Dictionary of Physical Metallurgy

Dictionary

EUGENIUSZ F. TYRKIEL
Associate Professor,
Institute of Materials Science,
Technical University, Warsaw

# of Physical Metallurgy

English
German
French
Polish
Russian

Elsevier Scientific Publishing Company
AMSTERDAM—OXFORD—NEW YORK
1977

Scientific Editor: ZBIGNIEW J. KOCH, M.A.

Groups 10, 30, 45, 70 & 80 have been reviewed by
L. KALINOWSKI, D.Sc., Associate Professor

The English terms and definitions have been checked by
MARGARET OTTO

Graphic Design by TADEUSZ PIETRZYK

Published in co-edition with Wydawnictwa Naukowo-Techniczne
Warsaw

Distribution of this book is handled by the following publishers:

   for the U.S.A. and Canada
Elsevier/North-Holland, Inc.
52 Vanderbilt Avenue
New York, N.Y. 10017

   for the East European Countries, China, Northern Korea, Cuba
   Vietnam and Mongolia
Wydawnictwa Naukowo-Techniczne
P.O. Box 359, 00-950 Warsaw, Poland

   for all remaining areas
Elsevier Scientific Publishing Company
335 Jan van Galenstraat
P.O. Box 211, Amsterdam, The Netherlands

Library of Congress Cataloging in Publication Data

Tyrkiel, Eugeniusz.
  Dictionary of physical metallurgy.

  Includes indexes.
  1. Physical metallurgy-Dictionaries-Polyglot.
2. Dictionaries, Polyglot. I. Title.
TN689.4.T98 1977   669'.9'03   77-8548
ISBN 0-444-99810-1

Copyright © 1977
by WYDAWNICTWA NAUKOWO-TECHNICZNE

All Rights Reserved. No part of this publication may be reproduced,
stored in a retrieval system, or transmitted, in any form or by any
means, electronic, mechanical, photocopying, recording or otherwise,
without prior written permission of the publisher.

PRINTED IN POLAND

PREFACE

The Technical Terminology Division of Wydawnictwa Naukowo-Techniczne, Publishers, has been concerned for over two decades with the compiling and editing of bi- and multilingual technological dictionaries in various branches of science and technology. The present one is the first to cope with modern physical metallurgy.

The idea of compiling this dictionary was first advanced by Mr. Zbigniew J. Koch (the editor of the dictionary), and materialized in the course of direct working talks between him and the author.

In accordance with the present state of the art and trends of development in physical metallurgy characterized by increasingly strong ties between this branch of science and physics, the scope of the dictionary has been extended to include, apart from terms related to classical metallurgy, also those lying in the sphere of the physics of metals. It is on the basis of comprehensive physical considerations that the essence of the metallic state as well as possibilities of alloy formation, their structure and properties, and the possibility of modifying the latter are interpreted. Physical methods are also increasingly used today for studying metals and their alloys. All this made the task of determining the scope of the dictionary difficult. In accordance with the general assumption that the dictionary should have a scientific character, neither the terms relating to e.g. equipment for heat treatment nor those relating to research apparatus have been included. Moreover, in view of the limited space, the names of chemical reagents used for etching metals and alloys have been omitted as they can easily be found in chemical dictionaries. Trade names have been included only if they have become common names. In the case of alloys of the same kind, differentiated only by the name of the metal present, only a few most typical examples are given as it is assumed that, if necessary, the reader will be able to derive the terms relating to other metals by analogy.

PREFACE

The dictionary contains, in a logical arrangement, some 2300 English terms with their definitions followed by equivalents in German (D), French (F), Polish (P) and Russian (R).
Emphasis was laid on providing the English entries and the equivalents in the other languages as fully as possible with the existing synonyms.

A term in any of the five languages can be easily found by using the alphabetical indexes placed at the end of the volume. Efforts were made to formulate the English definitions as concisely as possible, preferably in a single sentence, but in any case so as to explain the semantic range of the entry, and of course be suited to the requirements of physical metallurgy.

Synonyms are separated by commas. In some cases semicolons are used to separate equivalents which are not fully synonymous (their range is more limited or the meaning slightly different) although they belong to the same semantic range as the entry. When there are two synonymous forms of an entry, a full and an abbreviated one, the parts of the composite term which can be omitted are given in parentheses, for instance the term "microscopie (éléctronique) à balayage" contains two synonymous forms: "microscopie éléctronique à balayage" and "microscopie à balayage", similarly the entry "węzeł sieci(owy)" also has two synonymous forms: "węzeł sieciowy" and "węzeł sieci". The gender of nouns is indicated (only in the indexes) by the generally used abbreviations: *m* for masculine, *f* for feminine, *n* for neuter. The abbreviation *pl* stands for plural.

The dictionary was compiled on the basis of the contemporary literature on physical metallurgy in the relevant languages (handbooks, monographs, encyclopaedias, technical journals, standards, etc.). It is intended for scientists, engineers, technicians and students concerned with physical metallurgy as well as translators of scientific literature. The author as well as the publisher are fully aware that the dictionary, as the first in the field of physical metallurgy, is not free from errors and gaps.
All comments on the usefulness of the dictionary and on the gaps and errors noticed will therefore be gratefully acknowledged, and will serve to improve the next edition.

EUGENIUSZ F. TYRKIEL

Warsaw, May 1977

CONTENTS

Group 05 CHEMICAL ELEMENTS 1
Group 10 ELECTRON THEORY OF METALS 7
Group 15 THERMODYNAMICS OF ALLOYS 15
Group 20 CRYSTAL STRUCTURE OF METALS 23
Group 25 CRYSTALLIZATION AND SOLID STATE TRANSFORMATIONS 33
Group 30 LATTICE DEFECTS 47
Group 35 PHASE CONSTITUTION OF ALLOYS 59
Group 40 STRUCTURE OF METALS AND ALLOYS 75
Group 45 PLASTIC DEFORMATION AND THERMALLY ACTIVATED PROCESSES . 89
Group 50 IRON-CARBON SYSTEM 101
Group 55 TYPES OF METALS AND ALLOYS 111
Group 60 FERROUS ALLOYS 125
Group 65 NON-FERROUS AND SPECIAL ALLOYS 145
Group 70 HEAT TREATMENT 159
Group 75 PROPERTIES OF METALS AND ALLOYS 181
Group 80 EXAMINATION OF METALS AND ALLOYS 199
Group 85 MATERIAL DEFECTS 217
GERMAN INDEX (DEUTSCHES WÖRTERVERZEICHNIS) 225
ENGLISH INDEX 263
FRENCH INDEX (INDEX FRANÇAIS) 313
POLISH INDEX (SKOROWIDZ POLSKI) 337
RUSSIAN INDEX (РУССКИЙ УКАЗАТЕЛЬ) 361

05 CHEMICAL ELEMENTS
CHEMISCHE ELEMENTE
ÉLÉMENTS
PIERWIASTKI CHEMICZNE
ХИМИЧЕСКИЕ ЭЛЕМЕНТЫ

-005 **(chemical) element** (Z = atomic number)
  D (chemisches) Element
  F élément
  P pierwiastek (chemiczny)
  R (химический) элемент

-010 **actinium**, Ac ($Z = 89$)
  D Actinium, Aktinium
  F actinium
  P aktyn
  R актиний

-015 **aluminium, aluminum** (*USA*), Al ($Z = 13$)
  D Aluminium
  F aluminium
  P glin
  R алюминий

-020 **americium**, Am ($Z = 95$)
  D Americium, Amerizium
  F américium
  P ameryk
  R америций

-025 **antimony**, Sb ($Z = 51$)
  D Antimon
  F antimoine
  P antymon
  R сурьма

-030 **argon**, Ar ($Z = 18$)
  D Argon
  F argon
  P argon
  R аргон

-035 **arsenic**, As ($Z = 33$)
  D Arsen
  F arsenic
  P arsen
  R мышьяк

-040 **astatine**, At ($Z = 85$)
  D Astat(in)
  F astate
  P astat
  R астатин

-045 **barium**, Ba ($Z = 56$)
  D Barium
  F baryum
  P bar
  R барий

-050 **berkelium**, Bk ($Z = 97$)
  D Berkelium
  F berkélium
  P berkel
  R беркелий

-055 **beryllium**, Be ($Z = 4$)
  D Beryllium
  F béryllium, glucinium
  P beryl
  R бериллий

-060 **bismuth**, Bi ($Z = 83$)
  D Wismut
  F bismuth
  P bizmut
  R висмут

-065 **boron**, B ($Z = 5$)
  D Bor
  F bore
  P bor
  R бор

-070 **bromine**, Br ($Z = 35$)
  D Brom
  F brome
  P brom
  R бром

-075 **cadmium**, Cd ($Z = 48$)
  D Cadmium, Kadmium
  F cadmium
  P kadm
  R кадмий

-080 **caesium, cesium** (*USA*), Cs ($Z = 55$)
  D Caesium, Zäsium, Cäsium
  F cæsium, césium
  P cez
  R цезий

-085 **calcium**, Ca ($Z = 20$)
  D Calcium, Kalzium
  F calcium
  P wapń
  R кальций

-090 **californium**, Cf ($Z = 98$)
  D Californium, Kalifornium
  F californium
  P kaliforn
  R калифорний

05-

-095 **carbon**, C (Z = 6)
  D Kohlenstoff
  F carbone
  P węgiel
  R углерод

-100 **cerium**, Ce (Z = 58)
  D Cer, Zer
  F cérium
  P cer
  R церий

-105 **chlorine**, Cl (Z = 17)
  D Chlor
  F chlore
  P chlor
  R хлор

-110 **chromium**, Cr (Z = 24)
  D Chrom
  F chrome
  P chrom
  R хром

-115 **cobalt**, Co (Z = 27)
  D Kobalt, Cobalt
  F cobalt
  P kobalt
  R кобальт

-120 **copper**, Cu (Z = 29)
  D Kupfer
  F cuivre
  P miedź
  R медь

-125 **curium**, Cm (Z = 96)
  D Curium
  F curium
  P kiur
  R кюрий

-130 **dysprosium**, Dy (Z = 66)
  D Dysprosium
  F dysprosium
  P dysproz
  R диспрозий

-135 **einsteinium**, Es (Z = 99)
  D Einsteinium
  F einsteinium
  P ajnsztajn
  R эйнштейний

-140 **erbium**, Er (Z = 68)
  D Erbium
  F erbium
  P erb
  R эрбий

-145 **europium**, Eu (Z = 63)
  D Europium
  F europium
  P europ
  R европий

05-

-150 **fermium**, Fm (Z = 100)
  D Fermium
  F fermium
  P ferm
  R фермий

-155 **fluorine**, F (Z = 9)
  D Fluor
  F fluor
  P fluor
  R фтор

-160 **francium**, Fr (Z = 87)
  D Francium, Frankium
  F francium
  P frans
  R франций

-165 **gadolinium**, Gd (Z = 64)
  D Gadolinium
  F gadolinium
  P gadolin
  R гадолиний

-170 **gallium**, Ga (Z = 31)
  D Gallium
  F gallium
  P gal
  R галлий

-175 **germanium**, Ge (Z = 32)
  D Germanium
  F germanium
  P german
  R германий

-180 **gold**, Au (Z = 79)
  D Gold
  F or
  P złoto
  R золото

-185 **hafnium**, Hf (Z = 72)
  D Hafnium
  F hafnium, celtium
  P hafn
  R гафний

-190 **helium**, He (Z = 2)
  D Helium
  F hélium
  P hel
  R гелий

-195 **holmium**, Ho (Z = 67)
  D Holmium
  F holmium
  P holm
  R гольмий

-200 **hydrogen**, H (Z = 1)
  D Wasserstoff
  F hydrogène
  P wodór
  R водород

-205 **indium**, In (Z = 49)
  D Indium
  F indium
  P ind
  R индий

-210 **iodine, iodin** (*USA*), I (Z = 53)
  D Jod
  F iode
  P jod
  R иод

-215 **iridium**, Ir (Z = 77)
  D Iridium
  F iridium
  P iryd
  R иридий

-220 **iron**, Fe (Z = 26)
  D Eisen
  F fer
  P żelazo
  R железо

-225 **krypton**, Kr (Z = 36)
  D Krypton
  F krypton
  P krypton
  R криптон

-230 **lanthanum**, La (Z = 57)
  D Lanthan
  F lanthane
  P lantan
  R лантан

-235 **lawrencium**, Lr (Z = 103)
  D Lawrencium, Laurentium
  F lawrencium
  P lorens
  R лоуренций

-240 **lead**, Pb (Z = 82)
  D Blei
  F plomb
  P ołów
  R свинец

-245 **lithium**, Li (Z = 3)
  D Lithium
  F lithium
  P lit
  R литий

-250 **lutetium**, Lu (Z = 71)
  D Lutetium
  F lutétium
  P lutet
  R лютеций

-255 **magnesium**, Mg (Z = 12)
  D Magnesium
  F magnésium
  P magnez
  R магний

-260 **manganese**, Mn (Z = 25)
  D Mangan(ium)
  F manganèse
  P mangan
  R марганец

-265 **mendelevium**, Md (Z = 101)
  D Mendelevium
  F mendélévium
  P mendelew
  R менделевий

-270 **mercury**, Hg (Z = 80)
  D Quecksilber
  F mercure
  P rtęć
  R ртуть

-275 **molybdenum**, Mo (Z = 42)
  D Molybdän
  F molybdène
  P molibden
  R молибден

-280 **neodymium**, Nd (Z = 60)
  D Neodym
  F néodyme
  P neodym
  R неодим

-285 **neon**, Ne (Z = 10)
  D Neon
  F néon
  P neon
  R неон

-290 **neptunium**, Np (Z = 93)
  D Neptunium
  F neptunium
  P neptun
  R нептуний

-295 **nickel**, Ni (Z = 28)
  D Nickel
  F nickel
  P nikiel
  R никель

-300 **niobium, columbium**, Nb (Z = 41)
  D Niob
  F niobium
  P niob
  R ниобий

-305 **nitrogen**, N (Z = 7)
  D Stickstoff
  F azote
  P azot
  R азот

-310 **nobelium**, No (Z = 102)
  D Nobelium
  F nobélium
  P nobel
  R нобелий

-315 **osmium,** Os (Z = 76)
  D Osmium
  F osmium
  P osm
  R осмий

-320 **oxygen,** O (Z = 8)
  D Sauerstoff
  F oxygène
  P tlen
  R кислород

-325 **palladium,** Pd (Z = 46)
  D Palladium
  F palladium
  P pallad
  R палладий

-330 **phosphorus,** P (Z = 15)
  D Phosphor
  F phosphore
  P fosfor
  R фосфор

-335 **platinum,** Pt (Z = 78)
  D Platin
  F platine
  P platyna
  R платина

-340 **plutonium,** Pu (Z = 94)
  D Plutonium
  F plutonium
  P pluton
  R плутоний

-345 **polonium,** Po (Z = 84)
  D Polonium
  F polonium
  P polon
  R полоний

-350 **potassium,** K (Z = 19)
  D Kalium
  F potassium
  P potas
  R калий

-355 **praseodymium,** Pr (Z = 59)
  D Praseodym
  F praséodyme
  P prazeodym
  R празеодим

-360 **promethium,** Pm (Z = 61)
  D Promethium
  F prométhium
  P promet
  R прометий

-365 **protactinium,** Pa (Z = 91)
  D Protactinium, Protaktinium
  F protactinium
  P protaktyn
  R протактиний

-370 **radium,** Ra (Z = 88)
  D Radium
  F radium
  P rad
  R радий

-375 **radon,** Rn (Z = 86)
  D Radon
  F radon
  P radon
  R радон

-380 **rhenium,** Re (Z = 75)
  D Rhenium
  F rhénium
  P ren
  R рений

-385 **rhodium,** Rh (Z = 45)
  D Rhodium
  F rhodium
  P rod
  R родий

-390 **rubidium,** Rb (Z = 37)
  D Rubidium
  F rubidium
  P rubid
  R рубидий

-395 **ruthenium,** Ru (Z = 44)
  D Ruthen(ium)
  F ruthénium
  P ruten
  R рутений

-400 **samarium,** Sm (Z = 62)
  D Samarium
  F samarium
  P samar
  R самарий

-405 **scandium,** Sc (Z = 21)
  D Scandium, Skandium
  F scandium
  P skand
  R скандий

-410 **selenium,** Se (Z = 34)
  D Selen
  F sélénium
  P selen
  R селен

-415 **silicon,** Si (Z = 14)
  D Silicium, Silizium
  F silicium
  P krzem
  R кремний

-420 **silver,** Ag (Z = 47)
  D Silber
  F argent
  P srebro
  R серебро

-425 **sodium**, Na (Z = 11)
    D Natrium
    F sodium
    P sód
    R натрий

-430 **strontium**, Sr (Z = 38)
    D Strontium
    F strontium
    P stront
    R стронций

-435 **sulfur, sulphur**, S (Z = 16)
    D Schwefel
    F soufre
    P siarka
    R сера

-440 **tantalum**, Ta (Z = 73)
    D Tantal
    F tantale
    P tantal
    R тантал

-445 **technetium**, Tc (Z = 43)
    D Technetium
    F technétium
    P technet
    R технеций

-450 **tellurium**, Te (Z = 52)
    D Tellur
    F tellure
    P tellur
    R теллур

-455 **terbium**, Tb (Z = 65)
    D Terbium
    F terbium
    P terb
    R тербий

-460 **thallium**, Tl (Z = 81)
    D Thallium
    F thallium
    P tal
    R таллий

-465 **thorium**, Th (Z = 90)
    D Thorium
    F thorium
    P tor
    R торий

-470 **thulium**, Tm (Z = 69)
    D Thulium
    F thulium
    P tul
    R тулий

-475 **tin**, Sn (Z = 50)
    D Zinn
    F étain
    P cyna
    R олово

-480 **titanium**, Ti (Z = 22)
    D Titan
    F titane
    P tytan
    R титан

-485 **tungsten**, W (Z = 74)
    D Wolfram
    F tungstène
    P wolfram
    R вольфрам

-490 **uranium**, U (Z = 92)
    D Uran
    F uranium
    P uran
    R уран

-495 **vanadium**, V (Z = 23)
    D Vanadium, Vanadin
    F vanadium
    P wanad
    R ванадий

-500 **xenon**, Xe (Z = 54)
    D Xenon
    F xénon
    P ksenon
    R ксенон

-505 **ytterbium**, Yb (Z = 70)
    D Ytterbium
    F ytterbium
    P iterb
    R иттербий

-510 **yttrium**, Y (Z = 39)
    D Yttrium
    F yttrium
    P itr
    R иттрий

-515 **zinc**, Zn (Z = 30)
    D Zink
    F zinc
    P cynk
    R цинк

-520 **zirconium**, Zr (Z = 40)
    D Zirkon(ium)
    F zirconium
    P cyrkon
    R цирконий

Group 10

ELECTRON THEORY OF METALS
ELEKTRONENTHEORIE DER METALLE
THÉORIE ÉLECTRONIQUE DES MÉTAUX
ELEKTRONOWA TEORIA METALI
ЭЛЕКТРОННАЯ ТЕОРИЯ МЕТАЛЛОВ

-005 **physical metallurgy, structural metallurgy**
The scientific discipline dealing with the structure and the properties of metals and their alloys.
D Metallkunde, Metallogie, Metallehre
F métallurgie physique, métallurgie structurale
P metaloznawstwo (fizyczne)
R (физическое) металловедение

-010 **solid state physics, physics of solids**
The scientific discipline dealing with the structure and properties of solids on a strictly physical basis.
D Festkörperphysik
F physique de l'état solide, physique des solides
P fizyka ciała stałego
R физика твёрдого тела

-015 **metal physics, physics of metals**
The scientific discipline dealing with the structure and properties of metals and their alloys on a strictly physical basis.
D Metallphysik
F physique du métal, physique des métaux
P fizyka metali
R металлофизика, физика металлов

-020 **band theory**
The theory which interprets the electrical properties of solids on the assumption that there exist in crystals some energy states which can be occupied by them.
D Energiebändertheorie
F théorie des bandes
P teoria pasmowa (ciała stałego)
R зонная теория

-025 **electron theory of metals**
The theory which explains the properties of metals on the basis of their electronic structure.
D Elektronentheorie der Metalle, Metallelektronentheorie
F théorie électronique des métaux
P elektronowa teoria metali
R электронная теория металлов

-030 **electronic structure**
The mode of distribution of electrons over the different shells and sub-shells in the atom of a given element.
D Elektronenstruktur
F structure électronique
P budowa elektronowa, struktura elektronowa
R электронное строение, электронная структура

-035 **electron cloud**
The outer part of an atom made up of electrons.
D Elektronenwolke
F nuage électronique, nuage d'électrons
P chmura elektronowa
R электронное облако, облако электронов

-040 **atom(ic) core, atomic trunk**
An atom devoid of valency electrons (as in the space lattice of metal crystals).
D Atomrumpf, Atomrudiment
F tronc de l'atome
P rdzeń atomowy, ząb atomowy
R корпус атома, остов атома

-045 **electron gas**
The valency electrons of the atoms in a metal which can move freely in the metal as in a potential field created by the ions situated in the lattice points.
D Elektronengas
F gaz électronique, gaz d'électrons
P gaz elektronowy
R электронный газ

-050 **bound electron**
   An electron which is held by the nucleus by means of electrostatic attraction forces.
   D gebundenes Elektron
   F électron lié
   P elektron związany
   R связанный электрон

-055 **free electron**
   An electron which has freed itself from the attraction by the atomic nucleus and can move freely.
   D freies Elektron
   F électron libre
   P elektron swobodny
   R свободный электрон, несвязанный электрон

-060 **(electronic) work function**
   The energy required to remove an electron from the top levels of an energy distribution in a crystal to infinity.
   D Austrittsarbeit
   F travail d'extraction, travail de sortie
   P praca wyjścia
   R работа выхода

-065 **valency electron, outer(-shell) electron, external electron**
   An electron belonging normally to the outer electronic shell, the totality of these electrons determining chemical and optical properties of the given element.
   D Valenzelektron, Außenelektron, äußeres Elektron
   F électron de valence, électron périphérique
   P elektron walencyjny, elektron wartościowości, elektron zewnętrzny
   R валентный электрон, внешний электрон

-070 **inner(-shell) electron**
   An electron belonging normally to an electronic shell other than the outer shell.
   D Innenelektron, inneres Elektron
   F électron interne, électron intérieur
   P elektron wewnętrzny, elektron rdzeniowy
   R внутренний электрон

-075 **(relative) atomic mass, atomic weight**
   The mass of an atom of any element referred to a common standard, this standard being equal to one twelfth of the mass of an atom of carbon.
   D (relative) Atommasse, Atomgewicht
   F masse atomique (relative), poids atomique
   P masa atomowa (względna), ciężar atomowy
   R атомная масса, атомный вес

-080 **atomic diameter**
   The closest distance of approach of atom centres in the crystal lattice of a given element for a given coordination number.
   D Atomdurchmesser
   F diamètre atomique
   P średnica atomowa
   R атомный диаметр

-085 **atomic radius**
   A conventional quantity representing usually the half interatomic distance between the nearest neighbours in the space lattice of a given element.
   D Atomradius
   F rayon atomique
   P promień atomowy, promień atomu
   R атомный радиус

-090 **ionic radius**
   A conventional quantity representing usually the half distance between the centres of the nearest neighbours in the space lattice of an ionic crystal.
   D Ionenradius
   F rayon ionique
   P promień jonowy
   R ионный радиус

-095 **electronegativity**
   The ability of an element to add electrons to its atoms.
   D Elektronegativität
   F électronégativité
   P elektroujemność
   R электроотрицательность

-100 **ionization energy**
   The minimum energy required to ionize an atom, i.e. to remove an electron from the atom.
   D Ionisierungenergie
   F énergie d'ionisation
   P energia jonizacji
   R энергия ионизации

-105 **ionization potential, ionizing potential**
   The energy per unit charge required to remove an electron from an atom.
   D Ionisierungspotential
   F potentiel d'ionisation
   P potencjał jonizacji, potencjał jonizacyjny
   R потенциал ионизации, ионизационный потенциал

-110 **potential barrier, energy barrier**
   The amount of energy which opposes an electron contained in a periodic potential field, or a metal ion contained in the crystal lattice to be moved from its position and transferred to another stable position.
   D Potentialschwelle, Potentialwall, Potentialberg
   F barrière de potentiel, barrière énergétique
   P bariera potencjału, bariera energetyczna
   R потенциальный барьер, энергетический барьер

-115 **electron affinity**
   The tendency of atoms to accept electrons.
   D Elektro(nen)affinität
   F affinité électronique, électro-affinité
   P powinowactwo elektronowe
   R электронное сродство, электросродство

-120 **energy state, quantum state, electron state**
   The state of an electron in an atom as determined by the four quantum numbers.
   D Energiezustand, Quantenzustand
   F état énergétique, état quantique
   P stan energetyczny, stan kwantowy
   R энергетическое состояние, квантовое состояние

-125 **quantum number**
   Any one of the numbers which are necessary to define the state of an electron in an atom.
   D Quantenzahl
   F nombre quantique
   P liczba kwantowa
   R квантовое число

-130 **principal quantum number, total quantum number, shell quantum number**
   The quantum number which is a measure of the total energy of the electron in a given state.
   D Hauptquantenzahl, erste Quantenzahl
   F nombre quantique principal, nombre quantique total, premier nombre quantique
   P główna liczba kwantowa, pierwsza liczba kwantowa
   R главное квантовое число, первое квантовое число

-135 **second(ary) quantum number, azimuthal quantum number, subsidiary quantum number**
   The quantum number which is the measure of the angular momentum of the electron.
   D Nebenquantenzahl, zweite Quantenzahl, Azimutalquantenzahl, azimutale Quantenzahl, Drehimpulsquantenzahl
   F nombre quantique secondaire, nombre quantique azimutal, deuxième nombre quantique
   P poboczna liczba kwantowa, orbitalna liczba kwantowa, azymutalna liczba kwantowa, druga liczba kwantowa
   R побочное квантовое число, орбитальное квантовое число, второе квантовое число

-140 **magnetic quantum number, third quantum number**
   The quantum number which is a measure of the component of the angular momentum in the direction of an applied magnetic field.
   D magnetische Quantenzahl, Orientierungsquantenzahl, räumliche Quantenzahl
   F nombre quantique magnétique, troisième nombre quantique
   P magnetyczna liczba kwantowa, trzecia liczba kwantowa
   R магнитное квантовое число, третье квантовое число

-145 **spin quantum number, fourth quantum number**
   The quantum number which is related to the direction of the spin of the electron.
   D Spinquantenzahl
   F nombre quantique de spin, quatrième nombre quantique
   P spinowa liczba kwantowa, czwarta liczba kwantowa
   R спиновое квантовое число, четвёртое квантовое число

-150 **energy level**
 The energetic state of the electrons of an atom as determined by the first three quantum numbers.
 D Energieniveau
 F niveau énergétique, niveau d'énergie
 P poziom energetyczny
 R энергетический уровень

-155 **electron(ic) shell, principal shell**
 The set of electrons within an atom which have the same value of the principal quantum number.
 D Elektronenschale, Elektronenhülle, Hauptschale
 F couche électronique
 P powłoka elektronowa, warstwa elektronowa
 R электронная оболочка, основная оболочка

-160 **sub-shell**
 The set of electrons within an electron shell which have the same value of the second quantum number.
 D Unterschale, Nebenschale
 F sous-couche
 P podpowłoka (elektronowa), podwarstwa (elektronowa)
 R (электронная) подоболочка, (электронная) подгруппа, (электронный) подуровень

-165 **(Heisenberg) uncertainty principle**
 The principle stating that if Δx is the uncertainty in determining the position of a sub-atomic particle (e.g. of an electron) and Δp the uncertainty in measuring its momentum, then the product $\Delta x \Delta p$ has a minimum value equal to the Planck's constant.
 D Unbestimmtheitsprinzip, Unbestimmtheitsrelation, Ungenauigkeitsrelation, (Heisenbergsche) Unschärferelation, (Heisenbergsche) Unschärfebeziehung
 F relation d'indétermination de Heisenberg, principe d'indétermination de Heisenberg, principe d'incertitude de Heisenberg
 P zasada nieoznaczoności (Heisenberga), zasada nieokreśloności
 R соотношение неопределённости (Гейзенберга)

-170 **Pauli (exclusion) principle**
 The principle stating that an atom cannot contain two electrons in the same energy state, i.e. having the same values of all the four quantum numbers, or that any sub-shell can contain not more than two electrons, these electrons being of opposite spin.
 D (Paulisches) Ausschließungsprinzip, Pauli-Prinzip, Pauli-Verbot,
 F principe (d'exclusion) de Pauli, règle d'exclusion de Pauli
 P zasada Pauliego, zasada wykluczania, zakaz Pauliego
 R принцип (исключения) Паули, принцип несовместимости

-175 **Schrödinger equation**
 The fundamental equation of the quantum mechanics which, in general, determines the form of the wave function for various instances of motion and interaction of elementary particles, and which, applied to non localized electrons within the crystal enables to describe the allowed states of the valency electrons in the crystal.
 D Schrödinger-Gleichung
 F équation de Schrödinger
 P równanie Schrödingera
 R уравнение Шредингера

-180 **wave function**
 The principal quantity contained in the Schrödinger equation, characterizing the behaviour of a particle in a potential field.
 D Wellenfunktion
 F fonction d'onde(s)
 P funkcja falowa
 R волновая функция

-185 **wave vector**
 A vector, usually denoted by k, which is normal to a wave front in a crystal, its length being $2\pi/\lambda$, where λ is the wavelength of the wave motion involved.
 D Wellenvektor
 F vecteur d'onde
 P wektor falowy
 R волновой вектор

-190 **eigenfunction**
 Any wave function which, amongst other things, has the property of being a solution to the Schrödinger equation.
 D Eigenfunktion
 F fonction propre
 P funkcja własna
 R собственная функция

-195 **Bloch function**
An analytical expression representing the wave function of an electron in the crystal lattice.
D Bloch-Funktion
F fonction de Bloch
P funkcja Blocha
R функция Блоха

-200 **density of states**
The number of energy states, contained in a given energy interval, per unit volume.
D Zustandsdichte
F densité d'états
P gęstość stanów
R плотность состояний

-205 **k space, wave (vector) space**
The energetic space demarcated by the wave vectors.
D k-Raum
F espace k
P przestrzeń k, przestrzeń falowa
R пространство волновых векторов

-210 **Fermi energy**
The highest energy associated with the energy states which are occupied at the temperature of absolute zero.
D Fermienergie
F énergie (de) Fermi
P energia Fermiego
R энергия Ферми

-215 **Fermi surface**
The three-dimensional constant-energy surface in the k space on which all states have the Fermi energy at the temperature of absolute zero.
D Fermi(ober)fläche
F surface de Fermi
P powierzchnia Fermiego
R поверхность Ферми

-220 **Fermi level**
The energy corresponding to a Fermi surface.
D Fermigrenze
F niveau (de) Fermi
P poziom Fermiego
R уровень Ферми

-225 **Fermi sphere**
The Fermi surface of spherical shape, occurring when the wave vector has a low value.
D Fermikugel
F sphère de Fermi
P sfera Fermiego, kula Fermiego
R сфера Ферми

-230 **Brillouin zone**
A zone in the wave vector space within which the wave vector and the pertinent electron energy can be varied continuously.
D Brillouin-Zone
F zone de Brillouin
P strefa Brillouina
R зона Бриллюэна

-235 **energy band**
A set of so closely spaced energy levels of electrons in crystals that the transition from one level to another appears to be continuous.
D Energieband, Energiebereich
F bande d'énergie
P pasmo energetyczne
R энергетическая зона, энергетическая полоса

-240 **allowed band**
A zone of allowed states, i.e. an energy band comprising energy levels which can — as opposed to an energy gap — by occupied by electrons of a given substance.
D erlaubtes Band
F bande permise
P pasmo dozwolone
R зона дозволенных энергий

-245 **filled band**
An energy band within which each energy level is occupied by two electrons having opposite spin quantum numbers.
D (vollständig) besetztes Band, (völlig) aufgefülltes Band
F bande remplie, bande pleine
P pasmo zapełnione, pasmo obsadzone
R заполненная зона

-250 **(electron) energy gap, forbidden energy band**
An energy interval between two energy bands within which no energy levels exist that could by occupied by electrons of the given substance in a given state.
D verbotenes Band, verbotener Bereich, Energielücke
F bande interdite
P strefa energii wzbronionych, przerwa energetyczna, pasmo energii wzbronionych, pasmo wzbronione
R запрещенная зона, энергетическая щель

-255 **valency band, valence band**
The highest energy band filled with valency electrons.
D Valenzband
F bande de valence
P pasmo walencyjne
R валентная зона

-260 **conduction band**
An empty or incompletely filled energy band in which electrons can be raised on higher energy levels under the action of an electric field.
D Leitfähigkeitsband, Leit(ungs)band
F bande de conduction
P pasmo przewodnictwa
R зона проводнимости

-265 **conduction electron**
A valency electron contained in the conduction band.
D Leitungselektron
F électron de conduction
P elektron przewodnictwa
R электрон проводимости

-270 **thermal agitation, thermal vibration**
The oscillating motion of atoms in the crystal lattice, its amplitude increasing with the increase in temperature.
D thermische Schwingungen
F agitation thermique
P drgania cieplne, drgania termiczne
R тепловые колебания

-275 **electronic specific heat**
That part of the specific heat which comes not from the lattice vibrations but from the electronic contribution.
D spezifische Elektronenwärme, elektronische spezifische Wärme
F chaleur spécifique électronique
P ciepło właściwe elektronowe
R электронная теплоёмкость

-280 **Debye (characteristic) temperature**
The temperature, characteristic for a given substance, at which the whole spectrum of frequencies of atomic vibrations in the crystal lattice becomes narrowed close to its upper limit.
D Debyetemperatur, Debyesche Temperatur, (Debyesche) charakteristische Temperatur
F température (caractéristique) de Debye
P temperatura (charakterystyczna) Debye'a
R (характеристическая) температура Дебая, дебаевская температура, характеристическая температура

-285 **Mössbauer effect**
The phenomenon of emission and absorption of gamma rays occurring in such a way that the recoil is taken over by the whole crystal and not by an individual nucleus.
D Mößbauer-Effekt
F effet Mössbauer
P zjawisko Mössbauera
R эффект Мёссбауэра

-290 **Hall effect**
The occurrence of a transverse electric field and of a potential difference in a strip of metal or semiconductor through which an electric current flows and which is subjected to the action of a magnetic field perpendicular to the direction of the current.
D Hall-Effekt
F effet Hall
P zjawisko Halla
R эффект Холла

-295 **interaction energy**
The energy associated with the interaction between pairs of atoms in an alloy.
D Wechselwirkungsenergie
F énergie d'interaction
P energia oddziaływania
R энергия взаимодействия

-300 **interaction potential**
The potential energy representing the interaction between two atoms in the crystal lattice against the distance between these atoms.
D Wechselwirkungspotential
F potentiel d'interaction
P potencjał oddziaływania
R потенциал взаимодействия

-305 **phonon**
An energy quantum associated with vibrations of the crystal lattice or with oscillating motions within the atomic nucleus.
D Phonon
F phonon
P fonon
R фонон

-310 **cohesion**
Attraction between atoms or molecules within a given substance owing to which they are held together.
D Kohäsion
F cohésion
P kohezja, spójność
R когезия

-315 **cohesive force**
    The force of attraction between the atoms or molecules within a given substance.
    D Kohäsionskraft
    F force de cohésion
    P siła kohezji, siła spójności
    R сила сцепления, сила связи

-320 **chemical bond, chemical binding**
    The forces holding together the atoms, ions, or molecules within a given substance.
    D chemische Bindung
    F liaison chimique
    P wiązanie chemiczne
    R химическая связь

-325 **ionic bond, (hetero)polar bond, electrovalent bond**
    The chemical bond realized by the electrostatic attraction between ions of opposite signs which have been created by the passage of some valency electrons from the atom which is less electronegative to that one which is more electronegative.
    D Ionenbindung, heteropolare Bindung
    F liaison ionique, liaison hétéropolaire, liaison d'électrovalence
    P wiązanie jonowe, wiązanie heteropolarne, wiązanie elektrowalencyjne
    R ионная связь, (гетеро)полярная связь, электровалентная связь

-330 **atomic bond, covalent bond, homopolar bond**
    The chemical bond between atoms of a given element realized by the formation of valency-electron pairs which are shared in common by both atoms constituting the molecule.
    D Atombindung, kovalente Bindung, homöopolare Bindung
    F liaison covalente, liaison de covalence, liaison homopolaire
    P wiązanie atomowe, wiązanie kowalencyjne, wiązanie kowalentne, wiązanie homeopolarne
    R атомная связь, ковалентная связь, гомеополярная связь

-335 **van der Waals bond, molecular bond**
    The chemical bond existing between the molecules of liquefied or solidified inert gases.
    D van-der-Waalssche Bindung
    F liaison de van der Waals
    P wiązanie van der Waalsa, wiązanie (między)cząsteczkowe
    R вандерваальсов(ск)ая связь, молекулярная связь, связь Ван-дер-Ваальса

-340 **metallic bond**
    The chemical bond realized by the attraction between the positive ions (i.e. atom cores) and the free electrons constituting the electron gas.
    D metallische Bindung
    F liaison métallique
    P wiązanie metaliczne
    R металлическая связь

-345 **metallic state**
    The state of a substance manifesting itself by metallic properties.
    D metallischer Zustand
    F état métallique
    P stan metaliczny
    R металлическое состояние

-350 **metallic properties**
    The properties which result from the existence of the metallic bond and which are, therefore, peculiar to metals (good electrical and thermal conductivity, malleability, opacity, metallic lustre).
    D metallische Eigenschaften
    F propriétés métalliques
    P własności metaliczne
    R металлические свойства

-355 **metal**
    A crystalline material which is held together by metallic bonds.
    D Metall
    F métal
    P metal
    R металл

-360 **metallic element, metal**
    An element having a positive temperature coefficient of resistivity.
    D metallisches Element, Metall
    F élément métallique, métal
    P pierwiastek metaliczny, metal
    R металлический элемент, металл

-365 **metalloid, nonmetal, non-metal(lic element)**
    An element which does not possess metallic properties.
    D Metalloid, Nichtmetall, nichtmetallisches Element
    F métalloïde, élément non métallique
    P metaloid, niemetal, pierwiastek niemetaliczny
    R металлоид, неметалл, неметаллический элемент

-370 **semimetal, semi-metal**
    An element which possesses partly metallic properties and partly the properties of non-metals.
    D Halbmetall
    F semi-métal, demi-métal
    P półmetal, pierwiastek półmetaliczny
    R полуметалл, металл второго рода

-375 **semiconductor**
    A substance having a negative temperature coefficient of resistivity.
    D Halbleiter
    F semi-conducteur
    P półprzewodnik
    R полупроводник

-380 **intrinsic semiconductor**
    A semiconductor the electric properties of which do not rely on the presence of impurities.
    D Eigenhalbleiter
    F semi-conducteur intrinsèque
    P półprzewodnik samoistny
    R собственный полупроводник

-385 **extrinsic semiconductor, impurity semiconductor**
    A semiconductor the electric properties of which depend on the presence of impurities.
    D Fremdhalbleiter
    F semi-conducteur extrinsèque, semi-conducteur à impuretés
    P półprzewodnik domieszkowy, półprzewodnik niesamoistny
    R примесный полупроводник

-390 **superconductor**
    A metal, alloy, or intermetallic compound which is able to have no measurable electrical resistance below a certain critical temperature.
    D Supraleiter
    F supraconducteur
    P nadprzewodnik
    R сверхпроводник

-395 **atomic size effect, atomic size factor**
    The relation between the atomic diameters of two elements forming a binary alloy and the range of their solid solubility.
    D Atomgrößeneffekt
    F facteur de dimension, facteur de taille, facteur dimensionnel
    P czynnik wielkości atomu
    R масштабный фактор, масштабный эффект, размерный эффект, размерный фактор, объёмный фактор

-400 **electronegative valency effect, electrochemical factor**
    The relation between the difference in the electrochemical characteristics of the two components of a binary alloy and their tendency to form intermetallic compounds.
    D elektrochemischer Faktor
    F facteur électrochimique
    P czynnik elektrowartościowości ujemnej, czynnik elektrochemiczny
    R эффект электроотрицательной валентности, электрохимический фактор

-405 **relative valency effect**
    The relation between the mutual solid solubilities of metals of different valencies.
    D Valenzfaktor
    F facteur de valence
    P czynnik wartościowości względnej
    R эффект относительной валентности

15 THERMODYNAMICS OF ALLOYS
THERMODYNAMIK DER LEGIERUNGEN
THERMODYNAMIQUE DES ALLIAGES
TERMODYNAMIKA STOPÓW
ТЕРМОДИНАМИКА СПЛАВОВ

-005 **thermodynamics of solids**
    The scientific discipline dealing with the energetic interpretation of the structure of solids.
D Festkörperthermodynamik
F thermodynamique des solides
P termodynamika ciała stałego
R термодинамика твёрдого тела

-010 **thermodynamics of alloys, metallurgical thermodynamics**
    A branch of thermodynamics confined to metals and their alloys.
D Thermodynamik der Legierungen
F thermodynamique des alliages, thermodynamique métallurgique
P termodynamika stopów
R термодинамика сплавов

-015 **thermodynamic state**
    The state of a system as determined by the thermodynamic parameters.
D thermodynamischer Zustand
F état thermodynamique
P stan termodynamiczny
R термодинамическое состояние

-020 **thermodynamic parameter, thermodynamic variable, state variable**
    Any one of the three quantities — temperature, pressure, volume — which usually define the state of a system.
D Zustandsveränderliche, thermodynamische Variable
F variable d'état
P parametr termodynamiczny, parametr stanu
R термодинамический параметр, термодинамическая переменная, переменная состояния, (термодинамический) параметр состояния

-025 **equation of state**
    An equation expressing the relationship between the thermodynamic parameters, i.e. between temperature, pressure, and volume.
D Zustandsgleichung
F équation d'état
P równanie stanu
R уравнение состояния

-030 **thermodynamic quantity, thermodynamic function**
    A quantity which characterizes the system from the point of view of thermodynamics.
D thermodynamische Funktion, thermodynamische Größe
F fonction thermodynamique, grandeur thermodynamique
P funkcja termodynamiczna, wielkość termodynamiczna
R термодинамическая величина, термодинамическая функция

-035 **state quantity, state function, state property**
    A thermodynamic quantity which is unambiguously determined by the thermodynamic parameters and does not depend on the path through which the system has reached the given state.
D Zustandsfunktion, Zustandsgröße
F fonction d'état, grandeur d'état
P funkcja stanu
R функция состояния

-040 **extensive, quantity, extensive property, capacity property**
    A thermodynamic quantity the value of which depends on the mass of the system.
D extensive Größe, Quantitätsgröße, massenproportionale Größe
F grandeur extensive
P funkcja (termodynamiczna) ekstensywna, wielkość ekstensywna
R экстенсивная величина

-045 **intensive quantity, intensive property**
A thermodynamic quantity the value of which does not depend on the mass of the system.
D intensive Größe, Qualitätsgröße, massenunabhängige Größe
F grandeur intensive
P funkcja (termodynamiczna) intensywna, wielkość intensywna
R интенсивная величина

-050 **molar quantity, molar property**
A thermodynamic quantity referring to a mole of a substance.
D molare thermodynamische Funktion, molare thermodynamische Größe
F grandeur molaire
P funkcja (termodynamiczna) molowa
R молярная величина

-055 **integral (thermodynamic) quantity, integral thermodynamic function**
A thermodynamic quantity referring to the solution as a whole, i.e. not to a component of the solution.
D integrale thermodynamische Funktion, integrale thermodynamische Größe
F fonction thermodynamique integrale, grandeur (thermodynamique) integrale
P funkcja (termodynamiczna) całkowita
R интегральная (термодинамическая) величина

-060 **partial thermodynamic function, partial (thermodynamic) quantity**
A thermodynamic quantity referring to a component of a solution.
D partielle thermodynamische Funktion, partielle (thermodynamische) Größe
F fonction thermodynamique partielle, grandeur (thermodynamique) partielle
P funkcja (termodynamiczna) cząstkowa
R парциальная (термодинамическая) величина

-065 **excess (thermodynamic) quantity, excess (thermodynamic) function**
The difference between the value of a thermodynamic quantity of a real solution and that of an ideal solution.
D (thermodynamische) Zusatzfunktion, Überschußfunktion, Restgröße, Zusatzgröße, Exzeßfunktion
F fonction thermodynamique d'excès, fonction thermodynamique excédentive, grandeur thermodynamique excédentive, grandeur thermodynamique d'excès
P funkcja (termodynamiczna) nadmiarowa, funkcja (termodynamiczna) resztkowa
R избыточная (термодинамическая) величина

-070 **thermodynamic force, driving force**
The gradient of an intensive thermodynamic quantity acting as a stimulus for an irreversible process, e.g. for diffusion.
D treibende Kraft, Triebkraft
F force thermodynamique, force motrice
P siła termodynamiczna, bodziec termodynamiczny
R движущая сила, энергетический стимул

-075 **thermal gradient, temperature gradient**
The rate of change of temperature with distance within a given system.
D Temperaturgradient, Temperaturgefälle
F gradient thermique, gradient de température
P gradient temperatury
R температурный градиент, градиент температуры

-080 **concentration gradient**
The rate of change of the concentration of a given component with distance within a given system.
D Konzentrationsgradient, Konzentrationsgefälle
F gradient de concentration
P gradient stężenia
R концентрационный градиент, градиент концентрации

-085 **flux**
 The phenomenon of mass, energy, or electricity transport occurring within a system as a result of the action of driving forces.
 D Fluß
 F flux
 P przepływ (termodynamiczny)
 R поток

-090 **bond(ing) energy, binding energy**
 The difference between the energy of atoms after they have formed a bond and the sum of the energies of these atoms in an isolated state.
 D Bindungsenergie
 F énergie de liaison
 P energia wiązania
 R энергия связи

-095 **internal energy, intrinsic energy**
 The energy — contained within a system — associated with the interaction and motion of masses and electric charges of molecules, atoms, and elementary particles.
 D innere Energie
 F énergie interne
 P energia wewnętrzna
 R внутренняя энергия

-100 **bound energy, latent energy**
 The product of entropy and absolute temperature.
 D gebundene Energie
 F énergie liée
 P energia związana
 R связанная энергия

-105 **(Helmholtz) free energy, work function, thermodynamic potential at constant volume**
 That part of the internal energy of a system which can be transformed into work at constant temperature and volume, its value being equal to the difference between the internal energy and the bound energy.
 D (Helmholtzsche) freie Energie, thermodynamisches Potential bei konstantem Volumen
 F énergie libre (de Helmholtz), énergie disponible, potentiel thermodynamique à volume constant
 P energia (swobodna) Helmholtza, potencjał termodynamiczny w stałej objętości
 R свободная энергия (Гельмгольца), изохорно-изотермный потенциал

-110 **partial free energy**
 The free energy pertaining to a component of a solution.
 D partielle freie Energie
 F énergie libre partielle
 P energia swobodna cząstkowa
 R парциальная свободная энергия

-115 **excess free energy**
 The difference between the free energy of a real solution and that of an ideal solution.
 D freie Zusatzenergie, freie Überschußenergie
 F énergie libre d'excès, énergie libre excédentaire
 P energia swobodna nadmiarowa, energia swobodna resztkowa
 R избыточная свободная энергия

-120 **surface free energy**
 The amount of free energy associated with the existence of a free surface of a condensed phase, i.e. of a boundary surface between the condensed phase and the gaseous phase.
 D freie Oberflächenenergie
 F énergie libre superficielle
 P energia swobodna powierzchniowa
 R свободная поверхностная энергия

-125 **interfacial free energy**
 The amount of free energy associated with the existence of an interface between crystals of the same phase or between phases of different structure and composition.
 D freie Grenzflächenenergie
 F énergie libre interfaciale
 P energia swobodna powierzchni granicznej
 R свободная энергия поверхности раздела

-130 **free energy of mixing**
 The difference between the free energy of a solution and the sum of the free energies of its components in an unmixed state.
 D freie Mischungsenergie
 F énergie libre de mélange
 P energia swobodna tworzenia roztworu
 R свободная энергия смешения

-135 **free energy of formation**
   The difference between the free energy of a compound and the sum of the free energies of substances which have reacted to form this compound.
   D freie Bildungsenergie
   F énergie libre de formation
   P energia swobodna tworzenia (związku)
   R свободная энергия образования

-140 **free enthalpy, Gibbs free energy, Gibbs function, thermodynamic potential at constant pressure**
   That part of the enthalpy of system which can be transformed into work at a constant temperature and pressure, its value being equal to the difference between enthalpy and the bound energy.
   D freie Enthalpie, Gibbssche freie Energie, thermodynamisches Potential bei konstantem Druck
   F enthalpie libre,
      énergie libre de Gibbs,
      potentiel thermodynamique à pression constante
   P entalpia swobodna,
      energia swobodna Gibbsa,
      potencjał termodynamiczny pod stałym ciśnieniem,
      potencjał izotermiczno-izobaryczny
   R свободная энтальпия, свободная энергия Гиббса,
      изобар(но-изотерм)ный потенциал

-145 **excess enthalpy**
   The difference between the enthalpy of a real solution and that of an ideal solution.
   D Zusatzenthalpie, Überschußenthalpie, Restenthalpie
   F enthalpie d'excès, enthalpie excédentaire
   P entalpia nadmiarowa, entalpia resztkowa
   R избыточная энтальпия

-150 **partial enthalpy**
   The enthalpy pertaining to a component of a solution.
   D partielle Enthalpie
   F enthalpie partielle
   P entalpia cząstkowa
   R парциальная энтальпия

-155 **enthalpy of mixing**
   The difference between the enthalpy of a solution and the sum of the enthalpies of its components in an unmixed state.
   D Mischungsenthalpie
   F enthalpie de mélange
   P entalpia tworzenia roztworu
   R энтальпия смешения

-160 **enthalpy of formation**
   The difference between the enthalpy of a compound and the sum of the enthalpies of substances which have reacted to form this compound.
   D Bildungsenthalpie
   F enthalpie de formation
   P entalpia tworzenia (związku)
   R энтальпия образования

-165 **enthalpy of fusion, enthalpy of melting**
   The difference between the enthalpies of a substance in the liquid and solid state at melting point.
   D Schmelzenthalpie
   F enthalpie de fusion
   P entalpia topnienia
   R энтальпия плавления

-170 **bonding enthalpy, binding enthalpy**
   The difference between the enthalpy of atoms after they have formed a bond and the sum of the enthalpies of these atoms in an isolated state.
   D Bindungsenthalpie
   F enthalpie de liaison
   P entalpia wiązania
   R энтальпия связи

-175 **chemical potential**
   The partial molar free enthalpy of a given substance in a solution.
   D chemisches Potential
   F potentiel chimique
   P potencjał chemiczny
   R химический потенциал

-180 **common tangent principle, common tangent rule, common tangent construction**
A method of determining the compositions of coexisting phases at a given temperature by drawing a common tangent to the plots of the free enthalpy of these phases against composition.
D Prinzip der gemeinsamen Tangente, Methode der gemeinsamen Tangente, Tangentialschnitt--Methode
F principe de la tangente commune, methode de la tangente commune
P reguła wspólnej stycznej, metoda wspólnej stycznej
R правило общей касательной

-185 **spinodal point, spinode**
An inflection point in the curve representing free enthalpy against composition for a given solution.
D spinodaler Punkt
F point spinodal
P punkt spinodalny
R спинодальная точка, спинод, точка перегиба

-190 **spinodal curve, spinodal line, spinodal**
The locus of all inflection points in free enthalpy-composition curves for a given solution, plotted in the temperature-composition coordinate system.
D Spinodal(kurv)e
F (courbe) spinodale
P krzywa spinodalna, spinoda
R спинодальная кривая, спинодаль, линия точек перегиба, линия перегибов

-195 **entropy of mixing**
The difference between the entropy of a solution and the sum of the entropies of its components in an unmixed state.
D Mischungsentropie
F entropie de mélange
P entropia tworzenia roztworu
R энтропия смешения

-200 **entropy of formation**
The difference between the entropy of a compound and the sum of the entropies of substances which have reacted to form this compound.
D Bildungsentropie
F entropie de formation
P entropia tworzenia (związku)
R энтропия образования

-205 **vibrational entropy, thermal entropy**
Entropy which originates from thermal vibrations of atoms in a crystal lattice.
D Schwingungsentropie
F entropie de vibration, entropie vibrationnelle
P entropia wibracyjna
R вибрационная энтропия, колебательная энтропия

-210 **configurational entropy, positional entropy**
Entropy which originates from the multiplicity of possible configurations of solute atoms among solvent atoms in the crystal lattice of a solution.
D Vertauschungsentropie, Konfigurationsentropie
F entropie de configuration, entropie configurationnelle
P entropia konfiguracyjna, entropia pozycyjna
R конфигурационная энтропия

-215 **partial entropy**
The entropy pertaining to a component of a solution.
D partielle Entropie
F entropie partielle
P entropia cząstkowa
R парциальная энтропия

-220 **excess entropy**
The difference between the entropy of a real solution and that of an ideal solution.
D Zusatzentropie, Überschußentropie, Restentropie
F entropie d'excès, entropie excédentaire
P entropia nadmiarowa, entropia resztkowa
R избыточная энтропия

-225 **entropy of fusion, entropy of melting**
The difference between the entropies of a substance in the liquid and solid state at melting point.
D Schmelzentropie
F entropie de fusion
P entropia topnienia
R энтропия плавления

-230 **thermodynamic multiplicity, number of (microscopic) complexions, number of elementary complexions**
The number of ways of arranging microscopic particles (e.g. atoms in a crystal lattice) which make up a given macroscopic system.
D thermodynamische Wahrscheinlichkeit, Komplexionenzahl
F probabilité thermodynamique, nombre de complexions microscopiques
P prawdopodobieństwo termodynamiczne, liczba mikrostanów
R термодинамическая вероятность

-235 **energetic asymmetry**
A feature of the majority of real binary alloy systems consisting in that the curves representing the values of thermodynamic quantities of the system, or of its components, against composition are not symmetrical in relation to the ordinate going through the midpoint of the composition range.
D energetische Asymmetrie
F asymétrie énergetique
P asymetria energetyczna
R энергетическая асимметрия

-240 **ideal solution, perfect solution, Raoultian solution**
A solution the formation of which is not accompanied by any heat effect or volume change, the vapour pressure of the solute being proportional to its concentration.
D ideale Lösung, ideale Mischung
F solution idéale
P roztwór doskonały
R идеальный раствор, совершенный раствор

-245 **real solution, actual solution, nonideal solution**
A solution in which the vapour pressure of the solute is not proportional to its concentration, i.e. which does not obey Raoult's law.
D wirkliche Lösung, nichtideale Lösung
F solution réelle, solution imparfaite, solution non idéale
P roztwór rzeczywisty, roztwór niedoskonały
R реальный раствор, неидеальный раствор

-250 **dilute(d) solution**
A solution containing a very low percentage of the solute.
D verdünnte Lösung
F solution diluée
P roztwór rozcieńczony
R разбавленный раствор

-255 **regular solution**
A solution in which the enthalpy of mixing is not zero, whereas its entropy of mixing is equal to that of an ideal solution.
D reguläre Lösung, reguläre Mischphase
F solution regulière
P roztwór regularny
R регулярный раствор

-260 **semi-regular solution**
A solution differing from a regular solution in that its entropy of mixing is equal to that of an ideal solution multiplied by a coefficient the value of which differs somewhat from unity.
D semireguläre Lösung
F solution semi-regulière
P roztwór semiregularny
R семирегулярный раствор

-265 **athermal solution**
A solution in which the enthalpy of mixing is equal to zero, whereas its entropy of mixing differs from that of an ideal solution.
D athermische Lösung
F solution athermale
P roztwór atermiczny
R атермический раствор

-270 **Raoult's law**
A law stating that the partial vapour pressure of a substance in an ideal solution is, at constant temperature, directly proportional to its concentration in the solution, the proportionality factor being equal to the vapour pressure of this substance in the pure state.
D Raoultsches Gesetz
F loi de Raoult
P prawo Raoulta
R закон Рауля, закон Раула

-275 **Henry's law**
A law stating that the partial vapour pressure of a substance in a diluted solution is directly proportional to its concentration in this solution.
D Henrysches Gesetz
F loi de Henry
P prawo Henry'ego
R закон Генри

-280 **(thermodynamic) activity**
: The ratio of the vapour pressure of a substance in a real solution to its vapour pressure in a pure state, at the same temperature.
D (thermodynamische) Aktivität
F activité (thermodynamique)
P aktywność (termodynamiczna)
R (термодинамическая) активность

-285 **activity coefficient**
: The ratio of the activity of a substance in a solution, to its concentration in this solution, at a given temperature.
D Aktivitätskoeffizient
F coefficient d'activité
P współczynnik aktywności
R коэффициент активности

-290 **fugacity, fugitiveness**
: The value of the vapour pressure which would be possessed by a substance if its vapour were an ideal gas.
D Fugazität, Flüchtigkeit
F fugacité
P lotność, aktywność ciśnieniowa
R (термодинамическая) летучесть, фугитивность

-295 **interaction coefficient, activity factor**
: The coefficient representing the influence of each further component of a multicomponent solution on the activity of the given component of this solution.
D Wirkungskoeffizient, Wechselwirkungskoeffizient
F coefficient d'interaction
P współczynnik oddziaływania
R коэффициент взаимодействия

-300 **interaction parameter**
: Any one of the symbols $\varepsilon_2^{(i)}$ contained in the formula expressing the activity coefficient (γ) of a given substance (2) in a multicomponent solution
$\ln\gamma_2\ (N_2, N_3, N_4 ...) =$
$= N_2\varepsilon_2^{(2)} + N_3\varepsilon_2^{(3)} + N_4\varepsilon_2^{(4)} + ...$
where $\varepsilon_2^{(i)} = \dfrac{\partial \ln\gamma_2}{\partial N_i}$ and N_i is the mole fraction of the component i.
D Wechselwirkungsparameter, Wirkungsparameter, Interaktionsparameter
F coefficient d'effet, paramètre d'efficacité
P parametr oddziaływania
R параметр взаимодействия

-305 **Gibbs-Duhem equation**
: An equation expressing the relationship between partial molar quantities of a solution
$$\sum_{i=1}^{n} N_i\,\mathrm{d}\overline{X}_i = 0$$
where $\overline{X}_i$ denotes a partial molar quantity of the component i and N_i — the mole fraction of this component.
D Gibbs-Duhem-Gleichung
F équation de Gibbs-Duhem
P równanie Gibbsa-Duhema
R уравнение Гиббса-Дюгема

-310 **Hildebrand's equation**
: An equation expressing the relationship between the partial molar enthalpy of a component of a regular solution ($\Delta\overline{H}_i$) and the activity coefficient (γ_i) of this component in the solution.
$\Delta\overline{H}_i = RT\ln\gamma_i$
D Hildebrandsche Gleichung
F équation de Hildebrand
P równanie Hildebranda
R уравнение Гильдебранда

-315 **isoactivity line, isoactivity curve**
: A line, e.g. in the composition--temperature coordinate system, connecting points representing equal activity of a component in a given alloy system.
D Isoaktivitätslinie, Isoaktivitätskurve
F courbe d'isoactivité, ligne d'isoactivité
P izoaktywa, linia izoaktywności
R линия изоактивности

-320 **thermodynamic equilibrium**
: The state of a system when there is neither a temperature and pressure gradient, nor a gradient of the chemical potentials of individual components, within that system.
D thermodynamisches Gleichgewicht
F équilibre thermodynamique
P równowaga termodynamiczna
R термодинамическое равновесие

-325 **thermal equilibrium**
: The state of a system when there is no temperature gradient within it.
D Wärmegleichgewicht
F équilibre thermique
P równowaga cieplna, równowaga termiczna
R тепловое равновесие, термическое равновесие

-330 **chemical equilibrium**
The state of a heterogeneous system, when the chemical potential of each of its components is the same throughout all coexisting phases of the system.
D chemisches Gleichgewicht
F équilibre chimique
P równowaga chemiczna
R химическое равновесие

-335 **thermodynamic equilibrium constant**
The ratio, at equilibrium, of the product of activities of the reaction products to that of the activities of the reactants.
D thermodynamische Gleichgewichtskonstante
F constante d'équilibre thermodynamique
P stała równowagi termodynamicznej
R константа термодинамического равновесия

-340 **breakdown temperature, decomposition point**
The temperature at which a compound decomposes into simpler substances.
D Zersetzungstemperatur, Zersetzungspunkt
F température de décomposition
P temperatura rozkładu
R температура разложения

-345 **dissociation pressure**
The pressure of gaseous products resulting from thermal dissociation of a solid compound at a given temperature.
D Dissoziationsspannung, Dissoziationsdruck
F pression de dissociation, tension de dissociation
P prężność dysocjacji, prężność rozkładowa, ciśnienie dysocjacji
R давление диссоциации, упругость диссоциации

-350 **oxygen potential**
In the system metal-metal oxide, the free enthalpy of formation of the oxide at a given temperature.
D Sauerstoffpotential
F potentiel-oxygène
P potencjał tlenowy
R окислительный потенциал

Group 20

CRYSTAL STRUCTURE OF METALS
KRISTALLSTRUKTUR DER METALLE
STRUCTURE CRISTALLINE DES MÉTAUX
STRUKTURA KRYSTALICZNA METALI
КРИСТАЛЛИЧЕСКОЕ СТРОЕНИЕ МЕТАЛЛОВ

-005 **amorphous**
  Non-crystalline, i.e. having an irregular atomic arrangement in space.
  D amorph
  F amorphe
  P bezpostaciowy, amorficzny
  R аморфный

-010 **crystalline**
  Having a geometrically regular atomic arrangement in space.
  D kristallin(isch)
  F cristallin
  P krystaliczny
  R кристаллический

-015 **crystal**
  A solid composed of atoms or groups of atoms arranged in a definite pattern, periodic in three dimensions.
  D Kristall
  F cristal
  P kryształ
  R кристалл

-020 **metal(lic) crystal**
  A crystal held together by metallic bonds.
  D Metallkristall
  F cristal métallique
  P kryształ metaliczny
  R металлический кристалл

-025 **ionic crystal, polar crystal**
  A crystal held together by ionic bonds.
  D Ionenkristall
  F cristal ionique
  P kryształ jonowy
  R ионный кристалл, полярный кристалл

-030 **valence crystal**
  A crystal held together by atomic bonds.
  D Valenzkristall
  F cristal covalent
  P kryształ walencyjny
  R ковалентный кристалл

-035 **crystallography**
  The scientific discipline dealing with the study of the forms, structure and properties of crystals.
  D Kristallographie
  F cristallographie
  P krystalografia
  R кристаллография

-040 **crystal(lographic) structure**
  A real arrangement of atoms or groups of atoms in space according to the crystal lattice of the given substance.
  D Kristallstruktur, Kristallaufbau
  F structure cristallographique, structure cristalline
  P budowa krystalograficzna, struktura krystal(ograf)iczna
  R кристаллическая структура, кристаллическое строение

-045 **crystal(line) lattice, space lattice**
  A geometric pattern according to which the atoms of a crystalline substance are arranged in space.
  D Kristallgitter, Raumgitter
  F réseau cristallin, réseau spatial
  P sieć krystal(ograf)iczna, sieć przestrzenna
  R кристаллическая решётка, пространственная решётка

-050 **crystal class**
  A group of crystal symmetry, differing from another such group by its elements of symmetry.
  D Kristallklasse, Symmetrieklasse
  F classe cristalline, classe de symétrie
  P klasa krystalograficzna, klasa symetrii
  R класс симметрии, вид симметрии

-055 **crystal(line) symmetry**
  The quality possessed by crystals, by virtue of which they exhibit a repetetive arrangement of similar faces.
  D Kristallsymmetrie
  F symétrie cristalline
  P symetria kryształu
  R симметрия кристалла

-060 **crystal face**
Any one of the bounding surfaces of a crystal.
D Kristallfläche
F face du cristal
P ściana kryształu
R грань кристалла

-065 **rotation axis, axis of symmetry, rotatory reflection axis**
A line within a crystal, about which a rotation through 360 degrees replaces every element (point, line and plane) of the crystal by an equivalent element two, three, four or six times.
D Symmetrieachse, Dreh(ungs)achse, Deckachse, Gyre
F axe de symétrie, axe de rotation
P oś symetrii kryształu
R ось симметрии кристалла, гира

-070 **binary axis, two-fold rotation axis, diad axis, axis of two-fold symmetry**
The axis of symmetry of a crystal, about which a rotation through 360 degrees replaces twice every element of the crystal by an equivalent element.
D zweizählige Symmetrieachse, zweizählige Achse, Digyre
F axe (de symétrie) binaire
P oś symetrii dwukrotna
R двойная ось симметрии, ось симметрии второго порядка, дигира

-075 **ternary axis, three-fold rotation axis, triad axis, axis of three-fold symmetry**
The axis of symmetry of a crystal, about which a rotation through 360 degrees replaces three times every element of the crystal by an equivalent element.
D dreizählige Symmetrieachse, dreizählige Achse, Trigyre
F axe (de symétrie) ternaire
P oś symetrii trzykrotna
R тройная ось симметрии, ось симметрии третьево порядка, тригира

-080 **quaternary axis, four-fold rotation axis, tetrad axis, axis of four-fold symmetry**
The axis of symmetry of a crystal, about which a rotation through 360 degrees replaces four times every element of the crystal by an equivalent element.
D vierzählige Symmetrieachse, vierzählige Achse, Tetragyre
F axe (de symétrie) quaternaire
P oś symetrii czterokrotna
R четверная ось симметрии, ось симметрии четвёртого порядка, тетрагира

-085 **senary axis, six-fold rotation axis, hexad axis, axis of six-fold symmetry**
The axis of symmetry of a crystal, about which a rotation through 360 degrees replaces six times every element of the crystal by an equivalent element.
D sechszählige Symmetrieachse, sechszählige Achse, Hexagyre
F axe (de symétrie) sénaire
P oś symetrii sześciokrotna
R шестерная ось симметрии, ось симметрии шестого порядка, гексагира

-090 **unit cell**
A geometrical pattern, representing the arrangement of atoms in space, whose repetition at regular intervals in three dimensions, without change of orientation, builds up the lattice of a given crystal.
D Elementarzelle, Fundamentalzelle, Grundzelle
F maille cristalline, maille élémentaire
P elementarna komórka sieciowa, jednostkowa komórka sieciowa
R элементарная (кристаллическая) ячейка, единичная ячейка

-095 **primitive (unit) cell, simple cell**
A unit cell in which only the corners are occupied by atoms.
D primitive Elementarzelle, primitive Zelle
F maille primitive, maille simple
P komórka prymitywna, komórka prosta
R примитивная (элементарная) ячейка

-100 **lattice point**
The point at which three edges of a primitive unit cell intersect.
D Gitterpunkt
F nœud du réseau
P węzeł sieci(owy)
R узел (кристаллической) решётки

-105 **interstice, interstitial void**
The spaces between atoms which are — considered as spheres — situated at lattice points.
D Zwischengitterraum, Zwischengitterlücke, Gitterlücke
F interstice
P luka międzywęzłowa, międzywęźle
R междоузлие

-110 **lattice base**
The space arrangement of lattice sites, within the unit cell of a crystal lattice.
D Gitterbasis
F base du cristal
P baza sieci
R базис решётки

-115 **identity period, axial length, translation period**
The minimum distance over which a unit cell must be successively translated in anyone of the directions peculiar to a given crystal system, in order to create a crystal lattice.
D Identitätsperiode, Identitätsabstand, Translationsperiode
F période d'identité, période du réseau, distance d'identité
P okres identyczności, okres translacji, odcinek identyczności
R период решётки, период трансляции, осевая единица

-120 **lattice spacing, lattice parameter, lattice constant**
The length of an edge of a unit cell in a space lattice.
D Gitterabstand, Gitterparameter, Gitterkonstante
F paramètre cristallin, paramètre réticulaire, paramètre du réseau, constante réticulaire, constante de réseau
P parametr sieci (krystalicznej), stała sieci(owa)
R константа кристаллической решётки, постоянная решётки, параметр решётки

-125 **crystal(lographic) axis**
Any one of the three non-coplanar lines of reference intersecting at the centre of a crystal, their relative lengths within the crystal and arrangement to each other determining the system to which the crystal belongs.
D kristallographische Achse, Kristallachse
F axe cristallographique
P oś krystalograficzna
R кристаллографическая ось

-130 **translation**
A displacement parallel to a crystallographic axis.
D Translation, Gittertranslation
F translation
P translacja
R трансляция

-135 **(unit) lattice vector, unit cell vector**
A vector, the direction and length of which determines the position of the next lattice point.
D Gittervektor, Translationsvektor
F vecteur de base du réseau, vecteur de translation
P wektor sieci(owy), wektor translacji, wektor jednostkowy
R вектор решётки

-140 **lattice site, lattice position**
The location (of an atom) at a lattice point.
D Gitterplatz, Gitterstelle
F site réticulaire, site du réseau
P pozycja węzłowa
R узловое положение, узловая позиция

-145 **interstitial site, interstitial position**
The location (of an atom) in an interstice.
D Zwischengitterlage, Zwischengitterplatz
F site interstitiel
P pozycja międzywęzłowa
R междуузловое положение, междуузловая позиция

-150 **crystallographic direction, crystalline direction**
A direction traced in the space lattice by a straight line passing through two lattice points.
D kristallographische Richtung, Kristallrichtung
F direction cristallographique
P kierunek krystalograficzny
R кристаллографическое направление

-155 **crystal(lographic) plane, lattice plane**
   Any one of the planes which may be supposed to pass through the lattice points in a crystal.
   D kristallographische Ebene, Kristallebene, Netzebene
   F plan cristallographique, plan réticulaire
   P płaszczyzna krystalograficzna, płaszczyzna sieciowa
   R кристаллографическая плоскость, плоскость решётки

-160 **crystallographic index**
   Any one of the conventional sets of numbers used for crystallographic notation, i.e. for specifying crystallographic planes in a given crystal system.
   D kristallographischer Index
   F indice cristallographique
   P wskaźnik krystalograficzny
   R кристаллографический индекс

-165 **Miller indices**
   A set of three crystallographic indices which are the smallest integers proportional to the reciprocals of the intercepts, in terms of lattice parameters, cut by a given crystallographic plane on the three crystal axes.
   D Millersche Indizes
   F indices de Miller
   P wskaźniki Millera
   R миллеровские индексы

-170 **Miller-Bravais indices, hexagonal indices**
   A set of four crystallographic indices, analogous to Miller indices, used for specifying crystallographic planes in the hexagonal system.
   D Bravaissche Indizes
   F indices de Miller-Bravais
   P wskaźniki Millera-Bravais
   R индексы Миллера-Браве

-175 **atomic plane**
   Any one of the planes within a crystal which is populated by atoms.
   D Atomebene
   F plan atomique
   P płaszczyzna atomowa
   R атомная плоскость

-180 **stacking sequence**
   A pattern after which identical atomic planes are arranged in alternate position upon each other to build up a close-packed crystal structure.
   D Stapelfolge
   F empilage cristallographique
   P kolejność ułożenia (płaszczyzn atomowych)
   R последовательность упаковки, порядок упаковки

-185 **coordination number**
   The number of nearest equidistant atoms, surrounding any atom in a given crystal lattice.
   D Koordinationszahl
   F nombre de coordination, (nombre de) coordinence, indice de coordination
   P liczba koordynacyjna
   R координационное число

-190 **coordination shell, coordination zone**
   The totality of equidistant atoms surrounding any atom in a given crystal lattice.
   D Koordinationssphäre
   F zone de coordination
   P strefa koordynacyjna
   R координационная сфера, координационная зона

-195 **first-nearest neighbour, nearest neighbour (atom)**
   Any one of the group of nearest atoms equidistantly surrounding a given atom in a crystal lattice.
   D (erst)nächster Nachbar, nächstbenachbartes Atom, Nachbar erster Sphäre
   F voisin direct, voisin immédiat, voisin le plus proche, atome proche voisin, premier voisin
   P atom z pierwszej strefy koordynacyjnej
   R (первый) ближайший сосед

-200 **next-nearest neighbour, second-nearest neighbour**
   Any one of the group of second-nearest atoms equidistantly surrounding a given atom in the crystal lattice.
   D zweitnächster Nachbar, Nachbar zweiter Sphäre
   F second voisin
   P atom z drugiej strefy koordynacyjnej
   R второй сосед

-205 **(efficiency of) space filling**
   The ratio of the total volume of
   atoms, considered as rigid spheres
   belonging to one unit cell, to the
   volume of this unit cell.
   D Raumerfüllungsgrad,
   Raumerfüllung(szahl)
   F remplissage de l'espace,
   remplissage de volume
   P stopień wypełnienia, współczynnik
   wypełnienia
   R плотность упаковки,
   коэффициент упаковки,
   коэффициент заполнения,
   коэффициент компактности,
   степень компактности,
   объёмное заполнение

-210 **close(st) packing**
   Filling the space in a crystal lattice
   with atoms, considered as rigid
   spheres, as closely as possible.
   D dichte(ste) Kugelpackung
   F assemblage compact
   P wypełnienie zwarte
   R плотная упаковка

-215 **strong plane**
   A crystallographic plane of the
   greatest atomic density in a given
   system.
   D dichtest besetzte Gitterebene,
   dichtest gepackte Ebene
   F plan à assemblage compact,
   plan réticulaire le plus dense,
   plan à empilement dense
   P płaszczyzna najgęściej obsadzona
   R наиболее плотноупакованная
   плоскость

-220 **strong direction**
   A crystallographic direction along
   which there is the closest packing
   of atoms in a given system.
   D dichtest besetzte Richtung
   F direction (la plus) dense
   P kierunek najgęściej obsadzony
   R направление плотнейшей
   упаковки

-225 **interplanar spacing, plane spacing,
   grating constant, interplanar distance**
   The distance between parallel
   atomic planes in a crystal lattice.
   D Netzebenenabstand
   F distance interréticulaire,
   équidistance des plans réticulaires
   P odległość międzypłaszczyznowa
   R межплоскостное расстояние

-230 **interatomic spacing, interatomic
   distance**
   The distance between the centres
   of atoms in a crystal lattice.
   D Atomabstand
   F distance interatomique
   P odległość międzyatomowa
   R меж(ду)атомное расстояние

-235 **crystal(lographic) system**
   Any one of the six classes, into
   which all types of crystal structures
   are divided, on the basis of
   symmetry.
   D Kristallsystem, Syngonie
   F système cristallographique,
   système cristallin, syngonie
   P układ krystalograficzny, syngonia
   R кристалл(ограф)ическая система,
   сингония

-240 **cubic system, regular system,
   isometric system, tesseral system**
   A crystal system in which the unit
   cell is a cube.
   D kubisches Kristallsystem,
   reguläres Kristallsystem
   F système cubique
   P układ regularny, układ sześcienny
   R кубическая система

-245 **hexagonal system**
   A crystal system in which the unit
   cell is a regular hexagonal prism
   described by two lattice parameters
   a and c.
   D hexagonales Kristallsystem
   F système hexagonal, système sénaire
   P układ heksagonalny
   R гексагональная система

-250 **tetragonal system, quadratic system,
   pyramidal system**
   A crystal system in which the unit
   cell has two equal edges
   perpendicular to each other and
   to a third edge.
   D tetragonales Kristallsystem,
   quadratisches Kristallsystem
   F système quadratique,
   système tétragonal
   P układ tetragonalny
   R тетрагональная система

-255 **orthorhombic system**
   A crystal system in which the unit
   cell has three unequal edges
   mutually perpendicular.
   D orthorhombisches Kristallsystem
   F système orthorhombique
   P układ rombowy
   R (орто)ромбическая система

-260 **rhombohedral system, trigonal system**
A crystal system in which the unit cell has three equal edges mutually inclined at an angle, but not at right angles.
D rhomboedrisches Kristallsystem, trigonales Kristallsystem
F système rhomboédrique
P układ romboedryczny, układ trygonalny
R ромбоэдрическая система, тригональная система

-265 **monoclinic system, oblique system**
A crystal system in which the unit cell has three unequal edges, only one of them being perpendicular to the other two.
D monoklines Kristallsystem
F système monoclinique
P układ jednoskośny
R моноклинная система

-270 **triclinic system, anorthic system, asymmetric system**
A crystal system in which the unit cell has three unequal edges, no two of which are perpendicular.
D triklines Kristallsystem
F système triclinique
P układ trójskośny
R триклинная система

-275 **Bravais (space) lattice, point lattice, translation lattice**
Any one of the fourteen possible arrangements of points in space, which describe — from the view-point of symmetry — all the ways of disposition of lattice points occurring in crystals.
D Bravaisgitter, Translationsgitter
F réseau de Bravais, réseau de translation
P sieć Bravais, sieć punktowa, sieć translacyjna
R решётка Браве, трансляционная решётка

-280 **simple lattice**
A space lattice which contains atoms only at the corners of its unit cells.
D einfaches Gitter
F réseau simple
P sieć prosta
R простая решётка

-285 **sublattice**
Any one of the interpenetrating simple lattices which make up a more complex lattice.
D Teilgitter, Untergitter
F sous-réseau
P podsieć
R подрешётка

-290 **face-centred lattice**
A space lattice which apart from atoms at the corners, contains one atom at the centre of each face of its unit cell.
D flächenzentriertes Gitter
F réseau à faces centrées
P sieć ściennie centrowana, sieć płasko-centryczna
R гранецентрированная решётка, центрогранная решётка

-295 **body-centred lattice**
A space lattice which apart from atoms at the corners, contains one atom at the centre of its unit cell.
D raumzentriertes Gitter, innenzentriertes Gitter, körperzentriertes Gitter
F réseau centré
P sieć (przestrzennie) centrowana, sieć przestrzennie centryczna
R объёмноцентрированная решётка

-300 **cubic lattice**
A space lattice of the cubic crystal system.
D kubisches Gitter, würfeliges Gitter, reguläres Gitter
F réseau cubique
P sieć regularna, sieć sześcienna
R кубическая решётка

-305 **simple cubic lattice**
A cubic lattice which contains atoms only at the corners of its unit cell.
D einfaches kubisches Gitter
F réseau cubique simple
P sieć regularna prosta, sieć sześcienna prosta
R простая кубическая решётка, гексаэдрическая решётка

-310 **face-centred cubic lattice**
A cubic lattice which apart from atoms at the corners, contains one atom at the centre of each face of its unit cell.
D kubisch(es) flächenzentriertes Gitter
F réseau cubique à faces centrées
P sieć regularna ściennie centrowana, sieć regularna płasko-centryczna, sieć sześcienna zwarta
R кубическая гранецентрированная решётка, додекаэдрическая решётка, центрогранная кубическая решётка

-315 **body-centred cubic lattice**
A cubic lattice which apart from atoms at the corners, contains one atom at the centre of its unit cell.
D kubisch(es) raumzentriertes Gitter, kubisch(es) innenzentriertes Gitter, kubisch(es) körperzentriertes Gitter
F réseau cubique (à corps) centré, réseau cubique à maille centrée
P sieć regularna przestrzennie centrowana, sieć regularna przestrzennie centryczna, sieć sześcienna centrowana
R кубическая (объёмно)-центрированная решётка, октаэдрическая решётка

-320 **diamond cubic lattice, tetrahedral cubic lattice**
A space lattice composed of two face-centred cubic arrangements of atoms, either of which being displaced with respect to the other by a quarter of the diagonal of the unit cube.
D Diamantgitter
F réseau cubique tétraédrique, réseau du type diamant
P sieć regularna typu diamentu
R алмазная кубическая решётка, кубическая решётка типа алмаза

-325 **tetrahedral interstice, tetrahedral void**
The interstice which is surrounded by four atoms, situated at the corners of a tetrahedron.
D tetraedrische Gitterlücke, Tetraederlücke
F lacune tétraédrique, cavité tétraédrique
P luka czworościenna, luka tetraedryczna
R тетраэдрическая пустота, тетраэдрическая пора

-330 **octahedral interstice, octahedral void**
The interstice which is surrounded by six atoms, situated at the corners of a regular octahedron.
D oktaedrische Gitterlücke, Oktaederlücke
F lacune octaédrique, cavité octaédrique
P luka ośmiościenna, luka oktaedryczna
R октаэдрическая пустота, октаэдрическая пора

-335 **octahedral plane**
Any one of the sets of crystallographic planes in a cubic lattice, for which all three Miller indices are numerically equal.
D Oktaederebene
F plan octaédrique, plan octaédral, plan de (l')octaèdre
P płaszczyzna ośmiościanu
R плоскость октаэдра

-340 **hexagonal lattice**
A space lattice of the hexagonal crystal system.
D hexagonales Gitter
F réseau hexagonal
P sieć heksagonalna
R гексагональная решётка

-345 **simple hexagonal lattice**
A hexagonal lattice which contains atoms only at the corners of the hexagonal unit cell.
D einfach-hexagonales Gitter
F réseau hexagonal simple
P sieć heksagonalna prosta
R простая гексагональная решётка

-350 **hexagonal close-packed lattice, close-packed hexagonal lattice**
A hexagonal lattice which apart from twelve atoms at the corners of the hexagonal unit cell, contains one atom at the centre of each base and three coplanar atoms inside the cell at its mid-height, the axial ratio being approximately equal to 1.633.
D hexagonal(es) dichtgepacktes Gitter, Gitter mit hexagonal dichtester Kugelpackung
F réseau hexagonal (à assemblage) compact
P sieć heksagonalna, zwarta, sieć heksagonalna zwarcie wypełniona
R гексагональная плотноупакованная решётка, гексагональная компактная решётка

-355 **axial ratio, intercept ratio**
The ratio of the lattice parameter c to the lattice parameter a in the hexagonal or tetragonal crystal system.
D (kristallographisches) Achsenverhältnis
F rapport cristallographique
P stosunek osi(owy)
R отношение осей

-360 **basal plane**
A crystallographic plane in which the basis of the hexagonal unit cell lies.
D Basisebene
F plan de base
P płaszczyzna podstawowa, płaszczyzna bazowa
R базисная плоскость, плоскость базиса

-365 **prism(atic) plane**
Any one of the crystallographic planes in which a face of a hexagonal unit cell lies.
D Prismenebene
F plan prismatique, plan de face du prisme
P płaszczyzna słupa
R призматическая плоскость, плоскость грани призмы

-370 **tetragonal lattice**
A space lattice of the tetragonal crystal system.
D tetragonales Gitter
F réseau tétragonal, réseau quadratique
P sieć tetragonalna
R тетрагональная решётка

-375 **tetragonality**
The axial ratio in the tetragonal crystal system.
D Tetragonalität, tetragonale Verzerrung
F tétragonalité
P tetragonalność, stopień tetragonalności
R тетрагональность, степень тетрагональности

-380 **orthorhombic lattice**
A space lattice of the orthorhombic crystal system.
D rhombisches Gitter
F réseau orthorhombique
P sieć rombowa
R (орто)ромбическая решётка

-385 **base-centred orthorhombic lattice**
An orthorhombic lattice which apart from atoms at the corners, contains one atom at the centre of each of the two opposite bases of its unit cell.
D einseitig-flächenzentriertes rhombisches Gitter, rhombisch-einseitig--flächenzentriertes Gitter, rhombisch-grundflächenzentriertes Gitter
F réseau orthorhombique à bases centrées
P sieć rombowa jednostronnie centrowana, sieć rombowa jednostronnie centryczna, sieć rombowa o centrowanych podstawach
R односторонне-гранецентрированная ромбическая решётка, ромбическая базоцентрированная решётка

-390 **rhombohedral lattice, trigonal lattice**
A space lattice of the rhombohedral crystal system.
D rhomboedrisches Gitter
F réseau rhomboédrique
P sieć romboedryczna, sieć trygonalna
R ромбоэдрическая решётка

-395 **monoclinic lattice**
A space lattice of the monoclinic crystal system.
D monoklines Gitter
F réseau monoclinique
P sieć jednoskośna
R моноклинная решётка

-400 **triclinic lattice**
A space lattice of the triclinic crystal system.
D triklines Gitter
F réseau triclinique
P sieć trójskośna
R триклинная решётка

-405 **transition lattice**
An unstable intermediate crystallographic configuration, which occurs during solid state transformations, such as eutectoid decomposition or precipitation from a solid solution.
D Zwischengitter
F réseau de transition
P sieć przejściowa
R переходная решётка

-410 **reciprocal lattice**
A group of points arranged about a centre in such a way that the line joining each point to the centre is perpendicular to a family of planes in the crystal, and the length of this line is inversely proportional to their interplanar distance.
D reziprokes Gitter
F réseau réciproque, réseau polaire
P sieć odwrotna
R обратная решётка

-415 **crystal(lographic) orientation**
The position of important sets of lattice planes within a given crystal in relation to a reference system.
D Kristallorientierung
F orientation cristallographique, orientation cristalline
P orientacja krystalograficzna
R кристалл(ограф)ическая ориентировка, кристаллографическая ориентированность

-420 **random orientation, chaotic orientation**
A state characterized by a different, completely accidental crystallographic orientation of each grain, in a polycrystalline aggregate.
D regellose Orientierung
F orientation désordonnée
P orientacja chaotyczna, orientacja bezładna, orientacja przypadkowa
R беспорядочная ориентировка

-425 **preferred orientation**
The departure from randomness in crystallographic orientation in a polycrystalline metal.
D Vorzugsorientierung
F orientation privilégiée, orientation preférentielle
P orientacja uprzywilejowana, orientacja wyróżniona
R преимущественная ориентировка

-430 **single crystal, monocrystal**
A piece of a crystalline substance in which, as opposed to a polycrystalline aggregate, the whole mass has the same crystallographic orientation.
D Ein(zel)kristall, Monokristall
F monocristal, cristal unique
P monokryształ, kryształ pojedynczy
R монокристалл, одиночный кристалл

-435 **bicrystal**
A piece of metal (or other crystalline substance) composed of two parts differing in their crystallographic orientation.
D Bikristall, Zweikristall
F bicristal
P bikryształ
R бикристалл

-440 **polycrystal, polycrystalline aggregate**
An aggregate composed of many crystals with different orientations, as distinct from a single crystal.
D Polykristall, Vielkristall, Kristallhaufwerk, Kristallaggregat
F polycristal, aggrégat polycristallin
P polikryształ, ciało wielokrystaliczne
R поликристалл, поликристаллическое тело, многокристаллическое тело

-445 **anisotropy**
The dependence of the values of some physical properties of the material on the direction in which they are measured.
D Anisotropie
F anisotropie
P anizotropia
R анизотропия, анизотропность

-450 **isotropy**
The independence of the values of physical properties of the material, from the direction in which they are measured.
D Isotropie
F isotropie
P izotropia
R изотропия

-455 **quasi-isotropy**
The property of polycrystalline materials to be isotropic as a whole, though the randomly oriented individual grains which build up the polycrystalline aggregate are anisotropic.
D Quasiisotropie
F quasi-isotropie
P kwazyizotropia, quasi-izotropia
R квазиизотропия

-460 **crystal habit, habit (of a crystal)**
The general external shape of a crystal as determined by its internal crystallographic structure.
D Habitus (des Kristalls), Kristallhabitus, Kristalltracht, Kristallform
F faciès (d'un cristal), mode
P pokrój kryształu, postać kryształu, habitus
R габитус, облик кристалла, форма кристалла

-465 **habit plane**
   The crystallographic plane or system of planes of the parent phase along which the maximum section of the precipitated phase lies.
   D Habitusebene
   F plan limite, plan matrice, plan d'habitat
   P płaszczyzna habitus, płaszczyzna pokroju, płaszczyzna postaci
   R габитусная плоскость, плоскость габитуса

-470 **composition plane**
   A plane along which two crystals are coherently accreted.
   D Verwachsungsebene
   F plan d'accolement
   P płaszczyzna zrostu
   R плоскость срастания, плоскость сроста

-475 **isomorphism**
   The occurrence of different substances in the same type of crystal structure.
   D Isomorphie, Isomorphismus
   F isomorphi(sm)e
   P izomorfizm
   R изоморфизм

-480 **isomorphous crystals**
   Crystals of the same type of crystal structure.
   D isomorphe Kristalle
   F cristaux isomorphes
   P kryształy izomorficzne
   R изоморфные кристаллы

-485 **magnetic sublattice**
   In antiferromagnetic materials, any one of the interpenetrating space lattices in which all lattice points are occupied by the atoms having a parallel alignment of the spins.
   D magnetisches Untergitter
   F sous-réseau magnétique
   P podsieć magnetyczna
   R магнитная подрешётка

Group 25

CRYSTALLIZATION AND SOLID STATE TRANSFORMATIONS
KRISTALLISATION UND UMWANDLUNGEN IM FESTEN ZUSTAND
CRISTALLISATION ET TRANSFORMATIONS À L'ÉTAT SOLIDE
KRYSTALIZACJA I PRZEMIANY W STANIE STAŁYM
КРИСТАЛЛИЗАЦИЯ И ПРЕВРАЩЕНИЯ В ТВЕРДОМ СОСТОЯНИИ

-005 **freezing, solidification**
The passing of a substance from a liquid state to a solid state of aggregation.
D Erstarrung
F solidification
P krzepnięcie
R затвердевание

-010 **crystallization**
The precipitation of a crystalline solid from a solution.
D Kristallisation
F cristallisation
P krystalizacja
R кристаллизация

-015 **equilibrium solidification, equilibrium freezing**
A solidification process of an alloy conducted so slowly that diffusion in the liquid and solid eliminates all concentration gradients, thus maintaining equilibrium at all times.
D Gleichgewichtserstarrung
F solidification équilibrée, solidification d'équilibre
P krzepnięcie równowagowe, krystalizacja równowagowa
R равновесная кристаллизация

-020 **non-equilibrium solidification, non-equilibrium freezing**
A solidification process of an alloy conducted at such a rate that diffusion does not manage to eliminate concentration gradients, thus leading to the occurrence of segregation.
D Nichtgleichgewichtserstarrung
F solidification déséquilibrée, solidification de non équilibre, solidification hors d'équilibre
P krzepnięcie nierównowagowe, krystalizacja nierównowagowa
R неравновесное затвердевание, неравновесная кристаллизация

-025 **eutectic solidification**
Solidification of a liquid alloy at a constant temperature resulting in the formation of a eutectic.
D eutektische Erstarrung, eutektische Kristallisation
F solidification eutectique, solidification secondaire
P krzepnięcie eutektyczne, krystalizacja eutektyczna
R эвтектическое затвердевание, эвтектическая кристаллизация

-030 **controlled solidification**
A solidification process so conducted, that it proceeds in a desired manner as regards the direction of advance of the solidification front, the rate of this advance etc.
D kontrollierte Erstarrung
F solidification contrôlée
P krzepnięcie kierowane, krystalizacja kierowana
R управляемое затвердевание, управляемая кристаллизация

-035 **directed solidification, directional solidification, oriented solidification**
A solidification process so arranged that the solidification front advances in the desired direction.
D gelenkte Erstarrung, gerichtete Erstarrung
F solidification dirigée, solidification orientée
P krzepnięcie ukierunkowane, krystalizacja ukierunkowana, krystalizacja zorientowana
R направленное затвердевание, направленная кристаллизация, ориентированное затвердевание, ориентированная кристаллизация

-040 **unidirectional solidification, uniaxial solidification**
　　A solidification process in which the solidification front advances in one direction only.
　　D einachsige Erstarrung
　　F solidification unidirectionnelle
　　P krzepnięcie jednokierunkowe, krystalizacja jednokierunkowa
　　R одноосное затвердевание, одноосная кристаллизация

-045 **spontaneous crystallization**
　　Crystallization initiated by nuclei, formed only as a result of fluctuations within the phase in question, i.e. without interference of inclusions of another phase.
　　D spontane Kristallisation
　　F cristallisation spontanée
　　P krystalizacja samorzutna
　　R самопроизвольная кристаллизация, спонтанная кристаллизация

-050 **primary crystallization, primary solidification**
　　The precipitation of a crystalline solid phase from a liquid solution on cooling.
　　D Primärkristallisation, primäre Kristallisation, primäre Erstarrung
　　F cristallisation primaire, solidification primaire
　　P krystalizacja pierwotna, krystalizacja pierwszorzędowa
　　R первичная кристаллизация

-055 **mixed crystallization**
　　Crystallization resulting in the formation of a solid solution.
　　D Mischkristallbildung, Mischkristallisation
　　F cristallisation mixte
　　P powstawanie roztworu stałego
　　R образование твёрдого раствора

-060 **secondary crystallization**
　　The precipitation of a new crystalline phase from a solid solution on cooling.
　　D Sekundärkristallisation, sekundäre Kristallisation
　　F cristallisation secondaire
　　P krystalizacja wtórna, krystalizacja drugorzędowa
　　R вторичная кристаллизация

-065 **columnar crystallization**
　　Crystallization resulting in the occurrence of columnar crystals.
　　D Stengelkristallisation
　　F cristallisation basaltique
　　P krystalizacja słupkowa
　　R столбчатая кристаллизация

-070 **transcrystallization, fully-columnar crystallization**
　　Crystallization of an ingot resulting in the formation of such long columnar crystals that those growing from opposite sides meet.
　　D Transkristallisation
　　F transcristallisation
　　P transkrystalizacja
　　R транскристаллизация

-075 **equi-axed crystallization**
　　Crystallization resulting in the occurrence of equiaxed crystals.
　　D gleichachsige Kristallisation, globulitische Kristallisation
　　F cristallisation équiaxe
　　P krystalizacja równoosiowa
　　R равноосная кристаллизация, объёмная кристаллизация

-080 **electrocrystallization**
　　Crystallization taking place in the process of electrodeposition of metals.
　　D Elektrokristallisation
　　F électrocristallisation
　　P elektrokrystalizacja, krystalizacja katodowa
　　R электрокристаллизация

-085 **(single) crystal growing, growing single crystals**
　　The technique of producing single crystals.
　　D Einkristallzüchtung, Kristallzüchtung, Züchtung von Einkristallen, Einkristallherstellung
　　F préparation de monocristaux
　　P monokrystalizacja, hodowanie monokryształów
　　R выращивание монокристаллов

-090 **crystal pulling, pulling of crystals**
　　A method of producing single crystals by dipping a crystal seed into the molten metal in question just above its melting point and withdrawing it gently.
　　D Kristallziehen, Ziehen von Kristallen, Einkristallherstellung durch Ziehen
　　F tirage de (mono)cristaux
　　P wyciąganie monokryształu, wyciąganie monokryształów
　　R вытягивание монокристаллов

-095 **mother liquor, melt**
　　Liquid metal from which crystals form during solidification.
　　D Mutterlauge, Schmelze
　　F liquide mère
　　P ciecz macierzysta
　　R маточный расплав, маточный раствор

-100 **concentration fluctuations, composition(al) fluctuations**
Local limited departures of the composition of an alloy or a phase from its average composition.
D Konzentrationsschwankungen, Konzentrationsfluktuationen
F fluctuations de concentration
P fluktuacje stężenia
R концентрационные флюктуации, флюктуации концентрации, флюктуации состава

-105 **(crystal) nucleus, nucleus of crystallization**
A group of appropriately arranged atoms in a liquid metal at which crystallization starts and from which subsequent crystal growth continues.
D Kristall(isations)keim
F germe de cristallisation, centre de cristallisation
P zarodek krystalizacji
R зародыш кристаллизации, центр кристаллизации

-110 **homogeneous nucleus, spontaneous nucleus**
A crystal nucleus formed within the phase in question without any interference of inclusions of another phase, its occurrence being completely at random throughout the phase.
D arteigener Keim, spontaner Keim
F germe homogène, germe propre
P zarodek jednorodny, zarodek samorzutny
R гомогенный зародыш, спонтанный зародыш

-115 **heterogeneous nucleus**
A crystal nucleus in the form of a particle of foreign substance, a particle of another phase of the same substance, or a structural imperfection within the given phase.
D heterogener Keim
F germe hétérogène
P zarodek niejednorodny
R гетерогенный зародыш

-120 **foreign nucleus**
A crystal nucleus in the form of a particle of foreign substance within the given phase.
D artfremder Keim, Fremdkeim, Fremdkern, Impfkeim
F germe étranger
P zarodek obcy
R примесный зародыш

-125 **nucleant, nucleation catalyst, nucleating agent**
A foreign substance, in suspension within the given phase or in contact with this phase, which initiates heterogeneous nucleation.
D Keimbildner
F catalyseur de germination
P substancja zarodkotwórcza, katalizator zarodkowania
R возбудитель зарождения, катализатор зарождения

-130 **crystal seed**
A crystal of a foreign substance which, introduced into the melt, initiates nucleation and crystallization.
D Impfkristall
F cristal d'inoculation
P kryształ zaszczepiający
R затравочный кристалл, затравка

-135 **embryo**
A nucleus which is below the critical size, i.e. which is not yet capable of persisting and growing.
D Unterkeim, Embryo, Keimzentrum
F prégerme, germe subcritique, germe potentiel, embryon
P zarodek podkrytyczny
R докритический зародыш, эмбрион, дозародыш

-140 **critical (size) nucleus**
A nucleus of the minimum size necessary to be capable of persisting and growing.
D kritischer Kristallkeim
F germe critique
P zarodek krytyczny
R критический зародыш, зародыш критического размера, равновесный зародыш

-145 **critical radius (of a nucleus)**
The maximum radius of a nucleus below which the nucleus is not yet stable, i.e. not yet capable of persisting and growing.
D kritischer Keimradius
F rayon critique (d'un germe)
P promień krytyczny (zarodka)
R критический радиус (зародыша)

-150 **stable nucleus**
A nucleus the size of which is not smaller than that of the critical nucleus.
D wachstumsfähiger Kristallkeim
F germe stable
P zarodek trwały
R устойчивый зародыш

-155 **coherent nucleus**
   A nucleus built up in such a way that the interface between it and the parent phase is coherent.
   D kohärenter Keim
   F germe cohérent
   P zarodek koherentny
   R когерентный зародыш

-160 **oriented nucleus**
   A nucleus having a fixed crystallographic orientation with respect to the anisotropic substrate on which it forms.
   D orientierter Keim
   F germe orienté
   P zarodek zorientowany
   R ориентированный зародыш

-165 **substrate**
   The solid underlying material on which crystallization (e.g. electrocrystallization or crystallization from a gaseous phase) takes place.
   D Substrat
   F substrat
   P podłoże
   R подложка, основной материал

-170 **recrystallization nucleus**
   A nucleus in a cold-worked metal, initiating recrystallization.
   D Rekristallisationskeim
   F germe de recristallisation
   P zarodek rekrystalizacji
   R зародыш рекристаллизации, центр рекристаллизации

-175 **nucleation**
   The formation of crystal nuclei.
   D Keimbildung
   F germination, nucléation
   P zarodkowanie
   R зародышеобразование, образование зародышей, зарождение центров кристаллизации

-180 **homogeneous nucleation**
   Nucleation that occurs within a given phase without any interference of inclusions of another phase and which is completely random throughout the phase.
   D homogene Keimbildung, Eigenkeimbildung
   F germination homogène
   P zarodkowanie jednorodne, zarodkowanie homogeniczne
   R гомогенное образование зародышей, гомогенное зародышеобразование

-185 **heterogeneous nucleation**
   Nucleation that occurs at preferred sites within a given phase, e.g. at impurity particles within a melt or at lattice imperfections within a solid phase.
   D heterogene Keimbildung
   F germination hétérogène
   P zarodkowanie niejednorodne, zarodkowanie heterogeniczne
   R гетерогенное образование зародышей, гетерогенное зародышеобразование

-190 **spontaneous nucleation**
   Nucleation that occurs without any interference of inclusions of another phase and without any mechanical or electromagnetic stimuli. Often used synonymously with homogeneous nucleation.
   D spontane Keimbildung
   F germination spontanée
   P zarodkowanie samorzutne
   R самопроизвольное образование зародышей, самопроизвольное зародышеобразование

-195 **oriented nucleation**
   Nucleation taking place on an anisotropic substrate in such a way that the arising nuclei of a new phase have a fixed crystallographic orientation in relation to the substrate.
   D orientierte Keimbildung
   F germination orientée
   P zarodkowanie zorientowane
   R ориентированное зарождение

-200 **athermal nucleation**
   Nucleation which can take place without thermal activation or for which a very little activation energy is needed.
   D athermische Keimbildung
   F germination athermique
   P zarodkowanie atermiczne
   R атермическое образование зародышей, атермическое зарождение

-205 **dynamic nucleation**
   Nucleation initiated or intensified by mechanical or electromagnetic stimuli (e.g. vibrations, friction, pressure pulses, strong electric or magnetic fields etc.).
   D dynamische Keimbildung
   F germination dynamique
   P zarodkowanie dynamiczne
   R динамическое зародышеобразование, динамическое образование зародышей

-210 **tribonucleation**
   Dynamic nucleation initiated by friction.
   D Keimbildung durch Reibung, Tribonukleation
   F germination par frottement
   P zarodkowanie przez tarcie
   R трибозарождение

-215 **seeding (of crystallization), inoculation (of crystallization)**
   The introduction of an appropriate substance into the melt in order to initiate crystallization.
   D Animpfen (von Kristallbildung), Impfen (von Kristallbildung), künstliche Bekeimung
   F amorçage (de cristallisation), ensemencement
   P zaszczepianie krystalizacji
   R прививка кристаллизации, затравка кристаллизации

-220 **inoculation, modification**
   A technique used to improve the properties of some alloys (e.g. of grey cast iron) by introducing an inoculant into the liquid alloy before casting.
   D Modifizieren
   F inoculation, modification
   P modyfikowanie, modyfikacja
   R модифицирование

-225 **modification**
   A technique used to improve the properties of silumin by adding a small amount of e.g. sodium before casting.
   D Veredelung (des Silumins)
   F affinage (de l'alpax), traitement d'affinage
   P modyfikowanie (siluminu)
   R модифицирование (силумина)

-230 **inoculant, modifier**
   A substance added to a liquid alloy before casting in order to create a larger number of nuclei of a given phase with the view of obtaining finer grains and, sometimes, grains of a special shape.
   D Modifizierungsmittel, Modifikator
   F inoculant, inoculateur, modificateur
   P modyfikator
   R модификатор

-235 **demodifier**
   A substance which added to liquid grey cast iron hinders the occurence of graphite in nodular form.
   D Demodifizierungsmittel, Demodifikationsmittel
   F (agent) antinodulisant
   P demodyfikator
   R демодификатор

-240 **supercooling, su(pe)rfusion, undercooling**
   The cooling of a molten metal to a temperature below its normal freezing point without solidification.
   D Unterkühlung
   F surfusion
   P przechłodzenie
   R переохлаждение

-245 **undercooling**
   The cooling of a solid metal or alloy to a temperature below the equilibrium temperature for a given transformation, without transformation occurring.
   D Unterkühlung
   F surrefroidissement
   P przechłodzenie
   R переохлаждение

-250 **degree of supercooling, degree of undercooling**
   The difference between the theoretical and the actual freezing temperature.
   D Unterkühlungsgrad
   F degré de surfusion, taux de surfusion
   P stopień przechłodzenia
   R степень переохлаждения

-255 **constitutional supercooling, constitutional undercooling**
   Supercooling, or an increase in supercooling, brought about by a composition change rather than a temperature change.
   D konstitutionelle Unterkühlung
   F surfusion constitutionnelle
   P przechłodzenie stężeniowe
   R концентрационное переохлаждение

-260. **(equilibrium) distribution coefficient**
In a binary alloy, the equilibrium ratio of the solute concentration in a solid solution to its concentration in a coexisting liquid solution at a given temperature.
D (Gleichgewichts-) Verteilungskoeffizient
F coefficient de partage (à l'équilibre)
P (równowagowy) współczynnik rozdziału, teoretyczny współczynnik rozdziału
R (равновесный) коэффициент распределения, теоретический коэффициент распределения

-265 **effective distribution coefficient**
In a binary alloy, the ratio of the solute concetration in a solid solution, coexisting at a given temperature with a liquid solution, to its average concentration in the alloy.
D effektiver Verteilungskoeffizient, tatsächlicher Verteilungskoeffizient, wirksamer Verteilungskoeffizient
F coefficient de partage réel
P efektywny współczynnik rozdziału
R эффективный коэффициент распределения, фактический коэффициент распределения

-270 **nucleation rate, rate of nucleation**
The number of nuclei formed in a unit volume per unit time.
D Keimbildungsgeschwindigkeit, Keimbildungshäufigkeit, spezifische Keimzahl
F vitesse de germination
P szybkość zarodkowania
R скорость зародышеобразования, скорость зарождения центров кристаллизации

-275 **growth rate, rate of growth, speed of growth, rate of crystallization**
The increase in linear dimensions of a crystal in unit time.
D (Kristall-) Wachstumsgeschwindigkeit, Keimwachstumsgeschwindigkeit, (lineare) Kristallisationsgeschwindigkeit
F vitesse de croissance (des germes), vitesse de croissance du cristal
P szybkość wzrostu kryształów, szybkość krystalizacji
R скорость роста кристаллов, скорость кристаллизации

-280 **remaining liquid**
That portion of the melt which remains after the precipitation of primary crystals and which solidifies as a eutectic mixture.
D Restschmelze
F liquide résiduel, phase liquide restante
P resztka cieczy, resztka roztworu ciekłego
R остаточный расплав

-285 **crystal growth**
The formation and development of crystals from nuclei.
D Kristallwachstum
F croissance des cristaux
P wzrost kryształów
R рост кристаллов

-290 **grain growth, (grain) coarsening**
An increase in the average grain size of a metal or alloy.
D Kornwachstum, Kornvergröberung, Kornvergrößerung
F croissance du grain, grossissement du grain
P rozrost ziarn
R рост зерна, укрупнение зерна

-295 **solidification front**
An idealized boundary surface, moving in the direction of proceeding solidification, delimiting the already crystallized region from the melt.
D Erstarrungsfront, Kristallisationsfront
F front de solidification, front de cristallisation
P front krystalizacji
R фронт кристаллизации

-300 **principal crystallization axis**
The main direction along which a dendritic crystal grows.
D primäre Kristallisationsachse
F axe de cristallisation primaire
P główna oś krystalizacji, oś główna dendrytu
R ось первого порядка дендрита, главная кристаллографическая ось

-305 **secondary crystallization axis**
The direction perpendicular to the principal crystallization axis.
D sekundäre Kristallisationsachse
F axe de cristallisation secondaire
P wtórna oś krystalizacji, oś wtórna dendrytu
R ось второго порядка дендрита

-310 **dendritic growth**
The process of formation of dendritic crystals.
D dendritisches Wachstum
F croissance dendritique
P wzrost dendrytyczny
R дендритный рост

-315 **dendrite, dendritic crystal, fir-tree crystal, arborescent crystal, pine-tree crystal**
A crystal of ramified tree-like structure occurring in cast alloys.
D Dendrit, dendritischer Kristall
F dendrite, cristal dendritique
P dendryt, kryształ dendrytyczny
R дендрит, дендритный кристалл, ёлочный кристалл, разрывной кристалл

-320 **chemical dendrite**
A dendrite, its formation being controlled by chemical diffusion.
D chemischer Dendrit
F dendrite chimique
P dendryt chemiczny
R химический дендрит

-325 **thermal dendrite**
A dendrite, its formation being controlled by thermal diffusion.
D thermischer Dendrit
F dendrite thermique
P dendryt termiczny
R термический дендрит

-330 **growth step**
The incipience of a new atomic layer formed on the crystal surface during crystal growth.
D Wachstumsstufe
F marche de croissance
P stopień wzrostu, schodek wzrostu
R ступенька роста

-335 **growth spiral**
A growth step which is spiral in shape, developing at that point of the crystal surface where a dislocation emerges with an appreciable screw component more or less perpendicular to the surface.
D Wachstumsspirale, spiralförmige Wachstumsstufe
F spirale de croissance
P spirala wzrostu
R спиральная ступенька роста

-340 **epitaxy**
A fixed orientation relationship between a metal lattice and the lattice of a crystalline film formed thereon.
D Epitaxie
F épitaxie
P epitaksja
R эпитаксия

-345 **epitaxial growth**
The growth process of a crystalline film on a metal surface with the maintenance of a fixed orientation relationship between the crystal lattices of the two phases.
D epitaxisches Wachsen, epitaktisches Aufwachsen
F croissance épitaxiale, croissance épitaxique
P wzrost epitaksjalny
R эпитаксиальный рост, эпитаксиальное наращивание

-350 **epitaxial layer, epitaxial film**
A crystalline layer or film formed on a metal surface through epitaxial growth.
D Epitaxieschicht, epitaktische Schicht
F couche épitaxiale
P warstwa epitaksjalna
R эпитаксиальный слой, эпитаксиальная плёнка

-355 **idiomorphic crystal**
A crystal with well-developed faces peculiar to the given substance.
D idiomorpher Kristall
F cristal idiomorphe
P kryształ idiomorficzny
R идиоморфный кристалл

-360 **whisker, filamentary microcrystal**
Any one of the very fine filamentary single crystals of great tensile strength, grown on metal surfaces in such a way that they contain at the most one axial screw dislocation.
D Haarkristall, Nadelkristall, Fadenkristall, Whisker
F trichite, cristal filiforme, barbe, poil, whisker
P kryształ włoskowy, kryształ nitkow(at)y
R нитевидный кристалл, ус(ик)

-365 **unmixing**
  The decomposition of a solution into two phases.
  D Entmischung
  F démixion
  P rozkład roztworu, rozpad roztworu
  R распад (твёрдого) раствора, расслоение (жидкого) раствора

-370 **spinodal decomposition**
  The decomposition of a solution according to the spinodal curve.
  D spinodale Entmischung
  F décomposition spinodale
  P rozkład spinodalny, rozpad spinodalny
  R спинодальный распад

-375 **continuous precipitation**
  Precipitation of a new phase from a supersaturated solid solution proceeding simultaneously in all parts of the parent phase, although its rate may vary markedly from one region to another.
  D kontinuierliche Ausscheidung
  F précipitation continue
  P wydzielanie ciągłe
  R непрерывное выделение.

-380 **discontinuous precipitation**
  Precipitation of a new phase from a supersaturated solid solution implying the division of the parent phase into regions which have completely transformed into equilibrium phases and regions of untransformed parent phase.
  D diskontinuierliche Ausscheidung
  F précipitation discontinue
  P wydzielanie nieciągłe
  R прерывистое выделение

-385 **allotropy, polymorphism, polymorphy**
  The existence of some elements or intermetallic compounds in two or more crystallographically different forms, the change from one crystal system to another occurring at a definite temperature depending on pressure.
  D Allotropie, Polymorphie, Polymorphismus, Vielgestaltigkeit
  F allotropie, polymorphi(sm)e
  P alotropia, polimorfizm
  R аллотропия, полиморфизм

-390 **allotrope, polymorph, allotropic form, polymorphic form**
  Any one of the crystallographically different varieties of a given substance.
  D allotrope Modifikation, polymorphe Modifikation
  F allotrope, forme allotropique, variété allotropique, variété polymorphique
  P odmiana alotropowa, odmiana polimorficzna
  R аллотропическая модификация, аллотропическая разновидность, аллотропическая форма, полиморфная модификация

-395 **high-temperature allotrope, high-temperature form**
  An allotrope which is stable at elevated temperatures.
  D Hochtemperaturmodifikation
  F forme haute température
  P odmiana alotropowa wysokotemperaturowa
  R высокотемпературная модификация

-400 **segregation**
  Non-uniformity of the distribution or concentration of alloying components, inclusions and impurities in a metal or alloy, arising during freezing.
  D Seigerung
  F ségrégation
  P segregacja
  R ликвация, сегрегация

-405 **primary segregation**
  Segregation occurring in the pigs of a metal or alloy.
  D Primärseigerung, primäre Seigerung
  F ségrégation primaire
  P segregacja pierwotna
  R первичная ликвация

-410 **secondary segregation**
  Segregation occurring in the product made of pigs.
  D Sekundärseigerung, sekundäre Seigerung
  F ségrégation secondaire
  P segregacja wtórna
  R вторичная ликвация

-415 **macrosegregation, major segregation, long-range segregation**
Segregation of the chemical constituents of an ingot, visible with the naked eye or under slight magnification on a suitably prepared section of the ingot.
D Blockseigerung, Makroseigerung, makroskopische Seigerung, Zonenseigerung
F ségrégation majeure
P makrosegregacja, segregacja wlewka, segregacja strefowa
R макроликвация, ликвация по слитку, зональная ликвация

-420 **dendritic segregation, microsegregation, minor segregation, short-range segregation, coring**
Segregation of the chemical constituents of an alloy within the dendrites, visible under a microscope on a suitably prepared specimen.
D Mikroseigerung, mikroskopische Seigerung, dendritische Seigerung, Kristallseigerung
F ségrégation mineure, ségrégation dendritique
P mikrosegregacja, segregacja dendrytyczna
R микроликвация, дендритная ликвация

-425 **gravity segregation, gravitational segregation**
Segregation caused by differences of specific gravity of individual phases in a cast alloy.
D Schwereseigerung, Schwerkraftseigerung
F ségrégation par gravité
P segregacja grawitacyjna, segregacja ciężarowa
R ликвация по удельному весу

-430 **blowhole segregation, droplet segregation, sweating out**
A type of segregation where the low melting-point constituent of an alloy separates out in the form of drops.
D Gasblasenseigerung
F ségrégation en goutelettes, ségrégation gazeuse
P segregacja kroplista
R газовая ликвация, капельная ликвация, ликвация у газового пузыря, выпоты

-435 **normal segregation**
Segregation of low melting-point constituents of an alloy in those parts of a casting which solidify last.
D normale Seigerung
F ségrégation normale
P segregacja normalna, segregacja prosta
R нормальная ликвация, простая ликвация, прямая ликвация, нормальная сегрегация

-440 **inverse segregation, negative segregation**
Macrosegregation characterized by the presence of agglomerations of low melting-point constituents near the surface of the casting.
D umgekehrte Blockseigerung
F ségrégation (majeure) inverse
P segregacja odwrotna
R обратная ликвация

-445 **interdendritic segregation**
Segregation of a phase or some phases in the interdendritic spaces.
D interdendritische Seigerung, Seigerung in Dendritenzwischenräumen
F ségrégation interdendritique
P segregacja międzydendrytyczna
R междендритная ликвация, междендритная сегрегация

-450 **intercrystalline segregation, grain boundary segregation**
Precipitation of a phase at the grain boundaries of the parent phase.
D Korngrenzenseigerung
F ségrégation intercristalline, ségrégation aux joints de grains
P segregacja na granicach ziarn
R серегация на границах зёрен, серегация по границам зёрен, зернограничная сегрегация

-455 **allotropic transformation, allotropic change, polymorphic transformation, polymorphic change**
A transformation occurring at a definite temperature in some substances (e.g. iron, tin, sulfur etc.) which results in a different type of space lattice.
D allotrope Umwandlung, polymorphe Umwandlung
F transformation allotropique, transformation polymorphe, transformation polymorphique
P przemiana alotropowa, przemiana polimorficzna
R аллотропическое превращение, полиморфное превращение, полиморфный переход

-460 **phase transformation, phase transition, phase change**
The transformation of a metal from one phase into another (e.g. from the solid state into the liquid state, from one allotrope into another). In an alloy a transformation resulting in a change in the kind, in the number and kind or — sometimes — in the number and composition of phases existing or coexisting in a given system.
D Phasenumwandlung, Phasenumsetzung
F changement de phase, transition de phase, transformation de phase
P przemiana fazowa
R фазовое превращение, фазовый переход

-465 **first-order transformation, first-degree transformation, first-order transition, first-degree transition**
A transformation occurring with a discontinuity in the first derivatives of the free enthalpy.
D Umwandlung erster Ordnung, Umwandlung erster Art
F transition (de phase) de première espèce
P przemiana pierwszego rodzaju, przemiana pierwszego rzędu
R (фазовое) превращение первого рода, (фазовый) переход первого рода

-470 **second-order transformation, second-degree transformation, second-order transition, second-degree transition**
A transformation which is continuous with regard to the first derivatives of the free enthalpy but shows a discontinuity in its second derivatives.
D Umwandlung zweiter Ordnung, Umwandlung zweiter Art
F transition (de phase) de seconde espèce
P przemiana drugiego rodzaju, przemiana drugiego rzędu
R (фазовое) превращение второго рода, (фазовый) переход второго рода

-475 **solid (state) transformation, solid state transition**
Transformation occurring in the solid state.
D Umwandlung im festen Zustand, sekundäre Umwandlung
F transformation à l'état solide
P przemiana w stanie stałym
R превращение в твёрдом состоянии

-480 **temperature hysteresis, thermal hysteresis, thermal lag**
The lagging in the occurrence of a transformation behind the temperature change.
D Umwandlungsverzögerung
F hystérésis thermique, hystérésis de la transformation
P histereza cieplna, histereza przemiany
R температурный гистерезис, гистерезис превращения

-485 **recalescence**
Isothermal evolution of heat that occurs when iron or steel cools through a transformation point.
D Rekaleszenz
F recalescence
P rekalescencja
R рекалесценция

-490 **decalescence**
Isothermal absorption of heat that occurs when iron or steel is heated through a transformation point.
D Dekaleszenz
F decalescence
P dekalescencja
R декалесценция

-495 **reversible transformation, reversible transition**
   A transformation which on heating proceeds in the opposite direction than on cooling, the transformation temperature being theoretically the same in both cases.
   D reversible Umwandlung
   F transformation réversible
   P przemiana odwracalna
   R обратимое превращение

-500 **structural change**
   A change consisting in an alteration of the microstructure.
   D Gefügeumwandlung, Strukturumwandlung
   F transformation structurale, transformation de structure
   P przemiana strukturalna
   R структурное превращение

-505 **diffusional transformation, diffusional process**
   A transformation for which a displacement of atoms relatively to their neighbours over distances greater than the interatomic distance is required.
   D diffusionsartige Umwandlung, Diffusionsvorgang
   F transformation avec diffusion
   P przemiana dyfuzyjna
   R диффузионное превращение

-510 **diffusionless transformation, diffusion-free process, shear transformation**
   A transformation for which no atomic interchange is required, the atoms moving only a fraction of the interatomic distance relatively to their neighbours.
   D diffusionslose Umwandlung, Umklappumwandlung, Schiebungsumwandlung
   F transformation sans diffusion
   P przemiana bezdyfuzyjna
   R бездиффузионное превращение

-515 **isothermal transformation**
   A transformation taking place at constant temperature.
   D isothermische Umwandlung
   F transformation isotherme
   P przemiana izotermiczna
   R изотермическое превращение

-520 **athermal transformation**
   A transformation for which no activation energy is required.
   D athermische Umwandlung
   F transformation athermique
   P przemiana atermiczna
   R атермическое превращение

-525 **congruent transformation**
   A transformation of an alloy taking place at a fixed temperature without any change in the composition of either the phase which transforms or the phase which results from this transformation.
   D kongruente Umwandlung
   F transformation congruente
   P przemiana kongruentna
   R конгруэнтное превращение

-530 **homogeneous transformation**
   A transformation during which no regions of discontinuity, such as boundary surfaces between transformed regions and their surroundings, occur in the system.
   D homogene Umwandlung
   F transformation homogène
   P przemiana jednorodna
   R гомогенное превращение

-535 **heterogeneous transformation**
   A transformation during which regions of discontinuity, such as boundary surfaces between transformed regions and their surroundings, occur in the system, even if the initial and final conditions are both single phase.
   D heterogene Umwandlung, inhomogene Umwandlung
   F transformation hétérogène
   P przemiana niejednorodna
   R гетерогенное превращение

-540 **continuous transformation**
   A transformation which proceeds simultaneously in all parts of the system, although its rate may vary from one region to another.
   D kontinuierliche Umwandlung
   F transformation continue
   P przemiana ciągła
   R непрерывное превращение

-545 **discontinuous transformation**
   A transformation which implies the division of the system into regions which have completely transformed into equilibrium phases and regions which have not yet transformed.
   D diskontinuierliche Umwandlung
   F transformation discontinue
   P przemiana nieciągła
   R прерывистое превращение

-550 **massive transformation**
A very rapid transformation occurring on cooling in some solid solutions, resulting in a new single-phase structure with the same composition as the original phase.
D massive Umwandlung
F transformation massive, transformation en masse
P przemiana masywna
R массивное превращение

-555 **magnetic transformation**
Transition from a ferromagnetic to a paramagnetic state, or conversely.
D magnetische Umwandlung
F transformation magnétique
P przemiana magnetyczna
R магнитное превращение

-560 **magnetic transformation point, magnetic change point, Curie temperature, Curie point**
The temperature at which a magnetic transformation occurs, i.e. at which a ferromagnetic material becomes paramagnetic on heating.
D magnetischer Umwandlungspunkt, Curietemperatur, Curiepunkt
F température de Curie, point de Curie
P temperatura przemiany magnetycznej, temperatura Curie, punkt Curie
R температура магнитного превращения, температура Кюри, точка Кюри

-565 **Néel point, Néel temperature**
The temperature at which antiferromagnetic substances lose their antiparallel alignment and become paramagnetic.
D Néeltemperatur, Néel-Punkt
F température de Néel, point de Néel
P temperatura Néela, punkt Néela
R температура Нэля, температура Нееля, точка Нееля

-570 **(superconducting) transition temperature, superconducting critical temperature, transition point**
The temperature below which certain metals exhibit an abnormally high electrical conductivity.
D Sprungtemperatur, Sprungpunkt, (kritische) Übergangstemperatur
F température de transition, température critique, point de transition, point critique du supraconducteur
P temperatura przejścia w stan nadprzewodnictwa, temperatura przeskoku
R температура перехода в сверхпроводящее состояние

-575 **ordering**
A transformation occurring in certain solid šolutions during which a random arrangement of solvent and solute atoms is replaced by a regular or orderly arrangement of the different atoms on preferred lattice sites.
D Ordnungsvorgang, Unordnung-Ordnung-Umwandlung
F transformation d'ordonnancement, transition désordre-ordre, restauration de l'ordre, remise en ordre
P uporządkowanie
R упорядочение

-580 **disordering, order-disorder transformation, order-disorder transition**
A transformation which is the reverse of ordering.
D Entordnung, Ordnung-Unordnung-Umwandlung
F transformation ordre-désordre, transition ordre-désordre
P przemiana porządek-nieporządek
R разупорядочение, превращение порядок-беспорядок

-585 **disorder(ed) state**
A random arrangement of the different atoms, with respect to each other, in a crystal lattice.
D Unordnung, Unordnungszustand
F désordre
P nieuporządkowanie, stan nieuporządkowany, stan nieuporządkowania
R беспорядок, неупорядоченное состояние

-590 **order(ed state)**
A regular or orderly arrangement of the different atoms, with respect to each other, in a crystal lattice.
D Ordnung, Ordnungszustand
F ordre
P uporządkowanie, porządek, stan uporządkowany, stan uporządkowania
R порядок, упорядоченное состояние

-595 **short-range order**
The order restricted to the volume of an antiphase domain.
D Nahordnung
F ordre à courte distance, ordre à petite distance
P uporządkowanie bliskie, porządek bliski, uporządkowanie bliskiego zasięgu, porządek bliskiego zasięgu
R ближний порядок

-600 **long-range order**
The order extending over the whole crystal.
D Fernordnung
F ordre à longue distance, ordre à grande distance
P uporządkowanie dalekie, porządek daleki, uporządkowanie dalekiego zasięgu, porządek dalekiego zasięgu
R дальний порядок

-605 **degree of order(ing)**
A conventional quantity which is a measure for expressing the proximity of incompletely ordered solutions to the ideal order.
D Ordnungsgrad
F degré d'ordre
P stopień uporządkowania
R степень упорядочения, степень упорядоченности, степень порядка

-610 **degree of disorder(ing)**
A conventional quantity which is a measure for expressing deviations of incompletely ordered solutions from the ideal order.
D Unordnungsgrad, Fehlordnungsgrad
F degré de désordre
P stopień nieuporządkowania
R степень беспорядка

-615 **short-range order parameter**
The quantity σ defined by the relation $\sigma = 2q - 1$ where q denotes the probability that a B atom is the nearest neighbour of the given A atom in a binary solution which tends to be ordered.
D Nahordnungsparameter
F paramètre d'ordre à courte distance, paramètre d'ordre à petite distance
P parametr uporządkowania bliskiego, parametr porządku bliskiego
R параметр ближнего порядка

-620 **long-range order parameter**
The quantity s defined by the relation $s = \dfrac{p - N_A}{1 - N_A}$ where N_A is the atom fraction of the component A in the binary solution A—B which tends to be ordered, and p represents the probability that an A atom occupies the right site in the lattice of this solution.
D Fernordnungsparameter
F paramètre d'ordre à longue distance, paramètre d'ordre à grande distance
P parametr uporządkowania dalekiego, parametr porządku dalekiego
R параметр дальнего порядка

-625 **disordering temperature**
The temperature at which the order-disorder transformation occurs.
D Entordnungstemperatur
F température de transition ordre-désordre
P temperatura przemiany porządek-nieporządek
R температура разупорядочения

-630 **ordering energy**
The difference between the interaction energy of dissimilar pairs of atoms and the mean arithmetic of interaction energies of similar pairs of atoms in a binary alloy.
D Ordnungsenergie
F énergie de mise en ordre
P energia porządkująca
R энергия упорядочения

-635 **configurational specific heat**
An extra specific heat which is accounted for by the disordering of an alloy.
D konfigurationsbedingte Wärmekapazität
F chaleur spécifique additionnelle
P ciepło właściwe konfiguracyjne
R дополнительная удельная теплоёмкость

-640 **magnetic ordering**
Ordering occurring in some magnetic alloys during annealing in a magnetic field.
D magnetisches Ordnen
F mise en ordre magnétique
P uporządkowanie magnetyczne
R магнитное упорядочение

30 LATTICE DEFECTS
GITTERBAUFEHLER
DÉFAUTS RÉTICULAIRES
DEFEKTY SIECI KRYSTALICZNEJ
ДЕФЕКТЫ КРИСТАЛЛИЧЕСКОЙ РЕШЁТКИ

-005 **perfect crystal**
A crystal free from lattice defects.
D Idealkristall, perfekter Kristall, fehlerfreier Kristall
F cristal parfait
P kryształ doskonały
R совершенный кристалл, идеальный кристалл

-010 **real crystal, imperfect crystal**
A crystal really existing, i.e. not free from lattice defects.
D Realkristall
F cristal réel, cristal imparfait
P kryształ rzeczywisty
R реальный кристалл, несовершенный кристалл

-015 **lattice defect, crystal(lographic) imperfection**
An imperfection of a crystal lattice, i.e. a deviation from the normal arrangement of atoms in a crystal lattice.
D Gitter(bau)fehler, Kristallbaufehler, Gitterfehlstelle, Raumgitterfehler, Gitterdefekt
F défaut réticulaire, défaut cristallographique, imperfection réticulaire, défaut (du réseau) cristallin, imperfection cristalline
P defekt sieciowy, defekt sieci krystalicznej, defekt strukturalny, defekt struktury krystalicznej
R дефект (кристаллической) решётки

-020 **defect lattice**
A space lattice which shows a deviation from the normal arrangement of atoms characteristic of a given crystal system.
D fehlerhaftes Gitter
F réseau imparfait
P sieć zdefektowana
R дефектная решётка

-025 **lattice distortion**
A disturbance of the regular geometrical array of a space lattice caused e.g. by the presence of vacancies or foreign atoms.
D Gitterverzerrung
F distortion du réseau
P zniekształcenie sieci
R искажение решётки, искажённость решётки

-030 **point defect, zero-dimensional defect, point imperfection**
A lattice defect confined only to one point, i.e. consisting in a surplus atom or a missing atom.
D punktförmiger Gitter(bau)fehler, Punktdefekt, Punktfehler, nulldimensionaler Gitterfehler, nulldimensionale Fehlstelle, atomare Fehlstelle, Punktfehlstelle
F défaut ponctuel, imperfection ponctuelle
P defekt punktowy
R точечный дефект, нульмерный дефект

-035 **vacancy**
A missing atom in the crystal lattice.
D Leerstelle, Gitterleerstelle, Gitterlücke
F lacune (réticulaire)
P wakans
R вакансия

-040 **single vacancy**
A vacancy which has no other vacancy in its nearest proximity.
D Einfachleerstelle
F monolacune
P monowakans, wakans pojedynczy
R моновакансия, одиночная вакансия

-045 **divacancy**
A pair of two vacancies lying close to one another.
D Doppelleerstelle
F bilacune
P dwuwakans, wakans podwójny, diwakans, biwakans
R бивакансия, дивакансия

-050 **trivacancy**
    A group of three vacancies lying close together.
    D Dreifachleerstelle
    F trilacune
    P trójwakans, wakans potrójny
    R тривакансия

-055 **tetravancy**
    A group of four vacancies lying close together.
    D Vierfachleerstelle
    F quadrilacune
    P czterowakans, wakans poczwórny
    R тетравакансия

-060 **Frenkel defect, Frenkel disorder**
    A pair of point defects of the crystal lattice which arises when an atom leaves its normal lattice site and gets stuck in the interstice.
    D Frenkelsche Fehlstelle, Frenkel-Defekt, Frenkel-Fehlordnung, Frenkel-Paar
    F défaut de Frenkel, paire de Frenkel
    P defekt Frenkla
    R дефект Френкеля, пара Френкеля

-065 **Frenkel vacancy**
    A vacancy which forms part of a Frenkel defect.
    D Frenkelsche Leerstelle
    F lacune de Frenkel
    P wakans Frenkla
    R вакансия Френкеля

-070 **substitutional atom**
    A foreign atom which has replaced the atom of a given metal at a normal lattice site.
    D Substitutionsatom, substituiertes Atom
    F atome de substitution
    P atom podstawieniowy
    R замещенный атом, атом замещения

-075 **(interstitial) atom, intersticialcy**
    An atom which is trapped inside the crystal at a point intermediate between normal lattice sites.
    D Zwischengitteratom, Einlagerungsatom, eingelagertes Atom
    F (atome) interstitiel, atome d'insertion
    P atom międzywęzłowy
    R внедрённый атом, меж(до)узельный атом, атом внедрения

-080 **interstitial impurity**
    An impurity the atoms of which locate themselves in the interstices of the lattice of a given metal.
    D Zwischengitterverunreinigung
    F impureté interstitielle
    P zanieczyszczenie międzywęzłowe
    R примесь внедрения

-085 **dumbbell interstitial, split interstitial**
    A pair of interstitial atoms in a face-centred cubic lattice, which are situated equidistantly on both sides of a vacancy, along the direction passing through the centre of this vacancy.
    D aufgespaltenes Zwischengitteratom, Zwischengitterhantel
    F interstitiel dissocié, paire de semi-interstitiels
    P hantla
    R гантель, гантельная пара, расщеплённый междоузельный атом

-090 **structural vacancy, constitutional vacancy, equilibrium vacancy**
    Any one of the vacancies which are in thermal equilibrium at normal room temperature.
    D strukturelle Leerstelle
    F lacune de constitution
    P wakans strukturalny, wakans równowagowy
    R структурная вакансия

-095 **thermal vacancy**
    Any one of the vacancies resulting from an increase in temperature.
    D thermische Leerstelle
    F lacune thermique
    P wakans termiczny
    R термическая вакансия, тепловая вакансия

-100 **quenched-in vacancy**
    Any one of the thermal vacancies retained by quenching.
    D abgeschreckte Leerstelle
    F lacune trempée, lacune de trempe
    P wakans przechłodzony
    R закалённая вакансия, закалочная вакансия, вакансия зафиксированная закалкой

-105 **excess vacancy**
    Any one of the surplus vacancies which are in excess with respect to the amount of structural vacancies.
    D Überschußleerstelle
    F lacune excessive, lacune en excès
    P wakans nadmiarowy, wakans nierównowagowy
    R избыточная вакансия, неравновесная вакансия

-110 **quenching of vacancies**
Quenching a metal from an elevated temperature at which the vacancy concentration is greater, in order to immobilize and retain these vacancies at room temperature.
D Leerstellenübersättigung
F trempe de lacunes
P przechładzanie wakansów
R закалка вакансий

-115 **vacancy condensation**
The process of agglomeration of vacancies into larger groups.
D Leerstellenkondensation
F condensation de lacunes
P kondensacja wakansów
R конденсация вакансий

-120 **denuded zone, depleted zone**
A region within a crystalline material where many atoms in the space lattice are missing.
D verdünnte Zone, Verarmungszone
F zone lacunaire, zone de Seeger, zone appauvrie
P strefa rozrzedzona, strefa zubożona
R обеднённая зона

-125 **vacancy cluster**
A concentration of vacancies in a region within the crystal grain.
D Anhäufung von Leerstellen, Nest von Leerstellen, Leerstellenanhäufung
F amas de lacunes
P skupisko wakansów
R скопление вакансий

-130 **sink**
A region where a lattice defect can be annihilated.
D Senke
F puits
P ujście
R сток, поглотитель

-135 **annealing of defects**
The elimination of non-equilibrium defects by holding a metal at an elevated temperature and slow cooling.
D Ausheilen der Gitterfehler, Ausheilen der Defekte, Erholung
F recuit de défauts, guérison
P termiczna eliminacja defektów, wyżarzanie defektów
R отжиг дефектов

-140 **line defect, one-dimensional defect**
An imperfection of a crystal lattice extending along a line.
D linienförmiger Gitterfehler, eindimensionaler Gitterbaufehler, eindimensionale Fehlstelle
F défaut linéaire, imperfection linéaire
P defekt liniowy
R линейный дефект, одномерный дефект

-145 **dislocation**
A line defect of a crystal lattice consisting in a configuration of atoms, which exists at the boundary line between the slipped and unslipped area and which spreads across the crystal as the slipped region grows at the expense of the unslipped region.
D Versetzung, Dislokation
F dislocation
P dyslokacja
R дислокация

-150 **edge dislocation, Taylor(-Orowan) dislocation**
A dislocation, for which the Burgers vector is perpendicular to the dislocation line.
D Stufenversetzung, Kantenversetzung
F dislocation coin, dislocation de Taylor
P dyslokacja krawędziowa
R краевая дислокация, линейная дислокация

-155 **positive edge dislocation**
An edge dislocation, in which the extra half plane is situated in the upper part of the crystal, i.e. over the slip plane.
D positive Stufenversetzung
F dislocation coin positive
P dyslokacja krawędziowa dodatnia
R положительная краевая дислокация

-160 **negative edge dislocation**
An edge dislocation, in which the extra half plane is situated in the under part of the crystal, i.e. below the slip plane.
D negative Stufenversetzung
F dislocation coin negative
P dyslokacja krawędziowa ujemna
R отрицательная краевая дислокация

-165 **screw dislocation, Burgers dislocation**
   A dislocation, for which the Burgers vector is parallel to the dislocation line.
   D Schraubenversetzung
   F dislocation vis, dislocation de Burgers
   P dyslokacja śrubowa
   R винтовая дислокация

-170 **right-handed screw dislocation**
   A screw dislocation in which in order to go along the dislocation line downward, one has to move in the clockwise direction.
   D rechtsgängige Schraubenversetzung
   F dislocation vis à droite
   P dyslokacja śrubowa prawoskrętna
   R правовинтовая дислокация

-175 **left-handed screw dislocation**
   A screw dislocation in which in order to go along the dislocation line downward, one has to move in the counter-clockwise direction.
   D linksgängige Schraubenversetzung
   F dislocation vis à gauche, dislocation vis à pas inversé
   P dyslokacja śrubowa lewoskrętna
   R левовинтовая дислокация

-180 **mixed dislocation, compound dislocation**
   A dislocation for which the Burgers vector is neither perpendicular nor parallel to the dislocation line, i.e. a dislocation which can be regarded as the superposition of an edge dislocation and a screw dislocation.
   D gemischte Versetzung
   F dislocation mixte
   P dyslokacja mieszana, dyslokacja złożona
   R смешанная дислокация, составная дислокация

-185 **Burgers circuit**
   The path traversed in an atomic plane in the area enclosing the dislocation line, but at some distance from this line, starting from a lattice point and moving step by step by the same number of lattice vectors in each direction, the initial direction of the circuit and the final one having to intersect.
   D Burgersumlauf
   F circuit de Burgers
   P kontur Burgersa
   R контур Бюргерса

-190 **Burgers vector**
   The vector connecting the end point of the Burgers circuit with its starting point.
   D Burgersvektor, Burgers-Vektor
   F vecteur de Burgers
   P wektor Burgersa
   R вектор Бюргерса

-195 **dislocation line**
   The boundary line between the slipped and unslipped area in a crystal.
   D Versetzungslinie
   F ligne de dislocation
   P linia dyslokacji, linia dyslokacyjna
   R линия дислокации, дислокационная линия

-200 **extra half plane**
   An additional half plane of atoms inserted into a lattice, its presence involving the existence of an edge dislocation.
   D eingeschaltete Halbebene
   F demi-plan (atomique) supplémentaire
   P ekstrapłaszczyzna, półpłaszczyzna nadmiarowa
   R экстраплоскость, лишняя полуплоскость

-205 **dislocation core, dislocation kernel**
   The region near the centre of the dislocation line.
   D Versetzungskern, Kern der Versetzung, Zentrum der Versetzung
   F cœur de dislocation, noyau de dislocation
   P jądro dyslokacji, rdzeń dyslokacji
   R ядро дислокации

-210 **strength of dislocation**
   The length of the Burgers vector expressed as a multiple of the unit lattice vector.
   D Versetzungsstärke, Stärke der Versetzung
   F intensité de dislocation
   P moc dyslokacji
   R мощность дислокации

-215 **unit dislocation**
   A dislocation of unit strength, i.e. where the Burgers vector is equal to the unit lattice vector.
   D Einheitsversetzung
   F dislocation unitaire
   P dyslokacja jednostkowa, dyslokacja o mocy jednostkowej
   R единичная дислокация

-220 **superdislocation**
A pair of dislocations connected by a plane antiphase boundary.
D Superversetzung, Überversetzung
F superdislocation, dislocation de surstructure
P superdyslokacja
R супердислокация, сверхдислокация, сверхструктурная дислокация

-225 **stress field (of a dislocation)**
An elastic stress field surrounding a dislocation which produces forces on other dislocations and results in interaction between dislocations and solute atoms.
D Spannungsfeld der Versetzung
F champ de contraintes d'une dislocation
P pole naprężeń dyslokacji
R поле напряжений дислокации

-230 **self-energy of a dislocation (line)**
The surplus energy of a crystal, associated with the existence of a dislocation.
D Eigenenergie der Versetzung, Selbstenergie der Versetzung
F énergie interne d'une dislocation
P energia własna dyslokacji
R собственная энергия дислокации

-235 **dislocation density**
The number of dislocations intersected by unit area.
D Versetzungsdichte
F densité de dislocations
P gęstość dyslokacji
R плотность дислокации

-240 **(dislocation) jog**
An offset in the dislocation line, resulting from the intersection of two dislocations, or a step created in the dislocation line when a dislocation moves from one slip plane to another.
D Versetzungssprung, Sprung
F cran (de dislocation)
P próg (dyslokacji), uskok (dyslokacji)
R ступенька, порог

-245 **kink**
A step in the dislocation line which does not move the dislocation to another slip plane.
D Kink, Knick
F décrochement
P kolanko, przegięcie, załamanie
R перегиб, излом

-250 **primary dislocations**
Dislocations of the primary slip system.
D Primärversetzungen, primäre Versetzungen
F dislocations primaires
P dyslokacje pierwotne
R первичные дислокации

-255 **secondary dislocations**
Dislocations of secondary slip systems.
D Sekundärversetzungen, sekundäre Versetzungen
F dislocations secondaires
P dyslokacje wtórne
R вторичные дислокации

-260 **overshoot(ing)**
A phenomenon typical of single crystals of alloys, occurring during the progression of the primary slip when an abnormally high stress is needed to activate the conjugated slip system.
D Überschwingen, Überschuß, Überschießen
F rebondissement
P przeskok
R перелёт

-265 **glissile dislocation, glide dislocation, slip dislocation**
A dislocation which glides freely over the slip plane.
D Gleitversetzung, gleitfähige Versetzung
F dislocation glissile
P dyslokacja (po)ślizgowa
R скользящая дислокация, дислокация скольжения

-270 **pinned dislocation, anchored dislocation**
A dislocation which is immobilized at one point.
D verankerte Versetzung
F dislocation épinglée, dislocation ancrée
P dyslokacja zakotwiczona
R закреплённая дислокация

-275 **pinning point, anchoring point**
The point at which a dislocation is immobilized.
D Verankerungspunkt
F point d'ancrage, point d'épinglage
P punkt zakotwiczenia
R точка закрепления

-280 **sessile dislocation**
A dislocation which cannot glide.
D seßhafte Versetzung, sessile Versetzung, nichtgleitfähige Versetzung
F dislocation sessile
P dyslokacja osiadła, dyslokacja półutwierdzona, dyslokacja częściowo zakotwiczona, dyslokacja spoczynkowa
R сидячая дислокация

-285 **non-conservative motion**
A motion of dislocations by climbing.
D nicht-konservative Versetzungsbewegung, Kletterbewegung
F mouvement non conservatif, mouvement non conservateur
P ruch niekonserwatywny (dyslokacji)
R неконсервативное движение, неконсервативное перемещение, диффузионное движение

-290 **conservative motion**
A motion of dislocations by gliding.
D konservative Bewegung, Gleitbewegung
F mouvement conservatif, mouvement conservateur
P ruch konserwatywny (dyslokacji)
R консервативное движение, консервативное перемещение, скользящее движение

-295 **dislocation climb**
A motion of dislocations perpendicular to the slip plane.
D Klettern von Versetzungen
F montée des dislocations, ascension des dislocations
P wspinanie się dyslokacji
R переползание дислокаций, восхождение дислокаций

-300 **Peierls(-Nabarro) force**
The force required to move a dislocation over the potential barrier between successive positions of equilibrium.
D Peierls(-Nabarro)-Kraft
F force de Peierls(-Nabarro)
P siła Peierlsa(-Nabarro)
R сила Пайерльса(-Набарро)

-305 **perfect dislocation, complete dislocation**
A dislocation in which the Burgers vector is equal to a whole number of unit lattice vectors.
D vollständige Versetzung
F dislocation parfaite
P dyslokacja doskonała, dyslokacja całkowita
R полная дислокация

-310 **imperfect dislocation**
A dislocation in which the Burgers vector is not equal to a whole number of lattice vectors.
D unvollständige Versetzung
F dislocation imparfaite
P dyslokacja niedoskonała
R несовершенная дислокация, неполная дислокация

-315 **partial (dislocation), half-dislocation**
An imperfect dislocation in which the Burgers vector is shorter than the unit lattice vector.
D Teilversetzung, Halbversetzung
F dislocation partielle, demi-dislocation
P dyslokacja częściowa
R частичная дислокация, полудислокация

-320 **split(ting) of a dislocation, dissociation of a dislocation**
The breaking up of a perfect dislocation into two partial dislocations.
D Aufspaltung einer Versetzung
F dissociation d'une dislocation, décomposition d'une dislocation
P rozszczepienie dyslokacji, rozciągnięcie dyslokacji
R расщепление дислокации, диссоциация дислокации

-325 **extended dislocation**
A perfect dislocation which has split up into two partial dislocations.
D aufgespaltene Versetzung
F dislocation dissociée
P dyslokacja rozciągnięta, dyslokacja rozszczepiona
R растянутая дислокация, расщеплённая дислокация, вытянутая дислокация

-330 **Shockley (partial) dislocation**
A partial dislocation in the face-centred cubic lattice which is able to move by gliding.
D Shockley-Versetzung, Shockleysche Halbversetzung
F dislocation (partielle) de Shockley
P dyslokacja (częściowa) Shockleya
R дислокация Шокли

-335 **Frank (partial) dislocation, Frank sessile dislocation**
A partial dislocation in the face-centred cubic lattice which can move only by climbing.
D Frankversetzung, Franksche Halbversetzung
F dislocation (sessile) de Frank
P dyslokacja Franka (częściowo zakotwiczona)
R (сидячая) дислокация Франка

-340 **dislocation loop, dislocation ring**
A closed dislocation line.
D Versetzungsschleife, Versetzungsring
F boucle de dislocation, dislocation en anneau
P pętla dyslokacyjna, pętla dyslokacji
R дислокационная петля, дислокационное кольцо, дислокационный круг, петля дислокации

-345 **prismatic (dislocation) loop, prismatic dislocation, penny-shaped dislocation**
A dislocation loop which does not lie in a plane with its Burgers vector.
D prismatische Versetzung, prismatischer Versetzungsring
F boucle (de dislocation) prismatique, dislocation prismatique
P krawędziowa pętla dyslokacyjna, dyslokacja pryzmatyczna, dyslokacja krawędziowa nieregularna
R призматическая петля (дислокаций), призматическая дислокация

-350 **dislocation node**
A point at which two or more dislocations meet.
D Versetzungsknoten
F nœud de dislocation
P węzeł dyslokacji, węzeł dyslokacyjny
R узел дислокаций, дислокационный узел

-355 **helic(oid)al dislocation**
A dislocation in which the dislocation line is in the shape of a helix.
D Versetzungswendel, Versetzungsspirale
F dislocation en hélice, dislocation hélicoïdale, dislocation spirale
P dyslokacja helikoidalna, dyslokacja spiralna
R геликоидальная дислокация

-360 **stair-rod dislocation**
The resultant dislocation arising from the intersection of planes of two different partial dislocations, i.e. at a bend in the stacking fault.
D Winkelversetzung
F dislocation d'arête, dislocation tringle, dislocation anguleuse, dislocation stair-rod
P dyslokacja kątowa, dyslokacja narożnikowa
R вершинная дислокация

-365 **pole dislocation**
A dislocation of the dislocation forest, about which the two arms of a mobile dislocation rotate in opposite directions.
D Polversetzung
F dislocation-pôle, dislocation-poteau, dislocation-perche, dislocation-baton
P dyslokacja biegunowa
R полюсная дислокация

-370 **emissary dislocations**
Dislocations sent out by a noncoherent twin boundary.
D Vorläufer-Versetzungen
F dislocations émissaires
P dyslokacje emitowane
R испущенные дислокации

-375 **twinning dislocation**
Any one of the partial dislocations which create a twinned region.
D Zwillingsversetzung
F dislocation de macle
P dyslokacja bliźniakująca
R двойникующая дислокация

-380 **dislocation dipole**
A pair of closely spaced parallel dislocations of opposite sign.
D Versetzungsdipol
F dipôle de dislocations
P dipol dyslokacji, dipol dyslokacyjny
R дислокационный диполь

-385 **dislocation tangle**
A complex consisting of kinked and intertwined dislocations, interdispersed with prismatic dislocation loops, resulting from an interplay between point defects and dislocations.
D Versetzungsknäuel, Versetzungsansammlung
F écheveau de dislocations
P splot dyslokacji
R сплетение дислокаций, клубок дислокаций

-390 **dislocation network**
A three-dimensional system composed of intersecting dislocations.
D Versetzungsnetz(werk)
F réseau de dislocations
P sieć dyslokacji, sieć dyslokacyjna
R сетка дислокации, дислокационная сетка

-395 **dislocation wall**
A stable configuration of many dislocations of the same sign (i.e. only positive or only negative) arranged one under the other.
D Versetzungswand
F paroi de dislocations
P ściana dyslokacji
R стенка дислокаций, дислокационная стенка

-400 **dislocation pile-up, pile-up of dislocations**
An accumulation of successive dislocations, belonging to the same slip plane, which occurs when the first dislocation in a sequence of similar dislocations emitted by a dislocation source is held up by some extended obstacle.
D Versetzungsaufstauung, Aufstauung von Versetzungen, Dislokationsanhäufung
F empilement de dislocations
P spiętrzenie dyslokacji, skupienie dyslokacji
R скопление дислокаций

-405 **dislocation forest**
The dislocations threading through the active slip plane.
D Versetzungswald
F forêt de dislocations
P las dyslokacji
R лес дислокаций, дислокационный лес

-410 **(dislocation) tree, tree dislocation**
An individual dislocation of a dislocation forest.
D Waldversetzung
F arbre
P dyslokacja lasu, dyslokacja-drzewo
R дислокация леса

-415 **dislocation structure**
The structure of a metal or alloy exhibiting a dislocation pattern.
D Versetzungsstruktur
F structure de dislocation
P struktura dyslokacyjna
R дислокационная структура

-420 **dislocation multiplication**
The generating of new dislocations, attributable to the operation of a dislocation source, occurring when a crystal is being plastically deformed.
D Vervielfachung von Versetzungen, Vervielfältigung von Versetzungen, Versetzungsmultiplikation
F multiplication des dislocations
P generowanie dyslokacji, rozmnażanie dyslokacji, mnożenie dyslokacji
R размножение дислокаций, генерирование дислокаций

-425 **dislocation source, dislocation generator**
The mechanism by which new dislocations are generated in a crystal.
D Versetzungsquelle
F source (multiplicatrice) de dislocations
P źródło dyslokacji, generator dyslokacji
R источник дислокаций, генератор дислокаций, дислокационный источник

-430 **Frank-Read source, Frank-Read dislocation generator**
A dislocation source involving a dislocation line anchored at two points on the slip plane. As the stress applied is increased the part of this line between the anchoring points bows out, expands outwards in the slip plane, taking on the shape of a spiral around these points, until on further expansion it transforms into a free (i.e. non anchored) dislocation loop and a remaining portion linking the anchoring points, capable of repeating the above process.
D Frank-Read-Quelle
F source de Frank-Read, générateur de Frank-Read
P źródło Franka-Reada
R источник Франка-Рида

-435 **Koehler source**
A variant of the Frank-Read source, where the part of a dislocation line which has a screw orientation can pass from its slip plane on to an intersecting slip plane and, after travelling a short distance on the new plane, may resume its motion on another plane of the original slip system.
D Koehler-Quelle
F source de Koehler
P źródło Koehlera
R источник Кёлера

-440 **Cottrell atmosphere, impurity atmosphere, impurity cloud**
The accumulation of foreign atoms in the region around a dislocation.
D Cottrell-Atmosphäre, Cottrellsche Wolke, Verunreinigungsatmosphäre
F atmosphère de Cottrell, atmosphère dislocations-impuretés, nuage d'impuretés, nuage de Cottrell
P atmosfera Cottrella, atmosfera atomów obcych, atmosfera zanieczyszczeń, chmura zanieczyszczeń
R коттрелловская атмосфера, коттрелловское облако, атмосфера примесей, облако примесей, атмосфера Коттрелла, облако примесных атомов, облако Коттрелла

-445 **dislocation locking**
The generating of obstacles to the free motion of dislocations.
D Versetzungsblockierung, Blockierung der Versetzungen
F blocage de dislocations
P blokowanie dyslokacji
R блокировка дислокаций, блокирование дислокаций

-450 **Cottrell-Lomer barrier, Cottrell-Lomer lock**
A group of three sessile dislocations produced in the face centred cubic lattice by the glide of dislocations on intersecting (111) planes.
D Lomer-Cottrell-Versetzung
F verrou de Cottrell-Lomer, barrière de Lomer-Cottrell
P bariera Cottrella-Lomera
R барьер Ломера-Коттрелла, дислокация Ломера-Коттрелла

-455 **Suzuki effect**
The segregation of solute atoms in the stacking faults.
D Suzuki-Effekt
F effet Suzuki
P zjawisko Suzuki
R эффект Сузуки

-460 **Snoek effect**
Stress induced ordering of interstitial impurity atoms in a body-centred cubic lattice (e.g. carbon or nitrogen atoms in alpha iron).
D Snoek-Effekt
F effet Snoek
P zjawisko Snoeka
R эффект Снука

-465 **charged dislocation**
A dislocation in an ionic crystal.
D geladene Versetzung
F dislocation chargée
P dyslokacja naładowana
R заряженная дислокация

-470 **crowdion**
A special kind of linear lattice defect, consisting of an extra atom in a relatively short row of atoms along the direction [110] (i.e. $x+1$ atoms over a distance of x interatomic spacings).
D Crowdion
F crowdion
P kraudion, crowdion
R краудион

-475 **activation volume**
The product of the following three quantities: the Burgers vector of a glissile dislocation, the mean distance between tree dislocations, and the width of the stacking fault of tree dislocations.
D Aktivierungsvolumen
F volume d'activation
P objętość aktywowana
R активационный объём

-480 **disclination**
A line defect of the crystal lattice of a similar type as the dislocation, but created by rotational displacement and not by slipping.
D Disklination
F disclination
P dysklinacja
R дисклинация

-485 **radiation defect, irradiation-induced defect, irradiation-produced defect**
A lattice defect generated by irradiation.
D Bestrahlungsfehler
F défaut d'irradiation
P defekt radiacyjny, defekt popromienny
R радиационный дефект, дефект созданный облучением

-490 **(ir)radiation damage**
The existence of lattice defects which have been generated by irradiation.
D Strahlenschädigung, Strahlungsschädigung
F défectuosité due à l'irradiation
P zdefektowanie radiacyjne
R радиационная дефектность, радиационное повреждение

-495 **(ir)radiation hardening, radiation-induced strengthening**
Increasing the hardness and strength of a metal by means of irradiation-induced defects.
D Bestrahlungsverfestigung
F durcissement par irradiation
P utwardzanie radiacyjne, umacnianie radiacyjne
R радиационное упрочнение, упрочнение облучением

-500 **channeling**
The deep penetration of bombarding particles into the crystal, along channels created by closely packed atomic rows and planes.
D Kanalisierung
F canalisation
P efekt kanałowy
R каналирование, эффект каналирования

-505 **focused collision, focuson**
A collision process between bombarding particles and atoms in a crystal lattice occurring along a closely packed row of atoms.
D Fokusson, fokussierter Stoß
F collision focalisée, focuson
P fokuson
R фокусон, фокусирующее столкновение

-510 **displaced atom**
An atom which has been knocked out of its normal lattice site and has occupied an interstitial position.
D versetztes Atom, verlagertes Atom
F atome déplacé
P atom dyslokowany
R смещённый атом, дислоцированный атом

-515 **collision cascade**
A successive avalanche of collisions between the bombarding particles and atoms in a crystal lattice.
D Stoßkaskade, Stoßlawine
F cascade de collisions, collisions en cascade
P kaskada zderzeń, lawina zderzeń, łańcuch zderzeń
R каскад столкновений

-520 **displacement cascade, cascade of displacements**
An avalanche displacement of atoms in a crystal lattice resulting from a collision cascade.
D Verlagerungskaskade, Verlagerungslawine
F déplacement en cascade, cascade de déplacements
P kaskada przemieszczeń, lawina przemieszczeń
R каскад смещений

-525 **displacement spike**
The region of maximum intensity of the displacement cascade.
D Verlagerungsspitze
F pointe de déplacements
P szczyt przemieszczeń
R пик смещений

-530 **thermal spike**
The region of a very high temperature resulting from a collision cascade.
D thermischer Störungsbereich
F pointe thermique, pic thermique
P szczyt cieplny, szczyt temperaturowy
R температурный пик, термический пик, тепловой пик

-535 **plane defect, surface defect, planar defect, two-dimensional defect**
An imperfection of a crystal lattice extending over a surface.
D zweidimensionaler Gitterbaufehler, zweidimensionale Fehlstelle
F défaut de surface, imperfection de surface, défaut plan
P defekt powierzchniowy
R поверхностный дефект, двухмерный дефект

-540 **stacking fault**
A region within a crystal where the regular stacking sequence of close-packed atomic planes is disturbed.
D Stapelfehler
F faute d'empilement, défaut d'empilement
P błąd ułożenia
R дефект упаковки

-545 **intrinsic stacking fault**
A stacking fault associated with the removal of a close-packed layer of atoms.
D Stapelfehler I. Art
F faute d'empilement intrinsèque, défaut d'empilement intrinsèque
P błąd ułożenia wewnętrzny, błąd ułożenia pojedynczy
R дефект упаковки вычитания

-550 **extrinsic stacking fault**
A stacking fault associated with the insertion of an extra close-packed layer of atoms.
D Stapelfehler II. Art
F faute d'empilement extrinsèque, défaut d'empilement extrinsèque
P błąd ułożenia zewnętrzny, błąd ułożenia podwójny
R дефект упаковки внедрения

-555 **twin fault**
A stacking fault produced at the interface between two perfect crystallites which are in twin relation.
D Zwillingsstapelfehler
F faute d'empilement de maclage
P błąd ułożenia bliźniaczy
R двойниковый дефект (упаковки)

-560 **stacking fault energy**
The surplus energy per unit area of a crystal associated with the existence of a stacking fault.
D Stapelfehlerenergie
F énergie de faute(s) d'empilement, énergie de défaut d'empilement
P energia błędu ułożenia
R энергия дефектов упаковки, энергия дефекта упаковки

-565 **grain boundary**
An internal boundary which separates a crystal of one orientation from that of a differing orientation.
D Korngrenze
F joint de grain(s), joint intergranulaire
P granica ziarn(a)
R граница зёрен, граница зерна, межзёренная граница

-570 **equicohesive temperature**
The temperature at which the grain boundaries and grains have equal strength.
D äquikohäsive Temperatur
F température d'équicohésion
P temperatura ekwikohezyjna
R эквикогезивная температура, температура равносвязи

-575 **grain boundary energy**
The amount of energy associated with the existence of a grain boundary.
D Korngrenzenenergie
F énergie intergranulaire
P energia granic ziarn
R энергия границ зерен

-580 **grain boundary dislocation**
The disturbance left within a high-angle boundary by a glide dislocation that has passed through this boundary.
D Korngrenzenversetzung
F dislocation de joint de grains, dislocation d'interface
P dyslokacja granic ziarn
R пограничная дислокация, зернограничная дислокация

-585 **high-angle (grain) boundary, large angle boundary**
A grain boundary in which the misorientation angle is large and cannot be described directly by means of a dislocation model.
D Großwinkel(korn) grenze
F joint à grand angle
P granica szerokokątowa
R большеугловая граница

-590 **low-angle (grain) boundary, small angle boundary**
A boundary in which the misorientation angle is small and which can be described by means of a dislocation model.
D Kleinwinkel(korn)grenze
F joint d'angle faible
P granica wąskokątowa
R малоугловая граница

-595 **misorientation (angle), disorientation**
The angle between crystal lattices on either side of a boundary.
D Orientierungsunterschied, Desorientierung
F (angle de) désorientation
P kąt dezorientacji, kąt wzajemnej orientacji, dezorientacja
R угол разориентировки, разориентировка, дезориентация

-600 **twist boundary**
A low-angle boundary which consists entirely of a net of screw dislocations.
D Drehgrenze, Twist-Grenze
F joint de torsion, sous-joint de torsion
P granica skręcona, granica skręcenia
R граница кручения

-605 **tilt boundary**
A low-angle boundary which consists entirely of parallel edge dislocations.
D Kippgrenze, Tilt-Grenze
F (sous-)joint de flexion
P granica daszkowa, granica skośna, granica nachylona, granica pochylenia
R граница наклона, наклонная граница

-610 **interphase boundary, interphase interface, interface**
The surface of separation of two different phases.
D Phasengrenzfläche
F interface (de phase), limite interphases
P granica międzyfazowa, powierzchnia międzyfazowa, powierzchnia rozdziału faz
R межфазная граница, межфазная поверхность, поверхность раздела (фаз), граница раздела фаз

-615 **lattice misfit, lattice disregistry**
A lack of geometrical coincidence of two space lattices meeting at an interface.
D Gitterfehlanpassung
F discordance du réseau
P niedopasowanie sieciowe
R несоответствие решётки, несогласованность решётки

-620 **coherency**
The existence of a continuity of crystal structure across the interphase boundary.
D Kohärenz
F cohérence
P koherencja, sprzężenie
R когерентность

-625 **coherent boundary, coherent interface**
An interphase boundary having a special orientation relationship for which the atom spacing on one side of the boundary is very similar to that of the other side so that there is a continuity of rows and planes of lattice points across the boundary.
D kohärente Phasengrenze, kohärente Grenzfläche
F interface cohérente, joint cohérent
P granica (międzyfazowa) koherentna, granica (międzyfazowa) sprzężona
R когерентная граница

-630 **incoherent boundary, non-coherent boundary, incoherent interface**
An interphase boundary across which there is no continuity of rows and planes of lattice points, the transition region between the two phases being very disordered.
D nicht-kohärente Phasengrenze, nicht-kohärente Grenzfläche
F interface incohérente, joint incohérent
P granica (międzyfazowa) niekoherentna, granica (międzyfazowa) niesprzężona
R некогерентная граница

-635 **interface energy, interfacial energy**
The amount of energy associated with the existence of an interphase interface.
D Grenzflächenenergie
F énergie interfaciale, énergie d'interface
P energia granicy międzyfazowej
R энергия межфазовой границы, межфазовая энергия

-640 **surface energy**
The amount of energy required to create a unit area of the free surface of a condensed phase.
D Oberflächenenergie
F énergie superficielle, énergie de surface
P energia powierzchniowa
R поверхностная энергия

-645 **surface tension, interfacial tension**
The property of phase surfaces, resulting from unbalanced molecular cohesive forces at the surface, behaving as though they were covered by a thin elastic membrane in tension. This property is particularly pronounced in liquid phases.
D Oberflächenspannung, Grenzflächenspannung
F tension superficielle
P napięcie powierzchniowe
R поверхностное натяжение

-650 **sub(-)grain**
Any one of the individual regions within the sub-structure.
D Subkorn, Feinkorn, Unterkorn
F sous-grain
P podziarno
R субзерно

-655 **mosaic block**
A region in the mosaic structure within which the orientation does not vary.
D Mosaikblock
F bloc mosaïque
P blok mozaiki, krystalit podmikroskopowy
R мозаичный блок, блок мозаики, мозаичная область

-660 **subboundary, sub(-)grain boundary**
A low-angle boundary occurring as a result of recovery within grains delimited by high-angle boundaries.
D Unterkorngrenze, Feinkorngrenze, Subkorngrenze, sekundäre Grenze
F sous-joint (de grains)
P podgranica
R субграница, субзёренная граница

35 PHASE CONSTITUTION OF ALLOYS
KONSTITUTION DER LEGIERUNGEN
CONSTITUTION DES ALLIAGES
BUDOWA FAZOWA STOPÓW
ФАЗОВОЕ СТРОЕНИЕ СПЛАВОВ

-005 **system**
Any number and quantity of chemical substances considered separated from their surroundings.
D System
F système
P układ
R система

-010 **phase rule**
A thermodynamic rule stating that in any system in equilibrium $z = s - f + 2$, where z = number of degrees of freedom, s = number of components, f = number of phases.
D Phasengesetz, Phasenregel
F loi des phases, règle des phases
P reguła faz
R правило фаз

-015 **phase equilibrium, heterogeneous equilibrium**
The coexistence of phases in thermodynamic equilibrium.
D Phasengleichgewicht, heterogenes Gleichgewicht
F équilibre des phases, équilibre hétérogène, équilibre physico-chimique
P równowaga fazowa
R фазовое равновесие, гетерогенное равновесие

-020 **(phase) constitution**
The make-up of an alloy system, envisaged from the view-point of phases which form its structure.
D Konstitution
F constitution
P budowa fazowa
R фазовое строение

-025 **phase**
A part of an alloy system of uniform physical and chemical properties, separated from the remainder of the system by a boundary surface called an interface.
D Phase
F phase
P faza
R фаза

-030 **variance, number of degrees of freedom, degrees of freedom**
The number of variables which for a system at equilibrium may be arbitrarily specified — within certain limits — without the disappearance of a phase or the appearance of a new phase.
D Zahl der Freiheitsgrade, Anzahl der Freiheiten
F variance, nombre de degrés de liberté
P liczba stopni swobody
R число степеней свободы, вариантность, степень вариантности

-035 **component**
Any one of the independent chemical substances making up the various phases constituting an alloy system.
D Bestandteil, Komponente
F constituant, composant(e)
P składnik (układu)
R компонент

-040 **triple point**
A point in a temperature-pressure diagram of an one-component system, indicating the temperature and pressure at which three phases exist in equilibrium.
D Tripelpunkt
F point triple
P punkt potrójny
R тройная точка

-045 **equilibrium system**
An alloy system in a state of equilibrium.
D Gleichgewichtssystem
F système d'équilibre
P układ równowagi
R равновесная система

-050 **equilibrium state**
A state in which the properties of a system will not change with time, ad infinitum, unless acted upon by some constraint.
D Gleichgewichtszustand
F état d'équilibre
P stan równowagi
R равновесное состояние, состояние равновесия

-055 **stable equilibrium**
   The state of a system when its free enthalpy has the least value under given thermodynamic conditions.
   D stabiles Gleichgewicht
   F équilibre stable
   P równowaga trwała, równowaga stabilna
   R устойчивое равновесие, стабильное равновесие

-060 **metastable equilibrium**
   The state of apparent equilibrium of a system, i.e. one in which there is no tendency to spontaneous change, although the free enthalpy has not yet the least value for the given thermodynamic conditions, but which under certain circumstances can pass into stable equilibrium.
   D metastabiles Gleichgewicht
   F équilibre métastable, équilibre instable
   P równowaga metastabilna
   R неустойчивое равновесие, метастабильное равновесие

-065 **stable state, stable condition**
   The state of a system corresponding to the stable equilibrium.
   D stabiler Zustand
   F état stable
   P stan stabilny
   R устойчивое состояние, стабильное состояние

-070 **metastable state, metastable condition**
   The state of a system corresponding to the metastable equilibrium.
   D metastabiler Zustand
   F état métastable
   P stan metastabilny
   R метастабильное состояние

-075 **invariant system**
   A system with no degree of freedom.
   D nonvariantes System
   F système invariant
   P układ niezmienny, układ zerozmienny
   R нонвариантная система

-080 **monovariant system**
   A system with one degree of freedom.
   D univariantes System
   F système monovariant, système univariant
   P układ jednozmienny, układ o jednym stopniu swobody
   R моновариантная система, одновариантная система

-085 **bivariant system**
   A system with two degrees of freedom.
   D bivariantes System
   F système bivariant
   P układ dwuzmienny, układ o dwóch stopniach swobody
   R дивариантная система

-090 **single-phase system, homogeneous system, monophasis system**
   A system consisting of one phase.
   D Einphasensystem, einphasiges System, homogenes System, monophasiges System
   F système monophasé
   P układ jednofazowy
   R однофазная система, гомогенная система

-095 **heterogeneous system**
   A system in which there is more than one phase.
   D heterogenes System
   F système hétérogène
   P układ niejednofazowy
   R гетерогенная система, неоднородная система, негомогенная система

-100 **two-phase system, biphasic system**
   A system consisting of two phases.
   D zweiphasiges System, Zweiphasensystem
   F système biphasé
   P układ dwufazowy
   R двухфазная система

-105 **three-phase system**
   A system consisting of three phases.
   D Dreiphasensystem, dreiphasiges System
   F système triphasé
   P układ trójfazowy
   R трёхфазная система

-110 **polyphase system, multiphase system**
   A system consisting of several phases.
   D Mehrphasensystem, mehrphasiges System
   F système polyphasé
   P układ wielofazowy
   R многофазная система

-115 **un(it)ary system, one-component system, unicomponent system**
   A system consisting of one pure substance.
   D un(it)äres System, Einstoffsystem, einkomponentiges System, Einkomponentensystem
   F système un(it)aire, système à un composant
   P układ jednoskładnikowy
   R однокомпонентная система, унарная система

-120 **binary (system), two-component system**
   A system consisting of two components.
   D binäres System, Zweistoffsystem, zweikomponentiges System, Zweikomponentensystem
   F système binaire, système à deux composants
   P układ podwójny, układ dwuskładnikowy
   R двойная система, бинарная система, двухкомпонентная система

-125 **ternary (system), three-component system**
   A system consisting of three components.
   D ternäres System, Dreistoffsystem, dreikomponentiges System, Dreikomponentensystem
   F système ternaire, système à trois composants
   P układ potrójny, układ trójskładnikowy, układ trzyskładnikowy
   R тройная система, трёхкомпонентная система

-130 **pseudo(-)binary system, quasi-binary system**
   A ternary alloy system in which an intermetallic compound, formed by two of the three components, behaves as an independent component.
   D pseudobinäres System, quasibinäres System
   F système pseudobinaire, système quasibinaire
   P układ pseudopodwójny, układ kwazypodwójny, układ quasi-podwójny
   R псевдобинарная система, квазибинарная система

-135 **quaternary (system), four-component system**
   A system consisting of four components.
   D quaternäres System, Vierstoffsystem, vierkomponentiges System, Vierkomponentensystem
   F système quaternaire, système à quatre composants
   P układ poczwórny, układ czteroskładnikowy
   R четверная система, четырёхкомпонентная система

-140 **polynary system, multicomponent system, polycomponent system**
   A system consisting of several components.
   D Vielstoffsystem, Mehrstoffsystem, Vielkomponentensystem
   F système à composants multiples, système à constituants multiples
   P układ wieloskładnikowy
   R многокомпонентная система

-145 **condensed phase**
   A liquid or solid phase, as distinct from a gaseous phase.
   D kondensierte Phase
   F phase condensée
   P faza skondensowana
   R конденсированная фаза

-150 **metal(lic) phase**
   A phase which possesses the features of the metallic state, i.e. one in which the metallic bond exists exlusively, or at least prevails.
   D metallische Phase
   F phase métallique
   P faza metaliczna
   R металлическая фаза

-155 **nonmetallic phase**
   A phase which does not possess the features of the metallic state, i.e. one in which a bond other than the metallic bond exists or at least prevails.
   D nichtmetallische Phase
   F phase non métallique
   P faza niemetaliczna
   R неметаллическая фаза

-160 **high-temperature phase**
   A phase which is stable at high temperature.
   D Hochtemperaturphase
   F phase stable à haute température
   P faza wysokotemperaturowa
   R высокотемпературная фаза

-165 **high pressure phase**
A phase which is stable at high pressure.
D Hochdruckphase
F phase stable à haute pression
P faza wysokociśnieniowa
R фаза устойчивая при высоком давлении

-170 **stable phase**
A phase which, under given thermodynamic conditions, has the least value of the free enthalpy in a given system.
D stabile Phase
F phase stable
P faza stabilna
R стабильная фаза, устойчивая фаза

-175 **metastable phase**
A phase occurring in a state of apparent equilibrium, i.e. one which shows no tendency to spontaneous change but under certain conditions can transform into a stable form.
D metastabile Phase
F phase métastable
P faza metastabilna
R неустойчивая фаза, метастабильная фаза

-180 **equilibrium phase**
A phase the existence of which is in accordance with the requirements of the thermodynamic equilibrium.
D Gleichgewichtsphase
F phase d'équilibre
P faza równowagowa
R равновесная фаза

-185 **nonequilibrium phase**
A phase departing from a state of thermodynamic equilibrium.
D Ungleichgewichtsphase
F phase de non-équilibre, phase hors d'équilibre
P faza nierównowagowa
R неравновесная фаза

-190 **coexisting phases, conjugate phases**
Phases which are in thermodynamic equilibrium at a given temperature and a given pressure.
D koexistierende Phasen
F phases coexistentes
P fazy współistniejące, fazy termodynamiczne sprzężone
R сосуществующие фазы

-195 **transient phase, transition phase**
A metastable phase which is formed in the initial stage of some transformations, its existence being relatively short.
D Zwischenphase, Übergangsphase
F phase transitoire
P faza przejściowa
R переходная фаза

-200 **ordered phase, superlattice phase**
A solid phase in which the arrangement of atoms of the components with respect to each other in the crystal lattice is not random, but follows a determined periodically repeated pattern.
D geordnete Phase, geordnete Mischphase, Ordnungsphase, Überstrukturphase
F phase ordonnée
P faza uporządkowana
R упорядоченная фаза

-205 **disordered phase**
A solid phase in which the arrangement of atoms of the components with respect to each other in the crystal lattice is random.
D ungeordnete Phase, ungeordnete Mischphase
F phase désordonnée
P faza nieuporządkowana
R неупорядоченная фаза

-210 **intermediate phase, intermetallic phase**
A phase, occurring in an alloy system, in which the crystal structure differs from the crystal structure of either of the component elements.
D intermediäre Phase, intermetallische Phase
F phase intermédiaire, phase intermétallique
P faza pośrednia, faza międzymetaliczna
R промежуточная фаза, интерметаллическая фаза, интерметаллидная фаза

-215 **intermetallic compound**
An intermediate phase occurring at a definite atomic ratio.
D intermetallische Verbindung
F composé intermétallique
P związek międzymetaliczny
R интерметаллическое соединение, интерметаллидное соединение, интерметаллид

-220 **congruently melting compound, congruent(ly-melting) intermediate phase**
    An intermetallic compound which melts or freezes at a fixed temperature with no change in the composition of the solid or liquid phase.
    D kongruente intermetallische Verbindung
    F composé à fusion congruente
    P związek międzymetaliczny kongruentny, faza międzymetaliczna kongruentna
    R конгруэнтное интерметаллическое соединение, конгруэнтная интерметаллическая фаза

-225 **incongruently melting compound, incongruent(ly-melting) intermediate phase, peritectic compound**
    An intermetallic compound which after having reached, on heating, a certain definite temperature, splits up into a liquid phase and a solid phase of different compositions.
    D nichtkongruente intermetallische Verbindung, inkongruente intermetallische Verbindung
    F composé à fusion non congruente, composé à fusion incongruente
    P związek międzymetaliczny niekongruentny, związek międzymetaliczny inkongruentny, faza międzymetaliczna niekongruentna, faza międzymetaliczna inkongruentna
    R неконгруэнтное интерметаллическое соединение, неконгруэнтная интерметаллическая фаза

-230 **ionic compound, polar compound**
    A compound which is held together by the ionic bond.
    D heteropolare Verbindung
    F composé ionique, composé polaire
    P związek jonowy
    R ионное соединение, полярное соединение

-235 **(chemical) compound**
    A chemically pure substance composed of two or more elements united by chemical affinity in definite weight proportions.
    D chemische Verbindung
    F composé chimique, combinaison chimique
    P związek chemiczny
    R химическое соединение

-240 **stoichiometric composition**
    The chemical composition of an alloy corresponding to a definite chemical formula.
    D stöchiometrische Zusammensetzung
    F composition stœchiométrique
    P skład stechiometryczny
    R стехиометрический состав

-245 **daltonides**
    Intermetallic compounds, the composition of which is strictly stoichiometric.
    D daltonide Verbindungen, Daltonide, Daltonidverbindungen
    F composés stœchiométriques
    P daltonidy
    R дальтониды

-250 **berthollides**
    Intermetallic compounds the composition of which does not correspond strictly to a stoichiometric formula and which exist instead over a composition range.
    D berthollide Verbindungen, Berthollide, Berthollidverbindungen
    F composés non stœchiométriques
    P bertolidy
    R бертоллиды

-255 **equilibrium concentration**
    The percentage of a component in a phase concordant with the equilibrium diagram.
    D Gleichgewichtskonzentration
    F concentration d'équilibre
    P stężenie równowagowe
    R равновесная концентрация

-260 **Hägg's rule**
    A rule stating that the crystal structure of metalloid-metal compounds is simple or complex, according to whether the atomic radius ratio is less than or greater than 0.59.
    D Häggsche Regel
    F règle de Hägg
    P reguła Hägga
    R правило Хэгга

-265 **Laves phases, Laves compounds**
    A group of intermetallic phases of approximate general chemical formula AB_2, occurring in alloys formed by elements in which the atomic diameters are approximately in the ratio 1.2 to 1.
    D Laves-Phasen
    F phases de Laves
    P fazy Lavesa
    R фазы Лавеса

-270 **Zintl phases**
A group of intermetallic compounds formed by alkali metals, alkali earth metals, and rare earth metals with metals of the groups II, III, and IV of the periodic table. In these compounds — apart from the metallic bond — chemical valency effects also play an important part.
D Zintl-Phasen
F phases de Zintl
P fazy Zintla
R фазы Цинтля

-275 **sigma phase, σ phase**
A hard, brittle non-magnetic intermetallic phase, occurring in certain high chromium steels, its composition range being close to the formula FeCr.
D Sigma-Phase
F phase sigma, phase σ
P faza sigma, faza σ
R сигма-фаза

-280 **electron phase, electron compound, Hume-Rothery phase**
A phase having a fixed ratio between the number of valence electrons and the number of atoms.
D Elektronenkonzentrationsphase, Hume-Rothery-Phase
F composé intermétallique électronique, phase de Hume-Rothery
P faza elektronowa
R электронная фаза, электронное соединение, фаза Юм-Розери

-285 **electron concentration**
The ratio of the number of valency electrons to the number of atoms in a unit cell of a metallic phase.
D Valenzelektronenkonzentration
F concentration électronique, concentration des électrons
P stężenie elektronowe
R электронная концентрация

-290 **parent phase**
A phase from which another phase has been precipitated.
D Mutterphase, Matrix
F phase mère
P faza macierzysta
R маточная фаза, исходная фаза

-295 **solubility**
The ability of a component to dissolve in another component, i.e. to form a solution.
D Löslichkeit
F solubilité
P rozpuszczalność
R растворимость

-300 **(mutual) intersolubility, miscibility**
The ability of two (or more) components to dissolve in each other, i.e. to form solutions.
D Mischbarkeit
F solubilité réciproque, solubilité mutuelle, miscibilité
P rozpuszczalność wzajemna
R взаимная растворимость

-305 **liquid solubility, liquid miscibility**
The ability of two (or more) components to form a liquid solution.
D Mischbarkeit im flüssigen Zustand, Mischbarkeit in der Schmelze
F solubilité (à l'état) liquide, miscibilité (à l'état) liquide
P rozpuszczalność w stanie ciekłym
R растворимость в жидком состоянии, сплавляемость

-310 **solid solubility, solid miscibility**
The ability of two (or more) components to form a solid solution.
D Mischbarkeit im festen Zustand
F solubilité (à l'état) solide, miscibilité (a l'etat) solide
P rozpuszczalność w stanie stałym
R растворимость в твёрдом состоянии

-315 **complete miscibility, complete solubility**
The ability of two (or more) components to form a solution in the whole range of compositions.
D voll(ständ)ige Mischbarkeit, lückenlose Mischbarkeit
F solubilité continue, solubilité illimitée, miscibilité complète
P rozpuszczalność nieograniczona, rozpuszczalność ciągła
R полная растворимость, неограниченная растворимость

-320 **limited solubility, restricted solubility, partial solubility**
The ability of two (or more) components to form solutions only in a limited range of compositions.
D beschränkte Mischbarkeit, begrenzte Mischbarkeit, beschränkte Löslichkeit, begrenzte Löslichkeit
F miscibilité partielle, solubilité partielle, miscibilité limitée, miscibilité incomplète
P rozpuszczalność ograniczona
R частичная растворимость, ограниченная растворимость

-325 **Hume-Rothery rule**
A rule stating that if the difference between atomic diameters of component elements forming an alloy exceeds approximately 14–15%, solid solubility should become restricted.
D Hume-Rothersche Regel
F règle de Hume-Rothery
P reguła Hume-Rothery'ego
R правило Юм-Розери

-330 **solubility limit, limit of solubility, saturation point**
The maximum amount of a solute in a solution at a given temperature and pressure.
D Löslichkeitsgrenze
F limite de solubilité, limite de saturation
P rozpuszczalność graniczna, granica rozpuszczalności
R предельная растворимость, граница растворимости, предел растворимости

-335 **critical (solution) temperature**
The lowest temperature at which complete miscibility in a binary system occurs.
D kritische Temperatur
F température critique
P temperatura krytyczna
R критическая температура

-340 **immiscibility**
The incapability of two substances to form a solution.
D Unmischbarkeit
F insolubilité
P brak rozpuszczalności
R нерастворимость, несмешиваемость

-345 **miscibility gap, solubility gap**
The range of compositions and temperatures within which there is no miscibility in the given system.
D Mischungslücke
F lacune de miscibilité
P zakres mieszanin, zakres nierozpuszczalności
R область несмешиваемости

-350 **solution**
A perfectly homogeneous mixture, of variable composition, of two or more substances.
D Lösung
F solution
P roztwór
R раствор

-355 **solvent**
The major component of a solution.
D Lösungsmittel, Lösemittel, Löser
F solvant
P rozpuszczalnik
R растворитель

-360 **solute**
The minor component of a solution.
D Gelöste
F soluté
P ciało rozpuszczone, substancja rozpuszczona, pierwiastek rozpuszczony, składnik rozpuszczony
R растворённое вещество, растворённый элемент

-365 **liquid solution**
A solution in a liquid state of aggregation.
D flüssige Lösung
F solution liquide
P roztwór ciekły
R жидкий раствор

-370 **gaseous solution, gas mixture**
A solution in a gaseous state of aggregation.
D Gasgemisch
F solution gazeuse, mélange gazeux
P roztwór gazowy, mieszanina gazów
R газообразный раствор, газовая смесь, газообразная смесь

-375 **solid solution**
A phase, occurring in solid alloys, the chemical composition of which is capable of being continuously changed within certain limits, without abrupt changes in its properties.
D feste Lösung, Mischkristallphase
F solution solide
P roztwór stały
R твёрдый раствор

-380 **terminal solid solution, primary solid solution, terminal phase**
A solution which includes in its compositional range of stability the mole fraction 1 or 0.
D primäre feste Lösung
F solution solide primaire, solution solide terminale, solution solide finale
P roztwór stały podstawowy, roztwór stały pierwotny
R первичный твёрдый раствор

-385 **metallic solution**
A solution having metallic properties.
D metallische Lösung
F solution métallique
P roztwór metaliczny
R металлический раствор

-390 **continuous (series of) solid solutions, unbroken solid-solution series**
A solid solution in which the chemical composition is capable of being continuously changed over the whole composition range.
D lückenlose Mischkristallreihe, ununterbrochene Mischkristallreihe
F solution solide continue, série continue de solutions solides
P roztwór stały ciągły, roztwór stały nieograniczony
R неограниченный твёрдый раствор

-395 **saturated solid solution**
A solid solution which contains the maximum amount of the solute corresponding to equilibrium at a given temperature, i.e. which can coexist with an excess of the solute at this temperature.
D gesättigte feste Lösung
F solution solide saturée
P roztwór stały graniczny, roztwór stały nasycony
R насыщенный твёрдый раствор

-400 **supersaturated solid solution**
A solid solution in which the amount of the solute is greater than that corresponding to equilibrium at a given temperature.
D übersättigte feste Lösung, übersättigte Mischkristalle, Zwangslösung
F solution solide sursaturée
P roztwór stały przesycony
R пересыщенный твёрдый раствор

-405 **binary solution**
A solution consisting of two components.
D binäre Lösung
F solution binaire
P roztwór dwuskładnikowy
R бинарный раствор, двойной раствор, двухкомпонентный раствор

-410 **multicomponent solution, complex solution**
A solution consisting of several components.
D Mehrstofflösung
F solution complexe
P roztwór wieloskładnikowy
R многокомпонентный раствор

-415 **substitutional solid solution**
A solid solution in which atoms of the alloying element occupy spaces previously occupied by atoms of the basic metal.
D Substitutionsmischkristalle
F solution solide de substitution, solution solide substitutionelle
P roztwór stały różnowęzłowy, roztwór stały podstawieniowy
R твёрдый раствор замещения

-420 **interstitial solid solution**
A solid solution in which atoms of the alloying element occupy the interstices between atoms of the basic metal without undue distortion of the original lattice.
D Einlagerungsmischkristalle, Interstitionsmischkristalle, Zwischengittermischkristalle, Einlagerunsphase
F solution solide d'insertion, solution solide interstitielle
P roztwór stały międzywęzłowy, roztwór stały śródwęzłowy
R твёрдый раствор внедрения

-425 **Vegard's law**
A rule stating, that the lattice parameter of substitutional solid solutions vary linearly with the atomic percentage of the solute.
D Vegard-Regel, Vegardsche Regel
F loi de Vegard
P reguła Vegarda
R закон Вегарда

-430 **ordered solid solution**
A solid solution within which there is a regular or orderly arrangement of different atoms on preferred lattice sites.
D geordnete feste Lösung, geordnete Mischphase
F solution solide ordonnée
P roztwór stały uporządkowany
R упорядоченный твёрдый раствор

-435 **disordered solid solution, random solid solution**
A solid solution within which the distribution of different atoms among the lattice sites is completely random.
D ungeordnete Mischphase, ungeordnete feste Lösung
F solution solide désordonnée
P roztwór stały nieuporządkowany
R неупорядоченный твёрдый раствор

-440 **superlattice, superstructure**
A crystal structure existing in a solid solution after the ordering process has taken place.
D Überstruktur
F surstructure
P nadstruktura
R сверхструктура, сверхрешётка

-445 **melting, fusion**
The passing of a solid body into a liquid state.
D Schmelzen
F fusion
P topnienie
R плавление

-450 **isothermal melting**
A melting process during which all the solid substance liquefies on heating at a certain, definite temperature.
D isothermisches Schmelzen
F fusion isotherme
P topnienie izotermiczne
R изотермическое плавление

-455 **heterothermal melting**
A melting process during which on heating the liquefaction of the solid begins at a certain, definite temperature and terminates at a higher, also definite temperature; the system contains between these two temperatures both solid and liquid phases simultaneously.
D heterothermisches Schmelzen
F fusion hétérotherme, fusion non isotherme
P topnienie heterotermiczne, topnienie nieizotermiczne, topnienie w zakresie temperatur
R гетеротермическое плавление, плавление в интервале температур

-460 **congruent melting**
The melting of a homogeneous alloy during which the composition of the liquid phase is identical with that of the solid phase.
D kongruentes Schmelzen
F fusion congruente
P topnienie kongruentne
R конгруэнтное плавление

-465 **incongruent melting**
A peritectic melting of an intermetallic compound.
D nichtkongruentes Schmelzen, inkongruentes Schmelzen
F fusion incongruente, fusion non congruente
P topnienie niekongruentne, topnienie inkongruentne
R неконгруэнтное плавление

-470 **eutectic melting**
In a binary system, a three-phase melting process during which two crystalline phases melt at a constant temperature into a liquid phase.
D eutektisches Schmelzen
F fusion eutectique
P topnienie eutektyczne
R эвтектическое плавление

-475 **peritectic melting**
A sub-type of heterothermal melting process during which a solid phase, after having reached on heating a certain, definite temperature, splits up into a liquid phase and a solid phase of different compositions.
D peritektisches Schmelzen
F fusion péritectique
P topnienie perytektyczne
R перитектическое плавление

-480 **(phase) equilibrium diagram, constitution(al) diagram, thermal equilibrium diagram, phase diagram**
A diagram showing the stability ranges of liquid and solid phases of an alloy depending on temperature and chemical composition for a given pressure.
D Zustandsdiagramm, Zustandsschaubild, Konstitutionsdiagramm, Gleichgewichtsdiagramm, Phasendiagramm
F diagramme d'équilibre (des phases), diagramme de phases, diagramme d'état, diagramme de constitution
P wykres równowagi faz(owej), wykres równowagi
R (фазовая) диаграмма состояния, диаграмма фазового состояния, равновесная диаграмма состояния, диаграмма равновесия (фаз), диаграмма фазового равновесия

-485 **solidus**
A line in a binary constitutional diagram, or a surface in a ternary constitutional diagram, indicating the temperatures at which melting begins during heating and at which solidification is completed during cooling, under equilibrium conditions.
D Soliduskurve, Soliduslinie; Solidusfläche
F solidus
P solidus
R солидус

-490 **solidus curve, melting-point curve**
　　The solidus in a binary constitutional diagram.
　　D Soliduskurve, Soliduslinie
　　F ligne de solidus
　　P linia solidus(u)
　　R линия солидус(а), кривая солидус(а)

-495 **solidus surface**
　　The solidus in a ternary constitutional diagram.
　　D Solidusfläche
　　F surface de solidus
　　P powierzchnia solidus(u)
　　R поверхность солидус(а)

-500 **solidus temperature**
　　The temperature at which melting of an alloy begins during heating and at which solidification is completed during cooling, under equilibrium conditions.
　　D Solidustemperatur
　　F température de solidus
　　P temperatura solidusu
　　R температура начала плавления

-505 **retrograde solidus (curve)**
　　A solidus line, occurring in some binary alloy systems with limited solid solubility, which after running towards higher concentrations of the solute with falling temperature, changes its direction at a certain temperature towards lower concentrations of the solute.
　　D rückläufige Soliduskurve
　　F solidus retrograde
　　P solidus cofający się
　　R ретроградный солидус

-510 **liquidus**
　　A line in a binary constitutional diagram, or a surface in a ternary constitutional diagram, indicating the temperatures at which solidification begins during cooling and at which melting is completed during heating, under equilibrium conditions.
　　D Liquiduskurve, Liquiduslinie; Liquidusfläche
　　F liquidus
　　P likwidus
　　R ликвидус

-515 **liquidus curve, freezing-point curve**
　　The liquidus in a binary constitutional diagram.
　　D Liquiduskurve, Liquiduslinie
　　F ligne de liquidus
　　P linia likwidus(u)
　　R линия ликвидус(а), кривая ликвидус(а)

-520 **liquidus surface, freezing point surface**
　　The liquidus in a ternary constitutional diagram.
　　D Liquidusfläche
　　F surface de liquidus, surface-liquidus
　　P powierzchnia likwidus(u)
　　R поверхность ликвидус(а)

-525 **liquidus temperature**
　　The temperature at which solidification of an alloy begins during cooling and at which melting is completed during heating, under equilibrium conditions.
　　D Liquidustemperatur
　　F température de liquidus
　　P temperatura likwidusu
　　R температура ликвидуса, температура начала затвердевания

-530 **solvus (line), solvus curve, solubility curve**
　　A line in a binary equilibrium diagram, indicating the maximum solid solubility of one component in another against temperature.
　　D Löslichkeitskurve, Löslichkeitslinie, Sättigungskurve
　　F courbe limite de solubilité, courbe de saturation, ligne de précipitation
　　P krzywa rozpuszczalności granicznej, linia rozpuszczalności granicznej, solwus
　　R линия предельной растворимости, кривая растворимости

-535 **phase boundary (line)**
　　A line in a constitutional diagram delimiting the regions in which individual phases or phase mixtures exist.
　　D Phasengrenzlinie, Phasengrenze
　　F limite de phases
　　P granica fazowa, krzywa graniczna, linia graniczna
　　R фазовая граница, граничная кривая, граница между фазовыми областями

-540 **conjugate lines, conjugate curves**
　　Pairs of phase boundaries in binary constitutional diagrams, which delimit two-phase regions and indicate the compositions of coexisting phases.
　　D Koexistenzkurven
　　F courbes conjuguées
　　P krzywe termodynamicznie sprzężone
　　R сопряжённые кривые, сопряжённые линии

-545 **melting range, fusion range**
   The range of temperatures between the solidus and the liquidus of a given alloy.
   D Schmelzbereich, Schmelzintervall
   F intervalle de fusion
   P zakres temperatur topnienia
   R интервал (температур) плавления, область плавления

-550 **freezing range, solidification range, crystallization interval**
   The range of temperatures between the liquidus and the solidus of a given alloy.
   D Erstarrungsbereich, Erstarrungsintervall
   F intervalle de solidification
   P zakres temperatur krzepnięcia, zakres temperatur krystalizacji
   R интервал (температур) затвердевания, (температурный) интервал кристаллизации

-555 **transformation temperature, transformation point, transition temperature, critical temperature, critical point, transition point, change point**
   The temperature at which in a solid metal or alloy a transformation occurs (e.g. a phase transformation or a magnetic transformation).
   D Umwandlungstemperatur, Umwandlungspunkt
   F température de transformation, point de transformation, température critique, point critique, température de transition, point de transition
   P temperatura przemiany, temperatura krytyczna, punkt krytyczny, punkt przełomowy
   R температура превращения, критическая температура, критическая точка

-560 **transformation range, critical range, critical temperature range**
   The temperature interval, within which a transformation in an alloy takes place on heating or cooling.
   D Umwandlungsbereich, Umwandlungsintervall, kritisches Intervall
   F intervalle de transformation, intervalle critique, domaine (critique) de transformation, domaine critique
   P zakres temperatur przemiany
   R область превращения, критический интервал

-565 **eutectic transformation, eutectic reaction, eutectic change**
   A reversible isothermal change that occurs within a liquid alloy during cooling, consisting in the formation of at least two new solid phases from a liquid one.
   D eutektische Umwandlung, eutektische Reaktion
   F transformation eutectique
   P przemiana eutektyczna
   R эвтектическое превращение, эвтектическая реакция, эвтектический распад

-570 **peritectic transformation, peritectic reaction, peritectic change**
   A reversible isothermal change that occurs in a binary system, as a result of which a new solid phase of a definite composition is formed from a liquid and a solid phase during cooling.
   D peritektische Umwandlung, peritektische Reaktion
   F transformation péritectique, réaction péritectique
   P przemiana perytektyczna, reakcja perytektyczna
   R перитектическое превращение, перитектическая реакция

-575 **monotectic transformation, monotectic reaction, monotectic change**
   A reversible isothermal change in a binary system during which a liquid phase decomposes on cooling to yield a solid phase and a new liquid phase.
   D monotektische Umwandlung, monotektische Reaktion
   F transformation monotectique, réaction monotectique
   P przemiana monotektyczna, reakcja monotektyczna
   R монотектическое превращение, монотектическая реакция

-580 **metatectic transformation, metatectic reaction, metatectic change**
   A reversible isothermal change in a binary system during which a solid solution of a definite composition splits up on cooling into a mixture of one solid and one liquid phase in fixed proportions.
   D metatektische Umwandlung, metatektische Reaktion
   F transformation métatectique, réaction métatectique
   P przemiana metatektyczna
   R метатектическое превращение, метатектическая реакция

-585 **syntectic transformation, syntectic reaction, syntectic change**
    A reversible isothermal change in a binary system, involving the conversion of two immiscible liquid phases on cooling into a solid phase of definite composition.
    D syntektische Umwandlung, syntektische Reaktion
    F transformation syntectique, réaction syntectique
    P przemiana syntektyczna, reakcja syntektyczna
    R синтектическое превращение, синтектическая реакция

-590 **eutectoid(al) transformation, eutectoid(al) reaction, eutectoid change, eutectoid(al) decomposition**
    A reversible isothermal change that occurs within a solid alloy during cooling, consisting in the decomposition of a solid solution into two new solid phases.
    D eutektoide Umwandlung, eutektoide Reaktion
    F transformation eutectoïde, réaction eutectoïde
    P przemiana eutektoidalna
    R эвтектоидное превращение, эвтектоидный распад

-595 **peritectoid transformation, peritectoid reaction, peritectoid change**
    A reversible isothermal change that occurs in some binary alloys as a result of which a new solid phase of definite composition is formed from two solid phases during cooling.
    D peritektoide Umwandlung, peritektoide Reaktion
    F transformation péritectoïde, réaction péritectoïde
    P przemiana perytektoidalna, reakcja perytektoidalna
    R перитектоидное превращение, перитектоидная реакция

-600 **monotectoid transformation, monotectoid reaction, monotectoid change**
    A change analogous to a monotectic transformation, but in which all three phases involved are solid.
    D monotektoide Umwandlung, monotektoide Reaktion
    F transformation monotectoïde, réaction monotectoïde
    P przemiana monotektoidalna, reakcja monotektoidalna
    R монотектоидное превращение, монотектоидная реакция

-605 **eutectic temperature**
    The temperature at which a eutectic transformation takes place.
    D eutektische Temperatur
    F température eutectique, température d'eutexie
    P temperatura eutektyczna
    R эвтектическая температура

-610 **peritectic temperature**
    The temperature at which a peritectic transformation takes place.
    D peritektische Temperatur
    F température péritectique
    P temperatura perytektyczna
    R перитектическая температура

-615 **monotectic temperature**
    The temperature at which a monotectic transformation takes place.
    D monotektische Temperatur
    F température monotectique
    P temperatura monotektyczna
    R монотектическая температура

-620 **metatectic temperature**
    The temperature at which a metatectic transformation takes place.
    D metatektische Temperatur
    F température métatectique
    P temperatura metatektyczna
    R метатектическая температура

-625 **syntectic temperature**
    The temperature at which a syntectic transformation takes place.
    D syntektische Temperatur
    F température syntectique
    P temperatura syntektyczna
    R синтектическая температура

-630 **eutectoid temperature**
    The temperature at which a eutectoid transformation takes place.
    D eutektoide Temperatur
    F température eutektoïd(iqu)e
    P temperatura eutektoidalna
    R эвтектоидная температура

-635 **peritectoid temperature**
    The temperature at which a peritectoid transformation takes place.
    D peritektoide Temperatur
    F température péritectoïde
    P temperatura perytektoidalna
    R перитектоидная температура

-640 **eutectic line, eutectic horizontal, eutectic isothermal**
   A horizontal line, in a binary constitutional diagram, representing the temperature at which a eutectic transformation takes place.
   D Eutektikale, eutektische Horizontale, eutektische Linie
   F horizontale eutectique
   P linia eutektyczna, linia przemiany eutektycznej
   R эвтектическая линия, эвтектическая горизонталь

-645 **peritectic line, peritectic horizontal, peritectic isothermal**
   A horizontal line, in a binary constitutional diagram, representing the temperature at which a peritectic transformation takes place.
   D Peritektikale, peritektische Horizontale, peritektische Linie
   F horizontale péritectique
   P linia perytektyczna, linia przemiany perytektycznej
   R перитектическая линия, перитектическая горизонталь

-650 **monotectic line, monotectic horizontal, monotectic isothermal**
   A horizontal line in a binary constitutional diagram, representing the temperature at which a monotectic transformation takes place.
   D Monotektikale, monotektische Horizontale, monotektische Linie
   F horizontale monotectique
   P linia monotektyczna, linia przemiany monotektycznej
   R монотектическая линия, монотектическая горизонталь

-655 **eutectoid line, eutectoid horizontal, eutectoid isothermal**
   A horizontal line in a binary constitutional diagram, representing the temperature at which a eutectoid transformation takes place.
   D Eutektoidale, eutektoid(isch)e Horizontale, eutektoid(isch)e Linie
   F horizontale eutectoïde
   P linia eutektoidalna, linia przemiany eutektoidalnej
   R эвтектоидная линия, эвтектоидная горизонталь

-660 **eutectic point**
   A point in a phase equilibrium diagram, indicating the composition of the eutectic and the eutectic temperature.
   D eutektischer Punkt
   F point eutectique, point d'eutexie
   P punkt eutektyczny
   R эвтектическая точка

-665 **peritectic point**
   A point in a phase equilibrium diagram, indicating the composition of the peritectic and the peritectic temperature.
   D peritektischer Punkt
   F point péritectique
   P punkt perytektyczny
   R перитектическая точка

-670 **monotectic point**
   A point in a phase equilibrium diagram, indicating the composition of the monotectic and the monotectic temperature.
   D monotektischer Punkt
   F point monotectique
   P punkt monotektyczny
   R монотектическая точка

-675 **eutectoid point**
   A point in a phase equilibrium diagram, indicating the composition of the eutectoid and the eutectoid temperature.
   D eutektoider Punkt
   F point eutectoïd(iqu)e
   P punkt eutektoidalny
   R эвтектоидная точка

-680 **one-phase region, single-phase region, homogeneity range, one-phase area, homogeneous region**
   A region in a constitutional diagram representing the occurrence of a single phase.
   D Einphasenfeld, Einphasengebiet, homogenes Gebiet, Homogenitätsbereich
   F domaine monophasé
   P obszar jednofazowy
   R однофазная область, гомогенная область, область гомогенности, однофазное поле

-685 **two-phase region**
   A region in a constitutional diagram representing the coexistence of two phases.
   D Zweiphasenfeld, Zweiphasengebiet
   F domaine biphasé, domaine à deux phases
   P obszar dwufazowy
   R двухфазная область, двухфазное поле

-690 **equilibrium temperature**
  The temperature at which given phases can coexist in equilibrium.
  D Gleichgewichtstemperatur
  F température d'équilibre
  P temperatura równowagi, temperatura równowagowa
  R температура равновесия, равновесная температура

-695 **tie line, conode**
  A line connecting the compositions of coexisting phases in a binary alloy system.
  D Kon(n)ode
  F conode
  P konoda
  R конода

-700 **lever rule, lever law**
  The relation which enables calculation of the relative amounts of phases in a two-phase mixture in a binary alloy system, these amounts being inversely proportional to the portions of the tie line into which the latter is cut by the ordinate representing the composition of the given alloy.
  D Hebelbeziehung, Hebelgesetz
  F règle du bras de levier, règle de mélange, règle d'alliage, loi des segments proportionnels
  P reguła dźwigni, reguła odcinków
  R правило рычага, правило отрезков

-705 **Konovalov's rule**
  A rule stating that the solid solution is always richer than the melt with which it is in equilibrium in that component which raises the melting point when added to the system.
  D Konovalowscher Satz
  F règle de Konovalov
  P reguła Konowałowa
  R правило Коновалова

-710 **ternary (phase) equilibrium diagram**
  A phase equilibrium diagram for a three-component system.
  D ternäres Zustandsdiagramm, ternäres Gleichgewichtsdiagramm, ternäres Phasendiagramm
  F diagramme ternaire d'équilibre
  P wykres równowagi układu potrójnego
  R тройная диаграмма состояния

-715 **binary sub-system, terminal system**
  A binary system formed from two of the three components of a ternary system.
  D Randsystem
  F sous-système binaire
  P podukład podwójny, układ brzegowy
  R двойная подсистема

-720 **concentration triangle, Gibbs triangle, composition triangle**
  An equilateral triangle representing alloy compositions in a ternary system.
  D Konzentrationsdreieck, Dreieckschaubild
  F diagramme triangulaire
  P trójkąt stężeń, trójkąt składów
  R концентрационный треугольник, треугольник Гиббса-Роозебоома

-725 **triangulation (of the system)**
  The division of the concentration triangle of a complicated ternary system into a series of simple triangles.
  D Triangulation des Systems
  F triangulation du système
  P triangulacja układu
  R разбивка концентрационного треугольника

-730 **eutectic valley, eutectic trough**
  The trough in a ternary space diagram, bordered by two liquidus surfaces, stretching along the binary eutectic line towards the ternary eutectic point.
  D (binär-)eutektische Rinne
  F vallée eutectique
  P rynna eutektyczna
  R эвтектическое понижение

-735 **isothermal section, horizontal section**
  A section of a ternary space diagram at a constant temperature, i.e. the equilibrium diagram of a ternary system for a given temperature.
  D isothermer Schnitt
  F coupe à température constante, coupe isotherme
  P przekrój izotermiczny, przekrój poziomy
  R изотермический разрез, горизонтальный разрез

-740 **vertical section, isopleth section, polythermal section**
   A section of a ternary space diagram performed perpendicularly to the concentration triangle.
   D vertikaler Schnitt
   F coupe verticale
   P przekrój stężeniowy, przekrój pionowy, przekrój politermiczny
   R вертикальный разрез, политермический разрез

-745 **quasibinary section, pseudobinary section**
   A vertical section of a ternary space diagram performed at a constant percentage of one of the three alloying components.
   D quasibinärer Schnitt
   F coupe pseudobinaire
   P przekrój pseudopodwójny
   R псевдобинарный разрез

-750 **ternary (equilibrium) space diagram, ternary (phase equilibrium) spatial diagram**
   A phase equilibrium diagram for a three-component system, the axis of temperatures of this diagram being perpendicular to the plane in which the areas of concentrations of the individual components lie.
   D ternäres Raumzustandsdiagramm, ternäres Raumzustandsschaubild
   F diagramme ternaire spatial, diagramme ternaire prismatique
   P przestrzenny wykres równowagi układu potrójnego
   R пространственная диаграмма состояния тройной системы

40 STRUCTURE OF METALS AND ALLOYS
GEFÜGE DER METALLE UND LEGIERUNGEN
STRUCTURE DES MÉTAUX ET ALLIAGES
STRUKTURA METALI I STOPÓW
СТРУКТУРА МЕТАЛЛОВ И СПЛАВОВ

-005 **(metal) structure**
The mode of arrangement of crystals in a metal or alloy.
D Gefüge, Metallgefüge, Metallstruktur
F structure (de métal), structure (métallographique)
P struktura (metalu)
R структура (металла)

-010 **(micro)structural constituent, microconstituent**
A characteristic, metallographically discernible component of the microstructure of an alloy.
D Gefügebestandteil, Strukturbestandteil, Mikrogefügebestandteil
F constituant (métallographique), constituant micrographique, microconstituant
P składnik strukturalny
R структурная составляющая, микросоставляющая

-015 **matrix, groundmass**
The principal structural constituent of a heterogeneous alloy in which another structural constituent is embedded.
D Grundgefüge, Grundmasse, Matrix
F matrice
P osnowa struktury
R матрица, основа структуры

-020 **microstructural equilibrium**
The state of an alloy when the phases given by the equilibrium diagram are present and further changes in the spatial distribution and morphology — such as agglomeration — are very slow.
D Gefügegleichgewicht, Strukturgleichgewicht
F équilibre structural
P równowaga strukturalna
R микроструктурное равновесие

-025 **macrostructure**
The structure of a metal or alloy which, after appropriate treatment of its surface, may be seen with the naked eye or by means of a magnifying glass.
D Makrogefüge, Grobgefüge, Makrostruktur, Grobstruktur, Übersichtsgefüge
F macrostructure, structure macrographique
P makrostruktura, struktura makroskopowa
R макроструктура, макростроение

-030 **microstructure**
The structure of a metal or alloy as revealed by a microscope after the surface has been suitably prepared.
D Mikrogefüge, Kleingefüge, Mikrostruktur, Kleinstruktur
F microstructure, structure micrographique
P mikrostruktura, struktura mikroskopowa
R микроструктура, микростроение

-035 **substructure, subgrain structure, sub-boundary structure, internal structure**
A structure, appearing as a network of boundaries, which can be revealed within individual grains.
D Unterstruktur, Substruktur, Subgefüge
F sous-structure
P podstruktura, substruktura
R субструктура

-040 **fine structure**
The structure in the sense of arrangement of atoms in a space lattice, i.e. the structure which is revealed e.g. by X-ray diffraction methods.
D Feinstruktur, Feingefüge, Submikrostruktur, Submikrogefüge
F structure fine
P struktura podmikroskopowa, submikrostruktura
R тонкая структура

-045 **primary structure**
> The structure existing in an alloy immediately after solidification, i.e. before any solid state transformation occurs.
> D Primärgefüge, Primärstruktur, Erstarrungsgefüge, Erstgefüge
> F structure primaire
> P struktura pierwotna
> R первичная структура

-050 **(as) cast structure**
> The structure of a metal or alloy immediately after casting, i.e. before any further treatment.
> D Gußgefüge, Gußstruktur
> F structure de moulage, structure de fonderie, structure brute de coulée
> P struktura w stanie lanym
> R литая структура

-055 **secondary structure**
> The microstructure resulting from solid state transformations.
> D Sekundärgefüge, Sekundärstruktur
> F structure secondaire
> P struktura wtórna
> R вторичная структура

-060 **eutectic structure**
> A structure consisting entirely of a eutectic.
> D eutektisches Gefüge, eutektische Struktur
> F structure eutectique
> P struktura eutektyczna
> R эвтектическая структура

-065 **hypereutectic structure**
> A structure composed of a eutectic mixture and a pro-eutectic phase, the composition of this phase lying on the equilibrium diagram to the right of the eutectic point.
> D übereutektisches Gefüge
> F structure hypereutectique
> P struktura nadeutektyczna
> R заэвтектическая структура

-070 **hypoeutectic structure**
> A structure composed of a eutectic mixture and a pro-eutectic phase, the composition of this phase lying on the equilibrium diagram to the left of the eutectic point.
> D untereutektisches Gefüge
> F structure hypoeutectique
> P struktura podeutektyczna
> R доэвтектическая структура

-075 **eutectoid structure**
> A structure consisting entirely of a eutectoid.
> D eutektoides Gefüge
> F structure eutectoïde
> P struktura eutektoidalna
> R эвтектоидная структура

-080 **hypereutectoid structure**
> A structure composed of a eutectoid mixture and a pro-eutectoid phase, the composition of this phase lying on the equilibrium diagram to the right of the eutectoid point.
> D übereutektoides Gefüge
> F structure hypereutectoïde
> P struktura nadeutektoidalna
> R заэвтектоидная структура

-085 **hypoeutectoid structure**
> A structure composed of a eutectoid mixture and a pro-eutectoid phase, the composition of this phase lying on the equilibrium diagram to the left of the eutectoid point.
> D untereutektoides Gefüge
> F structure hypoeutectoïde
> P struktura podeutektoidalna
> R доэвтектоидная структура

-090 **dendritic structure**
> The structure of a solidified alloy consisting of dendrites.
> D Dendritengefüge, dendritisches Gefüge, Bäumchengefüge, Tannenbaumgefüge
> F structure dendritique
> P struktura dendrytyczna
> R дендритная структура

-095 **polyhedral structure, cell(ular) structure**
> A structure composed of grains which on the etched microsection have the appearance of irregular polygons.
> D Zellgefüge, Zellstruktur
> F structure polygonale, structure cellulaire, structure alvéolaire
> P struktura poliedryczna, struktura komórkowa
> R полиэдрическая структура, ячеистая структура

-100 **network structure, reticular structure, envelope structure**
A heterogeneous structure in which the grains of one structural constituent are enveloped in another structural constituent, these envelopes having the appearance of a network on the etched microsection.
D Netzgefüge, Netzstruktur
F structure réticulaire
P struktura siatkowa
R сетчатая структура

-105 **lamellar (micro)structure**
A heterogeneous structure composed of alternately arranged lamellae of coexisting phases.
D lamellares Gefüge
F structure lamellaire
P struktura płytkowa, struktura pasemkowa
R пластинчатая структура

-110 **(inter)lamellar spacing**
The distance between the lamellae of a given phase within a heterogeneous structural constituent having a lamellar structure.
D Lamellenabstand
F distance entre lamelles, distance interlamellaire
P odległość międzypłytkowa, odległość międzyblaszkowa
R межпластиночное расстояние

-115 **acicular structure**
A structure which appears on the etched microsection as composed of needle-like grains.
D nadeliges Gefüge, strahliges Gefüge
F structure aciculaire, structure en aiguilles
P struktura iglasta
R игольчатая структура

-120 **globular structure, granular structure**
A heterogeneous structure within which one phase occurs in the form of dispersed spherical grains.
D Globulargefüge, globulares Gefüge, körniges Gefüge, globulitisches Gefüge
F structure globulitique, structure globulisée, structure granulaire
P struktura globularna, struktura globulityczna
R глобулярная структура

-125 **columnar structure**
A structure composed of elongated parallel grains.
D stengeliges Gefüge
F structure colonnaire, structure basaltique
P struktura słupkowa
R столбчатая структура

-130 **Widmannstätten structure**
A heterogeneous structure occurring in overheated and therefore coarse-grained alloys, characterized by the formation of a new phase as plates lying along certain crystallographic planes of the parent solid solution.
D Widmannstättensches Gefüge
F structure de Widmannstätten
P struktura Widmannstättena
R видманштеттова структура

-135 **abnormal structure**
A defect structure, sometimes encountered in carburized steels, consisting in the occurrence of massive cementite along with free ferrite.
D anomales Gefüge
F structure anomale
P struktura anormalna
R аномальная структура

-140 **banded structure**
A heterogeneous segregated structure composed of parallel or almost parallel bands aligned in the direction of hot working.
D Zeilengefüge, Zeilenstruktur
F structure à bandes, structure en bandes, structure de bandes
P struktura pasmowa
R полосчатая структура, строчечная структура

-145 **homogeneous structure, single-phase structure**
A structure consisting of one phase.
D homogenes Gefüge, einphasiges Gefüge, monophasiges Gefüge
F structure homogène, structure monophasée
P struktura jednofazowa
R однофазная структура, гомогенная структура

-150 **two-phase structure, duplex structure**
A structure consisting of two phases.
D zweiphasiges Gefüge
F structure biphasée, structure duplex
P struktura dwufazowa
R двухфазная структура

-155 **heterogeneous structure**
A structure consisting of more than one phase.
D heterogenes Gefüge
F structure hétérogène
P struktura niejednofazowa
R гетерогенная структура

-160 **polyphase structure**
A structure consisting of several phases.
D mehrphasiges Gefüge
F structure polyphasée
P struktura wielofazowa
R многофазная структура

-165 **lineage structure**
The structure which forms during crystallization, in regions where the neighbouring branches of dendrites touch each other and which therefore bears — in its orientation — the impression of deviations of these branches from the perfect parallel alignment.
D Verzweigungsstruktur
F sous-structure de solidification orientée
P struktura dziedziczna
R линейчатая структура

-170 **mosaic structure, block structure**
A sub-structure composed of crystal fragments or blocks differing slightly in orientation.
D Mosaikstruktur
F structure mosaïque
P struktura mozaikowa
R мозаичная структура, блочная структура

-175 **chill zone**
The exterior fine-grained zone in an ingot, which owing to direct contact with the chill surface of the ingot mould solidifies first, this solidification being rapid.
D feindendritische Randzone
F zone de trempe, zone extérieure
P strefa zamrożona, strefa kryształów zamrożonych
R корковая зона, корковый слой, мелкозернистая корка

-180 **columnar (crystal) zone**
The second zone, after the chill zone, in an ingot, consisting of long columnar crystals more or less perpendicular to the chill surface of the ingot mould.
D Zone des gerichteten Dendritenwachstums
F zone (de cristallisation) basaltique, zone à structure basaltique
P strefa kryształów słupkowych
R зона столбчатых кристаллов, зона столбчатой кристаллизации, столбчатая зона

-185 **equiaxed zone, zone of equiaxed grains**
The interior zone of an ingot which solidifies last.
D gleichachsige Zone
F zone de cristaux équiaxes, zone (centrale de cristallisation) équiaxe
P strefa kryształów wolnych, strefa kryształów równoosiowych
R зона равноосных кристаллов

-190 **phase mixture, mechanical mixture (of phases)**
A structural constituent composed of two or more phases.
D heterogenes Gemenge (der Phasen), mechanisches Gemenge (der Phasen), Phasengemisch
F mélange (intime) (de phases)
P mieszanina faz
R механическая смесь (фаз)

-195 **mixed crystal**
A crystal of solid solution.
D Mischkristall
F cristal mixte
P kryształ roztworu stałego
R кристалл твёрдого раствора

-200 **eutectic (mixture)**
An intimate crystalline mixture of two or more phases in fixed proportions which solidifies and melts during eutectic transformation at a given constant temperature.
D Eutektikum, eutektisches Gemisch, eutektische Mischung
F (mélange) eutectique, eutexie
P eutektyka, mieszanina eutektyczna
R эвтектика, эвтектическая смесь

-205 **binary eutectic**
A eutectic composed of two phases.
D binäres Eutektikum
F eutectique binaire
P eutektyka podwójna
R двойная эвтектика

-210 **ternary eutectic**
    A eutectic composed of three phases.
    D ternäres Eutektikum
    F eutectique ternaire
    P eutektyka potrójna
    R тройная эвтектика

-215 **quaternary eutectic**
    A eutectic composed of four phases.
    D quaternäres Eutektikum
    F eutectique quaternaire
    P eutektyka poczwórna
    R четверная эвтектика

-220 **lamellar eutectic**
    A eutectic in which at least one of the phases present is lamellar in shape.
    D lamellares Eutektikum
    F eutectique lamellaire
    P eutektyka płytkowa
    R пластинчатая эвтектика

-225 **pseudo-eutectic**
    A eutectic formed from a supercooled liquid solution and therefore having a different composition from its equilibrium composition.
    D Quasieutektikum
    F pseudo-eutectique
    P pseudoeutektyka, kwazyeutektyka, quasi-eutektyka
    R квазиэвтектика

-230 **divorced eutectic**
    A eutectic structure in which one phase is either absent or present in a massive form.
    D entartetes Eutektikum
    F eutectique anormal
    P eutektyka anormalna
    R разбавленная эвтектика

-235 **discontinuous eutectic**
    A eutectic in which one of the phases must repeatedly renucleate, owing to the termination of growth of crystals of that phase.
    D diskontinuierliches Eutektikum
    F eutectique discontinu
    P eutektyka nieciągła
    R несплошная эвтектика

-240 **eutectoid (mixture)**
    An intimate crystalline mixture of two (or more) phases in fixed proportions, which results from the decomposition of a solid solution during cooling at a constant temperature.
    D Eutektoid, eutektoides Gemisch
    F (mélange) eutectoïde, agrégat eutectoïde
    P eutektoid, mieszanina eutektoidalna
    R эвтектоид, эвтектоидная смесь

-245 **peritectic**
    The structure that arises as a result of a peritectic transformation.
    D Peritektikum
    F péritectique
    P perytektyka
    R перитектика

-250 **peritectoid**
    The structure that arises as a result of a peritectoid transformation.
    D Peritektoid
    F péritectoïde
    P perytektoid
    R перитектоид

-255 **metatectic**
    A mixture of one solid and one liquid phase in fixed proportions arising from a solid solution as a result of the metatectic transformation.
    D Metatektikum
    F métatectique
    P metatektyka
    R метатектика

-260 **monotectic**
    A phase mixture quite analogous to the eutectic but with one of the solid phases replaced by a liquid phase.
    D Monotektikum
    F monotectique
    P monotektyka
    R монотектика

-265 **Sauveur's diagram**
    A diagram representing the percentage of individual structural constituents in iron-carbon alloys against the carbon content.
    D Gefügediagramm
    F diagramme de structure
    P wykres Sauveura, wykres strukturalny, trójkąt przystanków eutektycznych
    R структурная диаграмма, треугольник Таммана

-270 **primary crystals**
   The first dendritic crystals which form in an alloy during cooling below the liquidus temperature.
   D Primärkristalle, primäre Kristalle
   F cristaux primaires
   P kryształy pierwotne, kryształy pierwszorzędowe
   R первичные кристаллы

-275 **secondary crystals**
   Crystals which form in a solid alloy through the precipitation of a new phase from a solid solution.
   D Sekundärkristalle, sekundäre Kristalle
   F cristaux secondaires
   P kryształy wtórne, kryształy drugorzędowe
   R вторичные кристаллы

-280 **columnar crystals, directional crystals, fringe crystals**
   Elongated crystals which grow from a cooling surface, usually at right angles to that surface.
   D Stengelkristalle
   F cristaux basaltiques, cristaux bacillaires
   P kryształy słupkowe
   R столбчатые кристаллы, шестоватые кристаллы

-285 **equiaxial crystals, equi-axed crystals, equiaxed grains**
   Crystals, the dimensions of which are approximately equal in all directions and with random orientation, found for example in the interior of castings.
   D gleichachsige Kristalle, globulitische Kristalle
   F cristaux équiaxes
   P kryształy równoosiowe
   R равноосные кристаллы, ровноосные зёрна

-290 **spherulite, nodule**
   A grain which is spherical in shape.
   D Sphärolith
   F sphéroïde, sphérule
   P sferoid, sferolit
   R сферолит

-295 **parent crystal, original crystal**
   A crystal from which a new phase has precipitated.
   D Wirtskristall, Mutterkristall
   F cristal mère
   P kryształ macierzysty
   R матричный кристалл, материнский кристалл

-300 **precipitation**
   The separation of a new phase from a solid solution during cooling, caused by the decrease in solubility with decreasing temperature.
   D Ausscheidungsvorgang, Ausscheidung
   F précipitation
   P wydzielanie
   R процесс выделения

-305 **precipitate, precipitated phase, segregate**
   A phase which has separated from a solid solution during cooling owing to a decrease in solubility.
   D Ausscheidungen, ausgeschiedene Phase, Segregatphase, Segregat
   F précipité, phase précipitée
   P wydzielenia, faza wydzielona
   R выделения, сегрегаты

-310 **grain boundary precipitate**
   A precipitate which has separated at the grain boundaries of the parent phase.
   D Korngrenzenausscheidungen, Korngrenzensegregat
   F précipité intergranulaire
   P wydzielenia na granicach ziarn
   R пограничные выделения, межзёренные выделения

-315 **submicroscopic precipitate**
   A precipitate thus dispersed that it cannot be seen by means of an optical microscope.
   D submikroskopische Ausscheidungen
   F précipité submicroscopique
   P wydzielenia podmikroskopowe
   R ультрамикроскопические выделения

-320 **coherent precipitate**
   A precipitate having coherent boundaries with its parent phase.
   D kohärente Ausscheidungen
   F précipité cohérent
   P wydzielenia koherentne
   R когерентные выделения

-325 **incoherent precipitate**
   A precipitate having incoherent boundaries with its parent phase.
   D inkohärente Ausscheidungen
   F précipité incohérent
   P wydzielenia niekoherentne
   R некогерентные выделения

-330 **disperse(d) phase, dispersoid**
A phase occurring in an alloy in the form of finely scattered particles.
D disperse Phase, Dispersat, Dispersoid
F phase dispersée, dispersoïde
P faza dyspersyjna, faza rozproszona, faza zdyspergowana, dyspersoid
R дисперсная фаза, диспергированная фаза, мелкодисперсная фаза, дисперсоид

-335 **(degree of) dispersion**
The ratio of the total surface area of a dispersed phase, to the volume occupied by this phase.
D Dispersion, Dispersitätsgrad, Dispersität(sgröße)
F dispersité, degré de dispersion
P dyspersja, stopień dyspersji
R дисперсность, степень дисперсности

-340 **proeutectic (phase), pro-eutectic (phase)**
A solid phase which precipitates from a liquid solution during cooling, before the eutectic temperature is reached.
D voreutektische Phase
F phase proeutectique
P faza przedeutektyczna
R проэвтектическая фаза, избыточная фаза

-345 **proeutectoid (phase), pro-eutectoid (phase)**
A solid phase which precipitates from a solid solution during cooling, before the eutectoid temperature is reached.
D voreutektoidische Phase
F phase proeutectoïde
P faza przedeutektoidalna
R проэвтектоидная фаза, избыточная фаза

-350 **hardening phase, hardener**
A phase which due to its hardness increases the mean hardness of the alloy.
D härtesteigernde Phase
F phase durcissante
P faza utwardzająca
R упрочняющая фаза, упрочнитель

-355 **interdendritic spaces**
Regions between individual dendrites.
D Dendritenzwischenräume
F espaces interdendritiques
P przestrzenie międzydendrytyczne
R междендритные пространства, междендритные полости

-360 **(crystal) grain, allotriomorphic crystal, xenomorphic crystal, crystallite**
A crystal which has been hindered in assuming a regular geometrical external shape, as by interference of adjacent growing crystals in a polycrystalline material.
D Korn, Kristallkorn, Kristallit
F grain, cristallite
P ziarno (krystaliczne), krystalit
R зерно, кристаллит

-365 **coarse grain**
A grain having relatively large dimensions.
D Grobkorn, grobes Korn
F gros grain, grain grossier
P ziarno grube, ziarno duże
R крупное зерно

-370 **fine grain**
A grain having relatively small dimensions.
D Feinkorn, feines Korn
F grain fin
P ziarno drobne
R мелкое зерно

-375 **grain coarseness, coarse grainedness**
A term denoting the structural feature of a material made up of coarse grains.
D Grobkörnigkeit
F grossièreté du grain
P gruboziarnistość
R крупнозернистость

-380 **grain fineness**
A term denoting the structural feature of a material made up of fine grains.
D Feinkörnigkeit
F finesse du grain
P drobnoziarnistość
R мелкозернистость

-385 **grain refining, grain refinement**
A process in which the grain size of a metal or alloy is reduced.
D Kornverfeinerung, Kornfeinung, Gefügeverfeinerung
F affinage du grain, affinement du grain
P rozdrobnienie ziarna, rozdrobnienie struktury
R измельчение зерна, измельчение зернистости, измельчение структуры

-390 **coarse(-grained) structure**
    A structure consisting of coarse grains.
    D grobkörniges Gefüge, grobkristallines Gefüge
    F structure grossière, structure à gros grains
    P struktura gruboziarnista, struktura grubokrystaliczna
    R крупнозернистая структура, крупнозернистое строение, крупнокристаллическая структура

-395 **fine-grained structure**
    A structure consisting of fine grains.
    D feinkörniges Gefüge, feinkristallines Gefüge
    F structure fine, structure à grains fins
    P struktura drobnoziarnista, struktura drobnokrystaliczna
    R мелкозернистая структура, мелкозернистое строение, мелкокристаллическая структура

-400 **grain refiner, grain growth inhibitor**
    An alloying element added to a liquid metal or alloy in order to obtain a fine-grained structure in the casting, or to prevent the grain from coarsening during heat treatment of wrought structures.
    D Kornverfeiner, Feinkornzusatz, Wachstumshemmer
    F inhibiteur de croissance du grain
    P pierwiastek rozdrabniający ziarno, inhibitor rozrostu ziarna
    R измельчитель зерна

-405 **homogeneity**
    Uniformity in the composition or structure throughout the material.
    D Homogenität, Gleichartigkeit
    F homogénéité
    P jednorodność
    R однородность, гомогенность

-410 **heterogeneity, inhomogeneity**
    Non-uniformity in the composition or structure within the material.
    D Heterogenität, Inhomogenität
    F hétérogénéité, inhomogénéité
    P niejednorodność
    R неоднородность, негомогенность

-415 **chemical inhomogeneity, chemical non-uniformity**
    Differences in the chemical composition in various regions of the material or of the same phase.
    D chemische Heterogenität
    F hétérogénéité chimique
    P niejednorodność chemiczna
    R химическая неоднородность, концентрационная неоднородность

-420 **segregation ratio, index of segregation, degree of dendritic segregation**
    The ratio of the concentration of a component in the interdendritic spaces to its concentration in the centre of the dendrite.
    D Seigerungsgrad, Seigerungskennzahl, Seigerungskoeffizient, Segregationskoeffizient
    F taux de ségrégation, indice de ségrégation
    P stopień segregacji, stopień niejednorodności
    R степень дендритной ликвации, степень неоднородности

-425 **interparticle spacing**
    The mean distance between particles of the dispersed phase in a dispersion-strengthened metal.
    D mittlerer Teilchenabstand
    F espacement entre particules, distance interparticulaire
    P (średnia) odległość między cząstkami (dyspersoidu)
    R (средний) свободный путь (между частицами)

-430 **carbide**
    A compound of carbon with another element.
    D Karbid, Carbid
    F carbure
    P węglik
    R карбид

-435 **metal carbide**
    A compound of a metal with carbon.
    D Metallkarbid, Metallcarbid
    F carbure de métal, carbure métallique
    P węglik metalu
    R карбид металла

-440 **iron carbide**
    A compound of iron with carbon.
    D Eisenkarbid, Eisencarbid
    F carbure de fer
    P węglik żelaza
    R карбид железа

-445 **alloy carbide**
A carbide formed by an alloying element in steel.
D Legierungskarbid, Sonderkarbid
F carbure allié, carbure spécial
P węglik stopowy, węglik specjalny
R специальный карбид

-450 **tungsten carbide**
A compound of tungsten with carbon.
D Wolframkarbid
F carbure de tungstène
P węglik wolframu
R карбид вольфрама

-455 **chromium carbide**
A compound of chromium with carbon.
D Chromkarbid
F carbure de chrome
P węglik chromu
R карбид хрома

-460 **vanadium carbide**
A compound of vanadium with carbon.
D Vanadinkarbid
F carbure de vanadium
P węglik wanadu
R карбид ванадия

-465 **niobium carbide**
A compound of niobium with carbon.
D Niobkarbid
F carbure de niobium
P węglik niobu
R карбид ниобия

-470 **simple carbide**
A carbide formed by one metal and carbon.
D einfaches Karbid, binäres Karbid
F carbure simple
P węglik prosty
R простой карбид

-475 **double carbide, complex carbide**
A carbide formed by two metals and carbon.
D Doppelkarbid, Mischkarbid
F carbure mixte, carbure complexe
P węglik podwójny, węglik złożony
R двойный карбид, сложный карбид

-480 **undissolved carbides**
Carbides which have not been dissolved in austenite during the austenitizing process.
D Restkarbide
F carbures résiduels
P węgliki szczątkowe
R нерастворённые карбиды, остаточные карбиды

-485 **carbonitride**
A compound of a metal with carbon and nitrogen.
D Karbonitrid
F carbonitrure
P węgl(ik)oazotek
R карбонитрид

-490 **silicide**
A compound of a metal with silicon.
D Silizid
F siliciure
P krzemek
R силицид

-495 **hydride**
A compound of a metal with hydrogen.
D Hydrid, Metallhydrid
F hydrure (métallique)
P wodorek
R гидрид

-500 **tungstide**
An intermetallic compound of a metal with tungsten.
D Wolframid
F tungstènure
P wolframek
R вольфрамид

-505 **niobide**
An intermetallic compound of a metal with niobium.
D Niobid
F niobiure
P niobek
R ниобид

-510 **titanide**
An intermetallic compound of a metal with titanium.
D Titanid
F titanure
P tytanek
R титанид

-515 **non-metallic inclusions, sonims**
Extraneous microscopic substances of a non-metallic nature (oxides, silicates, etc.) present in a solid metal.
D nichtmetallische Einschlüsse
F inclusions non métalliques, sonims
P wtrącenia niemetaliczne
R неметаллические включения

-520 **exogenous inclusions**
Non-metallic inclusions which have been entrapped inadvertently in a metal during the metallurgical process.
D exogene Einschlüsse
F inclusions exogènes
P wtrącenia egzogeniczne
R экзогенные включения

-525 **indigenous inclusions**
Non-metallic inclusions which have separated within the molten or solid metal, owing to a change of composition or temperature.
D endogene Einschlüsse
F inclusions endogènes
P wtrącenia endogeniczne
R эндогенные включения

-530 **sulfide inclusions**
Non-metallic inclusions consisting of sulfides.
D Sulphideinschlüsse
F inclusions de sulphures, inclusions sulphurées
P wtrącenia siarczków, wtrącenia siarczkowe
R сульфидные включения, сернистые включения

-535 **oxide inclusions**
Non-metallic inclusions consisting of oxides.
D Oxideinschlüsse, oxidische Einschlüsse
F inclusions d'oxydes
P wtrącenia tlenków, wtrącenia tlenkowe
R окисные включения, включения окислов

-540 **slag inclusions**
Non-metallic inclusions consisting of particles of slag.
D Schlackeneinschlüsse
F inclusions de laitier, inclusions de scories
P wtrącenia żużlowe, zażużlenie
R шлаковые включения, шлаковины

-545 **nitride inclusions**
Non-metallic inclusions consisting of nitrides.
D Nitrideinschlüsse
F inclusions de nitrures
P wtrącenia azotków, wtrącenia azotkowe
R нитридные включения, включения нитридов

-550 **metallic inclusions**
Inclusions of an alloying element occurring in the matrix of an alloy when this alloying element is insoluble in the base metal of the given alloy.
D metallische Einschlüsse, metallische Einlagerungen
F inclusions métalliques
P wtrącenia metaliczne
R металлические включения

-555 **texture**
A preferred orientation of grains sometimes occurring in metals and their alloys.
D Textur, Gefügetextur
F texture
P tekstura
R текстура

-560 **deformation texture**
The texture produced by plastic deformation.
D Verformungstextur
F texture de déformation, texture d'écrouissage
P tekstura odkształcenia, tekstura deformacji, tekstura deformacyjna
R текстура деформации

-565 **fibre texture, fibrous texture**
The simplest deformation texture, produced e.g. by rolling or drawing, similar to the arrangement in naturally fibrous materials.
D Fasertextur
F texture de fibre, texture fibreuse
P tekstura włóknista
R волокнистая текстура

-570 **rolling texture**
The texture produced by the rolling process.
D Walztextur
F texture de laminage
P tekstura walcowania
R текстура прокатки

-575 **drawing texture**
The texture produced by the drawing process.
D Ziehtextur
F texture d'étirage
P tekstura ciągnienia
R текстура волочения

-580 **casting texture**
The texture occurring in a casting.
D Gußtextur
F texture de fonderie
P tekstura (pierwotna) odlewu, tekstura w stanie lanym
R текстура (металла) отливки, текстура кристаллизации

-585 **tensile texture**
The texture resulting from plastic deformation by tension.
D Zugtextur
F texture de traction
P tekstura rozciągania
R текстура растяжения, текстура вытяжки

-590 **compressive texture**
The texture resulting from plastic deformation by compression.
D Drucktextur
F texture de compression
P tekstura ściskania
R текстура сжатия

-595 **torsional texture**
The texture resulting from plastic deformation by torsion.
D Torsionstextur
F texture de torsion
P tekstura skręcania
R текстура кручения

-600 **recrystallization texture, annealing texture**
A texture produced by the recrystallization of a cold-worked metal.
D Rekristallisationstextur, Glühtextur
F texture de recristallisation, texture de recuit
P tekstura rekrystalizacji, tekstura wyżarzania
R текстура рекристаллизации, текстура отжига

-605 **cube texture**
A texture, sometimes occurring in rolled and annealed sheet metals, in which each grain is oriented in such a way that the lattice plane (100) coincides with the rolling plane, and the crystallographic direction [001] lies along the rolling direction.
D Würfeltextur, kubische Textur
F texture cubique
P tekstura regularna, tekstura sześcienna
R кубическая текстура

-610 **Goss texture, cube-on-edge texture**
A texture, sometimes occurring in rolled and annealed sheet metals, in which each grain is oriented in such a way that the lattice plane (110) coincides with the rolling plane, and the crystallographic direction [001] lies along the rolling direction.
D Goss-Textur, Gosstextur
F texture de Goss
P tekstura Gossa
R текстура Госса

-615 **domain**
Any one of the regions within a material in which the direction of magnetization is different, or which is perfectly ordered within itself.
D Domäne, Elementarbereich
F domaine
P domena
R домен

-620 **domain structure**
A structure composed of domains.
D Domänenstruktur, Bereichsstruktur
F structure à domaines, structure en domaines
P struktura domenowa
R доменная структура

-625 **antiphase domain**
Any one of the regions within an alloy which is perfectly ordered within itself but out of step with the others.
D Antiphasendomäne, Antiphasenbereich
F (domaine) antiphase, domaine ordonné
P domena antyfazowa, antyfaza
R антифазный домен, антифазовый домен

-630 **antiphase (domain) boundary**
The boundary between antiphase domains.
D Antiphasen(-Domänen)grenze
F limite d'antiphase, paroi d'antiphase, frontière d'antiphase
P granica antyfazowa
R антифазовая граница, антифазная граница, граница антифазных доменов, граница антифазовых доменов

-635 **magnetic domain**
   Any one of the regions within a magnetic material which is magnetically polarized to saturation but differs in the direction of magnetization from the others.
   D magnetische Domäne, magnetischer Bezirk, Weißscher Bereich, Weißscher Bezirk
   F domaine magnétique, domaine de Weiss
   P domena magnetyczna, domena Weissa
   R магнитный домен, магнитная область

-640 **closure domain**
   The end magnetic domain which is magnetized at right angle to the principal direction of magnetization.
   D Abschlußbereich, Abschlußbezirk
   F domaine de fermeture
   P domena zamykająca
   R замыкающий домен

-645 **Bloch wall, domain wall**
   The boundary which separates domains of different directions of magnetization from each other.
   D Bloch-Wand, Domänenwand, Domänengrenzfläche
   F paroi de Bloch, paroi de domaine
   P ściana Blocha
   R стенка Блоха, стенка домена, доменная стенка, граница Блоха

-650 **magnetic structure**
   The spatial arrangement of magnetic moments of atoms in a crystal lattice.
   D magnetische Struktur
   F structure magnétique
   P struktura magnetyczna
   R магнитная структура

-655 **parting fracture, parting rupture**
   A fracture occurring on the plane which is perpendicular to the principal stress.
   D Trenn(ungs)bruch, Spaltbruch
   F rupture par décohésion
   P przełom rozdzielczy, pęknięcie rozdzielcze
   R отрыв

-660 **shear(ing) fracture**
   A fracture occurring on the plane on which the maximum shear stress acts.
   D Verschiebungsbruch, Gleit(ungs)bruch, Scherbruch, Schiebungsbruch, Schiebebruch
   F rupture par glissement
   P przełom poślizgowy, pęknięcie poślizgowe
   R срез, сдвигающий разрыв

-665 **cleavage plane**
   A crystallographic plane along which a crystal fractures.
   D Spaltfläche
   F plan de clivage
   P płaszczyzna łupliwości
   R плоскость (ра)скола, плоскость скалывания, плоскость спайности

-670 **fracture (surface)**
   The surface of a metal object at the place of fracture or breaking.
   D Bruch(ober)fläche, Bruch
   F (surface de) cassure, surface de rupture
   P powierzchnia przełomu, przełom, złom
   R поверхность излома, излом

-675 **brittle fracture**
   A fracture which is not preceded by a plastic deformation.
   D spröder Bruch, Sprödbruch, verformungsloser Bruch
   F rupture fragile, cassure fragile
   P przełom kruchy, pęknięcie kruche
   R хрупкий излом, хрупкое разрушение

-680 **ductile fracture**
   A fracture which is preceded by a plastic deformation.
   D Verformungsbruch, zäher Bruch, duktiler Bruch
   F rupture ductile, cassure ductile, cassure résiliente
   P przełom plastyczny, przełom ciągliwy, pęknięcie plastyczne, pęknięcie ciągliwe
   R вязкий излом, тягучий излом, вязкое разрушение

-685 **crystalline fracture**
   A fracture surface occurring at brittle fractures of metals and usually having a bright, glittering crystalline appearance.
   D kristalliner Bruch
   F cassure cristalline
   P przełom krystaliczny
   R кристаллический излом

-690 **granular fracture**
A fracture surface having a rough sandstone-like appearance.
D körniger Bruch
F cassure granulaire
P przełom ziarnisty
R зернистый излом

-695 **coarse-grained fracture**
A brittle fracture characteristic of a coarse-grained material.
D grobkörniger Bruch
F cassure à grains gros
P przełom gruboziarnisty
R крупнозернистый излом, крупнокристаллический излом

-700 **fine-grained fracture, smooth fracture, even fracture**
A brittle fracture characteristic of a fine-grained material.
D feinkörniger Bruch
F cassure à grains fins
P przełom drobnoziarnisty
R мелкозернистый излом

-705 **intercrystalline fracture, intergranular fracture**
A fracture following the grain boundaries.
D interkristalliner Bruch, zwischenkristalliner Bruch, zwischenkörniger Bruch, Korngrenzenbruch
F rupture intercristalline, rupture intergranulaire, fracture intergranulaire
P przełom międzykrystaliczny
R межкристаллический излом, межкристаллитный излом, межзёренный излом, интеркристаллический излом

-710 **transcrystalline fracture, intracrystalline fracture, transgranular fracture**
A fracture going across the grains.
D transkristalliner Bruch, intrakristalliner Bruch, innerkristalliner Bruch, durchkörniger Bruch
F rupture transcristalline, rupture intragranulaire, fracture transcristalline, fracture transgranulaire
P przełom śródkrystaliczny
R внутрикристаллический излом, внутризеренный излом, транскристаллический излом

-715 **silky fracture**
A very fine-grained fracture having a smooth lustrous appearance.
D samtartiger Bruch
F cassure soyeuse
P przełom jedwabisty, przełom aksamitny
R шелковистый излом, бархатный излом, бархатистый излом, фарфоровидный излом

-720 **fibrous fracture, woody fracture**
A fracture occurring in metals which are sufficiently ductile for the grains to elongate before rupturing.
D faseriger Bruch, Holzfaserbruch, sehniger Bruch, Faserbruch
F cassure fibreuse, cassure à nerfs
P przełom włóknisty, przełom drzewiasty
R (древесно-)волокнистый излом

-725 **laminated fracture**
A fracture occurring in ductile metals which contain non-metallic inclusions stretched into layers by rolling.
D Schieferbruch
F cassure lamellaire
P przełom łupkowy, przełom warstwowy
R шиферный излом, слоистый излом

-730 **conchoidal fracture**
A fracture showing concentric rings, resembling those inside a shell.
D Muschelbruch, muscheliger Bruch
F cassure conchoïd(al)e
P przełom muszlowy
R ракови(ни)стый излом, раковинообразный излом

-735 **glassy fracture, vitreous fracture**
A fracture which is characteristic of brittle amorphous materials, its appearance resembling glass.
D glasiger Bruch
F cassure vitreuse
P przełom szklisty
R стекловидный излом

-740 **fish-scale fracture**
> A coarse-grained fracture, resembling fish scales or naphthaline flakes, occurring in quench-hardened high-speed steels which have been wrongly annealed before quench-hardening.
> **D** Naphtalinbruch, perlmutterartiger Bruch
> **F** cassure en écailles de poisson
> **P** przełom naftalinowy
> **R** нафталиновый излом, нафталинистый излом

-745 **fatigue fracture, cyclic fracture**
> A fracture caused by repeated applications of stresses, when their magnitude is lower than the yield stress of the material.
> **D** Ermüdungsbruch, Dauer(schwingungs)bruch
> **F** rupture par fatigue, cassure de fatigue, cassure par fatigue
> **P** przełom zmęczeniowy, pęknięcie zmęczeniowe
> **R** усталостный излом, усталостное разрушение

-750 **beach marks, clamshell mark(ing)s, conchoidal mark(ing)s, fatigue crescent, arrest lines**
> Curved markings roughly parallel, occurring on the surfaces of a fatigue fracture.
> **D** Rastlinien
> **F** lignes d'arrêts, lignes de repos
> **P** linie spoczynkowe
> **R** остановочные линии

-755 **delayed fracture**
> Fracture occurring after a time during which the metal was carrying a static stress, itself insufficient to cause immediate failure.
> **D** verzögerter Bruch
> **F** rupture différée, fracture retardée
> **P** przełom opóźniony, pęknięcie opóźnione
> **R** замедлённое разрушение, запоздывающее разрушение, задержанное разрушение

45 PLASTIC DEFORMATION AND THERMALLY ACTIVATED PROCESSES

PLASTISCHE VERFORMUNG UND THERMISCH AKTIVIERTE PROZESSE

DÉFORMATION PLASTIQUE ET PHÉNOMÈNES ACTIVÉS THERMIQUEMENT

ODKSZTAŁCENIE PLASTYCZNE I PROCESY AKTYWOWANE CIEPLNIE

ПЛАСТИЧЕСКАЯ ДЕФОРМАЦИЯ И ТЕРМИЧЕСКИ АКТИВИРУЕМЫЕ ПРОЦЕССЫ

-005 **elastic strain, elastic deformation**
A deformation which immediately vanishes after the removal of the stress.
D elastische Verformung, elastische Formänderung, elastische Deformation
F déformation élastique
P odkształcenie sprężyste
R упругая деформация

-010 **plastic strain, plastic deformation, permanent deformation**
A deformation which remains on removal of the stress.
D plastische Verformung, bleibende Verformung, bleibende Formänderung, plastische Formänderung, plastische Deformation
F déformation plastique, déformation permanente
P odkształcenie plastyczne, odkształcenie trwałe
R пластическая деформация, остаточная деформация

-015 **plastic working, mechanical working**
Changing the shape or dimensions of a metal object by means of an applied force which causes permanent deformation.
D plastische Verformung, bildsame Form(geb)ung, Verformen
F travail par déformation, transformation par corroyage
P obróbka plastyczna, przeróbka plastyczna
R обработка давлением

-020 **hot (plastic) working**
Plastic working at a temperature higher than the recrystallization temperature.
D Warmverformung
F corroyage (à chaud), transformation à chaud, deformation à chaud
P obróbka plastyczna na gorąco
R горячая обработка давлением

-025 **cold (plastic) working**
Plastic working at a temperature lower than the recrystallization temperature.
D Kaltverformung
F écrouissage, corroyage à froid, transformation à froid, deformation à froid
P obróbka plastyczna na zimno
R холодная обработка давлением

-030 **cold work**
The totality of phenomena which occur within a metal during cold plastic working, and their structural and mechanical consequences.
D Kaltreckung
F écrouissage
P zgniot
R наклёп, нагартовка

-035 **degree of deformation**
A per cent diminution of the transverse cross-sectional area occurring when a metal is being plastically deformed.
D Verformungsgrad
F degré de déformation, taux de déformation, degré de corroyage, taux de corroyage
P stopień odkształcenia (plastycznego)
R степень (пластической) деформации

-040 **degree of cold work**
　　Degree of deformation during cold working.
　　D Kaltverformungsgrad, Kaltreckungsgrad
　　F degré d'écrouissage, taux d'écrouissage
　　P stopień zgniotu
　　R степень наклёпа

-045 **strain hardening**
　　The increase in hardness produced by cold working of a metal.
　　D Verfestigung
　　F consolidation
　　P umocnienie
　　R упрочнение

-050 **work hardening**
　　Increasing the hardness and strength by cold working.
　　D Kalthärten, Kaltverfestigung
　　F durcissement par écrouissage, durcissement par déformation à froid
　　P utwardzanie przez odkształcanie plastyczne na zimno, utwardzanie zgniotem, utwardzanie przez zgniot, utwardzanie zgniotowe, umacnianie zgniotem
　　R деформационное упрочнение, упрочнение наклёпом, пластическое упрочнение

-055 **explosive hardening, shock-wave hardening**
　　Increasing the hardness of a metal by means of a sudden plastic deformation caused by an explosion wave shock.
　　D Explosivhärten, Stoßwellenbehandlung
　　F durcissement par explosion
　　P umacnianie wybuchowe, utwardzanie wybuchowe
　　R упрочнение взрывом, взрывное упрочнение, ударное упрочнение

-060 **stored energy**
　　The energy accumulated in a metal as a result of cold working.
　　D gespeicherte Energie
　　F énergie emmagasinée, énergie stockée
　　P energia zmagazynowana
　　R накопленная энергия, запасённая энергия

-065 **work-hardening coefficient, modulus of strain hardening, strain-hardening coefficient, strain-hardening index**
　　The ratio of the increase in true stress to the increase in strain occurring when a metal is being plastically deformed.
　　D Verfestigungskoeffizient
　　F taux de consolidation
　　P współczynnik umocnienia
　　R коэффициент упрочнения

-070 **strain-hardening curve, work-hardening curve**
　　A graph of the yield stress of a metal against the degree of work hardening which has been conferred on it.
　　D Verfestigungskurve
　　F courbe de consolidation, courbe d'écrouissage
　　P krzywa umocnienia
　　R кривая упрочнения

-075 **critical degree of deformation, critical strain, threshold strain**
　　The degree of deformation during cold working leading to a very coarse grain after primary recrystallization.
　　D kritischer Verformungsgrad, kritische Verformung
　　F écrouissage critique
　　P krytyczny stopień odkształcenia, odkształcenie krytyczne, zgniot krytyczny
　　R критическая деформация, критическая степень деформации

-080 **internal work hardening**
　　Cold work resulting from phase transformations which are accompanied by volume changes.
　　D Phasenverfestigung
　　F écrouissage interne
　　P zgniot fazowy, zgniot wewnętrzny
　　R фазовый наклёп, внутренний наклёп, фазовое упрочнение

-085 **shear(ing) stress**
　　A component of the externally applied stress which lies in a given slip plane.
　　D Schubspannung, Gleitspannung
　　F tension de glissement, tension tangentielle, tension de cisaillement, contrainte tangentielle, contrainte de cisaillement
　　P naprężenie styczne, naprężenie ścinające
　　R касательное напряжение, сдвиговое напряжение, напряжение сдвига, скалывающее напряжение

-090 **critical (resolved) shear stress**
　　The shear stress at which slip begins on a particular plane within an annealed single crystal.
　　D kritische Schubspannung, kritische Gleitspannung
　　F tension critique de glissement, tension critique de cisaillement, contrainte critique de glissement
　　P krytyczne naprężenie styczne, krytyczne naprężenie poślizgu
　　R критическое напряжение сдвига, критическое скалывающее напряжение

-095 **slip, glide**
　　A permanent shear displacement of one part of a metal crystal over the remaining of the crystal, occurring in a definite crystallographic direction and along a particular crystallographic plane, its magnitude being equal to a whole number multiple of the interatomic spacing.
　　D Gleitung, Gleitvorgang
　　F glissement
　　P poślizg
　　R скольжение, сдвиг

-100 **slip plane, gliding plane, glide plane**
　　A preferred crystallographic plane, along which blocks of atoms within a crystal slip past each other when the crystal is being plastically deformed.
　　D Gleitebene, Gleitfläche
　　F plan de glissement
　　P płaszczyzna poślizgu
　　R плоскость скольжения

-105 **slip direction**
　　The direction of slip in an individual crystal.
　　D Gleitrichtung
　　F direction de glissement
　　P kierunek poślizgu
　　R направление скольжения

-110 **slip system**
　　Any combination of a particular slip plane with a particular slip direction.
　　D Gleitsystem
　　F système de glissement
　　P system poślizgu
　　R система скольжения

-115 **primary slip system, primary glide system**
　　A slip system which is first activated during plastic deformation of a crystal.
　　D primäres Gleitsystem
　　F système de glissement primaire
　　P pierwotny system poślizgu
　　R первичная система скольжения

-120 **secondary slip system**
　　A slip system which is activated next to the first one.
　　D sekundäres Gleitsystem
　　F système de glissement secondaire
　　P wtórny system poślizgu
　　R вторичная система скольжения

-125 **conjugate slip system**
　　One of the possible secondary slip systems which is first activated.
　　D konjugiertes Gleitsystem
　　F système de glissement conjugué
　　P sprzężony system poślizgu
　　R сопряжённая система скольжения

-130 **simple slip, single slip, simple glide**
　　A slip process occurring in one slip system.
　　D Einfachgleitung, einfache Gleitung
　　F glissement simple
　　P poślizg pojedynczy, poślizg prosty
　　R одиночное скольжение, простое скольжение, одинарное скольжение, единичное скольжение, однократное скольжение

-135 **double slip, duplex slip**
　　A slip process occurring simultaneously in two slip systems.
　　D Doppelgleitung, zweifache Gleitung
　　F glissement double
　　P poślizg podwójny
　　R двойное скольжение

-140 **multiple slip, multiple glide, polyslip**
　　A slip process occurring simultaneously in more than two systems.
　　D Mehrfachgleitung, Vielfachgleitung
　　F glissement multiple
　　P poślizg wielokrotny
　　R множественное скольжение, сложное скольжение, многократное скольжение

-145 **cross slip**
Change over of a dislocation from one slip plane into another intersecting slip plane.
D Quergleitung
F glissement dévié, déviation (du glissement)
P poślizg poprzeczny
R поперечное скольжение

-150 **cross-slip plane**
The new slip plane into which a dislocation has moved from its primary slip plane during a cross slip.
D Quergleitebene
F plan de glissement dévié
P płaszczyzna poślizgu poprzecznego
R плоскость поперечного скольжения

-155 **easy glide**
The phenomenon of a very low rate of strain hardening occurring during the first stage of plastic deformation of single crystals.
D Easy-Gleitung, easy glide
F glissement facile, glissement léger
P poślizg łatwy
R лёгкое скольжение, неупрочняющее скольжение

-160 **grain boundary slip, grain boundary sliding**
A slip process which occurs along the grain boundary.
D Korngrenzengleitung
F glissement intergranulaire, glissement aux joints de grains
P poślizg po granicach ziarn, poślizg wzdłuż granic ziarn
R межзёренное скольжение, скольжение по границам (зёрен), пограничное скольжение, зернограничное скольжение

-165 **intracrystalline slip**
A slip process which occurs within a grain, as distinct from a grain-boundary slip.
D intrakristalline Gleitung
F glissement intracristallin
P poślizg śródkrystaliczny, poślizg wewnątrzkrystaliczny
R внутрикристаллическое скольжение

-170 **prismatic slip, prismatic gliding**
A slip in the hexagonal system in a prismatic plane.
D prismatische Gleitung
F glissement prismatique
P poślizg pryzmatyczny
R призматическое скольжение

-175 **pencil glide, pencil gliding**
A slip taking place simultaneously on many planes which form an almost arbitrarily corrugated surface; it occurs in metals having a body-centred cubic lattice.
D Stäbchengleitung
F glissement en pinceau
P poślizg ołówkowy
R карандашное скольжение

-180 **latent hardening**
Hardening caused by gliding in secondary slip systems, in which the resolved shear stress has not yet reached the critical value.
D latente Verfestigung
F durcissement latent, consolidation sur les systèmes latents
P umocnienie latentne
R латентное упрочнение, скрытое упрочнение

-185 **slip line**
Any one of the series of approximately parallel lines, visible on microscopic examination of a plastically deformed crystal, indicating the crystal planes along which slips have occurred.
D Gleitlinie
F ligne de glissement
P linia poślizgu
R линия скольжения, линия сдвига

-190 **slip band, glide band**
Any one of the bands or steps visible under the microscope on the polished and etched surface of a polycrystalline metal after plastic deformation, running parallel across each grain, but changing direction from grain to grain.
D Gleitband
F bande de glissement
P pasmo poślizgu
R полоса скольжения

-195 **flow lines, Lüders lines, Hartmann lines, stretcher strains, flow figures**
Lines which appear on the polished surface of certain materials, particularly iron and low-carbon steel, when stressed beyond the yield point.
D Hartmannsche Linien, Lüderssche Linien
F lignes d'écoulement, lignes de Lueders, bandes de Piobert(-Lueders)
P linie płynięcia, linie Lüdersa-Hartmanna
R линии Людерса-Чернова

-200 **deformation band**
Any one of the regions of well-defined but differing orientations occurring in individual grains after cold working.
D Verformungsband, Deformationsband
F bande de déformation
P pasmo odkształcenia
R полоса деформации

-205 **Neumann bands, Neumann lamellae, Neumann lines**
Straight narrow bands, related geometrically to the crystal lattice, occurring primarily in ferrite which has been deformed by impact.
D Neumannsche Linien
F bandes de Neumann, lignes de Neumann
P linie Neumanna
R линии Неймана

-210 **flow stress**
The shear stress which is necessary to initiate a plastic deformation of a solid metal.
D Fließspannung
F tension d'écoulement, contrainte d'écoulement
P naprężenie płynięcia plastycznego
R напряжение течения, деформирующие напряжение

-215 **plastic flow**
A flow process in which the gradient of strain rate is proportional to the difference between the shear stress and the yield stress.
D plastisches Fließen
F écoulement plastique
P płynięcie plastyczne
R пластическое течение

-220 **viscous flow**
A flow process in which the strain rate is proportional to stress.
D viskoses Fließen
F écoulement visqueux
P płynięcie lepkościowe, płynięcie lepkie
R вязкостное течение, вязкое течение

-225 **strain ageing**
Ageing induced by cold working.
D Reckalterung, mechanische Alterung
F vieillissement après écrouissage, vieillissement après déformation à froid
P starzenie po odkształceniu plastycznym na zimno, starzenie mechaniczne, starzenie po zgniocie
R механическое старение, деформационное старение

-230 **twin(ned) crystal, (crystal) twin**
A crystal grain in which the lattices of its two parts are related to each other in orientation, as an object and its mirror image across the plane named twinning plane.
D Zwilling, Kristallzwilling
F cristal maclé, grain maclé, macle
P kryształ bliźniaczy, bliźniak
R двойник, двойниковый кристалл

-235 **mechanical twin, deformation twin**
A twin produced by mechanical deformation.
D Verformungszwilling, Deformationszwilling
F macle mécanique, macle de déformation, macle d'écrouissage
P bliźniak deformacyjny, bliźniak mechaniczny
R деформационный двойник, механический двойник, двойник деформации

-240 **annealing twin, growth twin, recrystallization twin**
A twin produced as a result of annealing, following plastic deformation.
D Rekristallisationszwilling, Wachstumszwilling, Orientierungszwilling, Glühzwilling
F macle de recuit, macle de croissance, macle de recristallisation
P bliźniak rekrystalizacyjny, bliźniak wyżarzania
R двойник отжига, двойник роста, рекристаллизационный двойник, двойник рекристаллизации

-245 **twinning**
The process of generating twin crystals.
D Zwillingsbildung
F maclage
P bliźniakowanie
R двойникование

-250 **twin(ning) plane**
   The plane of symmetry between the two parts of a twinned crystal.
   D Zwillingsebene
   F plan de maclage, plan de macle
   P płaszczyzna bliźniacza, płaszczyzna bliźniakowania
   R двойниковая плоскость, плоскость двойникования

-255 **twin(ning) band**
   Any one of the bands with straight parallel sides occurring within the microstructure of metals with face-centred cubic lattice after plastic deformation and annealing; they testify to the existence of annealing twins.
   D Zwillingsstreifen, Zwillingslamelle, Zwillingsband
   F bande de macle, bande de maclage
   P pasmo bliźniacze
   R полоса двойникования, линии двойникования

-260 **twin axis**
   The axis about which one half of a twinned crystal should be rotated through 180 degrees in order to replace its other half.
   D Zwillingsachse
   F axe de macle
   P oś bliźniacza
   R ось двойникования

-265 **twin(ned) boundary**
   A boundary between two crystals which are mirror images of each other.
   D Zwillingsgrenze
   F joint de macle
   P granica bliźniaków
   R граница двойника, двойниковая граница

-270 **tin cry**
   A characteristic crackling sound emitted by tin and some tin-rich alloys caused by mechanical twinning during plastic deformation.
   D Zinngeschrei
   F cri de l'étain
   P chrzęst cynowy
   R оловянный треск, оловянный хруст

-275 **thermal activation**
   A process of initiating a physicochemical reaction by means of thermal energy, this energy being needed to overcome energy barriers.
   D thermische Aktivierung, thermische Anregung
   F activation thermique
   P aktywacja cieplna
   R термическая активация, термическое возбуждение

-280 **thermally activated process**
   A physicochemical process which occurs by increasing the temperature.
   D thermisch aktivierter Prozeß
   F phénomène activé thermiquement
   P proces aktywowany cieplnie
   R термически активируемый процесс, термически активированный процесс, термоактивируемый процесс

-285 **activation energy**
   The energy required for overcoming a potential barrier.
   D Aktivierungsenergie
   F énergie d'activation
   P energia aktywacji
   R энергия активации

-290 **diffusion**
   The mass transport in gases, liquids or solids, caused by migration of atoms or molecules.
   D Diffusion
   F diffusion
   P dyfuzja
   R диффузия

-295 **volume diffusion, bulk diffusion, lattice diffusion**
   A diffusion taking place through the bulk of the crystal grains.
   D Volumendiffusion, Gitterdiffusion
   F diffusion en volume, diffusion massique, diffusion volumique
   P dyfuzja objętościowa, dyfuzja przestrzenna, dyfuzja sieciowa
   R объёмная диффузия

-300 **surface diffusion**
   A diffusion taking place along the free surface of a given substance.
   D Oberflächendiffusion
   F diffusion en surface
   P dyfuzja powierzchniowa
   R поверхностная диффузия

-305 **vacancy diffusion**
A diffusion taking place through vacancies in the crystal lattice.
D Diffusion über Leerstellen, Diffusion über Gitterlücken
F diffusion par lacunes, diffusion lacunaire
P dyfuzja poprzez wakansy, dyfuzja wakansowa
R диффузия перемещением вакансий

-310 **grain boundary diffusion**
A diffusion taking place along grain boundaries.
D Korngrenzendiffusion
F diffusion intergranulaire
P dyfuzja po granicach ziarn
R пограничная диффузия, зернограничная диффузия, межзёренная диффузия, диффузия по границам зёрен

-315 **interstitial diffusion**
A diffusion consisting in the movement of atoms through the interstices.
D Zwischengitterdiffusion
F diffusion interstitielle, diffusion en insertion
P dyfuzja międzywęzłowa
R междуузельная диффузия

-320 **chemical diffusion**
A diffusion for which a gradient of chemical potential, and not a temperature gradient, constitutes the driving force.
D chemische Diffusion
F diffusion chimique
P dyfuzja chemiczna
R химическая диффузия, концентрационная диффузия

-325 **polyphase diffusion**
A diffusion by which a new phase is formed, other than a solid solution.
D mehrphasige Diffusion
F diffusion polyphasée
P dyfuzja reakcyjna
R реакционная диффузия, неоднофазная диффузия, реактивная диффузия, многофазная диффузия

-330 **single-phase diffusion**
A diffusion by which only a solid solution is formed.
D einphasige Diffusion, Einphasendiffusion
F diffusion monophasée
P dyfuzja jednofazowa, dyfuzja atomowa
R однофазная диффузия

-335 **heterodiffusion**
The movement of atoms of one element, through the crystal lattice of another element.
D Heterodiffusion, Fremddiffusion
F hétérodiffusion
P heterodyfuzja
R гетеродиффузия

-340 **up-hill diffusion**
A diffusion that takes place against the concentration gradient of a given element.
D Bergauf-Diffusion, negative Diffusion
F diffusion montante, diffusion à contresens, diffusion inverséé, diffusion négative
P dyfuzja wstępująca, dyfuzja ujemna
R восходящая диффузия

-345 **steady-state diffusion, stationary diffusion**
A diffusion taking place when the gradients which constitute the driving force do not vary in time.
D stationäre Diffusion
F diffusion (en régime) stationnaire
P dyfuzja ustalona
R установившаяся диффузия, стационарная диффузия

-350 **isothermal diffusion**
A diffusion that takes place at a constant temperature, i.e. without a temperature gradient.
D isotherme Diffusion
F diffusion isotherme
P dyfuzja izotermiczna
R изотермическая диффузия

-355 **enhanced diffusion, accelerated diffusion**
A diffusion taking place at a faster rate, as compared with that occurring in a perfect crystal lattice, due to lattice defects.
D beschleunigte Diffusion
F diffusion accélerée
P dyfuzja przyspieszona
R ускоренная диффузия

-360 **(dislocation) pipe diffusion**
An enhanced diffusion along dislocation pipes, i.e. channels formed by dislocation cores.
D Röhrendiffusion, Versetzungs(schlauch)diffusion
F diffusion dans les tubes
P dyfuzja kanalikowa
R трубочная диффузия

-365 **electrodiffusion, electromigration, electrotransport**
　　The process of mass transport by means of a movement of metal ions, occurring in a current-carrying metal.
　　D Elektrotransport, elektrische Überführung, Elektrodiffusion
　　F électrodiffusion, électromigration
　　P elektrodyfuzja, elektrotransport
　　R электродиффузия, электроперенос, электромиграция

-370 **thermal diffusion, thermomigration, Soret effect**
　　A diffusion for which the temperature gradient constitutes the driving force.
　　D Thermodiffusion, thermische Diffusion, Thermotransport
　　F diffusion thermique
　　P termodyfuzja, dyfuzja termiczna, zjawisko Soreta
　　R тепловая диффузия, термодиффузия, термическая диффузия, эффект Сорета, термоперенос, термомиграция

-375 **diffusion couple**
　　A welded pair of metals or alloys serving as a specimen in investigations of diffusion in metals.
　　D Diffusionspaar
　　F couple de diffusion
　　P złącze dyfuzyjne
　　R диффузионная пара

-380 **interdiffusion**
　　A diffusion of the atoms of each of the components of a diffusion couple into the crystal lattice of its partner.
　　D gegenseitige Diffusion
　　F interdiffusion
　　P dyfuzja wzajemna
　　R взаимная диффузия

-385 **Kirkendall effect**
　　Migration of the boundary between the component metals of a diffusion couple resulting from the different intrinsic diffusion coefficients of these metals.
　　D Kirkendall-Effekt
　　F effet Kirkendall
　　P zjawisko Kirkendalla
　　R эффект Киркендалля, эффект Киркендолла

-390 **diffusion coefficient, diffusivity**
　　A proportionality factor, representing the amount of substance in grammes diffusing in one second across a unit area under the influence of a unit driving force in a particular system.
　　D Diffusionskoeffizient, Diffusionskonstante
　　F coefficient de diffusion, constante de diffusion
　　P współczynnik dyfuzji
　　R коэффициент диффузии

-395 **intrinsic diffusion coefficient, partial diffusion coefficient, intrinsic diffusivity**
　　The diffusion coefficient pertaining to an element in the case of an interdiffusion.
　　D partielle Diffusionskonstante
　　F coefficient de diffusion intrinsèque
　　P cząstkowy współczynnik dyfuzji
　　R парциальный коэффициент диффузии, собственный коэффициент диффузии

-400 **self-diffusion**
　　The migration of atoms of a given substance through the space lattice of this substance.
　　D Selbstdiffusion
　　F autodiffusion
　　P samodyfuzja, dyfuzja własna
　　R самодиффузия

-405 **self-diffusion coefficient**
　　The proportionality factor, analogous to the diffusion coefficient, but relating to the process of self-diffusion.
　　D Selbstdiffusionskoeffizient
　　F coefficient d'autodiffusion
　　P współczynnik samodyfuzji
　　R коэффициент самодиффузии

-410 **diffusion front**
　　An idealized boundary surface, moving in the direction of diffusion, delimiting the region already swept by the diffusing substance from the region towards which the diffusion tends.
　　D Diffusionsfront
　　F front de diffusion
　　P front dyfuzji
　　R фронт диффузии, диффузионный фронт

-415 **diffusion zone**
The zone between the pair of metals in a diffusion couple, in which diffusion between the two has occurred.
D Diffusionszone
F zone de diffusion
P strefa dyfuzji, strefa dyfuzyjna
R диффузионная зона

-420 **recovery**
The restoration of the physical properties of a cold-worked metal without any noticeable change in microstructure.
D Erholung
F restauration
P zdrowienie, nawrót
R возврат, отдых

-425 **dynamic recovery**
Recovery already taking place during plastic deformation, i.e. a stress-aided recovery.
D dynamische Erholung
F restauration dynamique
P zdrowienie dynamiczne
R динамический отдых, динамический возврат

-430 **thermal recovery, static recovery**
Recovery taking place only as a result of providing thermal energy, i.e. without any action of external forces.
D statische Erholung, thermische Erholung
F restauration thermique, restauration statique
P zdrowienie termiczne, zdrowienie statyczne
R статический отдых, термический возврат

-435 **polygonization**
The formation of a sub-structure during recovery of a deformed metal.
D Polygonisation, Polygonisierung
F polygonisation
P poligonizacja
R полигонизация

-440 **recrystallization**
The replacement of a cold-worked structure by a new set of strain-free grains.
D Rekristallisation
F recristallisation
P rekrystalizacja
R рекристаллизация

-445 **primary recrystallization**
The formation of the first population of new grains which replace the structure of a deformed metal.
D primäre Rekristallisation, Primärrekristallisation, Bearbeitungsrekristallisation
F recristallisation primaire
P rekrystalizacja pierwotna
R первичная рекристаллизация, рекристаллизация обработки

-450 **secondary recrystallization**
A process during which some of the grains of a fine-grained recrystallized metal grow rapidly at the expense of the other grains when heated at a higher temperature.
D sekundäre Rekristallisation, Sekundärrekristallisation, Sammelrekristallisation, Grobkristallisation
F recristallisation secondaire, recristallisation de rassemblement
P rekrystalizacja wtórna
R вторичная рекристаллизация, собирательная рекристаллизация

-455 **self-annealing, spontaneous annealing**
Recrystallization taking place in time at room temperature, this phenomenon being shown by cold-worked metals of low melting point.
D spontane Rekristallisation
F recristallisation spontanée
P rekrystalizacja samorzutna
R самопроизвольная рекристаллизация

-460 **tertiary recrystallization**
The phenomenon of replacing the cube texture by the Goss texture sometimes occurring in thin sheets of silicon iron during annealing when the annealing atmosphere containing traces of oxygen has been replaced by very dry hydrogen or by high vacuum.
D tertiäre Rekristallisation
F recristallisation tertiaire
P rekrystalizacja trzeciorzędowa
R третичная рекристаллизация

-465 **dynamic recrystallization**
Recrystallization occurring during deformation at elevated temperatures.
D dynamische Rekristallisation
F recristallisation dynamique
P rekrystalizacja dynamiczna
R динамическая рекристаллизация

-470 **recrystallization temperature**
　　The lowest temperature at which the distorted grain structure of a cold-worked metal is replaced — as a result of prolonged annealing — by a new stress-free equiaxed grain structure.
　　D Rekristallisationstemperatur, Rekristallisationsgrenze
　　F température de recristallisation
　　P temperatura rekrystalizacji
　　R температура (начала) рекристаллизации, порог рекристаллизации

-475 **homologous temperature**
　　The quantity constituting a definite fraction of the absolute melting temperature of a given substance.
　　D homologe Temperatur
　　F température homologue
　　P temperatura homologiczna
　　R гомологическая температура, сходственная температура, соответственная температура

-480 **growth front**
　　An idealized boundary surface, moving in the direction of proceeding grain growth, delimiting the region where the grain growth process has already occurred from the region which has not yet undergone this process.
　　D Wachstumsfront
　　F front de croissance
　　P front rozrostu
　　R фронт роста

-485 **recrystallization diagram**
　　A spatial diagram representing the relationship between the degree of deformation, temperature and grain size after primary recrystallization.
　　D Rekristallisationsdiagramm, Rekristallisationsschaubild
　　F diagramme de recristallisation
　　P wykres rekrystalizacji
　　R диаграмма рекристаллизации

-490 **grain-boundary migration**
　　A motion of the grain boundary in a direction which is inclined to the grain boundary.
　　D Korngrenzenwanderung
　　F migration des joints de grain
　　P migracja granic ziarn
　　R миграция границ (зёрен), движение границ (зёрен)

-495 **creep**
　　Continuous deformation of metals occurring in time under steady load.
　　D Kriechen
　　F fluage
　　P pełzanie
　　R ползучесть, крип

-500 **primary creep, initial creep, transient creep, unsteady creep**
　　The first stage of creep during which the rate of creep decreases.
　　D primäres Kriechen, Übergangskriechen
　　F fluage primaire, fluage transitoire, fluage de transition
　　P pełzanie nieustalone
　　R неустановившаяся ползучесть, нестационарная ползучесть, затухающая ползучесть, переходная ползучесть, первичная ползучесть

-505 **secondary creep, steady-rate creep, steady-state creep, constant-rate creep**
　　The second stage of creep during which the rate of creep is constant.
　　D sekundäres Kriechen, stationäres Kriechen
　　F fluage secondaire, fluage stationnaire
　　P pełzanie ustalone
　　R установившаяся ползучесть, стационарная ползучесть

-510 **tertiary creep, accelerating creep, increasing-rate creep, accelerated creep**
　　The third stage of creep during which the rate of creep increases until the specimen fractures.
　　D tertiäres Kriechen, beschleunigtes Kriechen
　　F fluage tertiaire, fluage accéléré
　　P pełzanie przyspieszone, pełzanie progresywne
　　R прогрессирующая ползучесть, ускоренная ползучесть

-515 **creep curve**
　　A graph of the deformation of a material against time, the load and temperature being constant.
　　D Kriechkurve
　　F courbe de fluage
　　P krzywa pełzania
　　R кривая ползучести

-520 **Nabarro creep, Herring-Nabarro (diffusion) creep, diffusional creep**
  A special kind of steady-state creep, attributable to the mass transport through vacancy diffusion, occurring in specimens which are either very small or consisting of very fine grain material.
  D (Herring-)Nabarro-Kriechen, Diffusionskriechen
  F fluage Nabarro, fluage par diffusion
  P pełzanie dyfuzyjne
  R диффузионная ползучесть

-525 **stress relaxation, elastic relaxation**
  The vanishing or diminution of stresses in a deformed metal as a result of creep.
  D Spannungsrelaxation, Relaxation der Spannungen, elastische Relaxation
  F relaxation des contraintes
  P relaksacja naprężeń
  R релаксация напряжений

-530 **microcreep**
  Creep consisting in a glide movement of dislocations which drag solute atoms along with them in a viscous manner.
  D Mikrokriechen
  F microfluage
  P mikropełzanie
  R микроползучесть

-535 **work softening**
  A phenomenon occurring during deformation of single crystals, where the state of strain hardening reached at a low temperature is unstable at a higher temperature and a recovery process sets in, which tends to reduce the strain hardening to what it would have been if all the working had been accomplished at this higher temperature.
  D Entfestigung
  F déconsolidation
  P zanik umocnienia
  R деформационное разупрочнение

-540 **Portevin-Le Chatelier effect**
  The occurrence of serrations in a stress-strain curve of some alloys resulting from the fact that in the given range of temperature the time required for the diffusion of solute atoms to dislocations is much less than the time required for an ordinary tension test.
  D Portevin-Le Chatelier-Effekt
  F phénomène Portevin-Le Chatelier
  P zjawisko Portevina-Le Chateliera
  R эффект Портевена-Ле Шателье

Group 50

IRON-CARBON-SYSTEM
EISEN-KOHLENSTOFF-SYSTEM
SYSTÈME FER-CARBONE
UKŁAD ŻELAZO-WĘGIEL
СИСТЕМА ЖЕЛЕЗО-УГЛЕРОД

-005 **free carbon, graphitic carbon**
 The carbon that is present in iron-carbon alloys in the free state, i.e. as graphite or temper carbon.
 D freier Kohlenstoff
 F carbone libre, carbone graphitique
 P węgiel wolny
 R свободный углерод, несвязанный углерод

-010 **combined carbon**
 The carbon that is present in iron-carbon alloys as a metallic carbide, as distinct from that existing in a free state as graphite or temper carbon.
 D gebundener Kohlenstoff
 F carbone combiné
 P węgiel związany
 R связанный углерод

-015 **iron-carbon system**
 All alloys which can be formed by iron and carbon.
 D Eisen-Kohlenstoff-System
 F système fer-carbone
 P układ żelazo-węgiel
 R система железо-углерод

-020 **iron-carbon (phase) diagram**
 A phase equilibrium diagram of the iron-carbon system.
 D Eisen-Kohlenstoff(-Zustands) diagramm
 F diagramme d'équilibre fer-carbone
 P wykres (równowagi) żelazo-węgiel
 R диаграмма (состояния) железо-углерод

-025 **austenite**
 A structural constituent of iron-carbon alloys; a solid solution of carbon in gamma iron.
 D Austenit
 F austénite
 P austenit
 R аустенит

-030 **primary austenite**
 The austenite that separates from the melt before eutectic transformation begins.
 D Primäraustenit, primärer Austenit
 F austénite primaire
 P austenit pierwszorzędowy, austenit pierwotny
 R первичный аустенит

-035 **transformed austenite**
 The structural constituents insulting from the transformation of austenite, according to the iron-carbon diagram.
 D umgewandelter Austenit
 F austénite transformée
 P austenit przemieniony
 R превращённый аустенит

-040 **former austenite grain, prior austenite grain, prior austenitic grain**
 The austenite grain existing before austenite transformed on cooling, according to the iron-carbon diagram, which leaves its impression on the microstructure appearing after the transformation.
 D früheres Austenitkorn, ehemaliges Austenitkorn
 F ancien grain d'austénite, grain austénitique originel
 P ziarno byłego austenitu
 R исходное аустенитное зерно

-045 **ferrite**
 A structural constituent of iron-carbon alloys; a solid solution of carbide in alpha iron.
 D Ferrit
 F ferrite
 P ferryt
 R феррит

-050 **delta ferrite, high-temperature ferrite**
 A solid solution of carbon in alpha iron separating directly from the melt.
 D Deltaferrit, Delta-Ferrit
 F ferrite delta
 P ferryt wysokotemperaturowy, ferryt pierwotny
 R высокотемпературный феррит, первичный феррит

-055 **pro-eutectoid ferrite, hypoeutectoid ferrite, free ferrite**
   The ferrite that occurs in hypoeutectoid steels as a separate structural constituent, i.e. not in association with cementite in pearlite.
   D voreutektoider Ferrit, freier Ferrit
   F ferrite proeutectoïde, ferrite libre
   P ferryt przedeutektoidalny, ferryt wolny
   R проэвтектоидный феррит, свободный феррит, избыточный феррит

-060 **pearlitic ferrite**
   The ferrite that occurs in association with cementite in pearlite.
   D perlitischer Ferrit
   F ferrite perlitique
   P ferryt eutektoidalny
   R эвтектоидный феррит, феррит перлита

-065 **cementite**
   A hard structural constituent of the iron-carbon alloys, corresponding to the formula Fe_3C.
   D Zementit
   F cémentite
   P cementyt
   R цементит

-070 **primary cementite, hypereutectic cementite**
   The cementite which separates from the melt before the eutectic transformation begins.
   D Primärzementit, primärer Zementit
   F cémentite primaire, cémentite proeutectique, cémentite hypereutectique
   P cementyt pierwszorzędowy, cementyt pierwotny
   R первичный цементит

-075 **eutectic cementite**
   The cementite which is associated with austenite in ledeburite.
   D eutektischer Zementit, ledeburitischer Zementit
   F cémentite eutectique, cémentite lédeburitique
   P cementyt eutektyczny
   R эвтектический цементит

-080 **pro-eutectoid cementite, hypereutectoid cementite**
   The cementite which separates from austenite in the temperature range between the eutectic temperature and the eutectoid temperature.
   D Sekundärzementit, voreutektoider Zementit, sekundärer Zementit, übereutektoider Zementit
   F cémentite secondaire, cémentite proeutectoïde, cémentite hypereutectoïde
   P cementyt drugorzędowy, cementyt przedeutektoidalny
   R вторичный цементит, проэвтектоидный цементит

-085 **pearlitic cementite**
   The cementite which is associated with ferrite in pearlite.
   D perlitischer Zementit
   F cémentite eutectoïde, cémentite perlitique
   P cementyt eutektoidalny
   R цементит перлита

-090 **tertiary cementite**
   The cementite which separates from ferrite below the eutectoid temperature.
   D Tertiärzementit, tertiärer Zementit
   F cémentite tertiaire
   P cementyt trzeciorzędowy
   R третичный цементит

-095 **free cementite, excess cementite**
   The cementite that occurs as a separate constituent, i.e. not in association with ferrite in pearlite nor in association with austenite in ledeburite.
   D freier Zementit
   F cémentite libre, cémentite de séparation
   P cementyt wolny
   R избыточный цементит, структурно свободный цементит

-100 **spheroidal cementite, spheroidized cementite, nodular cementite, globular cementite, divorced cementite**
   Coagulated cementite formed by subjecting pearlite, or pearlite with hypereutectoid cementite, to an appropriate heat treatment.
   D kugeliger Zementit, körniger Zementit
   F cémentite globulaire, cémentite sphéroïdale
   P cementyt kulkowy
   R сфероидальный цементит, зернистый цементит, шаровидный цементит

-105 **cementite network**
   Secondary cementite occurring as an assembly of shells enveloping the grains of pearlite; visible in a microsection as a network.
   D Zementitnetz
   F réseau de cémentite, liséré de cémentite
   P siatka cementytu
   R цементитная сетка

-110 **alloy(ed) cementite**
   The cementite which apart from iron and carbon contains an alloying element.
   D legierter Zementit
   F cémentite alliée
   P cementyt stopowy
   R легированный цементит

-115 **cohenite**
   The cementite contained in meteorites.
   D Cohenit
   F cohénite
   P kohenit
   R когенит

-120 **cementite lamella**
   Any one of the plate-shaped crystals of cementite contained in lamellar pearlite.
   D Zementitlamelle
   F lamelle de cémentite, plaquette de cémentite
   P płytka cementytu
   R пластинка цементита, цементитная пластинка

-125 **pearlite**
   A structural constituent of iron-carbon alloys; a eutectoid composed of ferrite and cementite.
   D Perlit
   F perlite
   P perlit
   R перлит

-130 **lamellar pearlite, banded pearlite, laminated pearlite**
   Pearlite the components of which, i.e. ferrite and cementite, occur in the form of alternately arranged lamellae.
   D lamellarer Perlit, streifiger Perlit
   F perlite lamellaire, perlite striée
   P perlit płytkowy, perlit blaszkowy, perlit pasemkowy
   R пластинчатый перлит

-135 **fine (lamellar) pearlite, dense lamellar pearlite**
   Lamellar pearlite in which the interlamellar spacing is relatively small (between 0.8 and 1.3 μm).
   D feinlamellarer Perlit, dichtstreifiger Perlit, feinstreifiger Perlit
   F perlite (lamellaire) fine
   P perlit drobnopłytkowy, perlit drobnoblaszkowy, perlit drobnopasemkowy
   R мелкопластинчатый перлит

-140 **coarse (lamellar) pearlite, coarsely lamellar pearlite**
   Lamellar pearlite in which the interlamellar spacing is relatively large (more than 1.6 μm).
   D grobstreifiger Perlit, lockerstreifiger Perlit
   F perlite (lamellaire) grossière
   P perlit grubopłytkowy, perlit gruboblaszkowy, perlit grubopasemkowy
   R груборпластинчатый перлит

-145 **sorbitic pearlite, curly pearlite**
   Lamellar pearlite in which the interlamellar spacing is extremely small (less than 0.3 μm).
   D sehr feinstreifiger Perlit
   F perlite sorbitique
   P perlit sorbityczny, perlit bardzo cienkopłytkowy, perlit bardzo cienkoblaszkowy
   R сорбитовый перлит, сорбитообразный перлит

-150 **spheroidal pearlite, spheroidite, globular pearlite, divorced pearlite, granular pearlite**
   A structural constituent of iron-carbon alloys composed of globular cementite embedded in a ferritic matrix.
   D körniger Perlit
   F perlite globulaire, perlite globularisée, perlite coalescée, sphéroïdite, perlite nodulaire
   P perlit ziarnisty, sferoidyt, perlit z cementytem kulkowym
   R зернистый перлит, глобулярный перлит, точечный перлит

-155 **degenerate pearlite**
   Pearlite in which the ferrite has combined with the pro-eutectoide ferrite, while pearlitic cementite has placed itself at grain boundaries.
   D entarteter Perlit
   F perlite dégénérée
   P perlit zdegenerowany
   R вырожденный перлит

-160 **ledeburite**
A structural constituent of iron-carbon alloys; a eutectic composed of cementite and austenite.
D Ledeburit
F lédeburite
P ledeburyt
R ледебурит

-165 **graphite**
An allotropic variety of carbon, crystallizing in the hexagonal system, occurring in iron-carbon alloys as one of their structural constituents.
D Graphit
F graphite
P grafit
R графит

-170 **primary graphite, kish (graphite)**
Graphite which separates from liquid cast iron.
D Primärgraphit, Schaumgraphit, Garschaumgraphit
F graphite primaire, graphite écumeux, graphite d'écume
P grafit pierwotny, grafit pierwszorzędowy, grafit szumowy
R первичный графит, графитная спель

-175 **eutectic graphite**
Graphite which is a constituent of the iron-graphite eutectic.
D eutektischer Graphit
F graphite eutectique
P grafit eutektyczny
R эвтектический графит

-180 **secondary graphite**
Graphite which separates from austenite on cooling, in the temperature range between the eutectic temperature and the eutectoid temperature.
D Sekundärgraphit, sekundärer Graphit
F graphite secondaire
P grafit wtórny, grafit drugorzędowy
R вторичный графит

-185 **flake graphite**
Graphite having the form of flakes or lamellae.
D Flockengraphit, Lamellengraphit
F graphite lamellaire, graphite en lamelles
P grafit płatkowy
R пластинчатый графит, чешуйчатый графит

-190 **spheroidal graphite, spherulitic graphite, spherolitic graphite**
Graphite having the form of spheroids.
D Kugelgraphit, kugeliger Graphit, sphärolithischer Graphit
F graphite sphéroïdal
P grafit sferoidalny, grafit kulkowy
R шаровидный графит, сферолитный графит, сфероидальный графит, глобулярный графит

-195 **point graphite**
Fine graphite having the shape of points in the microsection.
D punktförmiger Graphit
F graphite punctiforme
P grafit punktowy
R точечный графит

-200 **rosette graphite**
Graphite having the shape of rosettes in the microsection.
D Rosettengraphit
F graphite en rosettes, graphite étoilé
P grafit gwiazdkowy, grafit rozetkowy
R розеточный графит

-205 **interdendritic graphite**
Graphite situated in the interdendritic spaces of cast iron.
D interdendritischer Graphit, zwischendendritischer Graphit
F graphite inter-dendritique
P grafit międzydendrytyczny
R междендритный графит

-210 **nodular graphite**
Clusters of temper carbon.
D Knotengraphit, nestförmiger Graphit
F graphite nodulaire, graphite en nodules
P grafit gniazdowy
R гнездообразный графит

-215 **graphite spherule**
A ball-shaped grain of graphite.
D Graphitkugel, Graphitsphärolith
F sphéroïde de graphite, sphérolite de graphite
P sferoid grafitu
R сферолит графита

-220 **temper carbon, temper graphite, annealing carbon**
Graphite formed during the annealing of black heart malleable cast iron as a result of the decomposition of cementite.
D Temperkohle
F carbone de recuit, graphite de recuit, graphite de malléabilisation
P węgiel żarzenia
R углерод отжига

-225 **graphitization**
    The decomposition of iron carbide in iron-carbon alloys, leading to the liberation of free carbon, i.e. of graphite.
    D Graphitisierung
    F graphitisation
    P grafityzacja
    R графитизация

-230 **ferrite halo**
    The ferritic region enveloping a graphite spherule in some spheroidal graphite cast irons, or a cluster of temper carbon in some malleable cast irons.
    D Ferrithof
    F liséré ferritique, auréole de ferrite
    P otoczka ferrytyczna
    R ферритный венец

-235 **steadite, phosphide eutectic**
    A ternary eutectic composed of iron phosphide, cementite, and ferrite.
    D Steadit, (ternäres) Phosphideutektikum
    F stéadite, eutectique phosphoreux (ternaire)
    P steadyt, (potrójna) eutektyka fosforowa
    R стедит, фосфористая эвтектика

-240 **manganese sulfide**
    A compound of manganese and sulfur (MnS).
    D Mangansulfid
    F sulfure de manganèse
    P siarczek manganu, siarczek manganawy
    R сульфид марганца, сернистый марганец

-245 **iron sulfide**
    A compound of iron and sulfur (FeS).
    D Eisensulfid
    F sulfure de fer
    P siarczek żelaza(wy)
    R сульфид железа, сернистое железо

-250 **iron phosphide**
    A compound of iron and phosphorus (Fe_3P).
    D Eisenphosphid
    F phosphure de fer
    P fosforek żelaza
    R фосфид железа

-255 **cementite eutectic**
    The eutectic occurring in the binary iron-iron carbide system.
    D (Eisen-)Zementit-Eutektikum
    F eutectique lédeburitique
    P eutektyka cementytowa, eutektyka ledeburytyczna
    R цементитная эвтектика, ледебуритная эвтектика

-260 **graphite eutectic**
    The eutectic occurring in the binary iron-graphite system.
    D (Eisen-)Graphit-Eutektikum
    F eutectique fer-graphite
    P eutektyka grafitowa
    R графитная эвтектика

-265 **undercooled austenite, unstable austenite**
    A temporarily subsisting austenite which has been brought by cooling to a temperature below the eutectoid temperature.
    D unterkühlter Austenit
    F austénite métastable
    P austenit przechłodzony
    R переохлаждённый аустенит, неустойчивый аустенит

-270 **retained austenite**
    An austenite which has not been transformed into martensite during quenching and subsists in quenched steel along with martensite.
    D Restaustenit
    F austénite résiduelle, austénite retenue
    P austenit szczątkowy
    R остаточный аустенит

-275 **pearlitic transformation, pearlite reaction, split transformation of austenite**
    The transformation of austenite resulting in the formation of pearlite or a pearlite-type structure.
    D Perlitumwandlung
    F transformation perlitique
    P przemiana perlityczna
    R (аустенитно-)перлитное превращение

-280 **martensitic transformation, martensitic change, martensite transformation**
    The transformation of undercooled austenite resulting in the formation of martensite.
    D Martensitumwandlung
    F transformation martensitique, transformation austénite-martensite
    P przemiana martenzytyczna
    R (аустенитно-)мартенситное превращение

-285 **secondary martensitic transformation**
 The martensitic transformation taking place at sub-zero temperatures after the stabilization of retained austenite at room temperature.
 D sekundäre Martensitumwandlung
 F transformation martensitique secondaire
 P przemiana martenzytyczna wtórna
 R вторичное мартенситное превращение

-290 **bainitic transformation, bainite transformation, bainite reaction**
 The transformation of undercooled austenite resulting in the formation of bainite.
 D Zwischenstufen-Umwandlung, Umwandlung in der Bainitstufe, bainitische Umwandlung
 F transformation bainitique
 P przemiana bainityczna
 R бейнитное превращение, промежуточное превращение

-295 **pearlite range**
 The temperature range within which undercooled austenite transforms into pearlite-type structures.
 D Perlitstufe
 F domaine perlitique
 P zakres przemiany perlitycznej, zakres perlityczny, obszar tworzenia się perlitu
 R перлитная область

-300 **martensite range**
 The temperature range within which undercooled austenite transforms into martensite.
 D Martensitstufe, Martensitbereich
 F domaine martensitique
 P zakres przemiany martenzytycznej, zakres martenzytyczny, obszar tworzenia się martenzytu
 R мартенситная область, интервал мартенситного превращения

-305 **bainite range**
 The temperature range within which undercooled austenite transforms into bainite.
 D Zwischenstufe
 F domaine bainitique
 P zakres przemiany bainitycznej, zakres bainityczny, obszar tworzenia się bainitu
 R промежуточная область, область бейнитного превращения

-310 **martensite**
 A structural constituent of iron-carbon alloys representing a product of the transformation of undercooled austenite, not accompanied by diffusion; it is a supersaturated solid solution of carbon in alpha iron.
 D Martensit
 F martensite
 P martenzyt
 R мартенсит

-315 **acicular martensite**
 Martensite which appears in microsections as having a needle-like structure.
 D nadeliger Martensit, Umklappmartensit, azikularer Martensit
 F martensite aciculaire
 P martenzyt iglasty
 R игольчатый мартенсит

-320 **coarse grained martensite**
 Martensite which appears in microsections as having the shape of coarse needles.
 D grobnadeliger Martensit, grobkörniger Martensit
 F martensite en gros grains
 P martenzyt gruboiglasty
 R крупноигольчатый мартенсит, крупнопластинчатый мартенсит

-325 **fine acicular martensite**
 Martensite which appears in microsections as having the shape of fine needles.
 D feinnadeliger Martensit
 F martensite aciculaire fine
 P martenzyt drobnoiglasty
 R мелкоигольчатый мартенсит, тонкоигольчатый мартенсит, мелкий мартенсит

-330 **cryptocrystalline martensite**
 A very fine martensite in which the needles are so minute that they are indiscernible under an optical microscope.
 D strukturloser Martensit, kryptokristalliner Martensit
 F martensite cryptocristalline
 P martenzyt skrytoiglasty, martenzyt skrytokrystaliczny
 R бесструктурный мартенсит, безигольчатый мартенсит, скрытокристаллический мартенсит

-335 **tetragonal martensite, alpha martensite**
Martensite in which the crystal lattice exhibits tetragonality, i.e. has an axial ratio different from unity.
D tetragonaler Martensit, Abschreckmartensit
F martensite tétragonale, martensite de trempe
P martenzyt tetragonalny
R тетрагональный мартенсит, мартенсит закалки

-340 **beta martensite, low-carbon martensite**
Martensite in which the crystal lattice almost does not exhibit tetragonality, i.e. has an axial ratio close to unity.
D kubischer Martensit
F martensite cubique
P martenzyt regularny
R кубический мартенсит

-345 **tempered martensite**
Martensite that has been heated to a relatively low temperature at which cementite does not yet occur, the carbon precipitated from the supersaturated solution forming a metastable transient carbide, its composition being different from Fe_3C.
D angelassener Martensit, Anlaßmartensit
F martensite (de) revenue
P martenzyt odpuszczony, martenzyt odpuszczania
R отпущенный мартенсит, мартенсит отпуска

-350 **secondary martensite**
Martensite formed from retained austenite during cooling after tempering.
D Sekundärmartensit, sekundärer Martensit
F martensite secondaire
P martenzyt wtórny
R вторичный мартенсит

-355 **massive martensite, lath martensite**
Martensite produced as a result of a transformation resembling the massive transformation, its morphology differing from the usual acicular martensite.
D massiver Martensit, Schiebungsmartensit, plattenförmiger Martensit, lattenförmiger Martensit
F martensite massive
P martenzyt masywny
R массивный мартенсит, реечный мартенсит

-360 **strain induced martensite, deformation martensite**
Martensite formed as a result of plastic deformation of austenite.
D Verformungsmartensit, Schleifmartensit
F martensite d'écrouissage, martensite de déformation
P martenzyt zgniotowy
R мартенсит наклёпа, мартенсит деформации, деформационный мартенсит

-365 **hardenite**
A fine martensite of exact eutectoid composition.
D Hardenit
F hardénite
P hardenit
R гарденит

-370 **martensite plate, martensitic plate**
A commonly encountered individual crystal of martensite.
D Martensitplatte
F plaquette martensitique, plaquette de martensite
P płytka martenzytu
R мартенситная пластинка, пластинка мартенсита

-375 **martensite needle**
A typical crystal of martensite as seen on the microsection.
D Martensitnadel
F aiguille martensitique, aiguille de martensite
P igła martenzytu
R игла мартенсита, мартенситная игла

-380 **sorbite**
A structural constituent of iron-carbon alloys, consisting of a fine aggregate of ferrite and cementite, produced as a result of an appropriate heat traetment.
D Sorbit
F sorbite
P sorbit
R сорбит

-385 **quenching sorbite, fine pearlite**
Sorbite which is obtained during quench hardening of steel, when the cooling rate of austenite is somewhat lower than that giving rise to troostite.
D Abschrecksorbit
F sorbite de trempe
P sorbit hartowania, drobny perlit
R сорбит закалки

-390 **temper sorbite**
Sorbite that occurs as the product of the tempering of martensite, when the tempering temperature is somewhat higher than that giving rise to temper troostite.
D Anlaßsorbit
F sorbite de revenu
P sorbit odpuszczania
R сорбит отпуска

-395 **troostite**
A structural constituent of the iron-carbon alloys, consisting of an extremely fine aggregate of ferrite and cementite, produced as a result of appropriate heat treatment.
D Troostit
F troostite
P troostyt
R троостит

-400 **quenching troostite, nodular troostite, very fine pearlite**
Troostite which is obtained during quench hardening of steel when the cooling rate of austenite is lower than that giving rise to martensite but higher than that giving rise to sorbite.
D Abschrecktroostit
F troostite de trempe, troostite primaire
P troostyt hartowania, bardzo drobny perlit
R троостит закалки

-405 **temper troostite**
Troostite that occurs as the first product of the tempering of martensite.
D Anlaßtroostit
F troostite de revenu, troostite secondaire
P troostyt odpuszczania
R троостит отпуска

-410 **bainite, troosto-martensite, troostite-martensite**
A structural constituent of iron-carbon alloys, of an acicular or feathery structure, produced during the isothermal transformation of austenite, undercooled to the temperature range stretching from about 450°C to M_s (martensite start temperature).
D Bainit, Zwischenstufengefüge
F bainite
P bainit
R бейнит, игольчатый троостит

-415 **lower bainite**
Bainite of an acicular structure, formed at the lower end of the temperature range stretching from about 450°C to M_s (martensite start temperature).
D Gefüge der unteren Zwischenstufe, unterer Bainit
F bainite inférieure
P bainit dolny
R нижний бейнит, игольчатый бейнит

-420 **upper bainite**
Bainite of a feathery structure, formed at the upper end of the temperature range stretching from about 450°C to M_s (martensite start temperature).
D Gefüge der oberen Zwischenstufe, oberer Bainit
F bainite supérieure
P bainit górny
R верхний бейнит, перистый бейнит

-425 **stabilization of austenite, austenite stabilization**
A decrease in the tendency of undercooled austenite to be transformed, occurring when this austenite has been held for some time at a constant temperature in the martensite range.
D Austenitstabilisierung, Stabilisation des Austenits
F stabilisation de l'austénite (résiduelle)
P stabilizacja austenitu
R стабилизация аустенита

-430 **destabilization of austenite**
The loss of capacity of an austenitic steel to withstand a transformation of austenite.
D Destabilisierung des Austenits
F déstabilisation de l'austénite (résiduelle)
P destabilizacja austenitu
R дестабилизация аустенита

-435 **carbide-forming element, carbide former**
An alloying element which combines with carbon to form carbides.
D Karbidbildner, karbidbildendes Element
F élément carburigène, générateur de carbures
P pierwiastek węglikotwórczy
R карбидообразующий элемент

-440 **eutectic carbides, ledeburitic carbides**
Carbides which are associated with austenite in ledeburite.
D eutektische Karbide, ledeburitische Karbide
F carbures eutectiques, carbures lédéburitiques
P węgliki eutektyczne, węgliki ledeburytyczne
R эвтектические карбиды

-445 **carbide network**
The arrangement of secondary carbides in steels at the grain boundaries of pearlite.
D Karbidnetz
F réseau de carbides
P siatka węglików
R карбидная сетка, сетка карбидов

-450 **austenite stabilizer, austenitizer, austenite former, austenite forming element, gamma-producing element**
Any alloying element which extends the range of stability of austenite.
D austenitbildendes Element, Austenitbildner, austenitstabilisierendes Element, gammagenes Element
F élément gammagène, stabilisateur d'austénite
P pierwiastek austenitotwórczy, pierwiastek stabilizujący austenit
R аустенитообразующий элемент

-455 **ferrite stabilizer, ferritizer, ferrite former, ferrite-forming element, alpha-producing element**
Any alloying element which extends the range of stability of ferrite.
D ferritbildendes Element, Ferritbildner, ferritstabilisierendes Element, ferritförderndes Element, alphagenes Element
F élément alphagène, stabilisateur de ferrite
P pierwiastek ferrytotwórczy, pierwiastek stabilizujący ferryt
R ферритообразующий элемент

-460 **graphitizer, graphite-forming element**
An alloying element which promotes carbon in iron-carbon alloys to occur in a free form, i.e. as graphite.
D Graphitbildner, karbidzerlegendes Element
F (élément) graphitisant
P grafityzator
R графитообразующий элемент

-465 **carbide stabilizer**
An alloying element which promotes carbon in iron-carbon alloys to persist in a combined form, i.e. as carbide.
D karbidstabilisierendes Element
F (élément) antigraphitisant, stabilisateur de carbure
P antygrafityzator
R карбидостабилизирующий элемент

-470 **nitride-forming element, nitride former**
An alloying element which combines with nitrogen to form nitrides.
D Nitridbildner, nitridbildendes Element
F élément nitrurigène
P pierwiastek azotkotwórczy
R нитридообразующий элемент, нитридообразователь

-475 **silicon ferrite, silicoferrite**
Ferrite in which silicon is dissolved.
D Siliziumferrit
F silicoferrite, ferrite au silicium
P krzemoferryt
R кремнистый феррит

-480 **carbon equivalent**
The value resulting from converting the percentages of alloying elements in steel, by means of an empirical formula, into an equivalent carbon content, this equivalence holding only for a certain definite application, e.g. for weldability purposes.
D Kohlenstoffäquivalent
F équivalent de carbone, équivalent en carbone
P równoważnik węglowy
R углеродный эквивалент

-485 **stabilizer**
An additional alloying element, e.g. titanium or niobium, which is introduced in small quantities to stainless steel to prevent intercrystalline corrosion.
D Stabilisierungselement, stabilisierendes Legierungselement
F (élément) stabilisant, addition stabilisante
P stabilizator, pierwiastek stabilizujący
R стабилизирующая присадка

-490 **eutectoid steel**
    Steel having in a fully annealed condition a eutectoid structure.
    D eutektoider Stahl
    F acier eutectoïde
    P stal eutektoidalna
    R эвтектоидная сталь

-495 **hypoeutectoid steel**
    Steel having in a fully annealed condition a hypoeutectoid structure.
    D untereutektoider Stahl, untereutektoidischer Stahl
    F acier hypoeutectoïde, acier sous-eutectoïde
    P stal podeutektoidalna
    R доэвтектоидная сталь

-500 **hypereutectoid steel, cementitic steel**
    Steel having in a fully annealed condition a hypereutectoid structure.
    D übereutektoider Stahl, übereutektoidischer Stahl
    F acier hypereutectoïde
    P stal nadeutektoidalna
    R заэвтектоидная сталь

-505 **eutectic white cast iron**
    White cast iron composed entirely of ledeburite.
    D eutektisches weißes Gußeisen
    F fonte lédeburitique, fonte blanche eutectique
    P żeliwo białe eutektyczne
    R эвтектический белый чугун

-510 **hypoeutectic white cast iron**
    White cast iron composed of transformed austenite (i.e. pearlite with secondary cementite) and ledeburite.
    D untereutektisches weißes Gußeisen
    F fonte blanche hypoeutectique
    P żeliwo białe podeutektyczne
    R доэвтектический белый чугун

-515 **hypereutectic white cast iron**
    White cast iron composed of primary cementite and ledeburite.
    D übereutektisches weißes Gußeisen
    F fonte blanche hypereutectique
    P żeliwo białe nadeutektyczne
    R заэвтектический белый чугун

55 TYPES OF METALS AND ALLOYS
TYPEN VON METALLEN UND LEGIERUNGEN
TYPES DES MÉTAUX ET ALLIAGES
RODZAJE METALI I STOPÓW
ТИПЫ МЕТАЛЛОВ И СПЛАВОВ

-005 **common metal, base metal**
A metal which is not resistant to corrosion or oxidation on exposure to air or moisture.
D unedles Metall, Unedelmetall
F métal commun
P metal pospolity, metal nieszlachetny
R неблагородный металл

-010 **noble metal, precious metal**
A metal which is not readily affected by corrosion or oxidation in the air or moisture.
D Edelmetall
F métal noble, métal precieux
P metal szlachetny
R благородный металл, драгоценный металл

-015 **non-ferrous metals**
Metals other than iron, cobalt, and nickel.
D Nichteisenmetalle, NE-Metalle, Buntmetalle
F métaux non ferreux
P metale nieżelazne
R цветные металлы

-020 **semi-noble metal**
A metal which is to some extent resistant to corrosion or oxidation in the air or moisture.
D halbedles Metall
F métal semi-noble
P metal półszlachetny
R полублагородный металл

-025 **transition metals, transition elements**
A group of chemical elements their free atoms having a characteristic electronic structure, in which there are electrons in the outer shell though the foregoing shell is not completely filled.
D Übergangsmetalle, Übergangselemente, T-Metalle
F métaux de transition, éléments de transition
P metale przejściowe, metale grup przejściowych
R переходные металлы, металлы переходных групп

-030 **alkali metals**
The monovalent metals lithium, sodium, potassium, rubidium, caesium, and francium which constitute the first group of the periodic system.
D Alkalimetalle
F (métaux) alcalins
P metale alkaliczne
R щелочные металлы

-035 **alkaline earth metals**
The bivalent metals calcium, strontium, barium, and radium which constitute the second group of the periodic system.
D Erdalkalimetalle, Erdalkalien
F (métaux) alcalino-terreux
P metale ziem alkalicznych
R щёлочноземельные металлы

-040 **rare earth metals**
A group of trivalent metals, very similar in their chemical properties, with atomic numbers ranging from 57 to 71.
D seltene Erdmetalle, Seltenerdmetalle, Selten-Erd-Metalle
F (métaux des) terres rares
P metale ziem rzadkich
R редкоземельные металлы

-045 **rare metals**
Metals which are not encountered in nature in large agglomerates, their occurrence being restricted to a highly dispersed state, i.e. to minute or even trace quantities.
D seltene Metalle, Spurenmetalle
F métaux rares
P metale rzadkie
R редкие металлы, рассеянные металлы

-050 **light metals**
Metals with the specific gravity lower than 4.5 g/cm^3.
D Leichtmetalle
F métaux légers
P metale lekkie
R лёгкие металлы

-055 **heavy metals, high-density metals**
　　Metals with the specific gravity greater than 4.5 g/cm³.
　　D Schwermetalle
　　F métaux lourds
　　P metale ciężkie
　　R тяжёлые металлы

-060 **low-melting-point metal**
　　A metal which melts at a relatively low temperature.
　　D niedrigschmelzendes Metall
　　F métal à bas point de fusion
　　P metal łatwotopliwy, metal niskotopliwy
　　R легкоплавкий металл

-065 **refractory metal, high-melting-point metal**
　　A metal which melts at a relatively high temperature.
　　D hochschmelzendes Metall
　　F métal refractaire, métal à haut point de fusion
　　P metal trudnotopliwy, metal wysokotopliwy
　　R тугоплавкий металл

-070 **reactive metal**
　　A metal which is readily susceptible to chemical change.
　　D reaktives Metall, reaktionsfreudiges Metall
　　F métal réactif
　　P metal chemicznie aktywny, metal reaktywny
　　R реакционноспособный металл

-075 **radioactive metal**
　　Any one of the group of metals with high atomic weights which break down spontaneously into other elements giving off (α, β, or γ) radiation.
　　D radioaktives Metall
　　F métal radioactif
　　P metal radioaktywny
　　R радиоактивный металл

-080 **fissi(onab)le metal**
　　A metallic element whose atomic nucleus has a tendency to split into two or more particles (these particles being nuclei of other elements) with the emission of much energy.
　　D spaltbares Metall, spaltfähiges Metall
　　F métal fissile
　　P metal rozszczepialny
　　R делящийся металл

-085 **nuclear metal**
　　A metal used for special purposes in nuclear energetics.
　　D Reaktormetall
　　F métal nucléaire
　　P metal reaktorowy
　　R реакторный металл

-090 **pure metal**
　　A metal containing no detectable impurities or, in a broader sense, a metal considered as a metallic element, as distinct from an alloy.
　　D Reinmetall, reines Metall
　　F métal pur
　　P metal czysty; czysty metal
　　R чистый металл

-095 **grade of purity**
　　A measure of the purity of metals expressed in percentage.
　　D Reinheitsgrad
　　F grade de pureté
　　P czystość, stopień czystości
　　R степень чистоты

-100 **fineness**
　　A measure of the purity of noble metals expressed mainly in parts per mille.
　　D Feingehalt
　　F aloi
　　P próba (metali szlachetnych)
　　R проба (благородного металла)

-105 **super-pure metal, ultra-pure metal, hyper-pure metal**
　　A metal of such high grade of purity that any further purification does not change its properties.
　　D Reinstmetall
　　F métal ultra-pur
　　P metal ultraczysty
　　R сверхчистый металл, металл сверхвысокой чистоты

-110 **native metal**
　　A naturally occurring metal.
　　D gediegenes Metall
　　F métal natif
　　P metal rodzimy
　　R самородный металл

-115 **primary metal, virgin metal, new metal**
　　A metal obtained directly from ore, i.e. previously unused, as distinct from a metal remelted from scrap.
　　D Primärmetall
　　F métal neuf, métal vierge
　　P metal pierwotny, metal nowy, metal nie przetapiany
　　R первичный металл

-120 **secondary metal, reused metal**
Metal obtained by remelting scrap.
D Sekundärmetall, Umschmelzmetall
F métal refondu
P metal przetopiony, metal wtórny
R вторичный металл

-125 **refined metal**
A metal which has been subjected to a process of purification.
D raffiniertes Metall, Feinmetall
F métal affiné
P metal rafinowany
R рафинированный металл

-130 **electrodeposited metal, electrolytic metal**
A metal obtained by means of electrodeposition from an electrolyte containing a salt of the metal to be deposited.
D Elektrolytmetall, elektrolytisch niedergeschlagenes Metall
F métal électrolytique
P metal elektrolityczny
R электролитический металл

-135 **vacuum(-melted) metal**
A metal obtained by melting in a vacuum furnace and therefore almost free from gaseous impurities.
D vakuumgeschmolzenes Metall
F métal fondu sous vide, métal élaboré sous vide
P metal topiony w próżni, metal próżniowy
R вакуумированный металл

-140 **zone-refined metal**
A metal with a high grade of purity, which has been attained through the elimination of impurities by means of zone melting.
D zonengeschmolzenes Metall
F métal de zone fondue
P metal oczyszczany strefowo
R металл очищенный зонной плавкой

-145 **(vacuum-)evaporated metal**
A metal deposited in the form of thin layers by means of evaporation in a vacuum and subsequent condensation of the vapour.
D aufgedampftes Metall
F métal déposé par evaporation
P metal napylany, metal osadzany z pary
R металл конденсированный в вакууме

-150 **melt**
A metal or alloy in a liquid state of aggregation.
D Schmelze, Metallschmelze
F métal fondu; alliage fondu
P metal ciekły, metal roztopiony; stop ciekły
R расплав, жидкий расплав, расплавленный металл

-155 **compact metal**
A metal constituting a dense uniform mass, i.e. without an accumulation of discontinuities such as pores or non-metallic inclusions.
D kompaktes Metall
F métal compact
P metal lity
R компактный металл

-160 **(metallic) alloy**
A material exhibiting metallic properties, composed of two or more elements of which the major one at least is metal.
D Legierung, Metallegierung
F alliage (métallique)
P stop (metalu), stop metali
R (металлический) сплав

-165 **metal(lic) system, alloy system**
All alloys which can be formed by the given components in the whole composition range, i.e. from 0 to 100% of each component.
D Metallsystem, metallisches System, Legierungssystem
F système métallique
P układ metaliczny, układ stopowy
R металлическая система, легированная система.

-170 **alloying**
The addition to a metal of one or more different elements in order to form an alloy.
D Legieren, Zusammenlegieren
F formation d'un alliage
P tworzenie stopu, wytwarzanie stopu; wprowadzanie składnika stopowego, stopowanie
R легирование, сплавление

-175 **alloying power**
The ability of an element to form an alloy with a given metal.
D Legierbarkeit, Legierungsfähigkeit
F aptitude à former un alliage
P zdolność stopotwórcza
R легируемость, способность сплавляться

-180 **component of an alloy, alloy(ing) component**
An element or compound contained in an alloy, deliberately introduced or preserved there in order to obtain certain desired properties.
D Legierungsbestandteil, Legierungskomponente
F composant d'un alliage, constituant d'un alliage
P składnik stopu
R компонент сплава, составляющая сплава, легирующий компонент

-185 **basis metal, base of an alloy, parent metal, major constituent**
The principal component of an alloy.
D Legierungsbasis, Legierungsträger, Grundmetall (einer Legierung)
F base d'alliage, métal de base
P podstawa stopu, metal podstawowy
R основа сплава, основной компонент (сплава)

-190 **alloying element, alloying addition**
Any of the components of an alloy other than the basis metal.
D Legierungselement, Legierungszusatz
F élément d'alliage, élément d'addition
P pierwiastek stopowy, dodatek stopowy
R легирующий элемент, легирующая добавка

-195 **major alloying element, principal alloying element**
The second — as regards percentage or importance — component of an alloy.
D Hauptlegierungselement, Hauptlegierungszusatz
F élément majeur d'addition, addition principale
P główny pierwiastek stopowy, dodatek stopowy główny
R основной легирующий элемент, главный легирующий элемент, основная легирующая добавка, главная легирующая добавка

-200 **tramp elements**
Those elements present in steel (or in other alloys) which are not intentionally added (e.g. those entrapped from non-typical scrap)
D zufällige Beimengungen
F additions inintentionnelles, éléments accidentels
P domieszki przypadkowe, domieszki niezamierzone
R случайные примеси

-205 **impurities**
Undesirable substances contained in a metal or alloy, their presence being inevitable.
D Verunreinigungen
F impuretés
P zanieczyszczenia
R загрязнения, примеси

-210 **alloy hardening**
Increasing the hardness and strength of a metal by forming an alloy on the basis of this metal.
D Legierungsverfestigung
F durcissement par formation d'un alliage
P utwardzanie przez tworzenie stopu, utwardzanie przez stopowanie, umacnianie przez tworzenie stopu
R упрочнение легированием

-215 **solution hardening, solution strengthening**
Increasing the hardness and strength of a metal by forming a solid solution of this metal with one or more alloying elements.
D Mischkristallhärtung, Mischkristallverfestigung, Lösungsverfestigung
F durcissement par solution
P utwardzanie roztworowe, umacnianie roztworowe
R упрочнение растворением, твёрдорастворное упрочнение

-220 **dispersion strengthening, dispersion hardening, dispersed phase hardening, particle reinforcement**
Increasing the hardness and strength of a metal by means of finely dispersed particles of another substance which is not soluble in the metal in question.
D Dispersionsverfestigung, Dispersionshärtung, Teilchenverfestigung, Teilchenhärtung
F durcissement par dispersion, durcissement par dispersoïdes, durcissement par des particules
P umacnianie dyspersyjne
R упрочнение дисперсными частицами, дисперсное упрочнение

-225 **chemical composition (of an alloy)**
The percentages of individual components of an alloy.
D chemische Zusammensetzung (einer Legierung)
F composition chimique (d'un alliage)
P skład chemiczny (stopu)
R химический состав (сплава)

-230 **concentration**
   The percentage of a given component in an alloy.
   D Konzentration
   F concentration
   P stężenie
   R концентрация

-235 **mass concentration**
   The concentration expressed as weight percentage.
   D Gewichtskonzentration
   F concentration pondérale, concentration en poids
   P stężenie ciężarowe, stężenie wagowe
   R весовая концентрация, концентрация по весу

-240 **molar concentration**
   The concentration expressed as mole fraction.
   D molare Konzentration
   F concentration molaire
   P stężenie molowe
   R молярная концентрация

-245 **bulk concentration**
   The average concentration of a component in a heterogeneous alloy, i.e. irrespective of the distribution of this component among the individual phases constituting the alloy.
   D Legierungskonzentration
   F concentration moyenne, concentration nominale
   P stężenie średnie, stężenie nominalne
   R средняя концентрация

-250 **trace element, oligo-constituent**
   An element contained in an alloy in a very minute proportion.
   D Spurenelement, Spurenbestandteil
   F élément-trace, oligo-élément, élément de trace
   P pierwiastek śladowy, mikrododatek
   R микроэлемент, (легирующая) микродобавка, микролегирующий элемент, микролегирующая добавка

-255 **hardening alloy element, hardener, hardening constituent**
   Any element which raises the hardness of an alloy.
   D härtesteigerndes Element
   F élément durcissant
   P pierwiastek utwardzający
   R упрочняющий элемент, упрочнитель

-260 **residual element**
   An element which usually occurs within a given metal and is difficult to eliminate, its origin being attributable to the ore or to the metallurgical process.
   D Begleitelement
   F élément résiduel, élément d'élaboration, élément associé
   P pierwiastek towarzyszący
   R сопутствующий элемент

-265 **native alloy, natural alloy**
   A naturally occurring alloy.
   D Naturlegierung
   F alliage naturel
   P stop naturalny
   R природный сплав

-270 **synthetic alloy, synthetized alloy**
   An alloy specially prepared from its components, as distinct from a naturally occurring alloy.
   D synthetische Legierung
   F alliage synthétique
   P stop syntetyczny
   R синтетический сплав

-275 **new alloy**
   An alloy newly made-up, as distinct from an alloy obtained by remelting scrap.
   D Neulegierung, Legierung erster Schmelzung
   F alliage de première fusion
   P stop pierwotny
   R первичный сплав, сплав первой плавки

-280 **secondary alloy, remelted alloy**
   An alloy obtained by remelting scrap or wasters.
   D Umschmelzlegierung, U-Legierung, Altmetallegierung
   F alliage de deuxième fusion
   P stop wtórny
   R вторичный сплав

-285 **master alloy, foundry alloy, tempering metal, rich alloy**
   An auxiliary alloy used to facilitate the addition of high-melting-point components to an alloy.
   D Vorlegierung
   F pré-alliage, alliage-mère, alliage d'addition
   P stop przejściowy, stop wstępny, zaprawa
   R лигатура, промежуточный сплав

-290 **electrodeposited alloy**
An alloy obtained by the process of electrodeposition.
D elektrolytische Legierung
F alliage électrolytique, alliage électrodéposé
P stop elektrolityczny, stop galwaniczny
R электролитический сплав, электроосаждённый сплав

-295 **binary (alloy), two-component alloy**
An alloy formed by two metals or by a metal and a non-metal.
D binäre Legierung, Zweistofflegierung, einfache Legierung
F alliage binaire
P stop podwójny, stop dwuskładnikowy
R двойной сплав, бинарный сплав, двухкомпонентный сплав

-300 **ternary (alloy), three-component alloy**
An alloy formed by three components.
D ternäre Legierung, Dreistofflegierung
F alliage ternaire
P stop potrójny, stop trójskładnikowy
R тройной сплав, трёхкомпонентный сплав

-305 **quaternary (alloy), four-component alloy**
An alloy formed by four components.
D quaternäre Legierung, Vierstofflegierung, vierkomponentige Legierung
F alliage quaternaire
P stop poczwórny, stop czteroskładnikowy
R четверной сплав

-310 **complex alloy, multicomponent alloy**
An alloy formed by several components.
D komplexe Legierung, Mehrstofflegierung
F alliage complexe, alliage à plusieurs composants
P stop wieloskładnikowy
R многокомпонентный сплав

-315 **pseudo-binary alloy, quasibinary alloy**
A ternary alloy in which an intermetallic compound formed by two of the three components exists, behaving as an independent component.
D quasibinäre Legierung
F alliage pseudobinaire
P stop pseudopodwójny
R псевдобинарный сплав, квазибинарный сплав

-320 **dilute alloy, low alloy**
An alloy in which the amount of the alloying element in relation to the basis metal is very small.
D verdünnte Legierung, niedrigprozentige Legierung, niedriglegierte Legierung
F alliage dilué
P stop niskoprocentowy
R низкопроцентный сплав, разбавленный сплав, низколегированный сплав

-325 **concentrated alloy, high alloy**
An alloy in which the amount of the alloying element in relation to the basis metal is great.
D hochprozentige Legierung, hochlegierte Legierung
F alliage concentré
P stop wysokoprocentowy
R высокопроцентный сплав, высоколегированный сплав

-330 **homogeneous alloy, single-phase alloy**
An alloy consisting of a single phase only.
D homogene Legierung, einphasige Legierung, monophasige Legierung
F alliage homogène, alliage monophasé, alliage à une phase
P stop jednofazowy
R однофазный сплав, гомогенный сплав

-335 **two-phase alloy**
An alloy consisting of two phases.
D zweiphasige Legierung
F alliage biphasé, alliage à deux phases
P stop dwufazowy
R двухфазный сплав

-340 **heterogeneous alloy**
An alloy containing more than one phase.
D heterogene Legierung
F alliage hétérogène
P stop niejednofazowy
R гетерогенный сплав, негомогенный сплав

-345 **polyphase alloy**
An alloy consisting of several phases.
D mehrphasige Legierung
F alliage polyphasé
P stop wielofazowy
R многофазный сплав

-350 **eutectic alloy**
An alloy having the lowest possible melting point for a given series of alloys or for a certain range of that series, its structure being a eutectic.
D eutektische Legierung
F alliage eutectique
P stop eutektyczny
R эвтектический сплав

-355 **hypoeutectic alloy**
An alloy, containing a eutectic, in which the content of the second component is less than the percentage corresponding to the eutectic.
D untereutektische Legierung
F alliage hypoeutectique
P stop podeutektyczny
R доэвтекический сплав

-360 **hypereutectic alloy**
An alloy, containing a eutectic, in which the content of the second component exceeds the percentage corresponding to the eutectic.
D übereutektische Legierung
F alliage hypereutectique
P stop nadeutektyczny
R заэвтектический сплав

-365 **eutectoid alloy**
An alloy having a eutectoid structure in a fully annealed condition.
D eutektoide Legierung
F alliage eutectoïde
P stop eutektoidalny
R эвтектоидный сплав

-370 **hypoeutectoid alloy**
An alloy, containing a eutectoid, in which the content of the second component is less than the percentage corresponding to the eutectoid.
D untereutektoid(isch)e Legierung
F alliage hypoeutectoïde
P stop podeutektoidalny
R доэвтектоидный сплав

-375 **hypereutectoid alloy**
An alloy, containing a eutectoid, in which the content of the second component exceeds the percentage corresponding to the eutectoid.
D übereutektoid(isch)e Legierung
F alliage hypereutectoïde
P stop nadeutektoidalny
R заэвтектоидный сплав

-380 **textured alloy**
An alloy in which a texture occurs.
D texturbehaftete Legierung
F alliage texturé
P stop steksturowany
R текстурованный сплав

-385 **precipitation-hardenable alloy, age-hardening alloy, age-hardenable alloy**
An alloy the hardness of which can be increased by the incipient or accomplished precipitation of a new, highly dispersed phase from a supersaturated solid solution.
D aus(scheidungs)härtbare Legierung, aushärtende Legierung, ausscheidungsfähige Legierung, Ausscheidungshärter
F alliage durcissant par précipitation, alliage durcissable par précipitation, alliage à durcissement structural
P stop utwardzalny wydzieleniowo
R дисперсионнотвердеющий сплав

-390 **dispersion-strengthened alloy, dispersion(-hardened) alloy**
An alloy the hardness and strength of which have been increased by means of finely dispersed particles of another substance, which is not soluble in the lattice of the alloy in question.
D dispersionsverfestigte Legierung, dispersionsgehärtete Legierung, DV-Legierung
F alliage durci par dispersion, alliage durci par dispersoïdes
P stop umocniony dyspersyjnie, stop dyspersyjny
R дисперсно-упрочнённый сплав

-395 **fusible alloy, low-melting(-point) alloy**
An alloy which melts at a low temperature or in a low temperature range.
D niedrigschmelzende Legierung, leichtschmelzende Legierung
F alliage fusible, alliage à bas point de fusion
P stop łatwotopliwy, stop niskotopliwy
R легкоплавкий сплав

-400 **refractory alloy, high-melting-point alloy**
>An alloy which starts to melt at a very high temperature.
>**D** hochschmelzende Legierung, hochschmelzbare Legierung
>**F** alliage réfractaire, alliage à haute température de fusion
>**P** stop trudnotopliwy, stop wysokotopliwy
>**R** тугоплавкий сплав, высокоплавкий сплав

-405 **light (metal) alloy**
>An alloy of a low-density metal base, especially an aluminium-base alloy.
>**D** Leichtmetallegierung
>**F** alliage léger
>**P** stop lekki
>**R** лёгкий сплав

-410 **ultra-light alloy**
>An alloy of a very low-density metal base, especially a magnesium-base alloy.
>**D** Magnesiumlegierung
>**F** alliage ultra-léger, alliage extra-léger
>**P** stop ultralekki
>**R** ультралёгкий сплав, сверхлёгкий сплав

-415 **hard alloy**
>An alloy which has a high hardness as its inherent property, i.e. not induced by any heat treatment.
>**D** Hartlegierung, Hartmetall
>**F** alliage dur
>**P** stop twardy
>**R** (сверх)твёрдый сплав

-420 **strong alloy, high-strength alloy**
>An alloy having a high mechanical strength.
>**D** hochfeste Legierung
>**F** alliage à haute résistance
>**P** stop o dużej wytrzymałości, stop wysokowytrzymały
>**R** высокопрочный сплав

-425 **super-strength alloy, ultra-high-strength alloy**
>An alloy having an extremely high mechanical strength.
>**D** ultrafeste Legierung
>**F** alliage à très haute résistance, alliage superrésistant
>**P** stop nadwytrzymały
>**R** сверх(высоко)прочный сплав

-430 **superplastic alloy**
>An alloy which is capable of undergoing unusually extensive strain-rate sensitive deformation under small forces without risk of fracture.
>**D** superplastische Legierung
>**F** alliage superplastique
>**P** stop nadplastyczny
>**R** сверхпластичный сплав

-435 **wear-resisting alloy**
>An alloy which is resistant to wear or abrasion.
>**D** verschleißfeste Legierung
>**F** alliage résistant à l'usure
>**P** stop odporny na ścieranie
>**R** износостойкий сплав, износоустойчивый сплав

-440 **low expansion alloy**
>An alloy having a low coefficient of thermal expansion.
>**D** Legierung mit geringer Wärmedehnung
>**F** alliage peu dilatable
>**P** stop o małej rozszerzalności cieplnej, stop o małym współczynniku rozszerzalności cieplnej
>**R** сплав с малым коэффициентом термического расширения, сплав с низким тепловым расширением

-445 **magnetic alloy**
>An alloy which can be magnetized.
>**D** magnetische Legierung
>**F** alliage magnétique
>**P** stop magnetyczny
>**R** магнитный сплав

-450 **permanent magnet alloy**
>A magnetically hard alloy which is suitable for making permanent magnets.
>**D** Dauermagnetlegierung
>**F** alliage pour aimants permanents
>**P** stop na magnesy trwałe
>**R** сплав для постоянных магнитов

-455 **ferromagnetic alloy**
>An alloy the magnetic permeability of which is considerably greater than that of a vacuum.
>**D** ferromagnetische Legierung
>**F** alliage ferromagnétique
>**P** stop ferromagnetyczny
>**R** ферромагнитный сплав

-460 **magnetically soft alloy**
   An alloy which owing to its low coercivity can be readily magnetized and demagnetized.
   D magnetisch weiche Legierung, weichmagnetische Legierung
   F alliage magnétiquement doux
   P stop magnetycznie miękki
   R магнитомягкий сплав

-465 **magnetically hard alloy, hard magnetic alloy, high coercive alloy**
   An alloy which owing to its high coercivity, once magnetized does not readily become demagnetized.
   D magnetisch harte Legierung, hartmagnetische Legierung
   F alliage magnétiquement dur
   P stop magnetycznie twardy
   R магнитотвёрдый сплав, высококоэрцитивный сплав

-470 **non(-)magnetic alloy**
   An alloy which is practically incapable of being magnetized.
   D nichtmagnetische Legierung, unmagnetische Legierung
   F alliage amagnétique
   P stop niemagnetyczny
   R немагнитный сплав

-475 **(high) permeability alloy**
   An alloy the permeability of which is greater than that of iron.
   D hochpermeable Legierung
   F alliage à perméabilité magnétique élevée, alliage à haute perméabilité magnétique
   P stop o dużej przenikalności magnetycznej
   R сплав с высокой (магнитной) проницаемостью

-480 **corrosion-resisting alloy, corrosion-resistant alloy, non-corrosive alloy**
   An alloy which is resistant to corrosion attacks.
   D korrosionsbeständige Legierung
   F alliage résistant à la corrosion
   P stop odporny na korozję
   R коррозионностойкий сплав, некорродирующий сплав

-485 **acid-resisting alloy**
   An alloy which is resistant to attack by acids.
   D säurebeständige Legierung
   F alliage résistant aux acides
   P stop kwasoodporny
   R кислотостойкий сплав, кислотоупорный сплав

-490 **alkali-resisting alloy**
   An alloy which is resistant to attack by alkalies.
   D laugenbeständige Legierung, alkalibeständige Legierung
   F alliage résistant aux alcalis
   P stop ługoodporny
   R щелочестойкий сплав, щелочеупорный сплав

-495 **oxidation-resisting alloy, oxidation-resistant alloy**
   An alloy which is capable of withstanding chemical attacks by an oxidizing atmosphere.
   D oxydationsbeständige Legierung
   F alliage résistant à l'oxydation
   P stop odporny na utlenianie
   R сплав стойкий к окислению

-500 **tarnish-resisting alloy, tarnish-resistant alloy, non-tarnishing alloy**
   An alloy capable of remaining lustrous in an ordinary atmosphere.
   D anlaufbeständige Legierung
   F alliage résistant au ternissement
   P stop odporny na matowienie (korozyjne)
   R сплав стойкий к тускнению

-505 **heat-resisting alloy, heat-resistant alloy**
   An alloy which is capable of resisting chemical attacks by air or combustion gases at elevated temperatures.
   D zunderbeständige Legierung, zunderfeste Legierung
   F alliage réfractaire
   P stop żaroodporny
   R жаростойкий сплав, жароупорный сплав, окалиностойкий сплав

-510 **high-temperature alloy, creep-resisting alloy, creep-resistant alloy**
   An alloy which is capable of withstanding creep at elevated temperatures.
   D warmfeste Legierung, Hochtemperaturlegierung, dauerstandfeste Legierung
   F alliage résistant au fluage (aux températures élevées)
   P stop żarowytrzymały
   R жаропрочный сплав, высокотемпературный сплав, термостойкий сплав

-515 **superalloy**
An alloy which is highly heat-resistant and maintains its mechanical strength at elevated temperatures (above 700°C).
D Superlegierung, hochwarmfeste Legierung
F superalliage, alliage super-réfractaire
P nadstop
R высокожаропрочный сплав, суперсплав

-520 **commercial alloy, industrial alloy, technical alloy**
An alloy used for industrial applications.
D technische Legierung, handelsübliche Legierung
F alliage technique, alliage industriel
P stop techniczny, stop przemysłowy
R промышленный сплав, технический сплав

-525 **heat-treatable alloy**
An alloy the properties of which can be improved by heat treatment.
D wärmebehandlungsfähige Legierung
F alliage à traitement thermique, alliage de traitement thermique
P stop obrabialny cieplnie
R термически обрабатываемый сплав, термообрабатываемый сплав

-530 **workable alloy, wrought alloy, forging alloy**
An alloy which is suitable for being subjected to plastic working.
D Knetlegierung, Schmiedelegierung
F alliage de forge, alliage (à laminer et) à tréfiler, alliage de forge et de laminage
P stop przerabialny plastycznie, stop do przeróbki plastycznej
R деформируемый сплав, ковкий сплав

-535 **cast(ing) alloy, castable alloy**
An alloy which owing to its good founding properties is used for castings.
D Gußlegierung
F alliage de fonderie, alliage de moulage
P stop odlewniczy
R литейный сплав

-540 **die-casting alloy**
An alloy which is suitable for use in die-casting processes.
D Druckgußlegierung
F alliage pour le moulage sous pression, alliage pour la coulée sous pression
P stop na odlewy ciśnieniowe
R сплав для литья под давлением

-545 **high-resistance alloy, (electrical) resistance alloy, (electrical) resistor alloy**
A heat-resisting alloy having a high electrical resistivity.
D Widerstandslegierung, Heizleiterlegierung
F alliage résistant
P stop oporowy, stop o dużej oporności właściwej
R реостатный сплав, высокоомный сплав

-550 **high-conductivity alloy**
An alloy which owing to its high electrical conductivity can be used for conductors.
D Leitlegierung
F alliage conducteur
P stop przewodowy
R проводниковый сплав, высокоэлектропроводный сплав

-555 **bearing alloy, antifriction alloy, bearing metal, antifriction metal**
An alloy which owing to its anti-frictional properties is used for bearings.
D Lager(schalen)legierung
F (alliage) antifriction, métal antifriction
P stop łożyskowy, stop panewkowy
R подшипниковый сплав, антифрикционный сплав

-560 **free cutting alloy**
An alloy to which some components have been added to improve its machinability by rendering the chips more breakable.
D Automatenlegierung
F alliage de décolletage
P stop automatowy
R автоматный сплав

-565 **coinage alloy, coin(age) metal, money metal**
An alloy used for making coins.
D Münzlegierung, Münzmetall
F alliage monétaire, alliage à monnaies
P stop monetowy
R монетный сплав

-570 **piston alloy**
   An alloy used for making pistons of internal combustion engines.
   D Kolbenlegierung
   F alliage à pistons
   P stop tłokowy
   R поршневой сплав

-575 **dental alloy**
   An alloy used for dentistry purposes.
   D Dentallegierung, zahntechnische Legierung
   F alliage dentaire
   P stop dentystyczny
   R зубопротезный сплав, сплав для зубных протезов, одонтологический сплав

-580 **jewellery alloy**
   An alloy used for jewellery purposes.
   D Juwelierlegierung, Schmucklegierung
   F alliage de bijouterie
   P stop jubilerski
   R ювелирный сплав

-585 **(electrical) contact alloy**
   An alloy used for contacts in electrical circuits.
   D Kontaktlegierung
   F alliage pour contacts électriques
   P stop stykowy
   R контактный сплав, электроэрозионностойкий сплав, сплав для контактов

-590 **hard-(sur)facing alloy**
   A hard alloy deposited by fusion on the surface of some machine parts or cutting tools, in order to render them hard and wear resistant.
   D (Hartmetall-)Auftraglegierung, Aufschweißlegierung
   F alliage de rechargement dur
   P stop (twardy) do napawania
   R наплавляемый твёрдый сплав, сплав для наплавки, сплав для наварки

-595 **pyrophoric alloy, sparking alloy, ignition alloy**
   An alloy which readily emits sparks when scratched or struck.
   D pyrophore Legierung
   F alliage pyrophorique
   P stop piroforyczny
   R пирофорный сплав

-600 **seal-in alloy**
   An alloy which owing to its coefficient of thermal expansion being approximately equivalent to that of glass, is used in the form of wire for embedding in, or passing through, glass in incandescent lamps and vacuum tubes.
   D Einschmelzlegierung
   F métal soudable au verre
   P stop wtopieniowy
   R сплав для спайки со стеклом, сплав для спаев со стеклом

-605 **decorative alloy, ornamental alloy**
   An alloy used for decorative purposes.
   D Zieratlegierung
   F alliage d'ornementation
   P stop zdobniczy, stop architektoniczny, stop dekoracyjny
   R декоративный сплав

-610 **printing metal, type metal**
   A ternary alloy of lead, antimony, and tin, used in the printing industry for casting types.
   D Typometall
   F alliage d'imprimerie, métal typographique
   P stop drukarski
   R типографский сплав

-615 **foundry type metal, printer's metal**
   A printing metal used to cast type for hand composition.
   D Letternmetall, Schriftmetall
   F alliage pour les caractères d'imprimerie
   P stop czcionkowy, metal czcionkowy
   R словолитный сплав, шрифтовой сплав

-620 **monotype metal**
   A printing metal used for die-casting individual type characters in monotype machines.
   D Monometall
   F métal monotype
   P stop monotypowy, metal monotypowy
   R монотипный сплав

-625 **linotype metal, slug-casting metal**
A printing metal used for die-casting entire lines of type characters in linotype machines.
D Linometall
F métal linotype
P stop linotypowy, metal linotypowy
R линотипный сплав

-630 **stereotype metal**
A printing metal used for casting printing plates.
D Stereometall
F métal stéréotype
P stop stereotypowy, metal stereotypowy
R стереотипный сплав

-635 **electrotype metal**
A printing metal used as a backing metal for electrotype printing plates.
D Elektrotypie-Legierung, Galvano-Legierung
F alliage d'électrotypie
P stop galwanotypowy, stop drukarski do galwanotypii
R электротипный сплав, гальванотипный сплав

-640 **spacing metal**
A printing metal used for filling in blank spaces in type-matter.
D Ausschlußmetall; Reglettenmetall
F métal typographique pour blancs
P stop justunkowy, metal justunkowy
R пробельный сплав

-645 **anatomical alloy, anatomical metal**
An alloy which owing to its compatibility with body fluids and tissues can be used in surgery for replacement of bones.
D Osteoplastiklegierung
F alliage ostéoplastique
P stop anatomiczny
R анатомический сплав

-650 **pseudo-alloy**
An alloy obtained by a method other than melting the components together or an alloy the components of which are not miscible in a solid state and tend to separate on freezing.
D Pseudolegierung
F pseudo-alliage
P pseudostop
R псевдосплав

-655 **sintered alloy**
An alloy obtained by compacting and sintering powder mixtures of relevant alloy components in a pure state or pre-alloyed powder mixtures.
D Sinterlegierung
F (alliage) fritté
P spiek
R металлокерамический сплав, порошковый сплав, спеканный сплав, спечённый сплав

-660 **sintered metal**
A metallic material obtained by compacting and sintering powder of a given metal.
D Sintermetall, gesintertes Metall
F métal fritté
P spiek metalowy, metal spiekany
R спечённый металл

-665 **sintered hard alloy, sintered carbides, cemented carbides**
A sintered alloy consisting of hard metal carbides, such as those of tungsten, tantalum, titanium, or molybdenum, embedded in a ductile binding metal, e.g. nickel or cobalt.
D Sinterhartmetall, gesintertes Hartmetall, gesinterte Karbide, Sinterkarbid
F carbure(s) fritté(s), alliage (fritté) dur
P spiek węglikowy, spiek twardy, węgliki spiekane
R спечённый твёрдый сплав, сплав карбидов металлов, спечённый карбид

-670 **infiltration alloy**
A metallic composite consisting of a porous, pressed or sintered metal body the pores of which are either completely or partially filled by infiltration of a melt of another metal or alloy.
D Tränklegierung, infiltrierte Legierung
F alliage formé par infiltration
P stop nasycany, spiek nasycany
R пропитанный сплав, пропитываемый сплав

-675 **cermet, ceramal**
> A material obtained by compacting and sintering ceramic-metal powder mixtures.
> **D** Metallkeramik, metallkeramische Legierung, Cermet
> **F** cermet
> **P** cermetal, spiek metaloceramiczny
> **R** металлокерамический материал, металлокерамика, кермет

-680 **composite (material)**
> A solid material which is made by physically combining two or more existing materials to produce a multiphase system with different properties from the starting materials.
> **D** Verbundwerkstoff, Zwitterwerkstoff, Kompositwerkstoff
> **F** (matériau) composite
> **P** materiał złożony, kompozyt
> **R** композиционный материал, композитный материал, комбинированный материал, композит

60 FERROUS ALLOYS

EISENLEGIERUNGEN
ALLIAGES FERREUX
STOPY ŻELAZA
ЖЕЛЕЗНЫЕ СПЛАВЫ

-005 **meteoric iron**
Native iron, usually alloyed with nickel and small amounts of other elements, found in mineral aggregates of cosmic origin which have reached the earth from inteplanetary space.
D Meteoreisen
F fer météorique
P żelazo meteorytowe, żelazo meteoryczne
R метеоритное железо

-010 **Armco iron**
Commercial iron containing less than 0.1% total impurities, noted for its high ductility and corrosion-resistance.
D Armco-Eisen
F fer Armco
P żelazo armco
R армко железо, железо Армко

-015 **commercial(ly pure) iron, industrial iron**
An iron-base alloy containing such a small percentage of carbon and other elements that its structure consists substantially of ferrite or ferrite with tertiary cementite.
D technisches Eisen, technisches Reineisen, Weicheisen
F fer commercial
P żelazo techniczne, żelazo technicznie czyste
R техническое железо, железо технической чистоты, промышленное железо

-020 **carbonyl iron**
Iron obtained by the decomposition of iron carbonyl.
D Carbonyleisen, Karbonyleisen
F fer (ex-)carbonyle
P żelazo karbonylkowe
R карбонильное железо

-025 **electrolytic iron**
Commercial iron obtained by means of electrodeposition.
D Elektrolyteisen
F fer électrolytique
P żelazo elektrolityczne
R электролитическое железо

-030 **Swedish iron**
Commercial iron of high purity produced from high-grade ores with the application of charcoal as fuel in the metallurgical processes.
D schwedisches Eisen
F fer suédois
P żelazo szwedzkie
R шведское железо

-035 **sponge iron, spongy iron, iron sponge**
Iron obtained by the reduction of iron oxide with carbon, without melting.
D Schwammeisen, Eisenschwamm
F fer en éponge, fer spongieux, éponge de fer
P żelazo gąbczaste
R губчатое железо

-040 **wrought iron**
A ferrous material, aggregated from a solidifying mass of pasty particles of highly refined metallic iron, with which a minutely and uniformly distributed quantity of slag is incorporated without subsequent fusion.
D Schweißeisen
F fer corroyé
P żelazo zgrzewne
R сварочное железо

-045 **bloomery iron, bloomary iron**
Wrought iron obtained directly from the ore within a single metallurgical process in a hearth by reduction of the ore with charcoal which simultaneously serves as fuel.
D Renn(feuer)eisen
F fer catalan
P żelazo dymarkowe, żelazo dymarskie
R сыродутное железо

-050 **finery(-fire) iron, charcoal(-hearth) iron, ball iron, fined iron**
Wrought iron obtained initially in balls by refining melted pig iron, falling in drops on to the hearth of the furnace, with strongly oxidizing charcoal combustion gases, these balls being subsequently aggregated into larger lumps and forged together.
D Frisch(feuer)eisen, Herdfrischeisen
F fer affiné, fer au (charbon de) bois
P żelazo świeżarskie, żelazo fryszerskie
R кричное железо

-055 **puddled iron**
Wrought iron produced in a puddling furnace.
D Puddeleisen
F fer puddlé
P żelazo pudlarskie
R пудлинговое железо

-060 **ingot iron**
Iron which is produced in a molten state, as opposed to wrought iron.
D Flußeisen
F fer coulé
P żelazo zlewne
R литое железо

-065 **sintered iron**
A material or product obtained by compacting and sintering iron powder.
D Sintereisen
F fer fritté
P spiek żelazny, żelazo spiekane
R металлокерамическое железо

-070 **ferrous alloy, iron-base alloy**
An alloy having iron as its major component.
D Eisenlegierung, Schwarzmetall
F alliage (à base) de fer
P stop żelaza, stop o podstawie żelazowej
R чёрный сплав, чёрный металл, железный сплав

-075 **steel**
A malleable iron-carbon alloy containing less than 2% carbon, which is present in the combined state as iron carbide, usually in addition to some other elements, obtained nowadays by melting.
D Stahl
F acier
P stal
R сталь

-080 **type of steel, category of steel**
The characteristic of a steel defined by its principal alloying elements.
D Stahlart, Stahltyp
F sorte d'acier
P rodzaj stali
R сорт стали

-085 **grade of steel**
The characteristic of a steel of a given type, defined by its chemical composition.
D Stahlsorte, Stahlmarke
F nuance d'acier
P gatunek stali, marka stali
R марка стали

-090 **designation of steel, steel designation**
A symbol, composed of letters and/or numbers, denoting a given grade of steel.
D Stahlbezeichnung, Markenbezeichnung
F désignation d'acier, symbole de nuance
P znak stali, cecha stali
R марочное обозначение

-095 **(straight) carbon steel, plain carbon steel, ordinary steel, unalloyed steel**
Steel which contains no deliberately introduced alloying elements.
D Kohlenstoffstahl, unlegierter Stahl
F acier au carbone, acier non allié, acier ordinaire
P stal węglowa, stal niestopowa, stal zwykła
R углеродистая сталь, нелегированная сталь, простая сталь

-100 **low-carbon steel, mild steel**
Steel containing a low percentage of carbon (up to 0.25%).
D niedriggekohlter Stahl, kohlenstoffarmer Stahl
F acier à bas carbone
P stal niskowęglowa
R низкоуглеродистая сталь, малоуглеродистая сталь

-105 **medium-carbon steel**
Steel containing from 0.25 to 0.6% carbon.
D mittelgekohlter Stahl
F acier à carbone moyen, acier à moyen carbone
P stal średniowęglowa
R среднеуглеродистая сталь

-110 **high-carbon steel**
   Steel containing a high percentage of carbon (more than 0.6%).
   D hochgekohlter Stahl, kohlenstoffreicher Stahl
   F acier à haute (teneur en) carbone, acier à carbone élevé
   P stal wysokowęglowa
   R высокоуглеродистая сталь

-115 **alloy steel, special steel**
   Steel which contains one or more deliberately introduced alloying elements.
   D legierter Stahl, Sonderstahl, Stahllegierung, Legierungsstahl
   F acier allié, acier spécial
   P stal stopowa, stal specjalna
   R легированная сталь, специальная сталь

-120 **low-alloy steel, mild-alloy steel**
   Steel containing low percentages of alloying elements (approximately up to 2% of each, up to 5% totally).
   D niedriglegierter Stahl, schwachlegierter Stahl
   F acier faiblement allié, acier légèrement allié, acier peu allié
   P stal niskostopowa
   R низколегированная сталь, малолегированная сталь, слаболегированная сталь

-125 **medium-alloy steel, moderately alloyed steel**
   Steel containing a moderate percentage of alloying elements.
   D mittellegierter Stahl
   F acier moyennement allié
   P stal średniostopowa
   R среднелегированная сталь

-130 **high-alloy steel, highly-alloyed steel**
   Steel containing a high percentage of alloying elements.
   D hochlegierter Stahl
   F acier fortement allié, acier hautement allié
   P stal wysokostopowa
   R высоколегированная сталь

-135 **tonnage steel, steel of commercial grade, commercial steel**
   Steel in which the content of sulfur as well as that of phosphorus exceeds 0.04%.
   D Massenstahl, Handelsbaustahl
   F acier ordinaire
   P stal zwykłej jakości
   R сталь обычного качества, сталь обыкновенного качества, рядовая сталь, сталь торгового качества

-140 **quality steel**
   Steel in which the content of sulfur as well as that of phosphorus does not exceed 0.04%.
   D Qualitätsstahl
   F acier de qualité
   P stal wyższej jakości, stal jakościowa
   R качественная сталь, сталь повышенного качества

-145 **high-quality steel, high-grade steel**
   Steel in which the content of sulfur as well as that of phosphorus does not exceed 0.035%.
   D Edelstahl
   F acier noble, acier fin
   P stal najwyższej jakości, stal wysokiej jakości
   R высококачественная сталь, сталь высокого качества

-150 **special(ty) steel**
   Steel having special physical or chemical properties.
   D Spezialstahl, Sonderstahl
   F acier spécial
   P stal specjalna
   R специальная сталь

-155 **binary steel**
   Steel which apart from carbon, contains one alloying element.
   D binärer Stahl, einfach legierter Stahl
   F acier binaire
   P stal trójskładnikowa
   R сталь легированная одним компонентом

-160 **ternary steel**
   Steel which apart from carbon, contains two alloying elements.
   D ternärer Stahl, zweifach legierter Stahl
   F acier ternaire
   P stal czteroskładnikowa
   R тройная сталь

-165 **quaternary steel**
   Steel which apart from carbon, contains three alloying elements.
   D quaternärer Stahl, dreifach legierter Stahl
   F acier quaternaire
   P stal pięcioskładnikowa
   R четверная сталь

-170 **complex steel**
   Steel containing several alloying elements.
   D mehrfach legierter Stahl
   F acier complexe
   P stal wieloskładnikowa
   R многокомпонентная сталь

-175 **chromium steel, chrome steel**
  Steel containing chromium as the principal alloying element.
  D Chromstahl
  F acier au chrome
  P stal chromowa
  R хромистая сталь, хромовая сталь

-180 **high-chromium steel**
  Steel containing a high percentage of chromium.
  D hochchromhaltiger Stahl
  F acier riche en chrome, acier à haute teneur en chrome
  P stal wysokochromowa
  R высокохромистая сталь

-185 **nickel steel**
  Steel containing nickel as the principal alloying element.
  D Nickelstahl
  F acier au nickel
  P stal niklowa
  R никелевая сталь

-190 **chromium-nickel steel**
  Steel containing chromium and nickel as the principal alloying elements.
  D Chromnickelstahl
  F acier au chrome-nickel
  P stal chromo(wo-)niklowa
  R хромоникелевая сталь

-195 **manganese steel**
  Steel containing manganese as the principal alloying element.
  D Manganstahl
  F acier au manganèse
  P stal manganowa
  R марганцевая сталь, марганцовистая сталь

-200 **silicon steel**
  Steel containing silicon as the principal alloying element.
  D Siliziumstahl
  F acier au silicium
  P stal krzemowa
  R кремнистая сталь

-205 **manganese-silicon steel**
  Steel containing manganese and silicon as the principal alloying elements.
  D Mangansiliziumstahl
  F acier mangano-siliceux
  P stal manganowo-krzemowa
  R марганцовокремнистая сталь

-210 **silicon iron**
  A silicon steel of very low carbon content, used mainly for its magnetic properties.
  D Siliziumeisen
  F fer au silicium
  P stal krzemowa niskowęglowa, żelazo krzemowe
  R кремнистое железо

-215 **tungsten steel**
  Steel containing tungsten as the principal alloying element.
  D Wolframstahl
  F acier au tungstène
  P stal wolframowa
  R вольфрамовая сталь, вольфрамистая сталь

-220 **molybdenum steel**
  Steel containing molybdenum as the principal alloying element.
  D Molybdänstahl
  F acier au molybdène
  P stal molibdenowa
  R молибденовая сталь

-225 **vanadium steel**
  Steel containing vanadium as the principal alloying element.
  D Vanadiumstahl, Vanadinstahl
  F acier au vanadium
  P stal wanadowa
  R ванадиевая сталь

-230 **titanium steel**
  Steel containing titanium as the principal alloying element.
  D Titanstahl
  F acier au titanium
  P stal tytanowa
  R титановая сталь, титанистая сталь

-235 **aluminium steel**
  Steel containing aluminium as the principal alloying element.
  D Aluminiumstahl
  F acier à l'aluminium
  P stal aluminiowa
  R алюминевая сталь

-240 **cobalt steel**
  Steel containing cobalt as the principal alloying element.
  D Kobaltstahl
  F acier au cobalte
  P stal kobaltowa
  R кобальтовая сталь

-245 **copper(-bearing) steel**
  Steel containing copper as an essential alloying element.
  D Kupferstahl
  F acier au cuivre
  P stal miedziowa
  R медистая сталь

-250 **lead-bearing steel, ledloy**
Steel to which some lead has been added to improve its machinability.
D bleihaltiger Stahl
F acier au plomb, acier plombifère
P stal ołowiowa, stal z dodatkiem ołowiu
R свинцовистая сталь

-255 **boron steel**
Steel containing boron as an alloying element.
D Borstahl, borlegierter Stahl
F acier au bore
P stal z dodatkiem boru
R борсодержащая сталь, бористая сталь

-260 **chromansil**
A chromium-manganese-silicon steel.
D Chrom-Mangan-Silizium-Stahl
F acier au chrome-manganèse-silicium
P chromansil
R хромансиль

-265 **standard steel**
Steel complying with the existing standards.
D Normstahl
F acier standardisé
P stal znormalizowana
R стандартная сталь

-270 **alternate steel**
Steel in which an alloying element that is in short supply has been replaced by a more common alloying element.
D Austauschstahl
F acier de remplacement, acier économique, acier succédané
P stal zastępcza, stal oszczędnościowa
R сталь-заменитель

-275 **(fully) annealed steel**
Steel which has been subjected to the full-annealing process.
D geglühter Stahl
F acier recuit
P stal wyżarzona
R отожжённая сталь

-280 **normalized steel**
Steel which has been subjected to the normalizing process.
D normalgeglühter Stahl
F acier normalisé, acier régénéré
P stal normalizowana
R нормализованная сталь

-285 **hardened steel, quench(-harden)ed steel**
Steel which has been subjected to the quench-hardening process.
D gehärteter Stahl
F acier trempé
P stal zahartowana, stal hartowana
R закалённая сталь

-290 **toughened steel**
Steel which has been subjected to the toughening process.
D vergüteter Stahl
F acier amélioré, acier traité
P stal ulepszona cieplnie
R термически улучшенная сталь

-295 **work-hardened steel**
Steel which has been hardened by cold working.
D kaltverfestigter Stahl
F acier écroui
P stal umocniona zgniotem
R нагартованная сталь

-300 **graphitic steel, graphitized steel**
A silicon steel in which cementite has partly decomposed.
D Graphitstahl, graphitischer Stahl, graphitisierter Stahl
F acier graphitisé, acier graphitique
P stal grafityzowana
R графитизированная сталь

-305 **stabilized steel**
A stainles steel which has been rendered resistant to intercrystalline corrosion by small additions of strong carbide-forming elements (e.g. niobium or titanium).
D stabilisierter Stahl
F acier stabilisé
P stal stabilizowana
R стабилиз(ир)ованная сталь

-310 **carburized steel**
Steel which has been subjected to the carburizing process.
D aufgekohlter Stahl
F acier cementé
P stal nawęglona
R науглероженная сталь, цементированная сталь

-315 **nitride(-hardened) steel, nitrided steel**
Steel which has been subjected to the nitriding process.
D nitrierter Stahl
F acier nitruré
P stal azotowana
R азотированная сталь

-320 overheated steel
Steel which has been heated to an unduly high temperature whereby it has become coarse-grained.
D überhitzter Stahl
F acier surchauffé
P stal przegrzana, stal przeżarzona
R перегретая сталь

-325 burnt steel
Steel which has been heated to such a high temperature that oxidation at grain boundaries has occurred or melting began.
D verbrannter Stahl
F acier brûlé
P stal przepalona
R пережжённая сталь

-330 forged steel, hammered steel
Steel which has been subjected to a forging process.
D geschmiedeter Stahl, Schmiedestahl
F acier forgé
P stal kuta
R кованая сталь

-335 clad steel
Steel which has been coated by plating it with a layer of another metal or alloy.
D Verbundstahl, plattierter Stahl
F acier plaqué
P stal platerowana
R плакированная сталь

-340 soft steel, mild steel
Steel of low hardness.
D Weichstahl, weicher Stahl
F acier doux
P stal miękka
R мягкая сталь

-345 dead soft steel, extra-soft steel
Steel of a very low hardness, i.e. fully annealed steel containing less than 0.15% C.
D sehr weicher Stahl
F acier extra-doux
P stal bardzo miękka
R весьма мягкая сталь, особо мягкая сталь

-350 hard steel
Steel of high hardness.
D Hartstahl, (natur)harter Stahl
F acier dur
P stal twarda
R твёрдая сталь

-355 high-strength steel, high-tensile steel
Steel having a high mechanical strength.
D hochfester Stahl
F acier à haute résistance mécanique, acier à résistance élevée
P stal o dużej wytrzymałości
R высокопрочная сталь

-360 hardenable steel
Steel which is suitable for being quench hardened.
D härtbarer Stahl
F acier trempant, acier susceptible de trempe
P stal hartująca się, stal hartowna
R закаливающаяся сталь

-365 water-hardening steel
Steel which can be quench hardened by water quenching.
D Wasserhärtestahl, Wasserhärter, wasserhärtender Stahl
F acier trempant à l'eau
P stal hartująca się w wodzie, stal do hartowania w wodzie
R сталь закаливаемая в воде, сталь закаливающаяся в воде

-370 oil-hardening steel
Steel which can be quench hardened by oil quenching.
D Ölhärtestahl, Ölhärter, ölhärtender Stahl
F acier trempant à l'huile
P stal hartująca się w oleju, stal do hartowania w oleju
R сталь закаливаемая в масле, сталь закаливающаяся в масле

-375 air-hardening steel, self-hardening steel
Steel which can be quench hardened by air quenching.
D Lufthärtestahl, Lufthärter, selbsthärtender Stahl
F acier trempant à l'air, acier autotrempant
P stal hartująca się w powietrzu, stal do hartowania w powietrzu, stal samohartowna
R воздушнозакаливающаяся сталь, воздушнозакаливаемая сталь, самозакаливающаяся сталь

-380 **shallow-hardening steel,
low-hardenability steel,
surface-hardening steel**
   Steel for which the hardness
   penetration resulting from quench
   hardening is small.
   D schwach einhärtender Stahl
   F acier peu trempant, acier de faible
   trempabilité
   P stal płytko hartująca się
   R неглубоко прокаливающаяся
   сталь, сталь с неглубокой
   прокаливаемостью, сталь
   небольшой прокаливаемости

-385 **deep-hardening steel
high-hardenability steel**
   Steel for which the hardness
   penetration resulting from quench
   hardening is great.
   D tief einhärtender Stahl,
   starkhärtender Stahl
   F acier très trempant, acier à bonne
   trempabilité
   P stal głęboko hartująca się
   R глубокопрокаливающаяся сталь,
   сталь с глубокой
   прокаливаемостью

-390 **non-hardenable steel**
   Steel which is practically
   incapable of being hardened by
   quenching.
   D nichthärtender Stahl
   F acier non trempable
   P stal nie hartująca się,
   stal niehartowna
   R незакаливающаяся сталь

-395 **temper resistant steel**
   Steel which after quench hardening
   retains its hardness to relatively
   high temperatures.
   D anlaßbeständiger Stahl
   F acier résistant au revenu
   P stal odporna na odpuszczanie
   R сталь устойчивая против отпуска,
   теплостойкая сталь

-400 **ag(e)ing steel**
   Steel in which the process of
   ageing can occur.
   D aushärtbarer Stahl
   F acier vieillissable
   P stal starzejąca się, stal ulegająca
   starzeniu
   R стареющая сталь

-405 **ag(e)ing-resisting steel, non-ag(e)ing
steel**
   Steel in which the process of
   ageing practically cannot occur.
   D nichtaushärtbarer Stahl
   F acier non vieillisable
   P stal odporna na starzenie
   R нестареющая сталь

-410 **non-deforming steel**
   Steel which does not warp during
   quenching.
   D verzugsfreier Stahl,
   stehenbleibender Stahl
   F acier indeformable
   P stal nie pacząca się
   R недеформирующаяся сталь

-415 **ferritic steel**
   An alloy steel in which during
   cooling the transformation of delta
   ferrite into austenite does not
   occur, the steel thus having down
   to room temperature a structure
   composed substantially of ferrite.
   D ferritischer Stahl
   F acier ferritique
   P stal ferrytyczna
   R ферритная сталь

-420 **austenitic steel**
   An alloy steel having, at room
   temperature, a structure composed
   substantially of austenite.
   D austenitischer Stahl
   F acier austénitique
   P stal austenityczna
   R аустенитная сталь

-425 **pearlitic steel**
   An alloy steel having at room
   temperature, after cooling in air,
   a structure composed of pearlite
   and ferrite.
   D perlitischer Stahl
   F acier perlitique
   P stal perlityczna
   R перлитная сталь

-430 **martensitic steel**
   An alloy steel having at room
   temperature, after cooling in air,
   a structure composed substantially
   of martensite.
   D martensitischer Stahl
   F acier martensitique
   P stal martenzytyczna
   R мартенситная сталь

-435 **bainitic steel**
   An alloy steel having at room
   temperature, after cooling in air,
   a structure composed of bainite.
   D bainitischer Stahl
   F acier bainitique
   P stal bainityczna
   R бейнитная сталь

-440 **ledeburitic steel**
A high-alloy steel in which in an as cast condition ledeburite occurs.
D ledeburitischer Stahl, Ledeburitstahl
F acier lédéburitique
P stal ledeburytyczna
R ледебуритная сталь

-445 **semi-austenitic steel**
An alloy steel in which on slow cooling only part of the austenite transforms while the remainder subsists down to the room temperature.
D halbaustenitischer Stahl
F acier semi-aust́enitique
P stal półaustenityczna
R полуаустенитная сталь

-450 **semi-ferritic steel**
An alloy steel in which on slow cooling only part of the delta ferrite transforms into austenite while the remainder subsists down to the room temperature.
D halbferritischer Stahl
F acier (se)mi-ferritique
P stal półferrytyczna
R полуферритная сталь

-455 **austenitic-ferritic steel**
An alloy steel having at room temperature, after slow cooling, a structure composed substantially of austenite and delta ferrite.
D austenitisch-ferritischer Stahl
F acier austéno-ferritique
P stal austenityczno-ferrytyczna
R аустенито-ферритная сталь

-460 **maraging steel**
A high-nickel steel, containing 18–25% Ni and small additions of other elements, in which a relatively ductile martensitic structure can be formed, this martensite being hardened by ageing at 450–500°C.
D martensitaushärtbarer Stahl, martensitaushärtender Stahl, Maragingstahl
F acier (du type) maraging
P stal martenzyczna starzona
R мартенситно-стареющая сталь

-465 **abnormal steel**
Steel in which some pearlite has broken down into massive cementite and free ferrite.
D anomaler Stahl
F acier anormal
P stal anormalna
R анормальная сталь

-470 **fine-grained steel**
Steel in which the austenite grain size does not increase until a temperature considerably superior to the critical range is reached.
D Feinkornstahl, überhitzungsunempfindlicher Stahl
F acier à grain fin, acier insensible à la surchauffe
P stal drobnoziarnista, stal skłonna do drobnoziarnistości, stal odporna na przegrzanie
R (наследственно) мелкозернистая сталь

-475 **coarse-grained steel**
Steel in which the austenite grain size increases just after the critical range on heating is passed.
D Grobkornstahl, überhitzungsempfindlicher Stahl
F acier à gros grain
P stal gruboziarnista, stal skłonna do gruboziarnistości
R (наследственно) крупнозернистая сталь

-480 **wear-resisting steel, wear-resistant steel, abrasion-resisting steel**
Steel which is resistant to abrasion or wear.
D verschleißfester Stahl
F acier résistant à l'usure
P stal odporna na ścieranie
R износостойкая сталь, износоустойчивая сталь

-485 **weldable steel**
Steel suitable for being joined by welding processes.
D schmelzschweißbarer Stahl; preßschweißbarer Stahl, feuerschweißbarer Stahl
F acier soudable
P stal spawalna; stal zgrzewalna
R свариваемая сталь, сваривающаяся сталь

-490 **extra-hard steel**
A tungsten-vanadium tool steel having a very high hardness after quench hardening.
D Diamantstahl, Riffelstahl
F acier diamant, acier extra-dur
P stal diamentowa
R алмазная сталь

-495 **magnetic steel**
Steel which can be magnetized.
D magnetischer Stahl
F acier magnétique
P stal magnetyczna
R магнитная сталь

-500 **magnetically soft steel, soft magnetic steel**
Steel which owing to its low coercivity can be readily magnetized and demagnetized.
D magnetisch weicher Stahl, weichmagnetischer Stahl
F acier magnétiquement doux
P stal magnetycznie miękka
R магнитомягкая сталь

-505 **magnetically hard steel, hard magnetic steel**
Steel which owing to its high coercivity does not readily become demagnetized once magnetized.
D magnetisch harter Stahl, hartmagnetischer Stahl
F acier magnétiquement dur
P stal magnetycznie twarda
R магнитотвёрдая сталь

-510 **non-magnetic steel**
Steel which is practically incapable of being magnetized.
D unmagnetischer Stahl, nichtmagnetisierbarer Stahl, unmagnetisierbarer Stahl, antimagnetischer Stahl
F acier amagnétique, acier non magnétique
P stal niemagnetyczna
R немагнитная сталь

-515 **stainless (steel), rustless steel, rust-resisting steel**
Steel which is immune against the process of rusting.
D nichtrostender Stahl, rostbeständiger Stahl, rostsicherer Stahl
F acier inox(ydable), inox(ydable)
P stal nierdzewna
R нержавеющая сталь

-520 **stainless iron, rustless iron, ferritic stainless steel**
A low-carbon chromium stainless steel, having a substantially ferritic structure.
D ferritischer nichtrostender Stahl
F acier inoxydable ferritique
P stal nierdzewna ferrytyczna
R ферритная нержавеющая сталь

-525 **resulfurized stainless steel**
A stainless steel to which some sulfur has been deliberately introduced in order to improve its machinability.
D aufgeschwefelter nichtrostender Stahl
F acier inoxydable au soufre, acier inoxydable resulphuré
P stal nierdzewna automatowa
R автоматная нержавеющая сталь

-530 **Schaeffler('s) diagram**
A diagram representing the relationship between the content of nickel and chromium (or other alloying elements expressed as nickel and chromium equivalents) in a stainless steel and the structure of this steel in a weld deposit.
D Schaeffler-Diagramm
F diagramme de Schaeffler
P wykres Schaefflera
R диаграмма Шефлера

-535 **corrosion-resisting steel**
Steel which is resistant to corrosion attacks.
D korrosionsbeständiger Stahl, chemisch beständiger Stahl
F acier résistant à la corrosion, acier inattaquable
P stal odporna na korozję
R коррозионностойкая сталь, коррозионноустойчивая сталь, коррозиеустойчивая сталь

-540 **weather-resisting steel**
Steel which resists atmospheric corrosion.
D witterungsbeständiger Stahl, schwerrostender Stahl
F acier résistant à la corrosion atmosphérique
P stal odporna na korozję atmosferyczną
R атмосферостойкая сталь

-545 **acid-resisting steel**
Steel which is resistant to attack by acids.
D säurebeständiger Stahl
F acier résistant aux acides
P stal kwasoodporna
R кислотостойкая сталь, кислотоупорная сталь

-550 **alkali-resisting steel**
Steel which is resistant to attack by alkalis.
D laugenbeständiger Stahl, alkalibeständiger Stahl, laugensicherer Stahl
F acier résistant aux alcalis
P stal ługoodporna
R щёлочестойкая сталь, щёлочеупорная сталь

-555 **low-temperature steel, steel tough at subzero**
 Steel which does not become brittle at subzero temperatures.
 D kaltzäher Stahl, Tieftemperaturstahl
 F acier tenace à froid, acier à haute tenacité à froid
 P stal mrozoodporna, stal kriotechniczna
 R хладостойкая сталь, хладоустойчивая сталь

-560 **heat-resisting steel, heat-resistant steel, oxidation-resistant steel**
 Steel which, owing to its chemical composition and structure, is capable of resisting chemical attacks by hot air or combustion gases at temperatures above the red heat.
 D zunderbeständiger Stahl, zunderfester Stahl, hitzebeständiger Stahl
 F acier réfractaire
 P stal żaroodporna
 R жаростойкая сталь, жароупорная сталь, окалиностойкая сталь

-565 **creep-resisting steel, creep-resistant steel**
 An alloy steel which, owing to its composition and structure, is capable of withstanding creep at elevated temperatures.
 D (hoch)warmfester Stahl
 F acier résistant au fluage (aux températures élevées), acier résistant à chaud
 P stal żarowytrzymała
 R жаропрочная сталь, крипоустойчивая сталь, теплопрочная сталь

-570 **wrought steel**
 Steel which is obtained as a solid pasty mass, i.e. without fusion.
 D Schweißstahl, Luppenstahl
 F acier corroyé
 P stal zgrzewna
 R сварочная сталь

-575 **puddled steel**
 Steel produced in a puddling furnace.
 D Puddelstahl
 F acier puddlé
 P stal pudlarska
 R пудлинговая сталь

-580 **ingot steel**
 Steel which is produced in a molten state, as opposed to wrought steel.
 D Flußstahl
 F acier coulé
 P stal zlewna
 R литая сталь

-585 **blister steel**
 Steel obtained by carburizing wrought-iron bars.
 D Zementstahl
 F acier poule, acier boursoufflé, acier à soufflures
 P stal cementowa
 R цементная сталь, томлёная сталь

-590 **converter steel**
 Steel produced in a refractory-lined container, named converter, by treating molten pig iron with a blast of air or oxygen under pressure.
 D Konverterstahl, Blasstahl
 F acier au convertisseur
 P stal konwertorowa
 R конвертерная сталь

-595 **Bessemer steel**
 Converter steel produced by the Bessemer process.
 D Bessemerstahl
 F acier Bessemer
 P stal bessemerowska
 R бессемеровская сталь

-600 **Thomas steel**
 Converter steel produced by the Thomas process.
 D Thomasstahl, Th-Stahl
 F acier Thomas
 P stal tomasowska
 R томасовская сталь

-605 **oxygen steel**
 Converter steel obtained by treating molten pig iron with a blast of oxygen under pressure.
 D Oxygenstahl, Sauerstoff-Blasstahl
 F acier à l'oxygène
 P stal konwertorowa świeżona tlenem, stal konwertorowa z procesu tlenowego
 R кислородно-конвертерная сталь

-610 **open-hearth steel, Siemens-Martin steel**
 Steel produced by the open-hearth process.
 D Siemens-Martin-Stahl, SM-Stahl, Martinstahl
 F acier Martin(-Siemens)
 P stal martenowska
 R мартеновская сталь

-615 **acid steel**
   Steel produced in a furnace having an acid (i.e. siliceous) lining, and under an acid slag.
   D saurer Stahl
   F acier acide
   P stal kwaśna
   R кислая сталь

-620 **basic steel**
   Steel produced in a furnace having a basic (magnesite or dolomite) lining, and under a lime slag.
   D basischer Stahl
   F acier basique
   P stal zasadowa
   R основная сталь

-625 **electric steel**
   Steel which has been refined in an electric furnace.
   D Elektrostahl, E-Stahl, Lichtbogenstahl
   F acier électrique
   P stal elektropiecowa, stal elektryczna
   R электросталь, электрическая сталь

-630 **crucible steel**
   Steel of high quality obtained after additional processing in a crucible furnace.
   D Tiegelstahl
   F acier au creuset
   P stal tyglowa
   R тигельная сталь

-635 **refined steel**
   Steel which has been subjected to a process of purification.
   D Raffinierstahl
   F acier raffiné
   P stal rafinowana
   R рафинированная сталь

-640 **vacuum-treated steel, vacuum-melted steel**
   Steel which has been purified by remelting in a vacuum.
   D Vakuumstahl, vakuumbehandelter Stahl
   F acier élaboré sous vide, acier traité sous vide, acier (re)fondu sous vide
   P stal próżniowa
   R вакуумированная сталь

-645 **damascene steel**
   A high-carbon crucible steel, produced mainly in the early Middle Ages for making swords, which owing to a special forging process acquires a structure characterized by a delicate alternation of higher and lower carbon areas, its mechanical properties being — after appropriate heat treatment — excellent.
   D Damaszenerstahl
   F acier damassé, damas
   P stal damasceńska
   R дамасская сталь, булатная сталь

-650 **rimming steel, unkilled steel**
   Steel which evolves gases during cooling and freezing.
   D unberuhigter Stahl
   F acier effervescent, acier mousseux, acier non calmé
   P stal nieuspokojona
   R кипящая сталь

-655 **killed steel**
   Steel which does not evolve gases during cooling and freezing, having been deoxidized prior to pouring.
   D beruhigter Stahl
   F acier calmé
   P stal uspokojona
   R спокойная сталь, успокоенная сталь

-660 **semi-killed steel, semi-rimming steel, balanced steel**
   Steel which has not been fully deoxidized before pouring, thus being intermediate between the rimming and the killed steel.
   D halbberuhigter Stahl
   F acier semi-calmé, acier semi-effervescent
   P stal półuspokojona
   R полуспокойная сталь

-665 **general-purpose steel**
   Steel consumed in great quantities for purposes where no special technological or physical properties are required.
   D allgemeiner Baustahl
   F acier de construction ordinaire
   P stal ogólnego przeznaczenia
   R сталь общего назначения

-670 **constructional steel**
   Steel designed for producing machines, machinery and mechanical installations.
   D Baustahl, Konstruktionsstahl
   F acier de construction
   P stal konstrukcyjna
   R конструкционная сталь

-675 **tool steel**
   Steel used for making tools.
   D Werkzeugstahl
   F acier à outils
   P stal narzędziowa
   R инструментальная сталь

-680 **machine(ry) steel**
   Constructional steel designed for producing machines and machinery.
   D Maschinen(bau)stahl
   F acier de construction mécanique
   P stal konstrukcyjna maszynowa
   R машиностроительная сталь, машиноподелочная сталь

-685 **structural steel**
   Steel used for building structures.
   D Hochbaustahl
   F acier de construction
   P stal (konstrukcyjna) budowlana
   R строительная сталь

-690 **carburizing steel, cementation steel, case-hardening steel**
   A low-carbon, unalloyed or alloy steel suitable for being subjected to the carburizing process.
   D Einsatzstahl
   F acier de cémentation
   P stal do nawęglania
   R цемент(ир)уемая сталь, сталь для цементации

-695 **nitriding steel**
   Steel suitable for being hardened by the nitriding process.
   D Nitrierstahl, nitrierbarer Stahl
   F acier de nitruration
   P stal do azotowania
   R азотируемая сталь

-700 **nitralloy (steel)**
   An alloy steel designed for nitriding.
   D legierter Nitrierstahl
   F acier allié de nitruration
   P stal stopowa do azotowania
   R (сталь-)нитраллой

-705 **toughening steel, heat treatable steel**
   Steel suitable for being subjected to the toughening process.
   D Vergütungsstahl
   F acier améliorable par traitement, acier de traitement
   P stal do ulepszania cieplnego
   R улучшаемая сталь

-710 **free-cutting steel, free-machining steel, easy-cutting steel, automatic steel, resulfurized steel**
   A steel containing an increased percentage of sulfur and phosphorus which improve its machinability by rendering the chips more breakable.
   D Automatenstahl
   F acier de décolletage
   P stal automatowa
   R автоматная сталь, легкообрабатываемая сталь

-715 **bearing steel**
   Steel used for producing rolling bearings.
   D Wälzlagerstahl, Lagerstahl
   F acier à roulement, acier pour roulements
   P stal łożyskowa, stal na łożyska toczne
   R подшипниковая сталь

-720 **ball-bearing steel**
   Steel used for producing ball bearings.
   D Kugellagerstahl
   F acier pour roulements à billes
   P stal na łożyska kulkowe
   R шарикоподшипниковая сталь

-725 **spring steel**
   Steel used for making springs.
   D Federstahl
   F acier à ressort, acier pour ressorts
   P stal sprężynowa, stal resorowa
   R пружинная сталь, рессорная сталь

-730 **rail steel**
   Steel used for the production of rails.
   D Schienenstahl
   F acier à rail
   P stal szynowa
   R рельсовая сталь

-735 **valve steel**
   Steel used for making valves of internal combustion engines.
   D Ventilstahl
   F acier à soupapes
   P stal zaworowa
   R клапанная сталь

-740 **boiler steel, pressure vessel steel**
   Steel used for making steam boilers.
   D Kessel(bau)stahl
   F acier à chaudières
   P stal kotłowa, stal do budowy kotłów
   R котельная сталь

-745 **rivet steel**
Steel used for making rivets.
D Nietstahl
F acier à rivets
P stal na nity
R заклёпочная сталь

-750 **cutlery steel**
Steel used for making knives and other cutlery products.
D Messerstahl
F acier de coutellerie, acier pour couteaux, acier pour taillanderie
P stal na noże, stal nożowa
R ножевая сталь

-755 **ship steel**
Steel used for building ships' hulls.
D Schiffbaustahl
F acier coque
P stal okrętowa
R корабельная сталь, судостроительная сталь, судовая сталь

-760 **tyre steel**
Steel used for making railway wheel tyres.
D Bandagenstahl
F acier à bandages
P stal obręczowa, stal na obręcze (kolejowe)
R бандажная сталь

-765 **armour plate**
An alloy steel in heavy plate form, usually heat treated, used for the protection of warships and tanks.
D Panzer(platten)stahl
F acier de blindage
P stal pancerna
R броневая сталь

-770 **transformer steel**
Magnetically soft steel used for the cores of transformers.
D Transformatorenstahl, Trafostahl
F acier pour transformateurs, acier pour tôles de transformateur
P stal transformatorowa
R трансформаторная сталь, трансформаторное железо

-775 **dynamo steel**
Magnetically soft steel used for dynamo sheets.
D Dynamostahl
F acier pour dynamos
P stal prądnicowa
R динамная сталь

-780 **(permanent) magnet steel**
Magnetically hard steel used for making permanent magnets.
D Magnetstahl, Dauermagnetstahl
F acier à aimants
P stal na magnesy (trwałe)
R сталь для постоянных магнитов

-785 **Hadfield('s manganese) steel**
An austenitic manganese steel, containing 10–15% Mn, used mainly for its wear resistance and toughness.
D (Hadfieldscher) Manganhartstahl
F acier Hadfield
P stal Hadfielda, stal manganowa nieścieralna
R сталь Гадфильда

-790 **deep-drawing steel**
Steel which is suitable for deep-drawing operations.
D Tiefziehstahl
F acier pour emboutissage profond
P stal do głębokiego tłoczenia, stal do głębokiego ciągnienia
R сталь для глубокой штамповки, сталь для глубокой вытяжки

-795 **reinforcing steel**
Steel used, in the form of wire or rods, for concrete reinforcement.
D Bewehrungsstahl, Betonstahl
F acier d'armature
P stal zbrojeniowa
R арматурная сталь

-800 **high-speed steel, rapid steel**
A high-alloy tool steel, containing tungsten, chromium, vanadium, and sometimes molybdenum or cobalt, which after appropriate heat treatment retains its hardness up to high temperatures and therefore is used for cutting tools for the high-speed machining of metals.
D Schnell(arbeits)stahl, SS-Stahl, Schnelldrehstahl
F acier (à coupe) rapide
P stal szybkotnąca
R быстрорежущая сталь

-805 **cold-work(ing) tool steel, cold-work steel, cold-forming tool steel**
A tool steel suitable for making tools which are designed to work at normal room temperature, i.e. without contact with hot metal.
D Kaltarbeitsstahl
F acier pour travail à froid, acier pour outils travaillant à froid
P stal (narzędziowa) do pracy na zimno
R инструментальная сталь для работы в холодном состоянии

-810 **hot-work tool steel, high temperature tool steel, hot-work(ing) die steel, hot-working steel**
A tool steel suitable for making tools (e.g. dies) which are designed to work in contact with hot metal.
D Warmarbeitsstahl
F acier pour travail à chaud, acier pour outils travaillant à chaud
P stal (narzędziowa) do pracy na gorąco
R (инструментальная) сталь для работы при высоких температурах, инструментальная сталь для горячей обработки, сталь для горячего деформирования и прессформ

-815 **die steel**
Tool steel used for making dies.
D Gesenkstahl
F acier à matrices, acier pour matrices
P stal matrycowa, stal na matryce
R штамповая сталь

-820 **silver steel**
A cold-drawn carbon or alloy tool steel, supplied in the form of ground rods.
D Silberstahl
F acier argenté
P stal srebrzanka
R серебрянка, серебристая сталь

-825 **cast steel**
Steel used in the production of castings.
D Gußstahl, Stahlguß
F acier moulé, acier de moulage
P staliwo
R лит(ейн)ая сталь

-830 **carbon cast steel**
Cast steel which contains no deliberately introduced alloying elements.
D unlegierter Gußstahl, unlegierter Stahlguß
F acier moulé non allié
P staliwo węglowe, staliwo niestopowe
R углеродистая литая сталь

-835 **alloy cast steel**
Cast steel to which alloying elements have been added to obtain special properties.
D legierter Gußstahl, legierter Stahlguß
F acier moulé allié
P staliwo stopowe
R легированная литая сталь

-840 **pig iron**
An iron-carbon alloy containing over 2% carbon in addition to other elements (Si, Mn, P, S), usually produced by the reduction of iron ore in a blast furnace, sometimes in an electric furnace.
D Roheisen
F fonte brute, fonte crue, fonte de première fusion
P surówka
R первичный чугун

-845 **Bessemer pig iron**
Pig iron which owing to its low phosphorus content is designed for being refined in the acid Bessemer process.
D Bessemerroheisen
F fonte Bessemer
P surówka bessemerowska
R бессемеровский чугун

-850 **Thomas pig iron**
Pig iron which owing to its chemical composition is suitable for being refined in the Thomas process.
D Thomasroheisen
F fonte Thomas
P surówka tomasowska
R томасовский чугун

-855 **open-hearth pig iron**
Pig iron which is designed for being refined in the open-hearth process.
D Siemens-Martin-Roheisen, Martinroheisen
F fonte Martin
P surówka martenowska
R мартеновский чугун

-860 **blast furnace pig iron**
Pig iron produced in a blast furnace.
D Hochofenroheisen
F fonte de haut fourneau
P surówka wielkopiecowa
R доменный чугун

-865 **electric furnace pig iron**
Pig iron smelted in an electric furnace.
D Elektroroheisen
F fonte électrique
P surówka elektropiecowa, surówka elektryczna
R электрочугун

-870 **coke pig iron**
Pig iron smelted with coke as fuel.
D Koksroheisen
F fonte au coke
P surówka koksowa
R коксовый чугун

-875 **charcoal pig iron**
    Pig iron smelted with charcoal as fuel.
    D Holzkohlenroheisen
    F fonte au (charbon de) bois
    P surówka drzewnowęglowa
    R древесноугольный чугун

-880 **steelmaking pig iron**
    Pig iron designed for being converted into steel.
    D Stahl(roh)eisen
    F fonte d'affinage
    P surówka przeróbcza, surówka stalownicza
    R передельный чугун

-885 **foundry pig iron**
    Pig iron used for the production of cast iron.
    D Gießereiroheisen
    F fonte de moulage
    P surówka odlewnicza
    R литейный чугун

-890 **phosphoric pig iron**
    A variety of foundry pig iron with a phosphorus content of 1.5%.
    D Phosphorroheisen, phosphorreiches Roheisen
    F fonte phosphoreuse
    P surówka fosforowa
    R фосфористый чугун

-895 **ha(e)matite pig iron**
    A variety of foundry pig iron with a phosphorus content of about 0.1%.
    D Hämatitroheisen
    F fonte hématite
    P surówka hematytowa
    R гематитовый чугун

-900 **medium-phosphoric pig iron**
    A variety of foundry pig iron with a phosphorus content of about 0.2%.
    D Halb-Hämatit-Roheisen
    F fonte semi-hématite
    P surówka półhematytowa
    R полугематитовый чугун

-905 **malleable pig iron**
    White pig iron designed for being converted into malleable cast iron.
    D Temperroheisen
    F fonte pour malléabilisation
    P surówka na żeliwo ciągliwe
    R неотожжённый ковкий чугун

-910 **white pig iron**
    Pig iron with a white fracture i.e. having all its carbon in the form of cementite.
    D weißes Roheisen
    F fonte blanche
    P surówka biała
    R белый чугун

-915 **grey pig iron**
    Pig iron with a grey fracture i.e. having all or almost all its carbon in the form of graphite.
    D graues Roheisen
    F fonte grise
    P surówka szara
    R серый чугун

-920 **mottled pig iron**
    Pig iron having its carbon in some regions in the form of cementite, and in others in the form of graphite.
    D meliertes Roheisen, halbiertes Roheisen
    F fonte truitée, fonte mélangée
    P surówka pstra, surówka połowiczna
    R половинчатый чугун

-925 **high-carbon pig iron**
    Pig iron containing a large amount of carbon in the form of graphite.
    D hochgekohltes Roheisen
    F fonte graphiteuse, fonte à haut carbone
    P surówka wysokowęglowa
    R высокоуглеродистый чугун

-930 **spiegel(eisen), specular pig iron**
    Pig iron with a high manganese content, used as a deoxidizer for cast steel or as a ferro-alloy for adding small quantities of manganese to cast iron.
    D Spiegeleisen
    F spiegel(eisen), fonte spéculaire
    P surówka zwierciadlista
    R зеркальный чугун

-935 **alloy pig iron**
    Pig iron containing deliberately introduced alloying elements.
    D legiertes Roheisen
    F fonte brute alliée
    P surówka stopowa
    R легированный чугун

-940 **synthetic pig iron**
    Pig iron obtained by smelting together the pure alloying constituents.
    D synthetisches Roheisen
    F fonte synthétique
    P surówka syntetyczna
    R синтетический чугун

-945 **cast iron**
A ferrous alloy produced by remelting pig iron and scrap, with or without additions of alloying elements, and casting it into moulds.
D Gußeisen, Guß
F fonte de deuxième fusion
P żeliwo
R вторичный чугун

-950 **grey cast iron**
Cast iron with a grey fracture in which only up to 0.8% carbon can be in the combined state, i.e. as cementite, the remaining carbon being in the free state, i.e. as graphite.
D graues Gußeisen, Grauguß
F fonte grise
P żeliwo szare
R серый чугун, графитистый чугун

-955 **white cast iron**
Cast iron with a white fracture having all its carbon in the combined state, i.e. as cementite.
D weißes Gußeisen, Weißguß
F fonte blanche
P żeliwo białe
R белый чугун

-960 **mottled cast iron**
A variety of cast iron intermediate between white cast iron and grey cast iron, its fracture being in some regions grey and in others white.
D meliertes Gußeisen, halbgraues Gußeisen
F fonte truitée, fonte mélangée
P żeliwo pstre, żeliwo połowiczne
R половинчатый чугун

-965 **ordinary cast iron**
Grey cast iron which has not been subjected to any treatment aimed at modifying the shape or size of graphite flakes, as opposed to inoculated cast iron.
D Handelsguß
F fonte (grise) ordinaire, fonte industrielle
P żeliwo zwykłe
R обыкновенный чугун, обычный чугун

-970 **inoculated cast iron**
Cast iron, the structure and properties of which have been modified by inoculation.
D modifiziertes Gußeisen
F fonte inoculée
P żeliwo modyfikowane
R модифицированный чугун

-975 **spheroidal (graphite) cast iron, ductile cast iron, nodular cast iron**
Inoculated cast iron in which the graphite occurs in the form of spheroids.
D Kugelgraphit-Gußeisen, Kugelgraphitguß, KG-Guß, KG-Gußeisen, sphärolitisches Gußeisen, Sphäroguß, duktiles Gußeisen
F fonte sphéroïdale, fonte à graphite sphéroïdal, fonte ductile, fonte (à graphite) nodulaire
P żeliwo sferoidalne
R чугун с шаровидным графитом, сфероидизованный чугун, магниевый чугун, высокопрочный чугун

-980 **chilled cast iron**
Cast iron which after casting has a hard skin with a structure of white or mottled cast iron, the interior consisting of grey cast iron.
D Schalen(hart)guß, Hart(schalen)guß
F fonte trempée
P żeliwo utwardzone, żeliwo zabielone
R отбелённый чугун

-985 **cast iron heredity**
The dependence of the properties of cast iron upon the type of pig iron from which it is produced.
D Gußeisenerblichkeit, Erblichkeit des Gußeisens
F hérédité de fonte
P dziedziczność żeliwa
R наследственность чугуна

-990 **plain cast iron**
Cast iron which contains no specially added alloying elements.
D unlegiertes Gußeisen
F fonte non alliée
P żeliwo niestopowe, żeliwo zwykłe
R нелегированный чугун, обыкновенный чугун

-995 **alloy cast iron**
Cast iron which contains one or more specially added alloying elements.
D legiertes Gußeisen
F fonte alliée, fonte spéciale
P żeliwo stopowe
R легированный чугун

-1000 **low-alloy cast iron**
 Cast iron containing a low
 percentage of alloying elements.
 D niedriglegiertes Gußeisen
 F fonte faiblement alliée
 P żeliwo niskostopowe
 R низколегированный чугун

-1005 **high-alloy cast iron, highly alloyed cast iron**
 Cast iron containing a high
 percentage of alloying elements.
 D hochlegiertes Gußeisen
 F fonte fortement alliée
 P żeliwo wysokostopowe
 R высоколегированный чугун

-1010 **silicon cast iron**
 Cast iron containing silicon as the
 principal alloying element.
 D Siliziumgußeisen
 F fonte au silicium
 P żeliwo krzemowe
 R кремнистый чугун

-1015 **chromium cast iron**
 Cast iron containing chromium as
 the principal alloying element.
 D Chromgußeisen
 F fonte au chrome
 P żeliwo chromowe
 R хромистый чугун

-1020 **aluminium cast iron**
 Cast iron containing aluminium as
 the principal alloying element.
 D Aluminiumgußeisen
 F fonte à l'aluminium
 P żeliwo aluminiowe
 R алюминиевый чугун, чугаль

-1025 **nickel cast iron**
 Cast iron containing nickel as the
 principal alloying element.
 D Nickelgußeisen
 F fonte au nickel
 P żeliwo niklowe
 R никелевый чугун

-1030 **manganese cast iron**
 Cast iron containing manganese
 as the principal alloying element.
 D Mangangußeisen
 F fonte au manganèse
 P żeliwo manganowe
 R марганцевый чугун,
 марганцовистый чугун

-1035 **semi-steel, steeled cast iron, ferro-steel, gun iron**
 Cast iron obtained from a charge
 containing an increased proportion
 of steel scrap.
 D Halbstahl
 F fonte aciérée
 P żeliwo staliste
 R сталистый чугун,
 малоуглеродистый чугун

-1040 **ferritic cast iron**
 Grey cast iron with a ferritic
 matrix.
 D ferritisches Gußeisen
 F fonte grise (à matrice) ferritique,
 fonte (à matrice) ferritique
 P żeliwo (szare) ferrytyczne
 R феррит(но-графит)ный чугун,
 феррито-графитовый чугун

-1045 **pearlitic cast iron**
 Grey cast iron with a pearlitic
 matrix.
 D perlitisches Gußeisen, Perlitguß,
 Perlitgußeisen
 F fonte grise (à matrice) perlitique,
 fonte (à matrice) perlitique,
 fonte à matrice eutectoïde
 P żeliwo (szare) perlityczne
 R перлит(но-графит)ный чугун

-1050 **austenitic cast iron**
 Alloy cast iron with an austenitic
 matrix.
 D austenitisches Gußeisen
 F fonte austénitique
 P żeliwo austenityczne
 R аустенитный чугун

-1055 **martensitic cast iron**
 Alloy cast iron with a martensitic
 matrix.
 D martensitisches Gußeisen
 F fonte (à matrice) martensitique
 P żeliwo martenzytyczne
 R мартенситный чугун

-1060 **acicular (grey) cast iron**
 Alloy cast iron which possesses
 a bainitic-martensitic or
 austenitic-martensitic structure
 with characteristic acicular
 crystals occurring in the matrix.
 D bainitisches Gußeisen
 F fonte aciculaire, fonte (à matrice)
 bainitique
 P żeliwo iglaste, żeliwo bainityczne
 R чугун с игольчатой структурой

-1065 **high-duty cast iron, high-grade cast iron, high-test cast iron**
Cast iron with outstanding mechanical properties.
D hochwertiges Gußeisen
F fonte résistante, fonte fine, fonte à haute résistance
P żeliwo wysokojakościowe
R высококачественный чугун

-1070 **armoured cast iron**
Grey cast iron reinforced with steel rods.
D stahlbewehrtes Gußeisen, armiertes Gußeisen
F fonte armée
P żeliwo zbrojone
R армированный чугун

-1075 **acid-resisting cast iron**
Cast iron which is resistant to attack by acids.
D säurebeständiges Gußeisen
F fonte résistante aux acides
P żeliwo kwasoodporne
R кислотостойкий чугун, кислотоупорный чугун

-1080 **alkali-resisting cast iron**
Cast iron which is resistant to attack by alkalis.
D laugenbeständiges Gußeisen, alkalibeständiges Gußeisen
F fonte résistante aux alcalis
P żeliwo ługoodporne
R щелочестойкий чугун, щелочеупорный чугун

-1085 **heat-resisting cast iron**
Alloy cast iron which, owing to its chemical composition and structure, is capable of resisting chemical attacks by air or combustion gases at temperatures above the red heat.
D zunderbeständiges Gußeisen, zunderfestes Gußeisen, feuerbeständiges Gußeisen
F fonte réfractaire
P żeliwo żaroodporne
R жаростойкий чугун, жароупорный чугун

-1090 **Ni-resist**
A corrosion-resistant austenitic cast iron containing nickel, copper and chromium.
D Niresist
F (fonte) Ni-résist
P niresist, nirezist
R нирезист

-1095 **nicrosilal**
A low-carbon heat-resistant cast iron containing nickel, silicon and chromium.
D Nicrosilal
F nicrosilal
P nicrosilal
R никросилал

-1100 **silal**
A high-silicon low-carbon cast iron which resists growth.
D Silal
F silal
P silal
R силал

-1105 **bearing cast iron**
Cast iron which, owing to its structure and properties, is used for the production of bearing shells.
D Lagergußeisen
F fonte antifriction
P żeliwo łożyskowe, żeliwo przeciwcierne
R антифрикционный чугун

-1110 **malleable (cast) iron**
Cast iron with good ductility, obtained by appropriate annealing of white cast iron.
D Temperguß
F (fonte) malléable
P żeliwo ciągliwe
R ковкий чугун

-1115 **black heart malleable cast iron**
Malleable cast iron with a black central part (heart), obtained from white cast iron by annealing in a neutral atmosphere.
D schwarzer Temperguß, Schwarzkern(temper)guß, Schwarzguß, amerikanischer Temperguß
F fonte (malléable) à cœur noir, fonte malléable americaine
P żeliwo ciągliwe czarne, żeliwo ciągliwe amerykańskie
R черносердечный ковкий чугун, американский ковкий чугун

-1120 **white heart malleable cast iron**
Malleable cast iron with a white central part (heart), obtained from white cast iron by annealing in an oxidizing atmosphere.
D weißer Temperguß, Weiß(kern)guß, europäischer Temperguß, deutscher Temperguß
F fonte (malléable) à cœur blanc, fonte (malléable) européenne
P żeliwo ciągliwe białe, żeliwo ciągliwe europejskie
R белосердечный ковкий чугун, светлосердечный ковкий чугун, европейский ковкий чугун

-1125 **ferritic malleable (cast) iron**
Malleable cast iron in which the process of graphitization has been performed completely, its matrix thus being ferritic.
D ferritischer Temperguß
F fonte malléable (à matrice) ferritique
P żeliwo ciągliwe ferrytyczne
R ферритный ковкий чугун

-1130 **pearlitic malleable (cast) iron**
Malleable cast iron in which the process of graphitization has been performed incompletely, its matrix thus being substantially pearlitic.
D perlitischer Temperguß
F fonte malléable (à matrice) perlitique
P żeliwo ciągliwe perlityczne
R перлитный ковкий чугун

-1135 **ferro-alloys**
Master alloys of iron and special constituents (e.g. Cr, Ni, Ti) or having considerable quantities of other elements (e.g. Si, Mn, P) which are usually present in cast iron or cast steel.
D Ferrolegierungen
F ferro(-alliage)s
P żelazostopy
R ферросплавы

-1140 **ferro-silicon**
A ferro-alloy with a high silicon content, used as a deoxidizer or alloying addition.
D Ferrosilizium
F ferro(-)silicium
P żelazokrzem
R ферросилиций

-1145 **ferro-chromium**
A ferro-alloy with a high chromium content, used as an alloying addition.
D Ferrochrom
F ferro-chrome
P żelazochrom
R феррохром

-1150 **ferro-manganese**
A ferro-alloy with a high manganese content, used as a deoxidizer or alloying addition.
D Ferromangan
F ferro-manganèse
P żelazomangan
R ферромарганец

-1155 **silico-manganese, ferro-manganese-silicon**
A ferro-alloy with a high content of silicon and manganese, used as a deoxidizer for steel.
D Silicospiegel, Ferrosilicomangan
F silico(-)spiegel, silicomanganèse
P (żelazo)krzemomangan, żelazokrzem zwierciadlisty
R силикошпигель, силикомарганец

-1160 **calcium-silicon**
A ferro-alloy with a high content of calcium and silicon, used as a deoxidizer in the production of steel and for the inoculation of cast iron.
D Kalzium-Silizium
F silico(-)calcium
P (żelazo)wapniokrzem
R силикокальций

-1165 **alsifer, ferro-silico-aluminium**
A ferro-alloy with a high content of silicon and aluminium, used as a deoxidizer.
D Alsimin
F alsifer
P żelazoaluminiumkrzem, glinokrzem
R силикоалюминий

65 NON-FERROUS AND SPECIAL ALLOYS
NICHTEISEN- UND SPEZIALLEGIERUNGEN
ALLIAGES NON-FERREUX ET SPÉCIAUX
STOPY NIEŻELAZNE I SPECJALNE
ЦВЕТНЫЕ И СПЕЦИАЛЬНЫЕ СПЛАВЫ

-005 **non-ferrous alloy**
   An alloy that does not contain iron as the major constituent.
   D Nichteisen(metal)legierung, NE-Legierung
   F alliage non-ferreux
   P stop nieżelazny
   R цветной сплав

-010 **copper(-base) alloy**
   An alloy having copper as its basis metal.
   D Kupferlegierung
   F alliage cuivreux, alliage (à base) de cuivre
   P stop miedzi, stop o podstawie miedziowej
   R медный сплав, сплав на медной основе

-015 **blister copper, converter copper, black copper**
   Crude copper produced in a converter.
   D Konverterkupfer, Blisterkupfer, Schwarzkupfer, Rohkupfer, Blasenkupfer
   F cuivre brut, cuivre blister, cuivre noir
   P miedź surowa, miedź konwertorowa, miedź czarna
   R конвертерная медь, черн(ов)ая медь

-020 **rafined copper**
   Copper which has been subjected to a process of purification.
   D Raffinatkupfer, Raffinadekupfer
   F cuivre (r)affiné
   P miedź rafinowana
   R рафинированная медь

-025 **fire-refined copper, lake copper**
   Copper which has been refined by remelting in a reverberatory furnace.
   D Garkupfer, Hüttenkupfer
   F cuivre affiné au feu, cuivre affiné thermiquement
   P miedź rafinowana ogniowo, miedź hutnicza
   R красная медь, медь рафинированная в пламенной печи

-030 **cathode copper**
   Raw electrolytically refined copper.
   D Katodenkupfer
   F cuivre de cathode
   P miedź katodowa
   R катодная медь

-035 **electrolytic copper, electrocopper**
   Cathode copper which has been remelted in a furnace.
   D Elektrolytkupfer, E-Kupfer
   F cuivre électro(lytique)
   P miedź elektrolityczna
   R электролитная медь, электролитическая медь

-040 **anode copper**
   Copper obtained from a fire-refining furnace and used for anodes in electrolytic refining.
   D Anodenkupfer
   F cuivre rouge
   P miedź anodowa
   R анодная медь

-045 **cement copper**
   Impure copper obtained by precipitating the metal from copper-bearing solutions.
   D Zementkupfer, Kupfersud
   F cuivre de cément, cuivre cémentaire
   P miedź cementacyjna
   R цемент(ацион)ная медь

-050 **dry copper, set copper**
   Copper containing a relatively large amount of cuprous oxide.
   D sauerstoffhaltiges Kupfer
   F cuivre non désoxyd(ul)é
   P miedź tlenowa, miedź nie odtleniona
   R нераскислённая медь

-055 **phosphorized copper**
   Copper which has been deoxidized by means of phosphorus.
   D phosphordesoxydiertes Kupfer
   F cuivre phosphoreux
   P miedź odtleniona fosforem
   R медь раскислённая фосфором

-060 **oxygen-free copper**
    Copper containing almost no oxygen.
    D sauerstoffreies Kupfer
    F cuivre sans oxygène
    P miedź beztlenowa
    R бескислородная медь

-065 **high conductivity copper, H.C.-copper**
    A high purity copper possessing very good electrical conductivity.
    D Leitkupfer, hochleitfähiges Kupfer
    F cuivre de haute conductibilité, cuivre pour conducteurs électriques, cuivre HC
    P miedź przewodowa, miedź elektrotechniczna
    R электротехническая медь

-070 **OFHC copper, oxygen-free high conductivity copper**
    Copper of a very high purity containing almost no oxygen and possessing a very good electrical conductivity.
    D OFHC-Kupfer, sauerstoffreies hochleitfähiges Kupfer
    F cuivre OFHC, cuivre de haute conductibilité à très basse teneur en oxygène
    P miedź beztlenowa o dużej przewodności elektrycznej, miedź OFHC
    R бескислородная медь высокой проводимости, бескислородная высокопроводящая медь

-075 **phosphor copper**
    A master alloy of copper and phosphorus, used as a deoxidizer for copper and its alloys.
    D Phosphorkupfer
    F cupro-phosphore
    P miedź fosforowa
    R фосфористая медь

-080 **copper-silicon**
    A master alloy of copper and silicon, used as a deoxidizer for copper and its alloys and as an alloying addition.
    D Silizium-Kupfer
    F cupro-silicium
    P miedź krzemowa
    R кремнистая медь

-085 **copper-manganese**
    A master alloy of copper and manganese, used as a deoxidizer for copper and nickel and their alloys and also as an alloying addition.
    D Mangankupfer
    F cupro-manganèse
    P miedź manganowa
    R марганцовистая медь

-090 **arsenical copper**
    Copper containing small amounts of arsenic.
    D Arsenkupfer
    F cuivre arsenical
    P miedź arsenowa
    R мышьяковистая медь

-095 **bronze**
    Originally a copper-rich alloy of copper and tin, now a copper-rich alloy other than a brass.
    D Bronze
    F bronze
    P brąz
    R бронза

-100 **casting bronze**
    Bronze used for the manufacture of products by casting.
    D Gußbronze
    F bronze de fonderie
    P brąz odlewniczy
    R литейная бронза

-105 **wrought bronze**
    Bronze which is suitable for being subjected to plastic working.
    D Knetbronze, Walzbronze
    F bronze malléable, bronze à laminer
    P brąz przerabialny plastycznie, brąz do przeróbki plastycznej
    R деформируемая бронза

-110 **tin bronze**
    A copper-rich alloy of copper and tin, sometimes containing small amounts of other elements.
    D Zinnbronze
    F bronze à l'étain, bronze ordinaire, cupro-étain
    P brąz cynowy
    R оловянная бронза, оловянистая бронза

-115 **high-tin bronze**
Bronze containing a high percentage of tin.
D hochzinnhaltige Bronze
F bronze à haute teneur en étain
P brąz wysokocynowy
R высокооловянная бронза, высокооловянистая бронза

-120 **low-tin bronze**
Bronze containing a low percentage of tin.
D niedrigzinnhaltige Bronze
F bronze à faible teneur en étain
P brąz niskocynowy
R низкооловянная бронза, малооловянная бронза, малооловянистая бронза

-125 **tin-free bronze**
A special bronze not containing tin.
D zinnfreie Bronze
F bronze sans étain
P brąz bezcynowy
R безоловянная бронза, безоловянистая бронза

-130 **special bronze, bronze alloy**
A bronze in which tin has been wholly or partly replaced by some other alloying component.
D Sonderbronze
F bronze spécial
P brąz specjalny
R специальная бронза

-135 **aluminium bronze**
A special bronze in which tin has been replaced by aluminium.
D Aluminiumbronze
F bronze d'aluminium, cupro-aluminium
P brąz aluminiowy, brązal, brąz glinowy
R алюминиевая бронза

-140 **lead bronze**
A special bronze in which tin has been replaced by lead.
D Bleibronze
F bronze au plomb, cupro-plomb
P brąz ołowiowy
R свинцов(ист)ая бронза

-145 **silicon bronze**
A special bronze in which tin has been replaced by silicon.
D Siliziumbronze
F bronze siliceux, bronze au silicium
P brąz krzemowy
R кремнистая бронза

-150 **beryllium bronze, beryllium copper**
A special bronze in which tin has been replaced by beryllium.
D Berylliumbronze
F bronze au glucinium, bronze au béryllium, cuivre au béryllium
P brąz berylowy
R бериллиевая бронза

-155 **manganese bronze, manganese copper**
A special bronze in which tin has been replaced by manganese.
D Manganbronze
F bronze au manganèse
P brąz manganowy
R марганцевая бронза, марганцовистая бронза

-160 **nickel bronze**
A special bronze in which tin has been partly replaced by nickel.
D Nickelbronze
F bronze au nickel
P brąz niklowy
R никелевая бронза

-165 **antimony bronze, antimonial bronze, antimonial copper**
A special bronze in which tin has been replaced by antimony.
D Antimonbronze
F bronze à antimoine
P brąz antymonowy
R сурьмянистая медь

-170 **tungsten bronze, wolfram bronze**
A special bronze in which tin has been wholly or partly replaced by tungsten.
D Wolframbronze
F bronze à tungstène
P brąz wolframowy
R вольфрамовая бронза

-175 **iron-bronze, ferro-bronze**
A special bronze in which tin has been replaced by iron and, usually, a little chromium.
D Eisenbronze
F bronze au fer
P brąz żelazowy
R железистая бронза

-180 **vanadium bronze**
Brass containing a little vanadium.
D Vanadinmessing
F bronze au vanadium
P mosiądz wanadowy
R ванадиевая латунь, ванадиевая бронза

-185 **leaded (tin-)bronze, plastic bronze**
A copper-base alloy containing tin and lead.
D Bleizinnbronze
F bronze au plomb-étain, bronze à l'étain-plomb
P brąz cynowo-ołowiowy
R оловянносвинцов(ист)ая бронза

-190 **cadmium copper, cadmium bronze**
A copper-cadmium alloy containing 0.5–1% Cd.
D Cadmiumbronze
F bronze au cadmium, cuivre au cadmium
P brąz kadmowy, miedź kadmowa
R кадмиевая бронза

-195 **chromium bronze**
A tin bronze containing chromium.
D Chrombronze
F bronze au chrome
P brąz chromowy
R хромистая бронза

-200 **chromium copper**
A copper-chromium alloy containing about 0.5% Cr.
D Chromkupfer
F cuivre au chrome
P miedź chromowa
R медь с добавкой хрома

-205 **zinc bronze, (Admiralty) gun metal, Government bronze, Admiralty bronze**
Tin bronze containing zinc and, in some cases, lead.
D Rotguß, Admiralitätsbronze
F bronze au zinc
P brąz cynowo-cynkowy, spiż
R оловянноцинковая бронза

-210 **phosphor tin-bronze**
Tin bronze containing phosphorus as one of its alloying components.
D Phosphor-Zinnbronze
F bronze phosphoreux
P brąz cynowo-fosforowy
R оловяннофосфористая бронза

-215 **phosphor bronze**
Tin bronze deoxidized by means of phosphor copper and containing traces of phosphorus.
D Phosphorbronze
F bronze désoxydé au phosphore
P brąz fosforowy, fosforobrąz
R фосфористая бронза

-220 **graphite bronze**
A sintered alloy of copper with graphite.
D Graphitbronze
F bronze à graphite
P brąz grafitowy
R графити(зи)рованная бронза, бронзографит

-225 **porous bronze, sintered bronze**
Bronze produced by sintering mixtures of copper and tin powders.
D Sinterbronze
F bronze poreux, bronze fritté
P brąz porowaty, brąz spiekany
R пористая бронза, металлокерамическая бронза

-230 **white bronze**
A copper-base alloy containing tin, nickel, and zinc, used chiefly in ornamental work.
D Weißbronze
F bronze blanc
P brąz biały
R белая бронза

-235 **bearing bronze**
Bronze used for bearings because of its anti-frictional properties.
D Lagerbronze
F bronze antifriction
P brąz łożyskowy, brąz panewkowy
R подшипниковая бронза

-240 **machine bronze**
Bronze used for the manufacture of machine parts.
D Maschinenbronze
F bronze de machines, bronze mécanique
P brąz maszynowy
R машинная бронза

-245 **ordnance metal, ordnance bronze**
A tin bronze containing about 10% Sn, used originally for making cannons.
D Geschützbronze, Kanonenbronze
F bronze à canon(s)
P brąz armatni
R орудийная бронза, пушечная бронза, артиллерийская бронза

-250 **bell bronze, bell metal**
Tin bronze containing 20–25% Sn, with, sometimes, a little zinc added as a deoxidizer, used mainly for making bells.
D Glockenbronze
F bronze de cloche, bronze à cloches, métal de cloche
P brąz dzwonowy
R колокольная бронза

-255 **(high-)conductivity bronze**
　　Bronze having a high electrical conductivity.
　　D Leit(ungs)bronze
　　F bronze conducteur, bronze téléphonique
　　P brąz przewodowy
　　R телефонная бронза

-260 **medal bronze, medal metal**
　　Bronze used for the manufacture of medals.
　　D Medaillenbronze
　　F bronze de médailles, bronze à médailles
　　P brąz medalierski
　　R медальная бронза

-265 **coinage bronze, coinage copper**
　　Bronze used for the manufacture of coins.
　　D Münz(en)bronze
　　F bronze de monnaies
　　P brąz monetowy
　　R монетная бронза

-270 **statuary bronze, statue bronze, architectural bronze**
　　Any one of the series of copper-base alloys, mostly containing tin, zinc and lead, used for casting statues, architectural ornaments and plaques.
　　D Statuenbronze, Architekturbronze, Kunstbronze
　　F bronze d'art
　　P brąz artystyczny, brąz architektoniczny, brąz dekoracyjny, brąz zdobniczy
　　R художественная бронза, декоративная бронза

-275 **valve bronze**
　　Bronze used for the manufacture of valves for some internal combustion engines.
　　D Ventilbronze
　　F bronze à soupapes
　　P brąz zaworowy
　　R клапанная бронза

-280 **speculum (alloy), mirror alloy**
　　Any one of the group of copper-base alloys, containing tin and zinc, and sometimes other elements, which owing to their capability of taking high polish are used for mirrors and reflectors.
　　D Spiegelbronze
　　F spéculum, spek
　　P brąz zwierciadłowy, stop zwierciadłowy
　　R зеркальная бронза

-285 **scrap bronze**
　　Bronze obtained by remelting scrap.
　　D Umschmelzbronze
　　F bronze de deuxième fusion
　　P brąz wtórny
　　R вторичная бронза, оборотная бронза, паспортная бронза

-290 **brass**
　　A copper-base alloy in which zinc is the principal alloying element.
　　D Messing
　　F laiton, cuivre jaune
　　P mosiądz
　　R латунь

-295 **malleable brass, wrought brass**
　　Brass which is suitable for being subjected to plastic working.
　　D Walzmessing, Preßmessing, Messing-Knetlegierung
　　F laiton malléable, laiton à laminer
　　P mosiądz przerabialny plastycznie, mosiądz do przeróbki plastycznej
　　R деформируемая латунь

-300 **casting brass**
　　Brass used for the manufacture of products by casting.
　　D Gußmessing, Gelbguß, Messing-Gußlegierung
　　F laiton de fonderie, laiton pour pièces moulées
　　P mosiądz odlewniczy
　　R литейная латунь

-305 **straight brass, common brass**
　　Brass which does not contain any other alloying components apart from copper and tin.
　　D unlegiertes Messing
　　F laiton ordinaire
　　P mosiądz zwykły, mosiądz dwuskładnikowy
　　R простая латунь

-310 **special brass, complex brass**
　　Brass which contains other alloying elements apart from zinc.
　　D Sondermessing
　　F laiton spécial, laiton complexe
　　P mosiądz specjalny, mosiądz wieloskładnikowy
　　R специальная латунь, спецлатунь

-315 **single-phase brass**
　　Brass consisting entirely of a single phase, e.g. solid solution or an electron phase.
　　D homogenes Messing
　　F laiton monophasé
　　P mosiądz jednofazowy
　　R однофазная латунь

-320 **two-phase brass**
Brass consisting of a mixture of two phases.
D heterogenes Messing
F laiton biphasé, laiton supérieur
P mosiądz dwufazowy
R двухфазная латунь

-325 **alpha brass, α-brass**
Brass consisting entirely of a homogeneous solid solution of zinc in copper, i.e. containing up to about 38% Zn.
D α-Messing
F laiton α, laiton de premier titre
P mosiądz α, mosiądz alfa
R α-латунь, альфа-латунь

-330 **alpha-beta brass, α+β-brass, yellow brass, Muntz metal**
Brass consisting of a mixture of two phases, the one being the solid solution of zinc in copper, the other an electron phase with the electron concentration equal to $\frac{3}{2}$ which at lower temperatures undergoes a process of ordering. The zinc content of this brass lies between about 38 and 45%.
D α+β-Messing, Muntzmetall
F laiton α+β, laiton de second titre, métal de Muntz
P mosiądz α+β
R α+β-латунь, альфа+бета-латунь

-335 **beta brass, β-brass**
Brass consisting entirely of an electron phase with the electron concentration equal to 3/2 which at lower temperatures undergoes a process of ordering. The zinc content of this brass lies between about 45 and 50%.
D β-Messing
F laiton β
P mosiądz β
R β-латунь, бета-латунь

-340 **aluminium brass**
Brass containing aluminium as the additional alloying element.
D Aluminiummessing
F laiton à l'aluminium
P mosiądz aluminiowy
R алюминиевая латунь

-345 **tin brass**
Brass containing tin as the additional alloying element.
D Zinn-Messing
F laiton à étain
P mosiądz cynowy
R оловянная латунь, оловянистая латунь

-350 **nickel brass**
Brass containing nickel as the additional alloying element.
D Nickelmessing
F laiton au nickel
P mosiądz niklowy
R никелевая латунь

-355 **manganese brass, manganese bronze, steel bronze, high-tensile brass**
A high-zinc brass containing up to 4% manganese and some other alloying elements in low proportions.
D Manganmessing, Stahlbronze, Ferrobronze
F laiton au manganèse, laiton (à) haute résistance
P mosiądz manganowy
R марганцовая латунь, марганцовистая латунь

-360 **leaded brass, leaded Muntz metal**
Brass containing lead as the additional alloying element.
D bleihaltiges Messing
F laiton au plomb
P mosiądz ołowiowy
R свинцов(ист)ая латунь

-365 **silicon brass**
Brass containing silicon as the additional alloying element.
D Siliziummessing
F laiton au silicium
P mosiądz krzemowy
R кремнистая латунь

-370 **free-cutting brass, free-cutting bronze**
Brass to which about 1–2% lead has been added in order to improve its machinability by rendering the chips more breakable.
D Automatenmessing, Schraubenmessing
F laiton de décolletage
P mosiądz automatowy
R автоматная латунь

-375 **machinery brass**
Brass used for the manufacture of machine parts.
D Maschinenmessing
F laiton mécanique
P mosiądz maszynowy
R машинная латунь

-380 **cartridge brass, shrapnel brass**
Brass which owing to its very good deep drawing capability is used for the manufacture of cartridge cases.
D Patronenmessing, Kartuschen-Messing
F laiton à cartouches
P mosiądz łuskowy
R гильзовая латунь, патронная латунь

-385 **naval brass**
An alpha-beta brass containing 1% Sn, added to improve the resistance to corrosion in sea water.
D heterogenes Marinemessing
F laiton marin
P mosiądz morski dwufazowy
R двухфазная морская латунь

-390 **Admiralty brass, Admiralty metal**
An alpha brass containing 1% Sn added to improve the resistance to corrosion in sea water.
D homogenes Marinemessing, Admiralitätslegierung
F laiton amirauté
P mosiądz morski jednofazowy
R однофазная морская латунь, адмиралтейская латунь, адмиралтейский металл

-395 **tombac (alloy), red brass**
Alpha brass containing 5–20% zinc.
D Rotmessing, Tombak
F tombac, simil(i)or
P tombak
R томпак, красная латунь

-400 **cupro-nickel**
An alloy of copper and nickel.
D Kupfernickel
F cupro(-)nickel
P miedzionikiel
R медноникелевый сплав

-405 **manganin**
A high-resistance copper-base alloy containing manganese and nickel.
D Manganin
F manganin(e)
P manganin
R манганин

-410 **nickelin**
A copper-base resistance alloy containing nickel and manganese or zinc.
D Nickelin
F cupronickel
P nikielina
R никелин

-415 **nickel silver, German silver, argentan, white copper**
A ternary alloy of copper, nickel, and zinc.
D Neusilber, Argentan, Alpaka
F maillechort, alpaca
P nowe srebro, argentan
R нейзильбер, новое серебро

-420 **Heusler('s) alloy**
Any one of the group of copper-base alloys, containing manganese and aluminium, which are strongly ferromagnetic.
D Heusler-Legierung
F alliage (de) Heusler
P stop Heuslera
R сплав Гейслера, гейслеров сплав

-425 **carbonyl nickel**
Nickel which has been purified by the carbonyl or Mond process, consisting in the formation of volatile nickel carbonyl and its subsequent decomposition.
D Carbonylnickel, Mond-Nickel
F nickel ex carbonyle, nickel Mond
P nikiel karbonylkowy, nikiel Monda
R карбонильный никель, мондникель

-430 **nickel(-base) alloy**
An alloy having nickel as its major component.
D Nickel(basis)legierung
F alliage (à base) de nickel
P stop niklu, stop o podstawie niklowej
R никелевый сплав, сплав на никелевой основе

-435 **Monel (metal)**
Any one of the group of nickel-base alloys containing copper as the principal alloying element.
D Monel(metall)
F monel
P monel, stop Monela
R монель(-металл)

-440 **permalloy**
Any one of the group of magnetically soft nickel-iron alloys which have a very high magnetic permeability.
D Permalloy
F permalloy
P permaloj
R пермаллой

-445 **inconel**
A heat-resisting nickel-base alloy containing chromium and iron, having a high resistance to corrosion.
D Inconel
F inconel
P inkonel
R инконель

-450 **constantan**
A resistance alloy composed of nickel and copper.
D Konstantan
F constantan
P konstantan
R константан

-455 **hastelloy**
Any one of the group of complex nickel-base alloys having high acid resistance and, in some cases, good mechanical properties at high temperatures.
D Hastelloy
F hastelloy
P hasteloj
R хастеллой

-460 **alumel**
A high-resistance nickel-base alloy containing manganese, aluminium, and silicon, used chiefly for thermocouples.
D Alumel
F alumel
P alumel
R алумель, алюмель

-465 **chromel**
A high-resistance nickel-chromium or nickel-chromium-iron alloy, used chiefly for electrical resistors and for thermocouples.
D Chromel
F chromel
P chromel
R хромель

-470 **nichrome, chrome-nickel**
Any one of the group of high-resistance nickel-chromium or nickel-chromium-iron alloys, used chiefly for electrical heating elements.
D Nichrom, Chromnickel
F nichrome
P chromonikielina, nichrom
R нихром, хром-никель

-475 **hipernik**
A nickel-iron alloy, containing almost equal amounts of these elements, having a high magnetic permeability.
D Hipernik
F hipernik
P hipernik
R гиперник, гайперник

-480 **nimonic alloy**
Any one of the group of complex nickel-base alloys having good resistance to oxidation and creep at high temperatures.
D Nimonic-Legierung
F nimonic
P nimonik
R нимоник

-485 **vitallium**
Any one of the group of complex cobalt-base alloys which have a good corrosion resistance, oxidation resistance, and high-temperature strength.
D Vitallium
F vitallium
P witalium
R виталлиум

-490 **permendur**
A cobalt-iron alloy, containing a little vanadium, used for its high saturation magnetization.
D Permendur
F permendur
P permendur
R пермендур

-495 **stellite**
Any one of the group of complex cobalt-base hard-facing alloys.
D Stellit
F stellite
P stellit
R стеллит

-500 **invar, nilvar**
An iron-nickel alloy, containing 36% Ni, having a very low coefficient of thermal expansion.
D Invar, Invarstahl, Invarlegierung
F invar, nilvar
P inwar
R инвар(-сталь)

-505 **elinvar, constant-modulus alloy**
An iron-nickel-chromium alloy having a longitudinal modulus of elasticity which does not change with temperature in the practically interesting temperature range.
D Elinvar, Elinvarlegierung
F élinvar
P elinwar
R элинвар

-510 **perminvar**
    Any one of the group of nickel-iron-cobalt alloys of high magnetic permeability, this permeability being nearly constant over a range of magnetizing force.
    D Perminvar
    F perminvar
    P perminwar
    R перминвар

-515 **platinite**
    An iron-nickel alloy having about the same coefficient of thermal expansion as that of platinum or glass.
    D Platinit
    F platinite
    P platynit
    R платинит

-520 **kovar, fernico**
    An iron-base alloy containing nickel and cobalt, its coefficient of thermal expansion being about the same as that of glass.
    D Kovar
    F kovar
    P kowar, ferniko
    R ковар, ферянко

-525 **kanthal (alloy)**
    An iron-base high-resistance alloy, containing about 25% Cr, 5% Al, and 3% Co, used for electric furnace heating elements.
    D Kanthal
    F kanthal
    P kantal
    R канталь

-530 **alnico, Alnico alloy**
    Any one of the group of magnetically hard carbon-free iron-base alloys, containing up to 25% Co, 20% Ni, and 15% Al, with a small addition of Ti.
    D Alnico, AlNiCo-Legierung
    F alnico, alliage Alnico
    P alniko
    R альнико

-535 **alsifer, sendust**
    A magnetically soft ternary alloy of iron, silicon, and aluminium.
    D Sendust, Alsifer
    F alsifer
    P alsifer
    R альсифер

-540 **thermalloy**
    A magnetically soft alloy composed of nickel, copper and iron, having a magnetic induction which varies considerably in the temperature range between -60 and $+50°C$.
    D Thermalloy
    F thermalloy
    P termaloj
    R термаллой

-545 **primary aluminium**
    Aluminium which after being obtained from the ore has not been subjected to a refining process.
    D Hüttenaluminium, Rohaluminium
    F aluminium ordinaire
    P aluminium hutnicze, aluminium pierwotne
    R первичный алюминий, алюминий-сырец

-550 **refined aluminium**
    Aluminium which after being obtained from the ore has been subjected to an additional refining process in order to increase its purity.
    D Raffinal
    F aluminium raffiné
    P aluminium rafinowane
    R рафинированный алюминий, катодный алюминий

-555 **duralumin, ulminium**
    A wrought age-hardenable aluminium-base alloy containing about 3.5% Cu, 0.5% Mn, and 0.5% Mg as alloying components, and silicon and iron as impurities.
    D Dural(umin)
    F duralumin
    P duraluminium, dural(umin)
    R дюраль, дюралюмин(ий), дураль, дуралюмин(ий)

-560 **alclad, aldural**
    A duralumin sheet coated with pure aluminium.
    D Alclad
    F alclad, védal
    P alclad
    R альклед, альклэд

-565 **silumin, alpax (alloy)**
    An aluminium-silicon alloy containing up to 13.5% Si, used mainly for castings.
    D Silumin
    F alpax, aladar
    P silumin
    R силумин

-570 **modified silumin, modified alpax**
Silumin, modified before casting by a small addition of e.g. sodium, having refined silicon grains in the eutectic and thus possessing improved mechanical properties.
D veredeltes Silumin
F alpax affiné
P silumin modyfikowany, silumin uszlachetniony
R модифицированный силумин

-575 **anticorodal**
Any one of the group of corrosion-resisting aluminium-base alloys containing silicon, manganese, magnesium, and titanium.
D Anticorodal
F anticorodal
P antykorodal
R антикорродаль

-580 **aludur**
An aluminium-base alloy containing 0.6% Mg and 0.88% Si.
D Aludur
F aludur
P aludur
R алудур, алюдур

-585 **aldrey**
An aluminium-base alloy containing about 0.5% Mg and 0.6% Si, and, occasionally, some iron or manganese.
D Aldrey
F aldrey
P aldrej
R ал(ь)дрей

-590 **Y-alloy**
An aluminium-base alloy containing about 4% Cu, 2% Ni, and 1.5% Mg.
D Y(psylon)-Legierung
F alliage Y
P stop Y
R сплав игрек, игрек-сплав, уай-сплав

-595 **lautal**
An aluminium-base alloy containing about 4% Cu, 2 to 7% Si, and a little manganese.
D Lautal
F lautal
P lautal
R лауталь

-600 **hydronalium**
An aluminium-base alloy containing up to 10% Mg and a little Mn, Si, and Zn, having good resistance to corrosion by sea water.
D Hydronalium
F duralinox, alumag
P hydronalium
R гидроналий

-605 **hiduminium**
Any one of the group of complex aluminium-base alloys having good mechanical properties at elevated temperatures.
D Hiduminium
F hiduminium
P hiduminium
R хидуминий

-610 **avional**
An aluminium-base alloy containing about 4% Cu and small quantities of Mg, Mn, and Si.
D Avional
F avional
P awional
R авиональ

-615 **pantal**
Any one of the group of multicomponent aluminium-base alloys containing up to 5% Si and up to 0.8% Mg.
D Pantal
F pantal
P pantal
R пантал

-620 **magnalium**
Any one of the group of aluminium-magnesium alloys containing up to 30% Mg.
D Magnalium
F magnalium
P magnal(ium)
R магналий

-625 **electron (alloy)**
A magnesium-base alloy containing aluminium and zinc.
D Elektron
F électron, élektron
P elektron
R электрон

-630 **white tin**
An allotrope of tin stable at temperatures above 13.2°C.
D weißes Zinn
F étain blanc
P cyna biała
R белое олово

-635 **grey tin**
    An allotrope of tin stable at temperatures below 13.2°C.
    D graues Zinn
    F étain gris
    P cyna szara
    R серое олово

-640 **white (bearing) metal**
    A tin-base or lead-base bearing alloy.
    D Lagerweißmetall, Weißmetall
    F régule, alliage anti-friction plastique, métal blanc antifriction
    P biały metal
    R баббит, белый металл, белый антифрикционный сплав

-645 **babbitt, Babbitt metal**
    A tin-base bearing alloy containing antimony and copper.
    D Babbitt(-Metall), Zinn-Weißmetall
    F métal Babbit, alliage de Babbit, métal blanc antifriction
    P bab(b)it
    R (высоко)оловянный баббит, (высоко)оловянистый баббит

-650 **crude lead**
    Lead which after being obtained from the ore has not been subjected to a purification process.
    D Werkblei, Rohblei, Hochofenblei, Hüttenblei
    F plomb brut, plomb d'œuvre
    P ołów surowy
    R черновой свинец, рабочий свинец, веркблей

-655 **refined lead**
    Lead which has been subjected to a purification process.
    D Weichblei, Feinblei, Raffinadeblei
    F plomb affiné
    P ołów rafinowany, ołów miękki, ołów czysty
    R рафинированный свинец, мягкий свинец

-660 **hard lead, antimonial lead, regulus metal**
    An alloy of lead and antimony containing up to 12% Sb.
    D Hartblei, Antimonblei
    F plomb dur(ci)
    P ołów twardy, ołów antymonowy
    R твёрдый свинец, сурьмянистый свинец

-665 **star antimony, star metal, regulus**
    Antimony which has been subjected to a purification process.
    D Regulus
    F antimoine affiné
    P antymon rafinowany
    R рафинированная сурьма

-670 **Wood's alloy, Wood's metal**
    A fusible alloy containing 50% Bi, 25% Pb, 12.5% Sn, and 12.5% Cd
    D Woodsche Legierung, Woodsches Metall, Wood-Metall
    F alliage de Wood, métal Wood
    P stop Wooda
    R сплав Вуда, металл Вуда

-675 **Rose's alloy, Rose's metal**
    A fusible alloy containing 50% Bi, 28% Pb, and 22% Sn.
    D Rosesche Legierung, Rosesches Metall
    F alliage de Rose, métal Rose
    P stop Rose'go
    R сплав Розе, металл Розе

-680 **cerrobend, Lipowitz alloy, Lipowitz's metal**
    A fusible alloy containing 50% Bi, 26.7% Pb, 13.3% Sn, and 10% Cd.
    D Lipowitz-Legierung, Lipowitz-Metall
    F cerrobend, alliage de Lipowitz
    P stop Lipowitza
    R церробенд

-685 **bahnmetall, Bahn metal, alkaline earth bearing alloy**
    A lead-base bearing alloy containing a little calcium and sodium.
    D Bahnmetall, Blei-Alkali-Lagermetall
    F alliage Bahnmetall
    P stop B
    R кальциевый баббит, банметалл, свинцовокальциевонатриевый сплав

-690 **spelter**
    Crude zinc obtained by smelting.
    D Hüttenzink
    F zinc thermique
    P cynk hutniczy
    R черновой цинк, нерафинированный цинк

-695 **rectified zinc**
    Zinc which has been purified by fractional distillation.
    D Feinzink
    F zinc rectifié, zinc affiné par distillation
    P cynk rektyfikowany, cynk redestylowany
    R дистилляционный цинк, редистиллированный цинк

-700 **hard zinc**
    Zinc containing up to 6% Fe.
    D Hartzink
    F zinc dur
    P cynk twardy
    R гартцинк

-705 **zamak (alloy), mazak**
A zinc-base alloy containing aluminium as the principal alloying element.
D Zamak
F zamak
P znal
R цинкоалюминиевый сплав

-710 **solder(ing alloy), brazing alloy**
An alloy having a relatively low melting point, used for joining metals by soldering.
D Lot, Lötlegierung, Lötmetall
F soudure, brasure
P lut(owie), stop lutowniczy, stop do lutowania
R припой

-715 **soft solder**
A soldering alloy which is fusible at temperatures below 450°C.
D Weichlot, Weißlot, Schnellot
F soudure tendre, brasure (tendre)
P lut miękki, lut łatwotopliwy, lut niskotopliwy
R мягкий припой, легкоплавкий припой

-720 **hard solder, brazing solder**
A soldering alloy which is fusible at temperatures above 450°C.
D Hartlot, Strenglot, Schlaglot
F soudure dure, soudure forte, brasure forte
P lut twardy
R твёрдый припой, крепкий припой, тугоплавкий припой

-725 **tin solder**
A tin-base soldering alloy.
D Zinnlot, Lötzinn
F soudure à l'étain
P lut cynowy
R оловянный припой

-730 **aluminium solder**
A zinc-base soldering alloy containing aluminium.
D Zinklot
F soudure au zinc
P lut cynkowy
R цинковый припой

-735 **lead solder**
A lead-base soldering alloy.
D Bleilot
F soudure au plomb
P lut ołowiany
R свинцовый припой

-740 **brazing brass, spelter solder**
Hard solder containing approximately equal parts of copper and zinc.
D Messinglot, Lötmessing
F laiton de brasure, laiton pour brasage
P lut mosiężny, mosiądz lutowniczy
R латунный припой

-745 **white solder**
A hard solder containing copper, zinc, and nickel.
D Stahllot, Argentanlot
F maillechort pour brasage
P lut mosiężnoniklowy
R латунно-никелевый припой, аргентановый припой

-750 **silver solder, silver brazing alloy**
Soldering alloy containing silver as one of its components.
D Silberlot
F brasure à l'argent
P lut srebrny
R серебряный припой

-755 **coin silver**
A silver-base alloy used for making coins.
D Münzsilber
F argent de monnaie
P srebro monetowe
R монетное серебро

-760 **fine silver**
Silver of a very high grade of purity.
D Feinsilber
F argent fin
P srebro czyste, srebro rafinowane
R рафинированное серебро

-765 **native gold**
A naturally occurring gold.
D Berggold
F or natif
P złoto rodzime
R самородное золото

-770 **fine gold**
Gold of a very high grade of purity.
D Feingold
F or fin, or affiné, or de coupelle
P złoto czyste, złoto rafinowane
R рафинированное золото

-775 **coin gold**
A gold-base alloy used for making coins.
D Münzgold
F or de monnaie
P złoto monetowe, złoto monetarne
R монетное золото

65-

-780 **jewellery gold**
   Gold or a gold-base alloy used for jewellery purposes.
   D Juweliergold, Schmuckgold
   F or de joaillerie
   P złoto jubilerskie
   R ювелирное золото

-785 **dental gold**
   Gold or a gold-base alloy used in dentistry.
   D Dentalgold
   F or dentaire
   P złoto dentystyczne
   R одонтологическое золото

-790 **standard gold, sterling gold**
   A gold-base alloy containing an officially predetermined percentage of copper.
   D Sterlinggold
   F or au titre (légal)
   P złoto standardowe
   R стандартное золото

-795 **high-carat gold alloy**
   A gold-base alloy in which the percentage of alloying elements is low.
   D hochkarätige Goldlegierung, Dukatengold
   F or à titre élevé
   P złoto wysokokaratowe, złoto dukatowe
   R высокопробное золото, червонное золото

-800 **red gold**
   A gold-base alloy containing 25% Cu.
   D Rotgold
   F or rouge
   P złoto czerwone
   R красное золото

-805 **white gold**
   Any one of the series of white jewellery gold-base alloys, used as substitutes for platinum.
   D Weißgold
   F or blanc
   P białe złoto
   R белое золото

-810 **electrum**
   A native or synthetic alloy of gold and silver.
   D Elektrum
   F électrum, or argental
   P elektrum
   R электрум

65-

-815 **imitation gold (alloy), mock gold**
   A copper-base alloy having the appearance of gold.
   D unechtes Gold, Gold-Imitation
   F or faux
   P imitacja złota
   R имитация золота

-820 **platinum (group) metals**
   The metals: ruthenium, rhodium, palladium, osmium, iridium, and platinum which have properties and appearance similar to platinum.
   D Platinmetalle
   F platinoïdes
   P platynowce
   R платиновые металлы, металлы платиновой группы

-825 **heavy platinum metals**
   The metals osmium, iridium, and platinum.
   D schwere Platinmetalle
   F platinoïdes lourds
   P platynowce ciężkie
   R тяжёлые платиновые металлы

-830 **light platinum metals**
   The metals ruthenium, rhodium, and palladium.
   D leichte Platinmetalle
   F platinoïdes légers
   P palladowce, platynowce lekkie
   R лёгкие платиновые металлы

-835 **platin(um-)iridium**
   A native alloy of platinum and iridium.
   D Platiniridium
   F platine iridié
   P platynoiryd
   R платинистый иридий, платина-иридий

-840 **rhodio-platinum, platinum-rhodium**
   An alloy of platinum and rhodium, containing 10% Rh, used primarily for thermocouple elements.
   D Platinrhodium
   F platine rhodié
   P platynorod
   R платинородий

-845 **osmiridium, iridosmine, iridosmium**
   A native alloy of osmium and iridium.
   D Osmiridium, Iridoosmium
   F iridosmium
   P osmoiryd, irydoosm
   R осмирид, осмистый иридий, осмиевый иридий, иридистый осмий

-850 **amalgam**
   A mercury-base alloy.
   D Amalgam
   F amalgame
   P amalgamat, ortęć
   R амальгама

-855 **zircaloy**
   Any one of the group of zirconium-base alloys containing some tin and a little iron, chromium, and nickel, used chiefly in nuclear reactors.
   D Zircaloy
   F zircaloy
   P cyrkaloj
   R циркаллой

-860 **beta-producing element**
   Any alloying element which extends the range of existence of the beta allotrope of titanium towards lower temperatures.
   D betagenes Element
   F élément betagène
   P pierwiastek betatwórczy, pierwiastek stabilizujący fazę β
   R β-стабилизатор

-865 **mischmetal, cerium standard alloy**
   An alloy of cerium with other rare earth metals.
   D Mischmetall, Cerit
   F mischmétal
   P miszmetal
   R мишметалл

-870 **non-sag molybdenum**
   Molybdenum which owing to a special treatment is prevented from forming equiaxed grains after recrystallization as a result of which the filament wire made of it does not hang down at elevated temperatures.
   D NS-Molybdän, non-sag-Molybdän
   F molybdène non-sag
   P molibden bezzwisowy
   R непровисающий молибден, беспровесный молибден

-875 **ferro-cerium**
   A cerium-base pyrophoric alloy containing about 30% Fe.
   D Ferrozer(ium), Zereisen, Auermetall
   F ferro-cérium
   P żelazocer, metal Auera, stop Auera
   R ферроцерий

-880 **bi(-)metal, duplex metal**
   A composite, usually a rolled product, consisting of two layers of different metals or alloys.
   D Bimetall
   F bimétal
   P bimetal
   R биметалл

70 HEAT TREATMENT

WÄRMEBEHANDLUNG
TRAITEMENT THERMIQUE
OBRÓBKA CIEPLNA
ТЕРМИЧЕСКАЯ ОБРАБОТКА

-005 **heat treatment**
A process in which a metal object or a portion thereof is intentionally submitted to thermal cycles and, if required, to other physical and/or chemical action in order to achieve desired properties.
D Wärmebehandlung
F traitement thermique
P obróbka cieplna
R термическая обработка, термообработка

-006 **thermal treatment, simple heat treatment**
Heat treatment in which a metal object is subjected to thermal cycles without any intentional chemical or additional physical action.
D thermische Behandlung, eigentliche Wärmebehandlung
F traitement thermique ordinaire, traitement uniquement thermique, traitement (purement) thermique, traitement thermique proprement dit
P obróbka cieplna zwykła
R собственно термическая обработка

-010 **surface treatment**
Treatment by means of which the properties of the surface or of the surface layer of a metal object are changed.
D Oberflächenbehandlung
F traitement de surface, traitement superficiel
P obróbka powierzchniowa
R поверхностная обработка

-015 **hardening**
Treatment aimed at increasing the hardness of a metal or alloy.
D Härte(steiger)n, härtesteigernde Behandlung, Härten, Härtung
F (traitement de) durcissement
P utwardzanie
R придавание твёрдости, упрочнение

-020 **hardening capacity, hardening power**
The ability of steel to become hardened, expressed as the highest possible hardness which can be acquired as a result of quench hardening.
D Aufhärtbarkeit, Härtefähigkeit
F pouvoir trempant, aptitude à la trempe, capacité de durcissement par trempe
P utwardzalność, zahartowalność
R восприимчивость к закалке

-025 **strengthening (treatment)**
Treatment aimed at increasing the mechanical strength of a metal or alloy.
D Verfestigungs(behandlung), Festigkeitssteigerung
F (traitement de) renforcement
P obróbka umacniająca, umacnianie, zwiększanie wytrzymałości
R упрочняющая обработка, упрочнение

-030 **heating**
Increasing the temperature of an object.
D Erwärmen, Erhitzen, Erwärmung, Erhitzung, Wärmen
F chauffage, échauffement
P nagrzewanie; grzanie
R нагрев, нагревание

-035 **heating through(out), temperature equalization**
Bringing the whole mass of an object under heating to the desired temperature.
D Durchwärmen, durchgreifendes Erwärmen
F chauffage à cœur, échauffement à cœur
P nagrzewanie na wskroś, nagrzewanie skrośne
R сквозной нагрев, прогрев

-040 **heating rate, rate of heating**
 The average increase in temperature as a function of time.
 D Erwärm(ungs)geschwindigkeit, Wärmgeschwindigkeit
 F vitesse de chauffage, vitesse d'échauffement
 P szybkość nagrzewania
 R скорость нагрева(ния)

-045 **preheating**
 Preliminary heating up to a temperature that is lower than the temperature at which the given heat treatment is to be performed.
 D Vorwärmen, Vorwärmung
 F préchauffage
 P podgrzewanie
 R подогрев, подгревание, предварительный нагрев

-050 **pulse heating**
 Heating by means of repeated discrete units of energy.
 D gepulstes Erwärmen
 F chauffage par impulsions, échauffement par impulsions
 P nagrzewanie tętniące
 R импульсный нагрев

-055 **soaking, holding**
 The portion of the thermal cycle during which the temperature is maintained constant.
 D Halten
 F maintien à température, maintien isotherme, maintien en température
 P wygrzewanie
 R выдержка

-060 **soaking time, holding time**
 The time during which the temperature is maintained constant.
 D Haltezeit, Haltedauer
 F dureé de maintien, temps de maintien
 P czas wygrzewania
 R время выдержки

-065 **overheating**
 The heating of a metal or alloy to an unduly high temperature whereby it acquires a coarse-grained structure.
 D Überhitzen, Überhitzung
 F surchauffe
 P przegrzanie
 R перегрев

-070 **cooling**
 Decreasing the temperature of an object.
 D Abkühlen, Abkühlung
 F refroidissement
 P chłodzenie, schładzanie
 R охлаждение

-075 **quenching**
 Cooling at rates faster than those produced by still air.
 D Abschrecken, Abschreckung
 F refroidissement rapide, refroidissement brusque, trempe
 P oziębianie, szybkie chłodzenie
 R быстрое охлаждение, резкое охлаждение, мгновенное охлаждение

-080 **slow cooling**
 Cooling during which the drop in temperature in unit time is small.
 D langsames Abkühlen, langsame Abkühlung
 F refroidissement lent
 P studzenie, chłodzenie powolne
 R медленное охлаждение, остуживание

-085 **liquisol quenching, splat cooling, splat quenching**
 The extremely rapid cooling of a liquid metal or alloy.
 D Liquisol-Abschreckung, Aufprallabschreckung, Klatschkühlung
 F trempe ultra-rapide, refroidissement par projection
 P chłodzenie ultraszybkie (cieczy)
 R закалка расплава, закалка из жидкого состояния

-090 **continuous cooling**
 A mode of cooling in which there is an uninterrupted decrease in temperature until the final temperature is reached.
 D kontinuierliches Abkühlen, kontinuierliche Abkühlung
 F refroidissement continu
 P chłodzenie ciągłe
 R непрерывное охлаждение

-095 **air cooling, cooling in (still) air, uncontrolled cooling**
A mode of cooling in which the free air of the natural environment is the cooling medium.
D Luftabkühlen, Luftabschrecken, Luftabkühlung, Luftabschreckung, Abkühlen an ruhender Luft
F refroidissement à l'air (libre), refroidissement à l'air calme
P chłodzenie w powietrzu
R охлаждение на воздухе, воздушное охлаждение, охлаждение в спокойном воздухе

-100 **water cooling, water quenching**
A mode of cooling in which water is the cooling medium.
D Wasserabschrecken, Wasserabschreckung
F refroidissement à l'eau
P chłodzenie w wodzie
R охлаждение в воде

-105 **oil cooling**
A mode of cooling in which oil is the cooling medium.
D Ölabschrecken, Ölabschreckung
F refroidissement à l'huile
P chłodzenie w oleju
R охлаждение в масле

-110 **furnace cooling**
A mode of cooling, giving the slowest decrease in temperature, that consists in leaving the object in the furnace in which it was heated to the desired temperature, until the temperature of this furnace, after switching off the power or heat supply, reaches the room temperature.
D Ofenabkühlen, Ofenabkühlung
F refroidissement au four, refroidissement dans le four
P chłodzenie z piecem, chłodzenie w piecu
R охлаждение в печи, охлаждение с печью

-115 **bath cooling**
Cooling by immersion into a liquid medium.
D Badabschrecken, Badabschreckung
F refroidissement au bain
P chłodzenie kąpielowe
R охлаждение в ванне

-120 **cooling rate, rate of cooling, quenching rate, quenching velocity**
Lowering the temperature of an object in unit time.
D Abkühlungsgeschwindigkeit, Abschreckgeschwindigkeit
F vitesse de refroidissement
P szybkość chłodzenia
R скорость охлаждения

-125 **quenching power**
The ability of a quenching medium to give a high cooling rate.
D Abkühlungsvermögen, Abkühlungswirkung
F pouvoir refroidissant, pouvoir refroidisseur, pouvoir de refroidissement
P zdolność chłodząca, zdolność chłodzenia
R охлаждающая способность

-130 **austenitizing**
The process of heating steel to a temperature at which its structure consists of austenite and maintaining it for some time at this temperature.
D Austenitisieren, Austenitisierung
F austénitisation
P austenityzowanie, austenityzacja
R аустенитизирование, аустени(ти)зация, обработка на аустенит

-135 **austenitizing temperature**
The temperature to which steel is heated and at which it is maintained with a view to austenitizing.
D Austenitisier(ungs)temperatur
F température d'austénitisation
P temperatura austenityzowania, temperatura austenityzacji
R температура аустенитизации

-140 **hardening temperature, quenching temperature**
The temperature from which a ferrous alloy object must be quenched in order to be quench-hardened.
D Härtetemperatur, Abschrecktemperatur
F température de trempe
P temperatura hartowania
R температура закалки, закалочная температура

-145 **hardening temperature range**
   The temperature range within which the hardening temperature should lie.
   D Härtebereich
   F intervalle de températures de trempe
   P zakres temperatur hartowania
   R область закалки

-150 **incubation period, incubation time**
   The length of time which must elapse before the beginning of the transformation of an undercooled phase becomes detectable.
   D inkubationsperiode, Inkubationszeit, Induktionsperiode, Brütezeit, Anlaufzeit
   F période d'incubation, durée d'incubation
   P okres inkubacji, okres inkubacyjny
   R инкубационный период, подготовительный период, период инертности

-155 **isothermal TTT diagram, isothermal (time-temperature-)transformation diagram**
   A diagram representing the course of the transformation of undercooled austenite against temperature and time under isothermal conditions.
   D isothermisches ZTU-Schaubild, isothermes ZTU-Diagramm, isothermisches Zeit-Temperatur-Umwandlungs-Schaubild, isothermes Zeit-Temperatur-Umwandlungs-Diagramm
   F diagramme de transformation isoauténitique, diagramme isotherme TTT, diagramme isotherme temps-temperature-transformation, diagramme de transformation isotherme, diagramme de transformation en conditions isothermes
   P wykres CTPi, izotermiczny wykres przemiany austenitu, wykres izotermicznej przemiany austenitu
   R диаграмма изотермического превращения аустенита, изотермическая диаграмма превращения аустенита

-160 **continuous TTT diagram, continuous cooling transformation diagram**
   A diagram representing the course of the transformation of undercooled austenite under conditions of continuous cooling.
   D kontinuierliches ZTU-Schaubild, kontinuierliches ZTU-Diagramm, kontinuierliches Zeit-Temperatur-Umwandlungs-Schaubild, kontinuierliches Zeit-Temperatur-Umwandlungs-Diagramm
   F diagramme de transformations anisothermes, diagramme de transformations en refroidissement continu, diagramme anisotherme TTT, diagramme de transformation en conditions anisothermes
   P wykres CTPc, anizotermiczny wykres przemiany austenitu, wykres przemiany austenitu przy chłodzeniu ciągłym
   R диаграмма анизотермического превращения аустенита, анизотермическая диаграмма превращения аустенита

-165 **martensite start temperature, M_s temperature**
   The temperature at which undercooled austenite starts to transform into martensite.
   D oberer Martensitpunkt, Martensitbildungstemperatur, M_s-Temperatur
   F température de début de la transformation martensitique, point M_s
   P temperatura (początku) przemiany martenzytycznej, temperatura M_s
   R температура начала мартенситного превращения, мартенситная точка, точка $M_н$

-170 **martensite finish temperature, M_f temperature**
   The temperature at which the transformation of undercooled austenite into martensite — during continuous cooling — ends.
   D unterer Martensitpunkt, Temperatur vollständiger Martensitumwandlung, M_f-Temperatur
   F température de fin de la transformation martensitique, point M_f
   P temperatura końca przemiany martenzytycznej, temperatura M_f
   R температура конца мартенситного превращения, точка $M_к$

-175 **quench hardening, hardening by quenching**
A thermal treatment, primarily applied to ferrous alloys, consisting in raising the temperature of an alloy above the critical point, holding it at that temperature for some time, and subsequently cooling it at a rate sufficient to increase the hardness substantially.
D Härten, Abschreckhärten, Umwandlungshärten
F trempe, durcissement par trempe
P hartowanie
R закалка (с полиморфным превращением)

-180 **critical cooling rate, critical quenching rate**
The minimum cooling rate sufficient to suppress the transformation of undercooled austenite into structures of the pearlitic type and to produce a martensitic structure.
D kritische Abkühlungsgeschwindigkeit
F vitesse critique de trempe
P krytyczna szybkość chłodzenia, krytyczna szybkość hartowania
R критическая скорость охлаждения, критическая скорость закалки

-185 **quenching medium, quenchant, hardening medium, quenching agent**
A liquid, gas, or sometimes solid used as the vehicle for removing heat from the hot object to be quenched.
D Abschreckmittel, Härtemittel
F milieu de trempe, milieu trempant
P ośrodek chłodzący, ośrodek hartowniczy, środowisko chłodzące
R охлаждающая среда, закалочная среда, закалочный охладитель

-190 **quenching liquid, quenching fluid, hardening liquid**
A liquid medium which takes away heat from the object to be quenched.
D Abschreckflüssigkeit, flüssiges Härtemittel
F liquide de trempe
P ciecz hartownicza, ciecz chłodząca
R закалочная жидкость

-195 **quench(ing) bath, hardening bath**
A liquid into which the objects to be quench hardened are immersed.
D Härtebad, Abschreckbad
F bain de trempe
P kąpiel hartownicza
R закалочная ванна

-200 **quenching intensity**
The quantity of heat transferred per unit area of the surface per unit time, per degree difference of temperature between the object which is quenched and the quenching medium, divided by the thermal conductivity of the material of the object.
D Abschreckintensität
F sévérité de trempe, drasticité de trempe
P intensywność chłodzenia, intensywność hartowania
R интенсивность охлаждения, резкость закалки

-205 **water hardening, water quenching**
A quench-hardening process implying cooling in water.
D Wasserhärten
F trempe à l'eau
P hartowanie w wodzie
R закалка в воде, водяная закалка

-210 **oil hardening, oil quenching**
A quench-hardening process implying cooling in an oil bath.
D Ölhärten
F trempe à l'huile
P hartowanie w oleju
R закалка в масле, масляная закалка

-215 **air hardening, self hardening, air quenching**
A quench-hardening process implying cooling in air.
D Lufthärten
F trempe à l'air
P hartowanie w powietrzu
R закалка на воздухе, воздушная закалка

-220 **air-blast quenching, gas quenching, dry quenching**
A quench-hardening process implying cooling with jets of cold compressed air.
D Härten im Luftstrom, Gebläselufthärten, Preßlufthärten
F trempe à l'air soufflé
P hartowanie w strumieniu powietrza
R закалка в струе воздуха, закалка струей воздуха, закалка в сжатом воздухе

-225 **hot quenching**
A quench-hardening process implying cooling in a medium with a temperature substantially higher than the room temperature.
D Warmbadhärten
F trempe au bain chaud
P hartowanie w ośrodku gorącym, hartowanie w gorącej kąpieli
R закалка в горячих средах

-230 **salt bath hardening, salt bath quenching, fused salt quenching**
A quench-hardening process implying cooling in a bath of fused salts.
D Salzbadhärten, Härten im Salzbad
F trempe au bain de sel
P hartowanie w kąpieli solnej
R закалка в соляной ванне, закалка в соляном расплаве

-235 **lead bath hardening, lead bath quenching**
A quench-hardening process implying cooling in a bath of molten lead.
D Bleibadhärten
F trempe au (bain de) plomb
P hartowanie w kąpieli ołowiowej
R закалка в свинцовой ванне

-238 **brine hardening, brine quenching**
A quench-hardening process implying cooling in a brine bath.
D Härten in Salzlösung, Härten in Salzwasser
F trempe à la saumure, trempe à l'eau salée, trempe dans une solution saline
P hartowanie w solance
R закалка в рассоле, закалка в соляном растворе

-240 **spray hardening, (pressure-)spray quenching**
A quench-hardening process in which the object to be hardened is cooled by means of a spray of water.
D Spritzhärten, Sprühhärten
F trempe par aspersion
P hartowanie natryskowe, hartowanie natryskiem
R струйчатая закалка, струйная закалка

-245 **immersion quenching, liquid quenching**
A quench-hardening process implying cooling by immersion in a bath.
D Tauchhärten, Badabschrecken
F trempe par immersion, trempe au bain
P hartowanie zanurzeniowe
R закалка погружением, закалка охлаждением в ванне, жидкостная закалка

-250 **contact hardening**
A surface-hardening process, applied for cylindrical objects, involving heating by means of electrical current which flows between two rollers moved along the surface to be hardened.
D Kontakthärten, Trockenhärten
F trempe par contact
P hartowanie kontaktowe
R контактная закалка

-255 **chill hardening**
A quench-hardening process in which the object to be hardened is placed in a cooled metal mould.
D Quettenhärten, Härten in abgekühlten Kokillen
F trempe en coquille refroidie
P hartowanie kokilowe, hartowanie w kokili chłodzonej
R закалка в кокилях, закалка в охлаждаемой кокили

-260 **rapid quenching, drastic quenching, fast quench**
A quench-hardening process in which the cooling rate is very high.
D schroffe Abschreckung
F trempe énergique, trempe brutale, trempe rapide
P hartowanie ostre
R резкая закалка, закалка с большой скоростью

-265 **mild quenching, slow quench**
A quench-hardening process in which the cooling rate is relatively low.
D Mildhärten, milde Abschreckung
F trempe douce
P hartowanie łagodne
R закалка с малой скоростью

-270 **conventional quenching, customary quenching**
A quench-hardening process implying continuous cooling in a medium with a temperature lower than the martensite start temperature.
D übliches Härten
F trempe ordinaire
P hartowanie zwykłe
R обычная закалка, обыкновенная закалка

-275 **local hardening, selective hardening, differential hardening**
A quench-hardening process by which only selected regions of an object are hardened.
D örtliches Härten
F trempe partielle, trempe localisée, trempe locale
P hartowanie miejscowe
R местная закалка, дифференциальная закалка

-280 **full hardening, through-hardening**
A quench-hardening process resulting in the hardening of the whole volume of the object.
D Durchhärten, Durchhärtung, durchgreifendes Härten
F trempe à cœur
P hartowanie na wskroś, hartowanie skrośne
R сквозная закалка,

-285 **surface hardening, skin hardening, shallow hardening**
A process of imparting increased hardness to the surface layer of steel or cast iron objects by rapidly heating this layer to the required hardening temperature and subsequently quenching.
D Oberflächenhärten, Randschichthärten
F trempe superficielle
P hartowanie powierzchniowe
R поверхностная закалка

-287 **bulk hardening, bulk quenching**
A quench-hardening process implying austenitizing of the whole volume of the object.
D Volumenhärten
F trempe de volume
P hartowanie objętościowe
R объемная закалка

-290 **flame hardening, torch hardening**
A surface-hardening process in which flame heating by means of high-intensity burners is utilized.
D Flamm(en)härten, Brennhärten
F trempe au chalumeau
P hartowanie płomieniowe, hartowanie palnikowe
R (газо)пламенная закалка

-295 **induction hardening, high frequency hardening**
A surface-hardening process in which induction heating is utilized.
D Induktionshärten, Induktionshärtung
F trempe électrique, trempe par induction
P hartowanie indukcyjne
R индукционная закалка, высокочастотная закалка, закалка ТВЧ

-300 **interrupted quenching, interrupted hardening, time quenching**
A quench-hardening process in which the object to be hardened is removed from the quenching medium before it has reached the temperature of this medium.
D gebrochenes Härten, unterbrochenes Härten
F trempe interrompue, trempe arrêtée
P hartowanie przerywane
R прерывистая закалка

-305 **martempering, marquenching, step quenching, graduated hardening**
A quench-hardening process which implies quenching of an austenitized ferrous alloy to a temperature slightly above the martensite start temperature, holding it at that temperature long enough, until the temperature throughout the mass is equalized without transformation of austenite, and subsequently cooling in air through the temperature range of martensite transformation.
D gestuftes Härten, Stufenhärten, Warmbadhärten
F trempe étagée (martensitique), trempe par étapes, martempering
P hartowanie stopniowe
R ступенчатая закалка

-310 **austempering**
> A heat treatment consisting in heating steel above the transformation range, so as to give rise to austenite, and quenching it in a molten salt or metal bath at a constant temperature between M_s and 450°C. The resulting structure is called bainite.
> **D** Zwischenstufenvergüten, Zwischenstufenhärten Austempering, Bainitisieren
> **F** trempe (étagée) bainitique, trempe isotherme, austempering
> **P** hartowanie z przemianą izotermiczną, hartowanie izotermiczne, hartowanie bainityczne
> **R** изотермическая закалка, бейнитирование

-315 **spin hardening**
> A flame-hardening process, applied to objects of a round section, in which the object to be surface-hardened is heated to the required hardening temperature by rotating in contact with stationary burners. After reaching this temperature it is then quenched by immersion or by water-spray.
> **D** Umlauf(mantel)härten, Drehungshärten
> **F** trempe superficielle par rotation
> **P** hartowanie obrotowe
> **R** закалка методом вращения

-320 **progressive hardening, scanning hardening**
> A flame-hardening process which implies a relative translatory motion between the object to be surface-hardened and the heating burner, the latter being followed by a water spray.
> **D** Vorschubhärten, Härten im Vorschub
> **F** trempe au défilé, trempe progressive, trempe de proche en proche
> **P** hartowanie posuwowe, hartowanie przesuwowe
> **R** закалка методом перемещения

-325 **progressive spin hardening**
> A flame-hardening process in which the cylindrical object to be surface-hardened is revolved on its axis while a ring of heating flames moves along it followed by a ring of quench sprays.
> **D** Umlaufvorschubhärten
> **F** trempe au defilé avec rotation
> **P** hartowanie posuwowo-obrotowe, hartowanie obrotowo-posuwowe
> **R** закалка методом вращения и перемещения

-330 **clean hardening, non-oxydizing hardening, bright hardening**
> A quench-hardening process carried out by heating the object to be hardened in a protective atmosphere so that a discolouration of its surface by oxidation is prevented.
> **D** Blankhärten
> **F** trempe brillante
> **P** hartowanie beznalotowe, hartowanie czyste
> **R** чистая закалка, светлая закалка

-335 **underhardening**
> A quench-hardening process in which a hypoeutectoid steel is quenched from a temperature at which ferrite as well as austenite exists.
> **D** Unterhärten
> **F** trempe incomplète
> **P** hartowanie niezupełne
> **R** неполная закалка, частичная закалка

-340 **sub-zero treatment, cold treatment, subquenching**
> Cooling of a previously quench-hardened steel to sub-zero temperatures in order to promote the transformation of retained austenite into martensite.
> **D** Tiefkühlen, Trieftemperaturbehandeln, Tieftemperaturhärten
> **F** traitement au froid, traitement par le froid
> **P** obróbka podzerowa, wymrażanie, hartowanie podzerowe
> **R** обработка холодом, низкотемпературная обработка

-345 **quenching from high temperatures**
  Austenitizing at a high temperature (usually above 1000°C) followed by water quenching, applied to some high-alloy steels in order to obtain an austenitic structure at room temperature.
  D Ablöschen
  F hypertrempe
  P przechładzanie
  R высокотемпературная закалка

-350 **depth of hardening, hardening depth, hardness penetration depth**
  The depth to which the hardening effect resulting from quench hardening penetrates.
  D Einhärtung(stiefe), Einhärtetiefe, Durchhärtungstiefe
  F pénétration de trempe, profondeur de trempe
  P głębokość zahartowania
  R глубина прокаливаемости

-355 **hardened zone, hardened case**
  The thickness of the quench hardened layer measured from the surface, usually to the point where there is still not less than 50 per cent martensite.
  D Härtezone, Härteschicht
  F zone trempée, couche trempée
  P strefa zahartowana, warstwa zahartowana
  R закалённая зона, закалённый слой

-360 **critical diameter, base diameter**
  The maximum diameter of a bar that can be quench hardened throughout in a given quenching medium.
  D kritischer Durchmesser
  F diamètre critique (de trempe)
  P średnica krytyczna
  R критический диаметр

-365 **tempering, drawing, letting down**
  A thermal treatment in which steel, previously quench hardened, is heated to a temperature below the critical range and then cooled slowly.
  D Anlassen, Anlaßglühen
  F revenu
  P odpuszczanie
  R отпуск

-370 **low-temperature tempering**
  Tempering during which the quench-hardened object is heated to a temperature of 150–250°C.
  D Anlassen bei niedrigen Temperaturen
  F revenu à basse température, revenu de détente
  P odpuszczanie niskie
  R низкотемпературный отпуск, низкий отпуск

-375 **high-temperature tempering**
  Tempering during which the quench-hardened object is heated to a temperature above 500°C.
  D Hochtemperaturanlassen, Anlassen bei hohen Temperaturen
  F revenu à haute température
  P odpuszczanie wysokie
  R высокотемпературный отпуск, высокий отпуск

-380 **tempering temperature**
  The temperature at which a quench-hardened object is held in order to be tempered.
  D Anlaßtemperatur
  F température de revenu
  P temperatura odpuszczania
  R температура отпуска

-385 **temper colour, heat tint**
  Any one of the colours occurring on the surface of polished steel during heating in air as a result of interference effects in the thin oxide films formed.
  D Anlauffarbe
  F couleur de revenu
  P barwa nalotowa
  R цвет побежалости

-390 **blue temper**
  The temper colour occurring at temperatures of about 300°C.
  D Blauanlauf
  F bleu
  P nalot niebieski
  R синий налёт, синий цвет побежалости

-395 **blue-heat range**
  The temperature range within which blue temper colour develops on the surface of polished steel.
  D Blauwärme, Blauhitze
  F température du bleu
  P temperatura niebieskiego nalotu
  R температура синего цвета побежалости

-400 **blu(e)ing**
A tempering process, performed within the blue-heat range in a suitable oxidizing atmosphere, applied to steel sheet, strip or finished parts (e.g. springs) in order to give them a better appearance (blue-black) and to improve their resistance to corrosion.
D Bläuen, Schwärzen, Oxydieren
F bleuissage
P czernienie, oksydowanie
R синение, воронение

-405 **secondary hardening**
Increase in hardness that occurs during high-temperature tempering of certain previously quench-hardened alloy steels containing one or more carbide-forming elements.
D Sekundärhärtung
F durcissement secondaire, trempe secondaire
P utwardzanie wtórne, hartowanie wtórne
R вторичная закалка, вторичное твердение

-410 **toughening**
A thermal treatment aimed at achieving high value of toughness at a given tensile strength by quench hardening and subsequently tempering at a sufficiently high temperature.
D Vergüten, Vergütung
F amélioration, traitement d'amélioration, trempe et revenu
P ulepszanie cieplne
R термическое улучшение

-415 **patenting**
A thermal treatment applied in the production of steel wire, consisting in heating it to a temperature above the transformation range and subsequently quenching in a molten lead bath or in compressed air.
D Patentieren
F patentage
P patentowanie
R патентирование, сорбитизация, одинарная обработка

-420 **annealing**
A thermal treatment consisting in heating a metal or alloy to a specified temperature, holding it at this temperature for some time, and then cooling it relatively slowly, thus approaching the structure corresponding to the equilibrium state.
D Glühen, Ausglühen
F recuit
P wyżarzanie
R отжиг

-425 **annealing temperature**
The temperature at which a metal or alloy is held for annealing.
D Glühtemperatur
F température de recuit
P temperatura wyżarzania
R температура отжига

-430 **bright annealing, clean annealing**
Annealing performed in a controlled atmosphere or vacuum, so as to prevent surface oxidation.
D Blankglühen
F recuit blanc, recuit brillant
P grzanie jasne, grzanie beznalotowe, wyżarzanie jasne, wyżarzanie beznalotowe
R ясный отжиг, светлый отжиг, безокислительный отжиг

-435 **box annealing, pack annealing, close annealing, pot annealing**
Annealing performed in an appropriately closed metal box or pot so as to minimize oxidation and prevent scaling.
D Kistenglühen, Kastenglühen, Topfglühen
F recuit en vase close, recuit en pot, recuit en caisse
P grzanie czyste, grzanie bezzgorzelinowe, wyżarzanie czyste, wyżarzanie bezzgorzelinowe
R чистый отжиг, отжиг в ящиках

-440 **black annealing**
A box-annealing process, applied e.g. to steel sheet before tinning, resulting in a blackening of the sheet surface.
D Dunkelglühen, Schwarzglühen
F recuit noir
P grzanie ciemne, wyżarzanie ciemne
R чёрный отжиг

-445 **blue annealing**
　　The annealing of hot-rolled steel plate and sheet at a temperature within the transformation range and subsequent cooling in air, aimed at softening them. It is accompanied by the formation of a bluish oxide film on the surface.
　　D Blauglühen
　　F recuit bleu
　　P wyżarzanie z niebieskim nalotem
　　R синий отжиг

-450 **isothermal annealing**
　　Annealing which involves the cooling of austenitized steel to a temperature below the critical range and holding it at that constant temperature until the transformation of undercooled austenite into a structure of the pearlitic type is completed.
　　D isothermisches Glühen, Perlitglühen, Perlitisieren, perlitisierendes Glühen
　　F recuit isotherme, (recuit de) perlitisation
　　P wyżarzanie z przemianą izotermiczną, wyżarzanie izotermiczne, wyżarzanie perlityzujące, perlityzowanie
　　R изотермический отжиг, перлитизирующий отжиг

-455 **stepped annealing**
　　An annealing process, applied to steel castings, performed in several steps, the time of cooling and the rate of cooling within each step not necessarily being equal.
　　D Stufenglühen
　　F recuit graduel
　　P wyżarzanie stopniowe
　　R ступенчатый отжиг

-460 **soft annealing, softening**
　　Annealing aimed at reducing the hardness of a metal or alloy.
　　D Weichglühen
　　F recuit d'adoucissement, recuit adoucissant, recuit doux, (traitement d')adoucissement
　　P wyżarzanie zmiękczające, zmiękczanie
　　R смягчающий отжиг

-465 **full annealing, dead annealing**
　　An annealing process, applied to steels, consisting of austenitizing and a very slow cooling (e.g. furnace cooling).
　　D vollständiges Ausglühen, Totglühen, Ausglühen
　　F recuit complet, recuit à mort
　　P wyżarzanie zupełne
　　R полный отжиг

-470 **under-annealing, partial annealing**
　　An annealing process, applied to steels, which involves soaking the steel at a temperature within the critical range and very slow cooling (e.g. furnace cooling).
　　D unvollständiges Ausglühen, unvollständiges Glühen
　　F recuit partiel, recuit incomplet
　　P wyżarzanie niezupełne
　　R неполный отжиг

-475 **normalizing, heat refining**
　　An annealing process, applied to steels, consisting in austenitizing the steel and then allowing it to cool in still air at room temperature.
　　D Normalglühen, Normalisieren, Feinglühen, Umkörnen
　　F recuit de normalisation, traitement de normalisation, recuit de regénération, normalisation, recuit d'affinage structural, (traitement d')affinage structural
　　P wyżarzanie normalizujące, normalizowanie
　　R нормализационный отжиг, нормализация

-480 **high-temperature annealing, grain growth annealing**
　　An annealing process, sometimes applied to steels, aimed at obtaining a coarse-grained structure and, consequently, at improving its machinability.
　　D Grobkornglühen, Hoch(temperatur)glühen
　　F recuit à haute température, recuit à gros grain
　　P wyżarzanie przegrzewające, przegrzewanie
　　R отжиг увеличивающий зерно, высокий отжиг, отжиг на крупное зерно

-485 **homogenizing, diffusion annealing**
  Annealing performed at a temperature a little below the solidus, aimed at giving greater chemical homogeneity to the alloy.
  D Homogen(isierungs)glühen, Diffusionsglühen
  F recuit d'homogénéisation, homogénéisation, recuit de diffusion
  P wyżarzanie ujednorodniające, ujednorodnianie, homogenizowanie
  R гомогенизационный отжиг, гомогенизация, гомогенизирующий отжиг, диффузионный отжиг

-490 **recrystallization annealing**
  Annealing, applied to cold-worked metals or alloys, performed at a temperature above the recrystallization temperature, its purpose being the elimination of the consequences of the cold-working without the interference of any phase transformation.
  D Rekristallisationsglühen
  F recuit de recristallisation, traitement de recristallisation
  P wyżarzanie rekrystalizujące, rekrystalizowanie
  R рекристаллизационный отжиг

-495 **isochronal annealing, isochronous annealing**
  The heating of a cold-worked metal at a constant rate, applied e.g. when calorimetric measurements of the stored energy are to be made.
  D isochrones Erholungsglühen, isochrone Erholung
  F recuit isochrone, guérison isochrone
  P wygrzewanie izochroniczne, wyżarzanie izochroniczne
  R изохрон(аль)ный отжиг

-500 **quench annealing, solution treatment, solution annealing**
  Annealing aimed at dissolving secondary phases in a solid solution.
  D Lösungsglühen
  F recuit de mise en solution
  P rozpuszczanie, roztwarzanie, wyżarzanie rozpuszczające
  R обработка на твёрдый раствор

-505 **flame annealing**
  Annealing involving heating by means of a high-temperature flame.
  D Flammenglühen
  F recuit au chalumeau
  P wyżarzanie płomieniowe
  R пламенный отжиг

-510 **spheroidizing (annealing)**
  Annealing of steel aimed at converting the cementite network or lamellae into a rounded or globular form.
  D Kugelkornglühen, Glühen auf kugeligen Zementit, Glühen auf kugelige Carbide
  F recuit de sphéroïdisation, recuit de coalescence
  P wyżarzanie sferoidyzujące, sferoidyzowanie
  R сфероидизирующий отжиг, сфероидизация

-515 **spheroidization of cementite, coalescence of cementite**
  The balling-up of the cementite lamellae and network, occurring in steels during spheroidizing.
  D Einformung des Zementits, Zusammenballung des Zementits
  F sphéroïdisation de la cémentite, globulisation de la cémentite, coalescence de la cémentite
  P sferoidyzacja cementytu
  R сфероидизация цементита

-520 **cyclic annealing, thermal cycling**
  A spheroidizing annealing involving soaking at temperatures repeatedly alternating from a little below the eutectoid temperature to a little above this temperature.
  D Pendelglühen
  F recuit oscillant, recuit alterné, cyclage thermique
  P wyżarzanie wahadłowe, wygrzewanie wahadłowe
  R маятниковый отжиг, колебательный отжиг, циклический отжиг, термоциклирование

-525 **stress relieving, stress-relief annealing**
A thermal treatment aimed at reducing residual stresses in a metal object by heating the object to a suitable temperature, holding it at this temperature for a specified time and then allowing it to cool down slowly.
D Spannungsarmglühen, Spannungsfreiglühen, Entspann(ungsglüh)en, Spannungsausgleichsglühung
F recuit de détente, traitement de détente, recuit de relaxation, traitement de relaxation
P wyżarzanie odprężające, odprężanie
R отжиг уменьшающий напряжения, низкий отжиг, отжиг для снятия напряжений

-530 **stabilizing (annealing)**
The overcoming of the susceptibility of austenitic stainless steels, containing a small amount of a strong carbide-forming element, such as titanium or niobium, to intercrystalline corrosion by soaking at an appropriate temperature in order to precipitate the maximum amount of carbon as titanium carbide or niobium carbide.
D Stabilglühen
F recuit de stabilisation, stabilisation
P stabilizowanie
R стабилизирование, стабилизация

-535 **stabilizing treatment**
A thermal treatment applied mainly to aluminium alloy castings, aimed at relieving internal stresses and attaining dimensional stability, consisting in holding the casting at about 200°C for several hours and then slowly cooling.
D Stabilisierungsglühen, Stabilisierung
F traitement de stabilisation, recuit de stabilisation, stabilisation
P stabilizowanie, wyżarzanie stabilizujące
R стабилизирующий отжиг, стабилизирование

-540 **natural stabilizing treatment**
A thermal treatment applied mostly to cast iron castings to relieve residual stresses, consisting in holding these castings for a sufficient length of time at normal room temperature.
D natürliches Stabilisieren
F stabilisation naturelle
P stabilizowanie naturalne, sezonowanie
R естественное стабилизирование

-545 **sensitization, sensitizing**
Holding an austenitic stainless steel for any length of time at a temperature at which chromium carbide particles precipitate on the grain boundaries, the austenite grains thus becoming depleted in chromium and susceptible to intercrystalline corrosion.
D Sensibilisieren, Sensibilisierungsglühung
F sensibilisation, recuit de sensibilisation
P wygrzewanie uwrażliwiające, wyżarzanie uwrażliwiające
R сенсибилизация

-550 **baking**
Heat treatment aimed at removing gases from a solid metal.
D Entgasen
F dégazage, traitement (thermique) de dégazage
P odgazowanie
R дегазирование

-555 **dehydrogenation**
Baking aimed at reducing the hydrogen content of steel, e.g. for preventing the occurrence of snowflakes.
D Wasserstofffreiglühen, Dehydrieren, Wasserstoffentzug; Flockenfreibehandlung
F déshydrogénation
P odwodorowywanie; wyżarzanie przeciwpłatkowe
R обезводороживание

-560 **precipitation hardening, ausageing**
A process in which an increase in the hardness of an alloy is produced by the incipient or accomplished precipitation of a new, highly dispersed phase from a supersaturated solid solution.
D Aus(scheidungs)härten, Aus(scheidungs)härtung
F durcissement par précipitation, durcissement structural, durcissement par ségrégation
P utwardzanie wydzieleniowe, utwardzanie dyspersyjne (wydzieleniowe), umacnianie wydzieleniowe
R дисперсионное твердение, дисперсионное упрочнение

-565 **age hardening**
The process of increasing the hardness of an alloy by ageing.
D Alterungshärtung
F durcissement par vieillissement
P utwardzanie przez starzenie
R упрочнение при старении

-570 **solution heat treatment**
Heating an alloy — which exhibits solid solubility of one constituent in another increasing with the rise of temperature — to a temperature at which it is homogeneous, holding it at that temperature for some time, and subsequently quenching in order to retain the single-phase structure as a supersaturated solid solution.
D (Lösungsglühen und) Abschrecken
F trempe (après recuit de mise en solution)
P przesycanie
R (истинная) закалка, закалка без полиморфного превращения)

-575 **ag(e)ing**
A change in the properties of an alloy occurring slowly at room temperatures and more rapidly at slightly elevated temperatures, usually after quenching or cold working.
D Altern, Alterung, Auslagern
F vieillissement, maturation
P starzenie
R старение

-580 **quench ageing**
Ageing of a supersaturated solid solution.
D Abschreckalterung
F vieillissement après trempe
P starzenie po przesycaniu
R старение после закалки

-585 **natural ageing**
Ageing occurring spontaneously at room temperature.
D natürliche Alterung, Kalt(aus)lagern, Kaltaushärtung
F vieillissement naturel, maturation
P starzenie naturalne, starzenie samorzutne
R естественное старение

-590 **artificial ageing**
Ageing occurring at elevated temperatures.
D künstliche Alterung, beschleunigte Alterung, Warm(aus)lagern, Warmaushärtung
F vieillissement accéléré, revenu durcissant
P starzenie przyspieszone, starzenie sztuczne
R искусственное старение, ускоренное старение, старение при нагреве

-595 **overageing**
Ageing at a temperature which is too high or for a time which is too long as compared with that required to obtain the optimum mechanical properties.
D Überaltern, Überalterung
F survieillissement, vieillissement poussé, surrevenu
P przestarzenie
R перестарение

-600 **pre-precipitation**
The phenomena occurring in the lattice of a supersaturated solid solution during ageing before the precipitation of a new phase begins.
D Vor-Ausscheidung, Vorausscheidung
F pré-précipitation
P przedwydzielanie, zjawiska przedwydzieleniowe
R предвыделение, явление предвыделения

-605 **Guinier-Preston zones**
The first clusters of the solute formed on definite crystallographic planes within a supersaturated solid solution before precipitation begins.
D Guinier-Preston-Zonen
F zones de Guinier-Preston
P strefy Guinier-Prestona
R зоны Гинье-Престона

-610 **reversion, retrogression**
The returning of naturally aged alloys, after a heating of short duration, to the condition in which they were before ageing, i.e. immediately after solution heat treatment.
D Rückbildung
F réversion
P nawrót
R возврат

-615 **maraging**
Ageing performed at 450—500°C, applied to steels containing 18—25% Ni and small additions of other elements and having a relatively ductile martensitic structure, its objective being the improvement of mechanical properties.
D Martensitaltern, Martensitaushärten
F maraging
P starzenie martenzytu
R старение мартенсита

-620 **magnetic ageing**
A change in the hysteresis and permeability of steel taking place in time, especially when heated, resulting from ageing effects.
D magnetisches Altern
F vieillissement magnétique
P starzenie magnetyczne
R магнитное старение

-625 **malleablizing**
The annealing of white cast iron aimed at converting it into malleable cast iron.
D Tempern
F malléabilisation, recuit de malléabilisátion
P uplastycznianie (żeliwa), wyżarzanie uplastyczniające
R томление (чугуна)

-630 **graphitizing, black heart process**
Malleablizing performed in a neutral atmosphere, i.e. implying the conversion of some or all of the combined carbon of the white cast iron into free carbon, thus resulting in the production of black heart malleable cast iron.
D Graphitisieren, Graphitisierung
F malléabilisation par graphitisation, recuit graphitisant, traitement de graphitisation, graphitisation
P wyżarzanie grafityzujące, grafityzacja
R графитизирующий отжиг, графитизация

-635 **decarburizing annealing, white heart process**
Malleablizing performed in an oxidizing atmosphere, i.e. implying the removal of carbon by oxidation, thus resulting in the production of white heart malleable cast iron.
D Glühfrischen
F malléabilisation par décarburation
P wyżarzanie odwęglające
R обезуглероживающий отжиг

-640 **growth of cast iron**
A permanent increase in volume of cast iron, mainly resulting from the decomposition of cementite caused by prolonged alternating temperature cycles in a range above 400°C.
D Wachsen des Gußeisens
F gonflement de la fonte
P pęcznienie żeliwa, puchnięcie żeliwa
R рост чугуна

-645 **ferritizing annealing**
The process of obtaining ferritic malleable cast iron.
D ferritisierende Glühbehandlung
F ferritisation
P ferrytyzacja
R ферритизация

-650 **thermo-chemical treatment, chemico-thermal treatment**
Heat treatment in which the chemical composition of a metal object, at least in the surface layer, is intentionally changed by the diffusion of one or more elements into or out of the surface.
D thermochemische Behandlung, chemisch-thermische Behandlung, (thermochemische) Diffusionsbehandlung
F traitement thermochimique (de diffusion), traitement thermique de diffusion
P obróbka cieplno-chemiczna, obróbka cieplno-dyfuzyjna
R химико-термическая обработка, термохимическая обработка

-655 **cementation, diffusion alloying**
Thermo-chemical treatment in which the chemical composition of the surface layer of a metal object is intentionally changed by the diffusion of one or more elements into the surface.
D Auflegieren im festen Zustand, Diffusionssättigung
F cémentation, traitement de cémentation
P nasycanie dyfuzyjne
R диффузионное легирование, поверхностное легирование, диффузионное насыщение

-660 **case**
The outer layer of a metal object within which the composition has been changed as a result of cementation.
D Einsatzschicht
F couche cémentée
P warstwa dyfuzyjna
R диффузионный слой

-665 **core, heart**
The interior part of a cemented metal object within which the composition, as opposed to the case, has not been changed by the cementation process.
D Kern
F cœur, âme
P rdzeń
R сердцевина

-670 **case hardening**
Hardening of the surface layer of a metal object by cementation and, mostly, a suitable thermal treatment.
D Einsatzhärten, Einsatzhärtung
F durcissement par cémentation, cémentation
P utwardzanie dyfuzyjne; węgloutwardzanie cieplne
R цементация

-675 **inert atmosphere, neutral atmosphere**
A gaseous atmosphere exerting no chemical action on the metal objects held in it at an elevated temperature.
D inerte Atmosphäre, inaktive Atmosphäre, Neutralgasatmosphäre
F atmosphère neutre, atmosphère inerte
P atmosfera obojętna
R инертная атмосфера

-680 **(re)active atmosphere**
A gaseous atmosphere which exerts a chemical action on the metal objects held in it at a given temperature.
D (re)aktive Atmosphäre
F atmosphère (ré)active
P atmosfera aktywna
R активная атмосфера

-685 **carrier gas**
That component of a reactive atmosphere which carries other components of this atmosphere as well as reduces their concentrations.
D Trägergas
F gaz support, gaz porteur
P gaz nośny
R газ-носитель

-690 **protective atmosphere**
An artificially produced gaseous atmosphere which protects metal objects against oxidation during heat treatment.
D Schutzgasatmosphäre
F atmosphère protectrice
P atmosfera ochronna
R защитная атмосфера

-695 **carburizing atmosphere**
A gaseous atmosphere capable of carburizing steel objects when held in this atmosphere at an appropriate temperature.
D Kohlungsatmosphäre
F atmosphère carburante
P atmosfera nawęglająca
R науглероживающая атмосфера

-700 **controlled atmosphere**
A gaseous atmosphere for carrying out heat treatment, produced either in the furnace itself or outside the furnace, its composition being automatically controlled.
D geregelte Atmosphäre, kontrollierte Atmosphäre
F atmosphère contrôlée
P atmosfera regulowana
R контролируемая атмосфера

-705 **endothermic atmosphere, endothermic gas, endogas**
A controlled atmosphere produced by the conversion of a fuel gas in a heated chamber.
D Endogasatmosphäre, Endogas
F atmosphère endothermique, gaz d'atmosphere endothermique
P atmosfera endotermiczna
R эндотермическая атмосфера, эндотермический газ, эндогаз

-710 **exothermic atmosphere, exothermic gas, exogas**
A controlled atmosphere produced by incomplete combustion of a fuel gas without heat supply.
D Exogasatmosphäre, Exogas
F atmosphère exothermique, gaz d'atmosphere exothermique
P atmosfera egzotermiczna
R экзотермическая атмосфера, экзотермический газ, экзогаз

-715 **carburizing, carburization, carbon cementation**
A thermo-chemical treatment consisting in enriching the surface layer of a steel object in carbon.
D Aufkohlen, Aufkohlung, Einsetzen, Zementieren
F carburation, cémentation par le carbone, cémentation (au carbone)
P nawęglanie
R науглероживание, цементация

-720 **powder carburizing, pack carburizing, box carburizing, solid carburizing**
Carburizing in a solid powdered medium.
D Pulveraufkohlen, Pulveraufkohlung
F cémentation à la poudre, carburation en caisse, cémentation en caisse, cémentation solide
P nawęglanie w proszkach, nawęglanie proszkowe
R науглероживание в порошке, науглероживание порошком, твёрдая цементация

-725 **liquid carburizing, (salt) bath carburizing**
Carburizing in a liquid medium.
D Badaufkohlen, Badeinsetzen, Salzbadaufkohlen
F carburation liquide, cémentation liquide, cémentation en bain de sels
P nawęglanie kąpielowe
R жидкостная цементация

-730 **gas carburizing**
Carburizing in a gaseous medium.
D Gasaufkohlen, Gasaufkohlung, Gaseinsetzen
F carburation gazeuse, cémentation gazeuse
P nawęglanie gazowe
R газовая цементация, науглероживание газами

-735 **selective carburizing**
Carburizing by which only certain parts of the surface of an object are carburized.
D örtliches Aufkohlen, partielle Aufkohlung
F carburation sélective, cémentation sélective
P nawęglanie miejscowe
R местное науглероживание, дифференциальное науглероживание, местная цементация

-740 **carburizer**
A medium consisting of carbonaceous matter in which the process of carburizing is carried out.
D Aufkohlungsmittel, Kohlungsmittel, Zementationsmittel, Aufkohlungsmedium
F cément (carburant), agent de cémentation, milieu cémentant
P karburyzator, (o)środek nawęglający
R карбюризатор

-745 **carburizing gas**
A gas which is the source of carbon for the surfaces to be carburized.
D Aufkohlungsgas, Kohlungsgas, Zementationsgas
F gaz carburant
P gaz nawęglający
R науглероживающий газ, цемент(ир)ующий газ

-750 **carbon potential, carbon level**
The carbon concentration in austenite in equilibrium with the components of the given carburizing atmosphere.
D Kohlenstoffpotential, Kohlenstoffpegel
F potentiel carbone, niveau de carbone
P potencjał węglowy
R углеродный потенциал

-755 **carbon transfer coefficient**
  The amount of carbon which penetrates from carburizer into the steel object under treatment in unit time at unit difference in concentration or activity of carbon between carburizer and steel.
  D Kohlenstoff-Übergangszahl, C-Übergangszahl
  F coefficient de transfert de carbone
  P współczynnik przenoszenia (węgla), współczynnik przejmowania (węgla)
  R коэффициент передачи (углерода)

-760 **direct quenching, direct hardening, pot quenching**
  The process of quenching carburized objects directly from the carburizing operation.
  D Direkthärten, Härten aus dem Einsatz, direktes Abschrecken, Direktabschrecken
  F trempe directe
  P hartowanie bezpośrednie, hartowanie z temperatury nawęglania
  R закалка с цементационного нагрева

-765 **regenerative quenching, double quenching, double hardening treatment**
  The double quenching of carburized objects aimed at refining both the case and the core and at hardening the case.
  D Doppelhärten
  F trempe double
  P hartowanie podwójne, hartowanie dwukrotne
  R двойная закалка

-770 **core refining**
  Heating a carburized object just above the critical range for the core and subsequently quenching in order to refine the coarse grain incurred by the core during the carburizing process.
  D Kernrückfeinen, Kernhärten
  F trempe du cœur, trempe de l'âme
  P hartowanie rdzenia (po nawęglaniu), normalizowanie rdzenia
  R закалка сердцевины (после цементации)

-775 **blank hardening**
  A method of determining the core properties of a steel to be carburized, consisting in quench hardening several discs of different thickness and measuring the hardness obtained.
  D Blindhärten, Blindhärtungsversuch, Blindhärteversuch
  F essai à blanc (de trempe), essai à temoin (de trempe)
  P hartowanie ślepe, próba ślepego hartowania
  R пробная закалка

-780 **carburized case**
  The outer layer of a carburized steel object within which the carbon content has been increased by the carburizing process.
  D aufgekohlte Randzone
  F couche carburée, couche cémentée
  P warstwa nawęglona
  R цемент(ир)ованный слой

-785 **nitriding, nitrogen (case) hardening**
  A thermo-chemical treatment consisting in saturating the surface layer of an object with nitrogen.
  D Nitrieren, Aufsticken
  F nitruration, cémentation par l'azote
  P azotowanie
  R азотирование, нитрирование

-790 **ammonia nitriding**
  Nitriding performed in an atmosphere of gaseous ammonia.
  D Gasnitrieren
  F nitruration gazeuse
  P azotowanie gazowe
  R газовое азотирование

-795 **bath nitriding, wet nitriding**
  Nitriding performed in a liquid medium.
  D Badnitrieren, Salzbadnitrieren
  F nitruration en bain de sel
  P azotowanie kąpielowe
  R жидкостное азотирование

-800 **ion nitriding, glow (discharge) nitriding, ionitriding**
  Nitriding performed in a gaseous atmosphere containing nitrogen, the latter being ionized by glow discharge.
  D Plasmanitrieren, Ionitrieren, Glimmnitrieren
  F nitruration (par bombardement) ionique, ionitruration, nitruration par décharge luminescente
  P azotowanie jonowe, azotowanie jarzeniowe
  R ионное азотирование, азотирование в тлеющем разряде

-805 **ion implantation**
  The introduction of ions into a solid by giving them a high kinetic energy, e.g. in an electric field, with a view to modify the properties of the surface layer of this solid.
  D Ionenimplantation
  F implantation d'ions
  P implantacja jonów
  R ионное внедрение

-810 **nitride**
  A compound of a metal and nitrogen.
  D Nitrid
  F nitrure
  P azotek
  R нитрид

-815 **iron nitride**
  A compound of iron and nitrogen.
  D Eisennitrid
  F nitrure de fer
  P azotek żelaza
  R нитрид железа

-820 **chromium nitride**
  A compound of chromium and nitrogen.
  D Chromnitrid
  F nitrure de chrome
  P azotek chromu
  R нитрид хрома

-825 **titanium nitride**
  A compound of titanium and nitrogen.
  D Titannitrid
  F nitrure de titane
  P azotek tytanu
  R нитрид титана

-830 **vanadium nitride**
  A compound of vanadium and nitrogen.
  D Vanadinnitrid
  F nitrure de vanadium
  P azotek wanadu
  R нитрид ванадия

-835 **nitrogen austenite**
  The solid solution of nitrogen in gamma iron.
  D Stickstoffaustenit, Nitroaustenit
  F nitrausténite, austénite à l'azote
  P austenit azotowy, azotoaustenit
  R азотистый аустенит

-840 **braunite**
  The eutectoid in the iron-nitrogen system.
  D Braunit
  F braunite
  P braunit
  R браунит

-845 **nitride(d) case**
  The outer layer of a nitrided steel object into which nitrogen has been introduced by diffusion.
  D Nitrierschicht
  F couche nitrurée
  P warstwa naazotowana
  R азотированный слой, нитрированный слой

-850 **denitriding, denitrogenizing**
  A thermo-chemical treatment, opposed to nitriding, sometimes applied in order to remove the effects of an improperly effectuated nitriding.
  D Denitrieren, Entsticken
  F dénitruration
  P odazotowanie
  R деазотирование

-855 **cyaniding, cyanide (case) hardening**
  A thermo-chemical treatment consisting in saturating the surface layer of an object with carbon and nitrogen.
  D Zyanieren, Zyanhärten
  F cyanuration
  P cyjanowanie, węgloazotowanie
  R цианирование

-860 **gas cyaniding, dry cyaniding, ammonia carburizing**
  Cyaniding performed in a gaseous atmosphere.
  D Gaszyanieren
  F cyanuration gazeuse
  P cyjanowanie gazowe, węgloazotowanie gazowe
  R газовое цианирование

-865 **liquid cyaniding**
  Cyaniding performed in a bath of molten cyanide.
  D Zyanbadhärten
  F cyanuration liquide, cyanuration en bain
  P cyjanowanie kąpielowe, węgloazotowanie kąpielowe
  R жидкостное цианирование

-870 **carbonitriding**
  Cyaniding in which the absorption of carbon predominates.
  D Carbonitrieren, Karbonitrieren
  F carbonitruration
  P cyjanowanie wysokotemperaturowe, węgloazotowanie wysokotemperaturowe, azotonawęglanie, nitronawęglanie
  R азотонауглероживание, высокотемпературное цианирование, нитроцементация

-875 **nitrocarburizing, ni-carbing**
Cyaniding in which the absorption of nitrogen predominates.
D Nitrokarburieren
F nitrocarburation
P cyjanowanie niskotemperaturowe, węgloazotowanie niskotemperaturowe
R углеродоазотирование, низкотемпературное цианирование

-880 **steam treating, steam treatment, bluing in steam**
A thermo-chemical treatment aimed at producing an oxide surface layer on mainly steel objects by holding them in superheated steam at an appropriate temperature.
D Heißdampfbehandlung
F oxydation à la vapeur
P utlenianie w parze wodnej
R обработка водяным паром

-885 **oxynitriding, nitroxydizing**
A thermo-chemical treatment consisting in saturating the surface layer of an object with oxygen and nitrogen.
D Nitrooxidieren
F nitrooxydation, oxynitruration
P tlenoazotowanie, azotopasywowanie
R нитрооксидирование

-890 **sulfonitriding, sulfinuzing**
A thermo-chemical treatment consisting in saturating the surface layer of an object with sulfur and nitrogen.
D Sulfonitrieren, Sulf-Inuzieren
F sulfonitruration, sulfinuzation
P siarkoazotowanie, azotowanie z nasiarczaniem
R сульфоазотирование

-892 **sulfocarbonitriding**
A thermo-chemical treatment consisting in saturating the surface layer of an object with sulfur, carbon and nitrogen.
D Sulfokarbonitrieren, Sulfocarbonitrieren
F sulfocarbonitruration
P siarkowęgloazotowanie
R сероуглеродоазотирование, сульфоцианирование

-895 **sulfurizing, sulfiding**
A thermo-chemical treatment consisting in saturating the surface layer of an object with sulfur.
D Sulfidieren, Aufschwefeln
F sulfuration, cémentation par le soufre
P siarkowanie dyfuzyjne, nasiarczanie
R сульфидирование, сульфуризация

-900 **boriding, boronizing**
A thermo-chemical treatment consisting in saturating the surface layer of an object with boron.
D Borieren
F boruration, cémentation par le bore
P borowanie dyfuzyjne, naborowywanie
R борирование

-905 **boride**
A compound of a metal with boron.
D Borid
F borure
P borek
R борид

-910 **siliconizing, siliciding, silicon impregnation**
A thermo-chemical treatment consisting in saturating the surface layer of an object with silicon.
D Silizieren, Silicieren
F siliciuration, cémentation par le silicium
P krzemowanie dyfuzyjne, nakrzemowywanie
R (диффузионное) силицирование

-915 **metallic cementation, metal impregnation, diffusion metallization**
A thermo-chemical treatment, applied to some metal objects, aimed at saturating the surface layer of the object with another metal in order to confer special properties on it.
D Diffusionsmetallisieren, Thermodiffusions-Metallüberziehen
F cémentation métallique
P metalizowanie dyfuzyjne
R (термо)диффузионная металлизация, диффузионное насыщение металлами

-920 **chromizing, chromium impregnation**
A thermo-chemical treatment consisting in saturating the surface layer of an object with chromium.
D Chromieren, Inkromieren, Inchromieren, Diffusionsverchromen
F chromage thermique, cémentation par le chrome, chromisation
P chromowanie dyfuzyjne, nachromowywanie
R диффузионное хромирование

-925 **calorizing, aluminizing, aluminium impregnation**
A thermo-chemical treatment consisting in saturating the surface layer of an object with aluminium.
D Aluminieren, Alitieren, Kalorisieren, Veraluminieren
F alitérage, calorisation, cémentation par l'aluminium
P aluminiowanie dyfuzyjne, glinowanie
R алитирование, калоризация, диффузионное алюминирование

-930 **sherardizing, zinc impregnation**
A thermo-chemical treatment consisting in saturating the surface layer of an object with zinc.
D Sherardisieren, Aufzinken
F shérardisation, cémentation par le zinc
P cynkowanie dyfuzyjne, szerardyzacja
R шерардизация, диффузионное цинкование

-935 **tungsten impregnation**
A thermo-chemical treatment consisting in saturating the surface layer of an object with tungsten.
D Wolframieren
F cémentation par le tungstène
P wolframowanie dyfuzyjne
R (диффузионное) вольфрамирование

-940 **chromoaluminizing**
A thermo-chemical treatment consisting in saturating the surface layer of an object with chromium and aluminium.
D Chromoalitieren
F chromoaluminisation
P chromoaluminiowanie
R хромоалитирование, хромоалюминирование

-945 **titanizing**
A thermo-chemical treatment consisting in saturating the surface layer of an object with titanium.
D Titanieren
F titanisation, cémentation par le titanium
P tytanowanie (dyfuzyjne)
R (диффузионное) титанирование

-947 **internal oxidation**
Formation of dispersed oxide particles within an alloy by oxydizing its less noble component with oxygen introduced by diffusion.
D innere Oxydation
F oxydation interne
P utlenianie wewnętrzne
R внутреннее окисление

-950 **thermo-mechanical treatment**
A combination of a thermal treatment with a plastic deformation of an alloy, aimed mainly at improving its mechanical properties.
D thermo(-)mechanische Behandlung
F traitement thermomécanique, TTM
P obróbka cieplno-plastyczna, obróbka cieplno-mechaniczna
R термомеханическая обработка

-955 **high-temperature thermo-mechanical treatment**
A thermo-mechanical treatment of steel during which the plastic deformation is effectuated at the austenitizing temperature, this deformation being followed immediately by quenching.
D thermomechanische Behandlung bei hohen Temperaturen
F traitement thermomécanique à haute température, TTMHT
P obróbka cieplno-plastyczna wysokotemperaturowa, wysokotemperaturowa obróbka cieplno-mechaniczna
R высокотемпературная термомеханическая обработка, ВТМО

-960 **ausforming**
   A thermo-mechanical treatment of steel consisting in undercooling austenite to a temperature of 400–500°C, deforming it at this temperature and subsequently quenching in order to obtain martensite.
   D Austenitformhärten
   F ausforming, austéniformage
   P obróbka cieplno-plastyczna niskotemperaturowa, niskotemperaturowa obróbka cieplno-mechaniczna
   R низкотемпературная термомеханическая обработка, НТМО, аусформинг

-965 **thermo-magnetic treatment, magnetic anneal(ing)**
   A combination of a thermal treatment with the action of a magnetic field, aimed mainly at changing some physical properties of an alloy.
   D thermisch-magnetische Behandlung, magnetothermische Behandlung
   F traitement thermomagnétique
   P obróbka cieplno-magnetyczna
   R термомагнитная обработка, ТМО, отжиг в магнитном поле

-970 **thermal cycle, heat treatment cycle**
   The change of temperature with time during a heat treatment process.
   D Temperatur-Zeit-Folge
   F cycle thermique
   P cykl obróbki cieplnej
   R режим термической обработки, режим термообработки

75

PROPERTIES OF METALS AND ALLOYS
EIGENSCHAFTEN DER METALLE UND LEGIERUNGEN
PROPRIÉTÉS DES MÉTAUX ET ALLIAGES
WŁASNOŚCI METALI I STOPÓW
СВОЙСТВА МЕТАЛЛОВ И СПЛАВОВ

-005 **mechanical properties**
Those properties of a material which determine its ability to carry loads and to undergo an elastic or plastic deformation.
D mechanische Eigenschaften
F propriétés mécaniques
P własności mechaniczne
R механические свойства

-010 **strength properties**
That group of mechanical properties which represent the ability of a material to sustain various static (short-time and permanent), dynamic, and periodically changing loads.
D Festigkeitseigenschaften
F propriétés de resistance
P własności wytrzymałościowe
R прочностные свойства

-015 **tensile properties**
Mechanical properties which are quantitatively determined by means of a tensile test.
D mechanische Eigenschaften (aus der Zugprobe)
F propriétés mécaniques en traction
P własności mechaniczne (określane z próby rozciągania)
R механические свойства (определённые испытанием на растяжение)

-020 **elasticity**
The ability of a material to regain its original shape and dimensions after removal of the stress.
D Elastizität
F élasticité
P sprężystość
R упругость

-025 **plasticity, malleability**
The capacity of a material for extensive permanent deformation under compressive stresses without fracture.
D Plastizität, plastische Verformbarkeit, Bildsamkeit
F plasticité, malléabilité
P plastyczność
R пластичность

-030 **ductility**
The capacity of a material for plastic deformation in tension.
D Duktilität, Dehnbarkeit, Dehnungsvermögen
F ductilité
P ciągliwość
R дуктильность, тягучесть

-035 **toughness**
The ability of a material to absorb energy by plastic deformation, i.e. a combination of strength and ductility, its practical measure being the area under the stress-strain curve.
D Zäh(fest)igkeit
F tenacité
P wiązkość
R вязкость

-040 **(mechanical) strength, ultimate strength**
The ability of a material to sustain loads, usually expressed as the nominal maximum stress, i.e. as the maximum load which can be sustained without fracture divided by the original cross-sectional area of the specimen.
D Festigkeit
F résistance mécanique
P wytrzymałość
R (механическая) прочность

-045 **superstrength, ultra-high strength**
An extremely high mechanical strength.
D Ultrafestigkeit
F super-résistance
P nadwytrzymałość
R сверхпрочность, сверхвысокая прочность

-050 **anelasticity**
Time-dependent non-linearity between stress and elastic strain.
D Anelastizität
F anélasticité
P niesprężystość, asprężystość, sprężystość opóźniona
R неупругость

-055 **superplasticity**
   The ability of metals and alloys to undergo unusually extensive strain-rate-sensitive deformation under small stresses without risk of fracture.
   D Superplastizität
   F superplasticité
   P nadplastyczność
   R сверхпластичность

-060 **strength characteristic, strength parameter**
   Any one of the characteristic quantities, determined experimentally, which represent the measure of the strength properties of a material.
   D Festigkeitskenngröße, Festigkeitskennzahl, Festigkeitskennwert
   F caractéristique de résistance (mécanique), caractéristique mécanique
   P wskaźnik wytrzymałości(owy)
   R показатель прочности

-065 **plasticity characteristic, plasticity parameter**
   Any one of the characteristic quantities, determined experimentally, which represent the measure of the plasticity of a material.
   D plastische Kenngröße
   F caractéristique de ductilité
   P wskaźnik plastyczności
   R показатель пластичности

-070 **tenacity, (ultimate) tensile strength, ultimate tensile stress**
   The greatest value of tensile stress applied to a test bar during the tensile test, referred to its initial cross-sectional area.
   D Zugfestigkeit
   F résistance à la traction
   P wytrzymałość na rozciąganie
   R прочность на растяжение, сопротивление растяжению, предел прочности (при растяжении), прочность при растяжении

-075 **(ultimate) compressive strength**
   The greatest value of the compressive stress applied to a test piece during the compression test, referred to its initial cross-sectional area.
   D Druckfestigkeit
   F résistance à la compression
   P wytrzymałość na ściskanie
   R прочность на сжатие, сопротивление сжатию

-080 **shear strength**
   The greatest value of the shearing stress in a test piece subjected to shear load until it fractures.
   D Scherfestigkeit, Schubfestigkeit
   F résistance au cisaillement
   P wytrzymałość na ścinanie
   R прочность на срез, сопротивление срезу, сопротивление сдвигу

-085 **torsional strength**
   The upper limit of shearing stress at which fracture or destructive plastic deformation of the test piece subjected to torsional loading occurs.
   D Drehfestigkeit, Torsionsfestigkeit, Verdrehungsfestigkeit
   F résistance à la torsion
   P wytrzymałość na skręcanie
   R прочность на кручение, прочность на скручивание, сопротивление кручению

-090 **bending strength**
   The upper limit of normal stress at which fracture or excessive plastic deformation of a beam in bending occurs.
   D Biegefestigkeit, Biegungsfestigkeit
   F résistance à la flexion
   P wytrzymałość na zginanie
   R прочность на изгиб, сопротивление изгибу

-095 **hot strength, high-temperature strength**
   The strength exhibited by a material at an elevated temperature.
   D Hochtemperaturfestigkeit
   F résistance mécanique à chaud, résistance mécanique aux températures élevées
   P wytrzymałość w podwyższonej temperaturze
   R высокотемпературная прочность, прочность при повышенных температурах

-100 **proportional(ity) limit, limit of proportionality**
   The greatest value of stress up to which there exists linearity between stress and strain.
   D Proportion(alität)sgrenze
   F limite de proportionalité
   P granica proporcjonalności
   R предел пропорциональности

-105 **elastic limit**
   The maximum stress up to which the strain can be considered as perfectly elastic.
   D Elastizitätsgrenze
   F limite d'élasticité vraie
   P granica sprężystości
   R предел упругости

-110 **(sharp) yield point, yield stress**
   The stress at which there is a rapid increase in plastic strain without an increase in stress.
   D Fließgrenze, Dehngrenze, (ausgeprägte) Streckgrenze, natürliche Streckgrenze, exakte Streckgrenze, physikalische Streckgrenze
   F limite élastique, limite apparente d'élasticité, limite d'écoulement, limite d'étirement
   P (wyraźna) granica plastyczności, naturalna granica plastyczności, fizyczna granica plastyczności
   R предел пластичности, (физический) предел текучести

-115 **yield strength, offset yield stress, proof stress**
   The stress corresponding to a small specified (e.g. 0.2 per cent) plastic strain.
   D 0,2-Dehngrenze
   F limite élastique conventionelle, limite conventionelle d'élasticité
   P umowna granica plastyczności
   R условный предел текучести

-120 **elongation at fracture, (percentage) elongation**
   The ratio of the increase in the gauge length of the specimen in the tensile test to its original gauge length, expressed as a percentage.
   D Bruchdehnung
   F allongement après rupture, allongement à la rupture, allongement (plastique) de rupture
   P wydłużenie plastyczne po rozerwaniu, wydłużenie (całkowite), wydłużenie względne
   R (остаточное) относительное удлинение, полное относительное удлинение

-125 **reduction of area (at fracture), percentage reduction of area, contraction of area**
   The ratio of the reduction in cross-sectional area at the place of fracture in the tensile test to the initial cross-sectional area of the specimen, expressed as a percentage.
   D Brucheinschnürung, Einschnürung, Bruchquerschnittsverminderung
   F striction (après rupture), striction à la rupture, coefficient de striction
   P przewężenie (po rozerwaniu)
   R (остаточное) относительное сужение

-130 **fracture stress**
   The true stress, i.e. the stress related to the actual cross-sectional area, at which fracture takes place in the tensile test.
   D Reißfestigkeit, wahre Zerreißfestigkeit
   F contrainte de rupture
   P naprężenie rozrywające
   R действительное сопротивление разрыву, разрушающее напряжение, истинный предел прочности

-135 **cohesive strength, disruptive strength**
   The stress which causes fracture of the test piece without any plastic deformation.
   D Kohäsionsfestigkeit, Trennungswiderstand
   F résistance au clivage
   P wytrzymałość rozdzielcza
   R сопротивление отрыву

-140 **specific strength, specific tenacity**
   The tensile strength of a material divided by its specific gravity.
   D relative Festigkeit
   F résistance relative
   P wytrzymałość względna, wykładnik konstrukcyjny
   R удельная прочность, относительная прочность

-145 **elastic constant, elastic coefficient, elastic compliance**
A value, characteristic for a given material, relating the stress to the elastic strain produced by this stress.
D Elastizitätskonstante, elastischer Koeffizient, elastische Konstante
F constante élastique, constante d'élasticité, coefficient élastique
P stała sprężystości, współczynnik sprężystości
R упругая постоянная, постоянная упругости, упругая константа, константа упругости

-150 **longitudinal modulus of elasticity, Young's modulus, elastic modulus**
The ratio of stress, below the elastic limit of the material in tension, to the corresponding strain.
D Elastizitätsmodul, Elastizitätsmaß, Dehnungsmodul, E-Modul, Youngscher Modul
F module d'élasticité longitudinale, module d'Young, module élastique
P moduł sprężystości liniowej, moduł sprężystości wzdłużnej, moduł sprężystości podłużnej, moduł Younga
R модуль продольной упругости, модуль Юнга

-155 **shear modulus (of elasticity), rigidity modulus, torsion modulus, modulus of rigidity**
The ratio of shearing stress, below the elastic limit of the material in shear, to the corresponding shear strain.
D Schubmodul, Gleitmodul, Torsionsmodul, Gleitmaß, Schermodul
F module de cisaillement, module de glissement, module de torsion, module d'élasticité transversale, module de cission élastique
P moduł sprężystości postaciowej, moduł sprężystości poprzecznej, moduł Kirchhoffa
R модуль поперечной упругости, модуль сдвига

-160 **bulk modulus (of elasticity), bulk elastic modulus, compression modulus of elasticity, Voight's modulus, volumetric modulus of elasticity**
The ratio of stress, below the elastic limit of the material in triaxial compression, to the corresponding reduction in volume.
D Kompressionsmodul, Raummodul, Druckmodul
F module de compression, module de volume, module de rigidité
P moduł sprężystości objętościowej, moduł ściśliwości
R модуль сжимаемости, модуль объёмного сжатия, модуль объёмной упругости, модуль всестороннего сжатия

-165 **coefficient of compressibility**
The reciprocal of the bulk modulus of elasticity.
D Kompressibilitätskoeffizient
F coefficient de compressibilité
P współczynnik ściśliwości
R коэффициент сжимаемости

-170 **Poisson's ratio**
The ratio of the transverse contraction of a specimen pulled in tension to its longitudinal elongation.
D Poisson-Konstante, Poisson-Zahl, Poissonsche Zahl
F coefficient de contraction transversale, coefficient de Poisson
P współczynnik Poissona, liczba Poissona
R коэффициент Пуассона

-175 **Bauschinger effect**
A reduction of the yield stress when a deformation in one direction is followed by a deformation in the opposite direction.
D Bauschinger-Effekt
F effet de Bauschinger
P zjawisko Bauschingera
R эффект Баушингера

-180 **elastic after-effect, delayed elasticity**
A phenomenon consisting in that after the load has begun (or ceased) to act, the elastic strain occurs (or disappears) at a rate which diminishes asymptotically to zero.
D elastische Nachwirkung, Nachwirkungseffekt
F retardation élastique
P opóźnienie sprężyste
R упругое последействие

-185 **dynamic strength, shock resistance**
　　The ability of a material to resist fracture under a sudden application of a load.
　　D Schlagfestigkeit, dynamische Festigkeit
　　F résistance aux chocs
　　P wytrzymałość dynamiczna, wytrzymałość na obciążenia dynamiczne
　　R динамическая прочность, ударная прочность, сопротивление удару

-190 **notch (impact) toughness, impact strength**
　　The ratio of the energy absorbed in fracturing a notched bar-specimen in the impact test, to the cross-sectional area of this specimen.
　　D Kerbschlagzähigkeit, Kerbschlagfestigkeit
　　F résilience
　　P udarność
　　R ударная вязкость

-195 **notch sensitivity**
　　The dependence of the nominal strength on the presence of local stress concentrations caused by notches.
　　D Kerbempfindlichkeit
　　F sensibilité à l'entaille
　　P wrażliwość na działanie karbu
　　R чувствительность к надрезу

-200 **fatigue limit, fatigue strength, cyclic strengh, endurance limit, endurance strength**
　　The greatest value of a periodically changing stress which a test piece of a given material will stand for an infinite or a sufficiently high, conventionally adopted, number of load cycles.
　　D Ermüdungsfestigkeit, Ermüdungsgrenze, Dauer(schwing)festigkeit, Schwingungsfestigkeit
　　F limite de fatigue, limite d'endurance, résistance à la fatigue
　　P wytrzymałość zmęczeniowa, granica zmęczenia, wytrzymałość na zmęczenie
　　R циклическая прочность, предел выносливости, предел усталости, выносливость, сопротивление усталости, усталостная прочность

-205 **damping capacity**
　　The ability of a metal to absorb vibratory energy by transforming it into heat.
　　D Dämpfungsvermögen, Dämpfungsfähigkeit
　　F capacité d'amortissement
　　P zdolność tłumienia drgań
　　R амортизационная способность

-210 **internal friction**
　　A mechanism by means of which the damping capacity manifests itself.
　　D innere Reibung
　　F frottement interne, frottement intérieur
　　P tarcie wewnętrzne
　　R внутреннее трение

-215 **creep resistance**
　　The ability of a metal or alloy not to creep or not to creep beyond a permissible limit.
　　D Kriechwiderstand
　　F tenue au fluage, résistance au fluage
　　P odporność na pełzanie
　　R крипоустойчивость, сопротивление ползучести

-220 **creep strength, creep limit**
　　The maximum stress which can be applied to a material at a given temperature, resulting — after a determined time — in a fixed amount of strain or in no measurable deformation.
　　D Kriechfestigkeit, Kriechgrenze
　　F résistance au fluage, limite de fluage
　　P wytrzymałość na pełzanie, granica pełzania
　　R предел ползучести, ползучепрочность

-225 **true limiting creep stress**
　　The maximum stress for which creep at a given temperature fades away with time.
　　D (wahre) Dauerstandfestigkeit, wahre Kriechgrenze, Dauerstandkriechgrenze
　　F résistance durable, limite de fluage vraie
　　P wytrzymałość trwała, rzeczywista granica pełzania
　　R истинный предел длительной прочности, абсолютный предел ползучести, истинный предел ползучести

-230 **limiting creep stress**
Stress which at a given temperature produces in a given time a conventionally fixed permanent strain.
D Zeitdehngrenze
F limite de fluage conventionelle
P umowna granica pełzania
R условный предел длительной прочности, условный предел ползучести, предел ограниченной длительной прочности

-235 **creep rupture strength, stress rupture strength**
Stress which at a given temperature causes fracture of the specimen after a conventionally fixed time.
D Zeitstandfestigkeit, Zeitstandkriechgrenze, Zeitbruchgrenze
F limite de fluage conventionelle
P wytrzymałość czasowa, wytrzymałość długotrwała, czasowa granica pełzania
R длительная прочность

-240 **high-temperature creep resistance, high-temperature (creep) strength**
Creep resistance at elevated temperatures.
D Warmfestigkeit
F résistance au fluage aux températures élevées
P żarowytrzymałość; cieplowytrzymałość
R жаропрочность; теплопрочность

-245 **hardness**
The property of solid bodies whereby resistance is offered to plastic deformation when two bodies in contact over a small area are pressed together.
D Härte
F dureté
P twardość
R твёрдость

-250 **micro-hardness**
Hardness of microscopic areas within an alloy, e.g. of individual phases.
D Mikrohärte
F microdureté
P mikrotwardość
R микротвёрдость

-255 **hot hardness**
Hardness of a material as measured at elevated temperatures.
D Warmhärte
F dureté à chaud
P twardość na gorąco
R горячая твёрдость

-260 **red hardness**
The property of quench-hardened high-speed steels consisting in retaining hardness although the temperature is raised to the red heat.
D Rotglühhärte, Rotwarmhärte, Rotglühhärte
F dureté au rouge
P twardość w temperaturze czerwonego żaru
R красностойкость, твёрдость при красном калении

-265 **wear resistance, abrasion resistance**
The ability of a material to withstand wear or abrasion.
D Verschleißfestigkeit, Abriebfestigkeit, Abnutz(ungs)festigkeit
F résistance à l'usure, résistance à l'abrasion
P odporność na ścieranie, odporność na zużycie przez tarcie
R износостойкость, износоустойчивость, сопротивление износу, сопротивление истиранию

-270 **brittleness, shortness**
The tendency of a material to break without appreciable deformation.
D Sprödigkeit, Brüchigkeit
F fragilité
P kruchość
R хрупкость

-275 **embrittlement**
The appearance of brittleness or an increase in brittleness.
D Versprödung
F fragilisation
P pojawienie się kruchości; wzrost kruchości
R охрупчивание, появление хрупкости; повышение хрупкости

-280 **(brittle fracture) transition temperature, nil-ductility temperature, ductility transition temperature**
The temperature below which a given material becomes brittle.
D Sprödbruchtemperatur, Versprödungstemperatur, Übergangstemperatur
F température de transition (ductile-fragile), température de transition de ductilité
P temperatura przejścia w stan kruchości, temperatura pojawienia się kruchości, graniczna temperatura kruchości na zimno, próg kruchości
R температура охрупчивания, критическая температура хрупкости, температура перехода (в хрупкое состяние), температура вязко-хрупкого перехода

-285 **cold shortness**
Brittleness at normal or low temperatures.
D Kaltsprödigkeit, Kaltbrüchigkeit
F fragilité à froid
P kruchość na zimno
R хладноломкость

-290 **hot shortness, hot brittleness, red shortness, red brittleness**
Brittleness at elevated temperatures.
D Warmsprödigkeit, Warmbrüchigkeit, Rotbrüchigkeit, Warmversprödung
F fragilité à chaud, fragilité au rouge
P kruchość na gorąco
R тепловая хрупкость, горячеломкость, теплоломкость, красноломкость

-295 **low-temperature brittleness**
Brittleness at low or sub-normal temperatures.
D Tieftemperatursprödigkeit
F fragilité à basse température
P kruchość niskotemperaturowa
R низкотемпературная хрупкость

-300 **blue brittleness, blue shortness**
Reduced ductility of some ferrous alloys occurring in the temperature range 200–300°C, i.e. in the range in which a blue oxide film forms on the surface of the alloy.
D Blausprödigkeit, Blaubrüchigkeit
F fragilité au bleu
P kruchość na niebiesko
R синеломкость

-305 **caustic embrittlement, boiler embrittlement**
The tendency of steel to intergranular cracking when immersed in a caustic alkali solution.
D Laugensprödigkeit, Laugenrissigkeit
F fragilité caustique
P kruchość ługowa, kruchość alkaliczna
R щелочная хрупкость, каустическая хрупкость

-310 **rheotropic brittleness**
Low-temperature brittleness which can be prevented by straining the metal beyond the elastic limit.
D rheotrope Sprödigkeit
F fragilité rhéotropique
P kruchość reotropowa
R реотропная хрупкость

-315 **physical properties**
Those properties of a material which characterize it from the point of view of physics.
D physikalische Eigenschaften
F propriétés physiques
P własności fizyczne
R физические свойства

-320 **atomic volume**
The volume per atom in a crystal lattice.
D Atomvolumen
F volume atomique
P objętość atomowa
R атомный объём

-325 **molar volume**
The volume of one mole of a given substance.
D Molvolumen
F volume molaire
P objętość molowa
R мольный объём

-330 **atomic number**
The number of protons in the atomic nucleus of the given element, this number indicating the position of this element in the periodic table.
D Atomnummer, Ordnungszahl
F nombre atomique
P liczba atomowa, liczba porządkowa
R атомное число, атомный номер

-335 **mass number**
   The mass of an atom of an element expressed by means of the nearest integer in relation to the mass of an oxygen atom which is assumed to be equal to 16.
   D Massenzahl
   F nombre de masse
   P liczba masowa
   R массовое число

-340 **capacitive properties, volumetric properties**
   Those properties of a material which are determined by the number of atoms per unit volume (e.g. density).
   D volumetrische Eigenschaften
   F propriétés volumetriques
   P własności objętościowe
   R объёмные свойства

-345 **specific density**
   A mass of any volume of a substance divided by the volume.
   D Dichte
   F densité, masse spécifique
   P gęstość, masa właściwa
   R плотность, удельная масса

-350 **specific gravity, weight density**
   The ratio of the weight of any volume of a substance to the volume.
   D spezifisches Gewicht, Wichte
   F poids spécifique
   P ciężar właściwy
   R удельный вес

-355 **molecular mass, molecular weight**
   The weight of a molecule of a substance referred to the weight of a molecule of hydrogen (taken as 2) or oxygen (taken as 32).
   D Mol(ekular)gewicht
   F masse moléculaire, poids moléculaire
   P masa cząsteczkowa, ciężar cząsteczkowy, ciężar molowy
   R молекулярная масса, молекулярный вес

-360 **metallic ring**
   The ring emitted by a metal when the latter is struck.
   D metallischer Klang
   F son métallique
   P dźwięk metaliczny
   R металлический звук

-365 **thermal properties**
   Those physical properties of a material which are associated with its behaviour when the temperature changes or when it absorbs or evolves heat.
   D thermische Eigenschaften
   F propriétés thermiques
   P własności cieplne
   R тепловые свойства, термические свойства

-370 **atomic heat**
   The quantity of heat equal to the product of the specific heat of an element at constant volume and its atomic weight.
   D Atomwärme
   F chaleur atomique
   P ciepło atomowe
   R атомная теплоёмкость

-375 **molecular heat**
   The quantity of heat equal to the product of the specific heat of a substance at constant volume and its molecular mass.
   D Molwärme
   F chaleur moléculaire
   P ciepło cząsteczkowe, ciepło molowe
   R молекулярная теплоёмкость

-380 **(latent) heat of fusion, (latent) heat of melting**
   The quantity of heat necessary for converting unit mass of a substance, at its melting point, from solid into liquid at the same temperature.
   D (latente) Schmelzwärme, (gebundene) Schmelzwärme
   F chaleur (latente) de fusion
   P (utajone) ciepło topnienia
   R (скрытая) теплота плавления

-385 **heat of solidification**
   The quantity of heat which is evolved during converting unit mass of a substance, at its freezing point, from liquid into solid at the same temperature.
   D Erstarrungswärme
   F chaleur de solidification
   P ciepło krzepnięcia
   R теплота затвердевания

-390 **(latent) heat of vaporization**
  The quantity of heat necessary for converting unit mass of a substance, at its boiling point, from liquid into vapour at the same temperature.
  D (latente) Verdampfungswärme, (gebundene) Verdampfungswärme
  F chaleur (latente) de vaporisation
  P (utajone) ciepło parowania
  R (скрытая) теплота испарения, (скрытая) теплота парообразования

-395 **heat of sublimation**
  The quantity of heat necessary for converting unit mass of a substance from solid directly into vapour at constant temperature and pressure.
  D Sublimationswärme
  F chaleur de sublimation
  P ciepło sublimacji
  R теплота сублимации

-400 **heat of transformation, heat of transition, critical heat**
  The quantity of heat evolved or absorbed during transformation in solid state, related to the unit mass of the substance involved.
  D Umwandlungswärme
  F chaleur de transformation, chaleur critique
  P ciepło przemiany
  R теплота превращения

-405 **heat of solution**
  The quantity of heat absorbed or evolved during dissolution of unit mass of a substance in a large volume of a solvent.
  D Lösungswärme
  F chaleur de dissolution
  P ciepło rozpuszczania
  R теплота растворения

-410 **heat capacity, thermal capacity**
  The amount of heat absorbed by a substance which undergoes a temperature change equal to one degree.
  D Wärmekapazität
  F capacité calorifique
  P pojemność cieplna
  R теплоёмкость

-415 **specific heat**
  The quantity of heat necessary for raising the temperature of unit mass of a substance through 1°C.
  D spezifische Wärme
  F chaleur spécifique
  P ciepło właściwe
  R удельная теплоёмкость, удельная теплота

-420 **melting point, fusion point, fusing point**
  The temperature at which a solid substance passes into the liquid state.
  D Schmelzpunkt, Schmelztemperatur
  F température de fusion, point de fusion
  P temperatura topnienia
  R температура плавления, точка плавления

-425 **freezing point, freezing temperature, solidification temperature, solidifying point**
  The temperature at which a liquid substance passes into the solid state.
  D Erstarrungspunkt, Erstarrungstemperatur, Gefrierpunkt
  F température de solidification, point de solidification
  P temperatura krzepnięcia
  R температура затвердевания, точка затвердевания

-430 **boiling point, boiling temperature**
  The temperature at which a substance, under constant pressure, changes from the liquid state into a vapour.
  D Siedepunkt, Siedetemperatur
  F température d'ébullition, point d'ébullition
  P temperatura wrzenia
  R температура кипения, точка кипения

-435 **sublimation point, sublimation temperature**
  The temperature at which a solid passes into a vapour without the intermediate formation of a liquid.
  D Sublimationspunkt, Sublimationstemperatur
  F température de sublimation, point de sublimation
  P temperatura sublimacji
  R температура сублимации

-440 **thermal conductivity, heat conductivity, coefficient of thermal conduction**
   The quantity of heat which flows in unit time across unit area of a block of the given material of unit thickness when there exists unit temperature gradient between the faces of the block.
   D (spezifische) Wärmeleitfähigkeit, thermische Leitfähigkeit, Wärmeleitvermögen, Wärmeleitzahl
   F conductibilité thermique, conductivité thermique, coefficient de conductibilité thermique
   P przewodność cieplna, przewodnictwo cieplne właściwe, współczynnik przewodzenia ciepła, współczynnik przewodnictwa cieplnego
   R теплопроводность, коэффициент теплопроводности

-445 **thermal diffusivity**
   A quantity determining the rate of propagation of temperature within a given material, its value being equal to the thermal conductivity divided by the product of specific heat and density.
   D Temperaturleitfähigkeit, Temperaturleitzahl, Temperaturleitungsvermögen
   F diffusivité thermique
   P przewodność temperaturowa, dyfuzyjność cieplna, współczynnik przewodzenia temperatury
   R температуропроводность, коэффициент теплопроводности

-450 **thermal expansion coefficient, coefficient of thermal expansion, thermal coefficient of expansion**
   The relative change in the volume or length of a body per degree change in temperature at constant pressure.
   D Wärmeausdehnungskoeffizient, thermischer Ausdehnungskoeffizient, Wärmeausdehnungszahl
   F coefficient de dilatation thermique
   P współczynnik rozszerzalności (cieplnej)
   R коэффициент теплового расширения, коэффициент термического расширения

-455 **coefficient of linear expansion, linear coefficient of thermal expansion**
   The relative change in the length of a body per degree change in temperature at constant pressure.
   D linearer Ausdehnungskoeffizient
   F coefficient de dilatation linéaire, coefficient linéaire de dilatation thermique
   P współczynnik rozszerzalności (cieplnej) liniowej
   R коэффициент линейного расширения

-460 **coefficient of cubical expansion, volume coefficient of thermal expansion**
   The relative change in the volume of a body per degree change in temperature at constant pressure.
   D kubischer Ausdehnungskoeffizient
   F coefficient de dilatation cubique, coefficient de dilatation en volume
   P współczynnik rozszerzalności (cieplnej) objętościowej
   R коэффициент объёмного расширения, термический коэффициент объёмного расширения

-465 **electrical properties**
   Those physical properties of a material which are associated with the passage of electric current through it or with its behaviour in an electric field.
   D elektrische Eigenschaften
   F propriétés électriques
   P własności elektryczne
   R электрические свойства, электрофизические свойства

-470 **electric(al) resistivity, resistivity, specific resistance**
   The electrical resistance of a block of a given material having unit length and unit cross-sectional area.
   D spezifischer (elektrischer) Widerstand
   F résistivité (électrique), résistivité spécifique
   P oporność (elektryczna) właściwa, opór właściwy, rezystywność
   R удельное (электрическое) сопротивление, удельное электросопротивление

-475 **Matthiessen's rule**
A rule stating that the total electric resistivity of a crystalline metallic substance is the sum of the resistivity due to the thermal agitation of the metal ions of the lattice and the resistivity due to the presence of imperfections in the crystal.
D Matthiessensche Regel
F règle de Matthiessen
P reguła Matthiessena
R правило Ма(т)ти(с)сена

-480 **temperature coefficient of (electrical) resistivity, temperature coefficient of resistance**
A quantity which determines the change in the electrical resistivity of a given material resulting from a change in temperature.
D Temperaturkoeffizient des (elektrischen) Widerstandes
F coefficient de température de la résistivité électrique
P współczynnik temperaturowy oporności elektrycznej
R температурный коэффициент удельного сопротивления, температурный коэффициент электрического сопротивления, температурный коэффициент электросопротивления

-485 **electrical conductivity, specific conductance**
The conductance of a block of a given material having unit length and unit cross-sectional area.
D elektrische Leitfähigkeit
F conductibilité électrique
P przewodność (elektryczna) właściwa, konduktywność
R удельная электрическая проводность, удельная (электро)проводность

-490 **superconductivity**
The ability of some pure metals and alloys to have no measurable electrical resistance below a certain critical temperature.
D Supraleitfähigkeit, Supraleitung
F supraconduction, supraconductibilité
P nadprzewodnictwo
R сверхпроводность, сверхпроводимость

-495 **thermoelectric power, thermoelectromotive force, thermal electromotive force, thermopower**
The electromotive force produced when the junction of two dissimilar metals is heated.
D Thermokraft
F pouvoir thermoélectrique
P siła termoelektryczna
R термоэлектродвижущая сила

-500 **optical properties**
Those physical properties of a material which are associated with its behaviour in relation to the light incident upon it.
D optische Eigenschaften
F propriétés optiques
P własności optyczne
R оптические свойства

-505 **opacity**
The property of some substances not to transmit rays of a specified wavelength.
D Undurchsichtigkeit
F opacité
P nieprzeźroczystość
R непрозрачность

-510 **metallic lustre**
The brilliant appearance of fresh (i.e. not oxidized) metal surfaces resulting from their high ability to reflect light.
D Metallglanz, metallischer Glanz
F éclat métallique
P połysk metaliczny
R металлический блеск

-515 **(optical) reflectivity, coefficient of reflection**
The ratio of the light flux reflected from the surface of a material to the light flux incident upon it.
D Reflexionskoeffizient, Reflexionszahl, Reflexionsfaktor, Reflexionsgrad
F coefficient de réflexion, facteur de réflexion
P współczynnik odbicia
R коэффициент отражения

-520 **nuclear properties**
Those properties which characterize a material from the point of view of nuclear physics or nuclear engineering.
D kernphysikalische Eigenschaften, kerntechnische Eigenschaften
F propriétés nucléaires
P własności jądrowe
R ядерные свойства

-525 **half life, half-life time**
   The period of time in which the radioactivity of a substance decreases to half its original value.
   D Halbwertszeit
   F période de radioactivité, demi-vie radioactive
   P półokres rozpadu
   R период полураспада

-530 **neutron capture cross-section, neutron absorption cross-section**
   A measure of the ability of the atomic nucleus of a given element to capture neutrons without fissioning.
   D Neutronen-Einfangsquerschnitt, Absorptionsquerschnitt für Neutrone
   F section efficace de capture de neutrons, section efficace d'absorption de neutrons
   P przekrój czynny na wychwyt neutronów
   R эффективное сечение захвата нейтронов, сечение поглощения нейтронов

-535 **moderating power, slowing-down power**
   A measure of the ability of a given material to reduce the kinetic energy of an elementary particle passing through this material.
   D Bremsvermögen
   F pouvoir ralentisseur, pouvoir de ralentissement
   P zdolność spowalniania, zdolność hamowania
   R замедляющая способность

-540 **magnetic properties**
   Those physical properties of a material which are associated with its behaviour in a magnetic field.
   D magnetische Eigenschaften
   F propriétés magnétiques
   P własności magnetyczne
   R магнитные свойства

-545 **ferromagnetism**
   A phenomenon inherent to some materials having a very high permeability, consisting in a tendency to spontaneous magnetization caused by a parallel alignment of the magnetic moments of atoms.
   D Ferromagnetismus
   F ferromagnétisme
   P ferromagnetyzm
   R ферромагнетизм, ферромагнитность

-550 **ferromagnetic substance, ferromagnetic material, ferromagnetic**
   A material having permeability considerably greater than that of a vacuum.
   D ferromagnetischer Stoff, Ferromagnetikum
   F (substance) ferromagnétique
   P ferromagnetyk
   R ферромагнетик, ферромагнитное вещество

-555 **paramagnetic substance, paramagnetic material, paramagnetic**
   A material having permeability slightly greater than that of a vacuum.
   D paramagnetischer Stoff, Paramagnetikum
   F (substance) paramagnétique
   P paramagnetyk
   R парамагнетик, парамагнитное вещество

-560 **diamagnetic substance, diamagnetic material, diamagnetic**
   A material having permeability less than that of a vacuum, i.e. less than unity.
   D diamagnetischer Stoff, Diamagnetikum
   F (substance) diamagnétique
   P diamagnetyk
   R диамагнетик, диамагнитное вещество

-565 **antiferromagnetism**
   A phenomenon inherent to some materials having a low positive magnetic susceptibility, consisting in an anti-parallel alignment of magnetic moments of neighbouring atoms.
   D Antiferromagnetismus
   F antiferromagnétisme
   P antyferromagnetyzm
   R антиферромагнетизм

-570 **superparamagnetism**
   A phenomenon inherent to an assembly of appropriately fine ferromagnetic particles with fluctuating magnetization, consisting in that this assembly behaves like a paramagnetic substance with large paramagnetic susceptibility.
   D Superparamagnetismus
   F superparamagnétisme
   P superparamagnetyzm
   R суперпарамагнетизм

-575 **magnetic anisotropy**
> Anisotropy of magnetic properties of a material which manifests itself mainly in the existence of an easy direction of magnetization.
> **D** magnetische Anisotropie
> **F** anisotropie magnétique
> **P** anizotropia magnetyczna
> **R** магнитная анизотропия

-580 **ferrimagnetism**
> A variant of antiferromagnetism which is characterized by the unequal magnitude of anti-parallel magnetizations and, hence, by the existence of a resultant magnetization.
> **D** Ferrimagnetismus
> **F** ferrimagnétisme
> **P** ferrimagnetyzm
> **R** ферримагнетизм

-585 **preferred magnetic axis, easy direction of magnetization, axis of easy magnetization**
> A direction in a single crystal for which the energy required for magnetizing the crystal has the minimum value.
> **D** Richtung leicht(est)er Magnetisierbarkeit, magnetische Vorzugsrichtung, leichte Richtung
> **F** direction d'aimantation facile, axe d'aimantation facile, direction privilegiée d'aimantation, axe privilegié d'aimantation
> **P** kierunek łatwego magnesowania, oś łatwego magnesowania
> **R** направление лёгкого намагничивания, ось лёгкого намагничивания, направление легчайшего намагничивания, ось легчайшего намагничивания

-590 **anisotropy energy**
> The excess energy, referred to unit volume, which is needed for magnetizing a single crystal when the process of magnetization is performed not along the preferred magnetic axis but in a direction in which the magnetizing becomes the most difficult.
> **D** Anisotropieenergie
> **F** énergie d'anisotropie
> **P** energia anizotropii
> **R** энергия анизотропии

-595 **anisotropy constant, anisotropy coefficient**
> Either of the two temperature-dependent parameters which characterize the ferromagnetic material from the point of view of anisotropy energy.
> **D** Anisotropiekonstante, Anisotropiekoeffizient, Anisotropiefaktor
> **F** coefficient d'anisotropie
> **P** współczynnik anizotropii, stała anizotropii
> **R** константа анизотропии, фактор анизотропии

-600 **(intensity of) magnetization**
> The difference between the magnetic induction and the magnetizing force, divided by 4π.
> **D** Magnetisierung
> **F** intensité d'aimantation
> **P** namagnesowanie, magnetyzacja, natężenie namagnesowania, intensywność namagnesowania
> **R** намагниченность, интенсивность намагничения

-605 **saturation magnetization, saturation magnetic intensity**
> The highest obtainable value of magnetization in a given material.
> **D** Sättigungsmagnetisierung
> **F** aimantation à saturation, aimantation de saturation
> **P** magnetyzacja nasycenia
> **R** намагниченность насыщения

-610 **magnetic induction, magnetic flux density**
> The magnetic flux induced in a ferromagnetic material by an external magnetic field, related to the unit sectional area of this material.
> **D** magnetische Induktion
> **F** induction magnétique, densité de flux magnétique
> **P** indukcja magnetyczna
> **R** магнитная индукция, плотность магнитного потока

-615 **(magnetic) permeability, magnetic inductive capacity**
> The ratio of the magnetic induction to the corresponding magnetizing force.
> **D** magnetische Permeabilität
> **F** perméabilité magnétique
> **P** przenikalność magnetyczna
> **R** магнитная проницаемость

-620 **initial permeability**
The permeability as measured when both the magnetizing force and the magnetic induction are minute.
D Anfangspermeabilität
F perméabilité (magnétique) initiale
P przenikalność magnetyczna początkowa
R начальная (магнитная) проницаемость

-625 **magnetic susceptibility**
The ratio of the intensity of magnetization of an isotropic material to the magnetizing force.
D magnetische Suszeptibilität
F susceptibilité magnétique
P podatność magnetyczna
R магнитная восприимчивость, коэффициент намагниченности

-630 **remanence, residual magnetism**
The magnetism remaining in a ferromagnetic substance after the magnetizing force has been removed.
D Remanenz, Restmagnetismus, remanente Magnetisierung
F rémanence, aimantation rémanente, intensité d'aimantation rémanente, induction rémanente
P pozostałość magnetyczna, magnetyzm szczątkowy
R реманенц, остаточная намагниченность, остаточная (магнитная) индукция

-635 **coercivity, coercive force**
The magnetic field that must be applied to annul the remanence of a ferromagnetic material.
D Koerzitivkraft, Koerzitivfeldstärke
F force coercitive, champ coercitif
P natężenie powściągające, koercja, natężenie koercyjne
R коэрцитивная сила, магнитозадерживающая сила

-640 **magnetic hysteresis**
A lagging of the changes in the magnetism of a ferromagnetic material placed in a magnetic field behind the changes in the strength of this field.
D magnetische Hysterese
F hystérésis magnétique
P histereza magnetyczna
R магнитный гистерезис

-645 **magnetic after(-)effect**
A phenomenon occurring during the magnetization of ferromagnetic materials, consisting in a lagging of magnetization changes in the material behind the changes of the magnetic field.
D magnetische Nachwirkung
F traînage magnétique
P opóźnienie magnetyczne
R магнитное последействие

-650 **magnetostriction, magnetoelasticity**
A reversible phenomenon, inherent to ferromagnetic materials, consisting in an elastic changing of dimensions with magnetization.
D Magnetostriktion
F magnétostriction, striction magnétique
P magnetostrykcja
R магнетострикция

-655 **chemical properties**
Those properties of a material which characterize its behaviour against other chemical species.
D chemische Eigenschaften
F propriétés chimiques
P własności chemiczne
R химические свойства

-660 **(chemical) affinity**
The tendency of two substances to react chemically.
D (chemische) Affinität
F affinité (chimique)
P powinowactwo (chemiczne)
R (химическое) сродство

-665 **chemical resistance**
The resistance of a material to changes under the influence of chemical agents.
D chemische Beständigkeit
F résistance chimique
P odporność chemiczna
R химическая стойкость, химическая устойчивость

-670 **corrosion resistance**
The resistance of a metal or alloy to changes or destruction due to the direct action of chemical agents or to electrochemical reaction with the environment.
D Korrosionsbeständigkeit, Korrosionswiderstand
F résistance à la corrosion, résistance aux corrosifs, tenue à la corrosion
P odporność na korozję, odporność korozyjna
R коррозионная стойкость, сопротивление коррозии, антикоррозионная устойчивость

-675 **corrodibility**
   The susceptibility of a metal or alloy to a corrosion attack.
   D Korrosionsanfälligkeit, Korrodierbarkeit, Korrosionsempfindlichkeit
   F corrodabilité, susceptibilité de corrosion, susceptibilité à la corrosion
   P skłonność do korozji, podatność na korozję
   R корродируемость, коррозионная чувствительность, восприимчивость к коррозии

-680 **tarnish resistance**
   The ability of some metals to remain lustrous in a normal atmosphere, i.e. not to become covered and discoloured by a film of oxide or sulphide.
   D Anlaufbeständigkeit
   F résistance au ternissement
   P odporność na matowienie (korozyjne)
   R стойкость к тускнению

-685 **oxidation resistance, resistance to oxidation**
   The ability of a metal or alloy to withstand attack by an oxidizing atmosphere.
   D Oxydationsbeständigkeit
   F résistance à l'oxydation, inoxydabilité
   P odporność na utlenianie
   R сопротивление окислению

-690 **heat resistance, scale resistance, resistance to scaling**
   The ability of an alloy to resist chemical attacks by air or combustion gases at elevated temperatures.
   D Zunderbeständigkeit, Zunderfestigkeit
   F réfractairité, résistance à l'oxydation à haute température
   P żaroodporność, odporność na zgorzelinowanie
   R жаростойкость, жароупорность, окалиностойкость

-695 **acid resistance**
   Resistance to attack by acids.
   D Säurebeständigkeit
   F résistance aux acides
   P kwasoodporność
   R кислотоупорность, кислотостойкость

-700 **alkali resistance**
   Resistance to attack by alkalis.
   D Laugenbeständigkeit, Alkalibeständigkeit
   F résistance aux bases
   P ługoodporność, odporność na alkalia
   R щелочеупорность, щелочестойкость

-705 **electrochemical properties**
   Those properties of a metal which characterize the tendency of its atoms to accept electrons.
   D elektrochemische Eigenschaften
   F propriétés électrochimiques
   P własności elektrochemiczne
   R электрохимические свойства

-710 **electromotive series, electrochemical series, displacement series, galvanic series**
   A list of metals arranged in order of their standard electrode potentials.
   D elektrochemische Spannungsreihe
   F série electrochimique, série des forces electromotrices
   P szereg napięciowy (metali)
   R ряд напряжений

-715 **electrochemical potential**
   The potential difference existing at the interface between a metal dipped in a solution containing its ions and this solution.
   D elektrochemisches Potential, Galvanipotential, Potential des Metalles
   F potentiel électrochimique
   P potencjał elektrochemiczny
   R электрохимический потенциал

-720 **technological properties, processing properties**
   Those properties of a material which characterize its ability to be subjected to various processing treatments, e.g. forging, drawing, bending, machining, welding, quench hardening etc.
   D technologische Eigenschaften, Verarbeitungseigenschaften
   F propriétés technologiques
   P własności technologiczne
   R технологические свойства

-725 **forgeability, malleability**
   The capacity of a metal to be hot worked by forging.
   D Schmiedbarkeit
   F forgeabilité
   P kowalność, kujność
   R ковкость

-730 **hardenability**
The property that determines the depth and distribution of hardness induced by quenching.
D Härtbarkeit; Einhärtbarkeit
F trempabilité
P hartowność
R закаливаемость; прокаливаемость

-735 **weldability**
The capacity of a metal to be joined by a welding process and to assure satisfactory service of the joint produced.
D Schweißbarkeit
F soudabilité
P spawalność; zgrzewalność
R свариваемость

-740 **solderability**
The capacity of a metal to be joined by a soldering process and to assure satisfactory service of the joint produced.
D Lötfähigkeit
F aptitude au brasage
P lutowność
R паяемость

-745 **machinability**
The property of a material determining its suitability for being machined.
D Zerspanbarkeit
F usinabilité
P skrawalność, obrabialność
R обрабатываемость резанием

-750 **deep(-)drawability, deep drawing capability**
The ability of a metal to be subjected to deep drawing processes.
D Tiefziehfähigkeit
F emboutissabilité, aptitude à l'emboutissage profond
P (głęboko)tłoczność
R способность к глубокой вытяжке, штампуемость

-755 **deep drawing characteristic**
The depth of the impression of a hemispherical-ended plunger forced into sheet metal, measured at the incipient fracture.
D Tiefziehkenngröße, Tiefungswert, Erichsentiefung
F flèche d'emboutissage, profondeur d'emboutissage, indice Erichsen
P tłoczność, liczba tłoczności
R глубина вытяжки

-760 **susceptibility to overheating, susceptibility to grain growth, overheating sensitivity**
The proneness of steels to a coarsening of the austenite grain under the influence of an elevated temperature and time.
D Überhitzungsempfindlichkeit
F susceptibilité à la surchauffe, aptitude à la surchauffe, aptitude au grossissement du grain
P przegrzewność, wrażliwość na przegrzanie, skłonność do gruboziarnistości, skłonność do rozrostu ziarna
R чувствительность к перегреву, перегреваемость, склонность к перегреву

-765 **temperability**
The susceptibility of a quench-hardened steel to be tempered on heating.
D Anlaßbarkeit
F susceptibilité au revenu
P odpuszczalność
R способность к отпуску

-770 **work-hardening capacity**
The capacity of a metal or alloy to become hardened when it is subjected to plastic working.
D Verfestigbarkeit, Verfestigungsfähigkeit
F aptitude à la consolidation
P zdolność umacniania się
R упрочняемость, способность к упрочнению, восприимчивость к упрочнению

-775 **casting properties, founding properties**
That group of properties which determines the suitability of metals and alloys for casting.
D Gießeigenschaften
F propriétés de fonderie
P własności odlewnicze
R литейные свойства

-780 **castability, runnability**
The property of a molten metal or alloy determining its suitability for pouring into moulds, measured by the length of a standard spiral casting.
D Gießbarkeit, Gießfähigkeit, Vergießbarkeit
F coulabilité
P lejność
R жидкотекучесть, жидкоподвижность

-785 **liquidity, flowing power, fluidity**
The ability of a liquid metal to flow readily.
D Dünnflüssigkeit, Fließfähigkeit, Fließvermögen
F fluidité
P rzadkopłynność
R жидкоплавкость

-790 **fusibility**
The readiness with which metals pass from a solid into the liquid state.
D Schmelzbarkeit
F fusibilité
P topliwość
R плавкость

-795 **volume shrinkage, shrinkage, contraction**
The reduction in volume experienced by metals or alloys on cooling.
D Kontraktion, Schwinden, Schwindung, Schrumpfen, Schrumpfung
F retrait, contraction
P skurcz
R усадка

-800 **liquid shrinkage**
Shrinkage occurring in a metal as it cools from the pouring to the freezing temperature or, in the case of an alloy, to the liquidus.
D Flüssigkontraktion, Schrumpfen, Schrumpfung
F retrait avant la solidification
P skurcz w stanie ciekłym
R усадка в жидком состоянии, усадка до затвердевания

-805 **solidification shrinkage, freezing shrinkage**
Shrinkage occurring in metals during solidification at the freezing temperature or, in the case of an alloy, over the freezing range.
D Erstarrungskontraktion, Erstarrungsschrumpfung
F retrait de solidification
P skurcz przy krzepnięciu, skurcz krzepnięcia
R усадка при затвердевании

-810 **solid shrinkage**
Shrinkage occurring during the cooling of a metal from its freezing point to the room temperature or, in the case of an alloy, from the solidus to the room temperature.
D Festkontraktion, Schwinden, Schwindung
F retrait après solidification
P skurcz w stanie stałym
R усадка в твёрдом состоянии

-815 **sectional sensitivity**
The dependence of the structure and properties of a casting upon the thickness of its walls.
D Wanddickenempfindlichkeit
F sensibilité sectionnelle, sensibilité à l'épaisseur
P wrażliwość na grubość ścianki
R чувствительность к толщине стенки

-820 **working properties**
Those properties of a material which characterize its behaviour and reliability in definite practical applications.
D Gebrauchseigenschaften
F propriétés de service, propriétés d'usage, propriétés d'emploi, propriétés d'utilisation
P własności użytkowe
R служебные свойства, эксплуатационные свойства

-825 **thermal shock resistance**
The ability of a material to withstand sudden changes of temperature without cracking.
D Wärmestoßbeständigkeit
F résistance aux chocs thermiques
P odporność na nagłe zmiany temperatury, odporność na udary cieplne, odporność na uderzenia cieplne
R стойкость к тепловым ударам

-830 **thermal endurance, resistance to thermal cycling**
The ability of a material to withstand changes of temperature without signs of failure by thermal fatigue.
D Temperaturwechselbeständigkeit
F résistance au cyclage thermique
P odporność na zmęczenie cieplne, wytrzymałość na zmęczenie cieplne
R стойкость к термической усталости, термоциклическая прочность, разгаростойкость

-835 **heat stability, thermal resistance**
　　Immunity against the action of an elevated temperature.
　D thermische Stabilität, Temperaturbeständigkeit, Wärmebeständigkeit, Warmbeständigkeit
　F stabilité thermique, résistance thermique, résistance à la chaleur
　P odporność na działanie podwyższonej temperatury, odporność termiczna, stabilność termiczna
　R теплостойкость, теплоустойчивость, термическая стойкость, термостойкость, термическая устойчивость

-840 **tempering resistance, resistance to softening, retention of hardness**
　　The ability of some quench-hardened high-alloy steels (especially of high-speed steels) not to be softened by tempering on heating up to a relatively high temperature.
　D Anlaßbeständigkeit
　F résistance au revenu
　P odporność na odpuszczanie
　R устойчивость против отпуска

-845 **ageing resistance**
　　The ability of a solution heat treated alloy to withstand ageing.
　D Alterungsbeständigkeit
　F résistance au viellissement
　P odporność na starzenie
　R устойчивость против старения

-850 **macroscopic properties, gross properties**
　　Those properties of a material which are accessible for direct observation.
　D makroskopische Eigenschaften
　F propriétés macroscopiques, propriétés directement perceptibles
　P własności makroskopowe
　R макроскопические свойства

-855 **metallurgical properties**
　　Such properties of a metal or alloy as grain size, tendency to form or to stabilize carbides, tendency to segregation, ability to deoxidize etc.
　D metallurgische Eigenschaften
　F propriétés métallurgiques
　P własności metalurgiczne
　R металлургические свойства

-860 **structure-sensitive properties, structural properties**
　　Those properties of a metal which depend on the kind of microstructure or which are affected by the presence of lattice defects.
　D strukturempfindliche Eigenschaften, gefügeabhängige Eigenschaften
　F propriétés dépendantes de la structure, propriétés extrinsèques
　P własności wrażliwe na strukturę, własności zależne od struktury
　R структурночувствительные свойства

-865 **structure-insensitive properties**
　　Those properties of a metal which do not depend on the kind of microstructure or which are not affected by the presence of lattice defects.
　D strukturunempfindliche Eigenschaften, strukturunabhängige Eigenschaften, gefügeunabhängige Eigenschaften
　F propriétés indépendantes de la structure, propriétés intrinsèques
　P własności niewrażliwe na strukturę, własności niezależne od struktury
　R структурно нечувствительные свойства

80 EXAMINATION OF METALS AND ALLOYS
PRÜFUNGEN DER METALLE UND LEGIERUNGEN
EXAMENS DES MÉTAUX ET ALLIAGES
BADANIA METALI I STOPÓW
ИССЛЕДОВАНИЯ МЕТАЛЛОВ И СПЛАВОВ

-005 **metallography**
　　The branch of physical metallurgy dealing with the study of the structure of metals and alloys.
　　D Metallographie
　　F métallographie
　　P metalografia
　　R металлография

-010 **metallographic microscopy**
　　Examination of the structure of metals and alloys by means of a microscope.
　　D Metallmikroskopie, Metallmikrographie
　　F métallographie microscopique
　　P metalografia mikroskopowa
　　R металломикроскопия

-015 **high-temperature metallography**
　　Metallography dealing with the study of structures at elevated temperatures.
　　D Hochtemperatur-Metallographie
　　F métallographie à haute température
　　P metalografia wysokotemperaturowa
　　R высокотемпературная металлография

-020 **quantitative metallography**
　　Metallography dealing with the quantitative determination of the individual phases in the microstructure under examination.
　　D quantitative Metallographie
　　F métallographie quantitative
　　P metalografia ilościowa
　　R количественная металлография

-025 **electron metallography**
　　Metallography based upon the application of the electron microscope.
　　D Elektronenmetallographie
　　F métallographie électronique
　　P metalografia elektronowa
　　R электрономикроскопическая металлография

-030 **colour metallography**
　　Metallography in which colour etching is applied.
　　D Farbenmetallographie
　　F métallographie en couleurs
　　P metalografia barwna
　　R цветная металлография

-035 **metallographic examination, metallographic study**
　　Examination of the structure of a metal or alloy.
　　D metallographische Untersuchung, metallographische Prüfung, metallographische Analyse
　　F examen métallographique, essai métallographique
　　P badanie metalograficzne
　　R металлографическое исследование, металлографический анализ

-040 **microscopic examination, microscopy, microscopic analysis**
　　Examination by means of a microscope.
　　D mikroskopische Untersuchung, Mikroskopie
　　F examen microscopique, microscopie, examen micrographique
　　P badanie mikroskopowe, mikroskopia, analiza mikroskopowa
　　R микроскопическое исследование, микроскопический анализ, микроанализ

-045 **macroscopic examination, macrographic examination, macroexamination, macrography**
　　Examination with the naked eye or by means of a magnifying glass.
　　D makroskopische Untersuchung
　　F examen macroscopique, macroscopie, examen macrographique
　　P badanie makroskopowe
　　R макроскопическое исследование

-050 sample, specimen
A piece of material taken from a larger mass of the same material for examination purposes.
D Probestück, Probekörper
F échantillon, éprouvette, pièce d'essai
P próbka
R образец

-055 bulk specimen
A specimen of sufficiently large extension in all three dimensions, as distinct e.g. from a thin foil.
D massive Probe
F échantillon massif
P próbka masywna
R массивный образец

-060 sampling
Selecting and taking (e.g. by cutting or another technique) a representative piece of a given material in order to examine its structure or properties.
D Probe(ent)nahme
F échantillonnage, prélèvement d'échantillons
P pob(ie)ranie próbek
R взятие пробы, отбор проб

-065 mounting
An operation applied to metallographic specimens which are of awkward shape or small in cross section consisting in embedding them — for comfortable handling — in moulding plastics or fusible alloys.
D Einbettung
F enrobage, enrobement
P inkludowanie
R заливка в оправку

-070 mechanical polishing
Polishing performed by mechanical abrasion, usually by pressing the microsection against a revolving disc which is covered with a cloth suitably wetted and impregnated with a suspension of alumina or other polishing powder.
D mechanisches Polieren
F polissage mécanique
P polerowanie mechaniczne
R механическая полировка

-075 electrolytic polishing, electropolishing
Polishing performed through the electrolytic dissolution of projecting portions of the microsection by making the specimen the anode in an electrolytic cell.
D elektrolytisches Polieren, Elektropolieren, Polierelysieren
F polissage électrolytique
P polerowanie elektrolityczne
R электролитическая полировка, электрополировка

-080 chemical polishing
Polishing performed by subjecting the microsection to the action of a suitable solution without interference of electric current or mechanical abrasion.
D chemisches Polieren
F polissage chimique
P polerowanie chemiczne
R химическая полировка

-085 polish attack, polish etch
A process constituting a combination of polishing and etching.
D Ätzpolieren
F polissage avec attaque
P polerowanie z trawieniem
R полировка с травлением

-090 Beilby layer
A thin amorphous layer formed on the surface of metallographic specimens during mechanical polishing.
D Beilby-Schicht
F couche de Beilby
P warstewka Beilby'ego
R слой Бильби, слой Бейльби

-095 microsection
The flat, polished surface of a specimen, prepared for metallographic examination.
D Schliff, Metallschliff, Schliffffläche
F coupe métallographique, coupe micrographique, coupe polie
P szlif, zgład
R (микро)шлиф

-100 **etchant, etching reagent**
    A chemical reagent, mostly in the form of a solution which — applied to microsections — will reveal structural features or constituents, its action consisting in differentially attacking areas of different chemical composition or of different orientation.
    D Ätzmittel
    F réactif métallographique, réactif d'attaque
    P odczynnik do trawienia, odczynnik metalograficzny
    R травитель, травящий реактив, реактив для травления

-105 **staining metallographic (etching) reagent**
    An etchant which gives colour effects within the microstructure under examination.
    D Farbätzmittel
    F réactif d'attaque en couleurs
    P odczynnik do trawienia barwnego, odczynnik do trawienia kolorowego
    R красящий травитель

-110 **nital**
    An alcoholic solution of nitric acid used as an etchant for ferrous alloys.
    D Nital, alkoholische Salpetersäure
    F nital
    P nital
    R нитал, спиртовой раствор азотной кислоты

-115 **picral**
    An alcoholic solution of picric acid used as an etchant for some ferrous alloys.
    D Picral, alkoholische Pikrinsäure
    F picral
    P pikral
    R пикрал, спиртовой раствор пикриновой кислоты

-120 **etching**
    The revealing of the structure of a metal or alloy by selective attack of appropriate reagents on the microsection of the specimen.
    D Ätzen, Ätzung
    F attaque chimque, attaque métallographique
    P trawienie, wytrawianie
    R травление

-125 **etching time, etching period**
    The time during which a microsection is subjected to attack by an etchant.
    D Ätzdauer, Ätzzeit
    F durée d'attaque, temps d'attaque
    P czas trawienia, czas wytrawiania
    R продолжительность травления

-130 **overetching**
    Etching for a longer period of time or with stronger etchants than those required to obtain an optimum revealing of a given structure.
    D Überätzung
    F attaque trop poussée, attaque exagérée
    P przetrawienie
    R перетравление

-135 **macro(-)etching**
    Etching aimed at revealing the macrostructure.
    D Makroätzung
    F attaque macrographique
    P trawienie makrostruktury, makrotrawienie
    R травление для макроскопического исследования

-140 **micro-etching**
    Etching aimed at revealing the microstructure.
    D Mikroätzung
    F attaque micrographique
    P trawienie mikrostruktury, mikrotrawienie
    R травление для микроскопического исследования

-145 **grain boundary etching**
    Etching aimed at revealing grain boundaries.
    D Korngrenzenätzung
    F attaque des joints des grains
    P trawienie granic ziarn
    R травление по границам зёрен

-150 **grain surface etch, grain-contrast(ing) etch**
    Etching aimed at revealing the different orientation of individual grains.
    D Kornflächenätzung, Kornfelderätzung
    F attaque de la surface des grains
    P trawienie pól ziarn
    R травление зёрен

-155 **orientation contrast**
>The contrast between individual grains in an etched microsection of a metal or homogeneous alloy, caused by differences in their crystallographic orientation with respect to the microsection plane.
>D Orientierungskontrast
>F contraste d'orientation
>P kontrast orientacji
>R ориентационный контраст

-160 **deep etching**
>Etching with strong reagents, used to reveal macrostructure.
>D Tiefätzung
>F attaque profonde
>P trawienie głębokie
>R глубокое травление

-165 **colour etching**
>Etching with reagents which give different colour effects for various microconstituents.
>D Farb(en)ätzung
>F attaque en couleurs
>P trawienie barwne, trawienie kolorowe
>R цветное травление

-170 **chemical etching**
>Etching by chemical attack.
>D chemische Ätzung
>F attaque chimique
>P trawienie chemiczne
>R химическое травление

-175 **electrolytic etching**
>Etching by means of an electrochemical process, in which the specimen to be etched is made the anode in a suitable electrolyte.
>D elektrolytische Ätzung, anodische Ätzung, elektrochemische Ätzung
>F attaque électrolytique, attaque anodique
>P trawienie elektrolityczne
>R электролитическое травление, электротравление

-180 **cathodic (vacuum) etching, ion bombardment etching**
>A way of preparing specimens for metallographic examination using ionic bombardment in a vacuum instead of chemical etching.
>D katodisches Ionenätzen, katodische Vakuumätzung
>F attaque (par pulvérisation) cathodique, attaque ionique
>P trawienie jonowe (próżniowe), trawienie katodowe (próżniowe)
>R ионное травление, травление ионной бомбардировкой

-182 **gas etching**
>A way of preparing specimens for metallographic examination using ionized gases as the etching reagent. The etching effect is based on colour contrasts produced by layers of different thickness on the different phases.
>D Gasätzen
>F attaque gazeux, attaque par gaz ionisés
>P trawienie gazowe
>R газовое травление

-185 **heat tinting**
>Heating a metallographic specimen in air, aimed at revealing the microstructure by selective oxidation effects.
>D Anlaßätzung, Ätzanlassen, Anlaufätzung, Trockenätzung
>F attaque par oxydation, coloration thermique, coloration à chaud
>P trawienie przez utlenianie, trawienie przez odpuszczanie
>R тепловое травление на воздухе, травление окислением на воздухе, оксидное травление, тепловое окрашивание, травление цветами побежалости

-190 **thermal etching, (thermal) grooving**
>Heating a microsection to a fairly high temperature in a vacuum or in an inert atmosphere, resulting in the appearance of thermal grooves at the grain boundaries.
>D thermische Ätzung, Grabenbildung
>F attaque thermique
>P trawienie termiczne
>R тепловое травление, термическое травление

-195 **thermal groove**
>Any one of the grooves appearing at the grain boundaries of a microsection during heating.
>D Ätzgraben, Ätzfurche, thermischer Graben
>F sillon d'attaque, sillon intergranulaire
>P rowek trawienny
>R канавка термического травления

-200 **etch figures, etch(ing) pattern**
>The pattern which appears on the surface of a microsection after etching.
>D Ätzfiguren
>F figures d'attaque
>P figury trawienia
>R фигуры травления

-205 **etch(ing) pit**
Any one of the small cavities produced on a highly polished surface of a metal by etching, often used for studying the dislocation density, as they are forming when the etching of the surface proceeds slower than the etching of dislocations.
D Ätzgrübchen
F point d'attaque, cavité d'attaque
P jamka trawienia
R ямка травления, пятно травления

-210 **metallurgical microscope, metallographic microscope, metallograph**
A microscope designed for the examination of metallographic specimens.
D Metallmikroskop
F microscope métallographique, microscope métallurgique
P mikroskop metalograficzny
R металлографический микроскоп, металломикроскоп

-215 **light microscopy, optical microscopy**
Examination by means of a light microscope.
D Lichtmikroskopie
F microscopie optique
P mikroskopia świetlna, mikroskopia optyczna
R световая микроскопия, оптическая микроскопия

-220 **transmission microscopy**
Microscopy in which a beam of light or a stream of electrons passes through the specimen under examination.
D Durchstrahlungsmikroskopie, Durchlicht-Mikroskopie, Transmissionsmikroskopie
F microscopie par transparence, microscopie par transmission
P mikroskopia prześwietleniowa, mikroskopia typu prześwietleniowego, mikroskopia transmisyjna
R просвечивающая микроскопия, трансмиссионная микроскопия

-225 **reflection microscopy**
Microscopy in which a beam of light or a stream of electrons is reflected from the microsection under examination.
D Auflichtmikroskopie, Reflexions-Mikroskopie
F microscopie par réflexion, microscopie en réflexion
P mikroskopia odbiciowa, mikroskopia typu odbiciowego, mikroskopia refleksyjna
R отражательная микроскопия

-230 **dark-field microscopy**
A technique of microscopy in which dark-field illumination is used.
D Dunkelfeldmikroskopie
F microscopie sur fond noir
P badanie mikroskopowe w polu ciemnym
R темнопольная микроскопия

-235 **bright-field microscopy**
A technique of microscopy in which bright field illumination is used.
D Hellfeldmikroskopie
F microscopie sur fond clair
P badanie mikroskopowe w polu jasnym
R светлопольная микроскопия

-240 **hot-stage microscopy, thermal microscopy**
Microscopic examination at elevated temperatures.
D Hochtemperaturmikroskopie, Heiztischmikroskopie
F microscopie à haute température
P mikroskopia wysokotemperaturowa
R (высоко)температурная микроскопия

-245 **micrograph**
A graphic reproduction of a microstructure.
D Mikrographie, Gefügebild, Mikrobild
F micrographie
P mikrografia
R микрография

-250 **photomicrograph**
A photographic reproduction of a microstructure.
D Mikrophotogramm, Fotomikrographie, Mikrofotographie, Mikro(struktur)aufnahme, Gefügeaufnahme, Photomikrographie
F photomicrographie
P fotomikrografia
R микрофотография

-255 **photomicrography**
The art and practice of making photomicrographs.
D Photomikrographie, Mikrophotographie
F photomicrographie, microphotographie
P fotomikrografia, mikrofotografia
R микрофотография

-260 **polarized light microscopy**
Microscopic examination by means of polarized light.
D Polarisationsmikroskopie
F microscopie en lumière polarisée
P badanie mikroskopowe w świetle spolaryzowanym, mikroskopia polaryzacyjna
R поляризационная микроскопия

-265 **ultra-violet microscopy**
Microscopic examination by means of ultra-violet radiation.
D Ultraviolett-Mikroskopie, UV-Mikroskopie
F microscopie en ultraviolet
P badanie mikroskopowe w świetle nadfioletowym, badanie mikroskopowe w nadfiolecie, mikroskopia w nadfiolecie, mikroskopia nadfioletowa
R ультрафиолетовая микроскопия

-270 **phase-contrast microscopy**
Microscopic examination based on the use of the phase contrast.
D Phasenkontrastmikroskopie
F microscopie à contraste de phase
P badanie mikroskopowe metodą kontrastu fazowego, mikroskopia fazowo-kontrastowa
R фазоконтрастная микроскопия

-275 **phase contrast**
The contrast between individual regions in the image of the surface of a specimen, caused by slight phase changes in the light reflected from this surface.
D Phasenkontrast
F contraste de phase
P kontrast fazowy
R фазовый контраст

-280 **revealing (of) the microstructure**
Application of an appropriate procedure aimed at making the microstructure of a metal or alloy visible.
D Gefügeentwicklung, Sichtbarmachung des Gefüges
F mise en évidence de la microstructure
P ujawnienie struktury, ujawnianie struktury
R выявление микроструктуры

-282 **magnification**
The ratio of the linear dimensions of the image obtained by means of an optical system (e.g. a microscope) to the linear dimensions of the object under observation.
D Vergrößerung
F grossissement
P powiększenie
R увеличение

-285 **grain size**
The average grain diameter in a given microstructure.
D Korngröße
F grosseur du grain, taille du grain
P wielkość ziarna
R величина зерна, размер зерна

-290 **grain-size number, grain-size index, index number**
A conventional number serving to determine the average size of the grain section in a given microstructure by comparison with standard charts.
D Richtreihennummer, Richtzahl
F numéro du grain
P numer ziarna
R балл зерна, номер зерна

-295 **interference microscopy**
The microscopic examination of the surface topography of metal specimens by using optical interference.
D Interferenzmikroskopie
F microscopie interférentielle
P mikroskopia interferencyjna
R интерференционная микроскопия

-300 **interferogram**
A pattern obtained by means of interference microscopy.
D Interferogramm
F interférogramme
P interferogram
R интерферограмма

-305 **macrograph**
A graphic reproduction of a macrostructure.
D Makrographie
F macrographie
P makrografia
R макрография

-310 **photomacrograph**
A photographic reproduction of a macrostructure.
D Makroaufnahme
F photomacrographie, macrophotographie
P fotomakrografia
R макрофотография

-315 **photomacrography**
 The art and practice of making photomacrographs.
 D Makrophotographie
 F photomacrographie, macrophotographie
 P fotomakrografia, makrofotografia
 R макрофотография

-320 **sulfur print test, Baumann printing**
 A macroscopic test by means of which the distribution of sulfide inclusions in steel is determined.
 D Baumann-Abdruckverfahren, Baumann-Schwefelabdruckverfahren
 F essai Baumann, épreuve Baumann
 P próba Baumanna, próba na siarkę
 R проба на серу

-325 **sulfur print**
 An image obtained on photographic paper in the sulfur print test, indicating the distribution of sulfides in the steel specimen under examination.
 D Baumann-Abdruck, Schwefelabdruck
 F empreinte (de) Baumann, impression aux sels d'argent
 P odbitka Baumanna, odbitka na siarkę, odbitka siarkowa
 R отпечаток Баумана, баумановский отпечаток

-330 **electron microscopy**
 Examination by means of an electron microscope, i.e. a microscope in which a stream of electrons is used instead of a beam of light.
 D Elektronenmikroskopie, Übermikroskopie
 F microscopie électronique
 P mikroskopia elektronowa
 R электронная микроскопия, электронно-микроскопическое исследование

-335 **electron emission microscopy**
 Examination by means of an electron microscope in which the specimen under examination emits electrons.
 D Elektronen(emissions)mikroskopie
 F microscopie électronique à emission
 P mikroskopia elektronowa emisyjna
 R эмиссионная электронная микроскопия, автоэмиссионная микроскопия

-340 **transmission electron microscopy**
 Electron microscopy in which the stream of electrons passes through the specimen under examination.
 D Elektronen-Durchstrahlungsmikroskopie, Elektronen-Transmissionsmikroskopie
 F microscopie électronique par transparence, microscopie électronique par transmission
 P mikroskopia elektronowa prześwietleniowa, mikroskopia elektronowa transmisyjna
 R просвечивающая электронная микроскопия

-345 **replica**
 A specimen for transmission electron microscopy, thin enough to be transparent to electrons, representing the model or a copy of the etched microsection which is to be examined.
 D Reliefabdruck, Oberflächenabdruck, Prägeabdruck
 F réplique, empreinte
 P replika
 R слепок, реплика

-350 **thin foil, thin film**
 A metallographic specimen used in electron transmission microscopy, thin enough to transmit electrons.
 D dünne Folie, Dünnschliff, Dünnschliffolie
 F lame mince, coupe mince
 P cienka folia
 R тонкая плёнка, тонкая фольга

-355 **thinning**
 A technique of reducing the thickness (of specimens) used in preparing thin foils for transmission electron microscopy.
 D Dünnen, Verdünnen
 F amincissement
 P ścienianie
 R утонение

-360 **scanning (electron) microscopy, flying-spot microscopy**
A technique of microscopy in which X-rays are excited by an electron spot moving in a regular scan over the specimen under examination, and are observed by television, on a cathode-ray tube screen.
D Raster(elektronen)mikroskopie
F microscopie (électronique) à balayage
P mikroskopia elektronowa skaningowa, mikroskopia elektronowa analizująca, mikroskopia elektronowa rastrowa
R сканирующая (электронная) микроскопия, растровая электронная микроскопия

-365 **electron (photo)micrograph**
A photomicrograph obtained by means of an electron microscope
D elektronenmikroskopische Aufnahme
F micro(photo)graphie électronique
P (foto)mikrografia elektronowa
R электронная микрофотография, электронномикроскопический снимок

-370 **shadowing**
Reinforcing the contrast of a replica for electron microscopy by depositing on it heavy atoms under vacuum these atoms approach the replica at a shallow angle of incidence and thus accentuate the relief.
D Beschatten
F ombrage
P cieniowanie
R оттенение

-375 **decorating, decoration**
The technique of making individual dislocations visible, based on the property of dislocations to attract impurity atoms and provide preferred sites for the nucleation of precipitates.
D Dekorierung, Dekoration
F décoration
P dekorowanie (dyslokacji), dekoracja (dyslokacji)
R декорирование (дислокаций)

-380 **field-ion microscopy**
A technique of microscopy with an extremely high resolving power, in which instead of a beam of light a stream of ionised gas atoms is used, ionisation taking place at the tip of a pointed metal specimen biassed, relative to the phosphorescent screen, to a high positive potential.
D Feldionenmikroskopie
F microscopie par ionisation, microscopie ionique
P mikroskopia (polowo-)jonowa
R автоионная микроскопия

-385 **fractography**
The examination of fracture surfaces.
D Fraktographie, fraktographische Untersuchung
F fractographie, examen fractographique, analyse fractographique
P fraktografia, badanie fraktograficzne, badanie przełomu
R фрактография

-390 **microfractography**
The examination of fracture surfaces by means of a microscope.
D Mikrofraktographie
F microfracto(métallo)graphie, étude microfractographique
P mikrofraktografia
R микрофрактография

-395 **microfractograph**
A photographic reproduction of the microstructure of a fracture surface.
D Mikrofraktographie
F microfractographie
P mikrofraktografia
R микрофрактограмма

-400 **mechanical testing**
The techniques by means of which the mechanical properties of a metal are determined.
D Festigkeitsprüfung, Festigkeitsversuch
F essai mécanique
P badanie własności mechanicznych; badanie wytrzymałościowe
R механическое испытание, испытание механических свойств; испытание прочностных свойств

-405 **tensile test, tension test**
A fundamental mechanical test in which a special bar specimen is subjected to stretching in a longitudinal direction until it fractures.
D Zugversuch, Zugprobe
F essai de traction
P próba rozciągania
R испытание на растяжение

-410 **stress-strain diagram, stress-strain curve**
The plot, obtained usually from the tensile test, representing unit stress against unit strain.
D Spannungs-Dehnungs-Kurve, Spannungs-Verformungs-Kurve
F diagramme tension-allongement, courbe tension-déformation, courbe contrainte-déformation
P wykres rozciągania, wykres naprężenie-odkształcenie, krzywa naprężenie-odkształcenie
R кривая напряжение-деформация

-415 **hardness test**
A test consisting in forcing an indentor of specified shape into a flat surface of the material being tested.
D Härteversuch, Härteprüfung
F essai de dureté
P próba twardości
R испытание на твёрдость, дюрометрическое исследование, испытание твёрдости

-420 **fatigue test, endurance test**
A mechanical test in which the bar specimen is subjected to stresses varying cyclically between a maximum and minimum value, the purpose of this test being to determine the endurance limit.
D Dauer(schwing)versuch, Ermüdungsversuch
F essai de fatigue, essai d'endurance
P próba zmęczeniowa, próba na zmęczenie
R испытание на усталость, испытание на выносливость

-425 **fatigue curve, S-N curve, Wöhler curve**
A plot of the alternating stress (S) that causes a fatigue fracture of the specimen against the number (N) of reversals of stress after which this failure occurs.
D Wöhler-Kurve, Ermüdungskurve
F courbe de Wöhler, courbe de fatigue
P krzywa zmęczenia, krzywa Wöhlera
R кривая выносливости, кривая усталости, кривая Вёлера

-430 **impact test**
A mechanical test, executed mostly on standard notched bar-specimens in bending, consisting in fracturing the specimen at a high velocity by a single blow from a pendulum and serving to determine the energy absorbed in fracturing.
D Schlagversuch, Kerbschlagversuch
F essai de choc
P próba udarności
R ударное испытание, испытание на удар

-435 **test bar, bar specimen**
A specimen of regular cross-section specially prepared for testing.
D Probestab
F barreau d'épreuve
P próbka (prętowa)
R (прутковый) образец

-440 **spark test(ing)**
A rapid method for determining the approximate type of steel under examination by observing the nature, i.e. shape, volume, colour, and intensity, of the spark pattern emitted when it is abraded by a grindstone wheel.
D Funkenprobe, Funkenprüfung, Schleiffunkenanalyse
F essai aux étincelles
P próba iskrowa
R испытание на искру, искровая проба

-445 **fracture test**
A test consisting in breaking a piece of metal for the purpose of examining the fractured surface in order to determine the structure or composition of the metal or the presence of internal defects.
D Bruchprobe
F essai de rupture
P próba przełomu
R испытание на излом, проба на излом

-450 **end quench hardenability test, Jominy (end quench) test**
A test serving to determine the hardenability of steel by quench hardening the end surface of a cylindrical specimen and measuring the hardness distribution along the generating line of the cylinder.
D Stirnabschreckversuch, Stirnabschreckprobe, Jominy-Probe
F essai de trempabilité Jominy
P próba hartowania od czoła, próba Jominy'ego
R определение прокаливаемости торцовым методом

-455 **(end-quench) hardenability curve, hardenability line**
A plot of hardness distribution at different depths within the piece that has been quench hardened.
D Stirnabschreck-Härtekurve
F courbe (Jominy) de pénétration de trempe
P krzywa hartowności
R кривая прокаливаемости

-460 **hardenability band**
A plot representing the interval of hardness distribution at different depths within the piece of steel that has been quench hardened, based upon the data pertaining to several tests of the same grade of steel.
D Härtbarkeitsstreuband
F bande de trempabilité
P pasmo hartowności
R полоса прокаливаемости

-465 **spot analysis, spot test**
A rapid qualitative non-destructive test carried out by placing a drop of a chemical reagent on a cleaned surface of the alloy under examination.
D Tüpfelprobe, Tüpfelanalyse
F essai à la touche
P analiza kroplowa
R капельный анализ

-470 **chemical analysis**
The qualitative or quantitative determination of the chemical composition of a material.
D chemische Analyse
F analyse chimique
P analiza chemiczna
R химический анализ

-475 **spectral analysis, spectrum analysis**
A method of examination, which can be both qualitative and quantitative, based on the observation and analysis of the spectrum of the given substance.
D Spektralanalyse
F analyse spectrale
P analiza widmowa, analiza spektralna
R спектральный анализ

-480 **mass spectrography**
A highly accurate analytical method of determining the presence and content of various elements in the material under examination based on the accelerating and deflecting action of an electrostatic and magnetic field on a stream of positively charged ions, the deflection of the ions of different kinds — recorded photographically — being unambiguously related to their mass.
D Massenspektrographie
F spectrographie de masse
P spektrografia masowa
R масс-спектрография

-485 **mercurous nitrate test**
A test performed using mercurous nitrate by means of which the tendency of cold worked or stressed brass to season cracking can be detected.
D Quecksilbernitratversuch, Quecksilber(nitrat)probe
F essai au nitrate mercureux
P próba rtęciowa
R испытание на сезонное (само)растрескивание

-490 **physicochemical analysis**
Examination of physical properties of an alloy system against its chemical composition.
D physikalisch-chemische Analyse
F analyse physico-chimique
P analiza fizykochemiczna
R физико-химический анализ

-495 **singular point**
   The breaking point in the plot
   of physical properties against
   the chemical composition of a given
   alloy system, occurring at the
   composition which corresponds
   to an intermetallic compound.
   D singulärer Punkt
   F point singulier
   P punkt osobliwy, punkt szczególny
   R сингулярная точка, особая точка

-500 **calorimetric investigation,
   calorimetric study**
   The technique of determining the
   relative heat contents of a
   material or heat effects
   accompanying various reactions
   by means of a calorimeter.
   D kalorimetrische Untersuchung
   F étude calorimétrique
   P badanie kalorymetryczne
   R калориметрическое исследование,
   калориметрическое испытание

-505 **thermal analysis**
   An analysis of the process of
   heating or cooling of a metal or
   an alloy, as a basis for plotting
   constitutional diagrams.
   D thermische Analyse, Thermoanalyse
   F analyse thermique
   P analiza termiczna, analiza cieplna
   R термический анализ

-510 **simple thermal analysis**
   A technique of thermal analysis
   in which the absolute values of
   temperature changes against time
   are registered.
   D gewöhnliche Thermoanalyse
   F analyse thermique absolue
   P analiza termiczna bezpośrednia
   R простой термический анализ

-515 **differential thermal analysis**
   A technique of thermal analysis
   in which the differences between
   the temperature changes against
   time for the material under
   examination and those for a
   standard sample are registered.
   D Differential-Thermoanalyse
   F analyse thermique différentielle
   P analiza termiczna różnicowa
   R дифференциальный термический
   анализ,
   дифференциально-термический
   анализ

-520 **heating curve, heating diagram**
   A curve obtained by plotting
   temperature against time as
   a metal or alloy is heated under
   constant conditions.
   D Erwärmungskurve,
   Erhitzungskurve
   F courbe de chauffage, diagramme
   de chaufflage
   P krzywa nagrzewania,
   krzywa ogrzewania
   R кривая нагрева(ния)

-525 **cooling curve, cooling diagram**
   A curve obtained by plotting
   temperature against time as a
   metal or alloy is cooled under
   constant conditions.
   D Abkühlungskurve
   F courbe de refroidissement,
   diagramme de refroidissement
   P krzywa stygnięcia
   R кривая охлаждения

-527 **thermogram**
   A plot of temperature against
   time during heating or cooling,
   e.g. in thermal analysis.
   D Thermogramm
   F thermogramme
   P termogram
   R термограмма

-530 **thermal arrest**
   The horizontal section of a
   heating or cooling curve.
   D thermischer Haltepunkt
   F arrêt thermique, palier thermique
   P przystanek termiczny,
   przystanek cieplny
   R термическая остановка,
   термическзя площадка

-535 **quantitative thermal analysis,
   thermodynamical analysis
   (by calorimetry)**
   A variety of thermal analysis as
   applied to determining the values
   of thermodynamic quantities of
   an alloy system.
   D quantitative thermische Analyse,
   thermodynamische Analyse
   F étude thermodynamique
   par calorimétrie
   P analiza termodynamiczna
   R количественный
   термодинамический анализ

-540 **dilatometry, dilatometric analysis**
The study of changes in length of a metal or alloy sample in the solid state on heating or cooling.
D Dilatometer-Untersuchung, Dilatometrie
F analyse dilatométrique, essai dilatométrique, dilatométrie
P analiza dylatometryczna, badanie dylatometryczne, dylatometria
R дилатометрический анализ, дилатометрическое исследование, дилатометрия

-545 **simple dilatometry**
A method of dilatometric analysis based on measuring absolute values of length changes.
D absolute Dilatometer-Untersuchung, absolute Dilatometrie
F analyse dilatométrique absolue, essai dilatométrique, absolue, étude dilatométrique absolue, dilatométrie absolue
P analiza dylatometryczna bezwzględna, analiza dylatometryczna prosta, dylatometria prosta
R простой дилатометрический анализ, простая дилатометрия

-550 **differential dilatometry, differential dilatometric analysis**
A method of dilatometric analysis in which the difference between the linear change of the sample under study and the linear change of a standard sample is measured.
D Differentialdilatometrie
F analyse dilatométrique différentielle, essai dilatométrique différentiel, étude dilatométrique différentielle, dilatométrie différentielle
P analiza dylatometryczna różnicowa, dylatometria różnicowa
R дифференциальный дилатометрический анализ, дифференциальная дилатометрия

-555 **dilatometric curve, dilatometer curve**
A plot of linear changes of a solid specimen against temperature during heating and/or cooling, obtained by means of an instrument called a dilatometer.
D Dilatometerkurve, Dilatationskurve
F diagramme dilatométrique, courbe de dilatation, courbe dilatométrique
P krzywa dylatometryczna, dylatogram
R дилатометрическая кривая, дилатограмма

-560 **magnetic analysis**
Examination of metals and alloys based on the interrelation between their constitution, structure, or composition and their magnetic properties.
D magnetische Analyse
F analyse magnétique
P analiza magnetyczna, badanie magnetyczne
R магнитный анализ

-565 **thermomagnetic analysis**
Magnetic analysis extended to measuring the magnetic properties of a metal or alloy at various temperatures.
D thermomagnetische Analyse
F analyse thermomagnétique
P analiza termomagnetyczna
R термомагнитный анализ

-570 **X-ray examination, X-ray study**
Testing of materials by means of X-rays.
D röntgenographische Untersuchung
F examen aux rayons X, radiographie X
P badanie rentgenograficzne
R рентгенографическое исследование, рентгеноанализ, рентгенографирование

-575 **X-ray metallography**
Metallography based on the application of X-rays.
D Röntgenmetallographie
F radiométallographie par rayons X
P rentgenografia metali, metalografia rentgenowska
R рентгеновская металлография, рентгенография металлов

-580 **X-ray diffraction analysis, X-ray crystal analysis, X-ray structure analysis**
Examination of the crystal structure based on the diffraction of X-rays in a crystal lattice.
D röntgenographische Strukturanalyse, Röntgen(fein)strukturanalyse, röntgenographische Feinstrukturanalyse, Röntgenkristallstrukturanalyse
F analyse radiocristallographique, examen radiocristallographique, radiocristallographie
P rentgenografia dyfrakcyjna, rentgenowska analiza strukturalna, rentgenografia strukturalna
R рентгеноструктурный анализ, структурная рентгенография

-585 **X-ray phase analysis**
The qualitative and quantitative determination of phases contained in a given specimen by means of X-ray diffraction methods.
D röntgenographische Phasenanalyse
F analyse des phases par rayons X
P rentgenowska analiza fazowa
R рентгеновский фазовый анализ, рентгенографический фазовый анализ

-590 **Laue method**
A method of X-ray crystal analysis which uses a fixed single crystal as the specimen and employs a continuous spectrum of X-rays.
D Laue-Methode
F méthode de Laüe, méthode de Laue
P metoda Lauego, metoda kryształu nieruchomego
R метод неподвижного кристалла, метод Лауэ

-595 **powder method, Debye-Scherrer method**
A method of X-ray crystal analysis in which a powdered or finely polycrystalline specimen is used.
D Debye-Scherrer-Verfahren
F méthode des poudres, méthode de Debye et Scherrer
P metoda proszkowa, metoda Debye'a-Scherrera
R порошковый метод, метод Дебая

-600 **rotating-crystal method, revolving crystal method, Bragg method**
A method of X-ray crystal analysis in which a rotating single crystal is used as the specimen.
D Drehkristallmethode
F méthode du cristal tournant
P metoda kryształu obracanego, metoda Braggów
R метод вращающегося кристалла

-605 **oscillating crystal method**
A method of X-ray crystal analysis in which a single crystal rotating alternately in both directions, but through much less than a complete rotation, is used as the specimen.
D Schwingkristallmethode
F méthode du cristal oscillant
P metoda kryształu wahliwie obracanego, metoda kryształu kołysanego, metoda kryształu oscylującego
R метод качающегося кристалла, метод колебательного кристалла

-610 **diffraction pattern**
The image, registered photographically, representing the totality of intersections of X-ray or electron beams, diffracted by a crystal or by a polycrystalline aggregate, with a plane or other surface.
D Streudiagramm, Beugungsbild, Beugungsaufnahme
F diagramme de diffraction, cliché de diffraction
P dyfraktogram
R дифффрактограмма, диффракционная картина, рефлексограмма

-615 **X-ray diffraction pattern, X-ray diffraction photograph**
A diffraction pattern obtained by means of X-rays.
D Röntgenbeugungsaufnahme, Röntgenfeinstrukturdiagramm
F diagramme de diffraction des rayons X, cliché de diffraction de rayons X
P rentgenogram dyfrakcyjny, dyfraktogram rentgenowski
R диффракционная рентгенограмма

-620 **Laue pattern, Laue picture, Laue photogram, Laue photograph, Laue diagram**
A diffraction pattern obtained by means of the Laue method of X-ray crystal analysis.
D Lauediagramm, Laue-Aufnahme
F diagramme de Laüe, diagramme de Laue, cliché de Laue
P lauegram, rentgenogram Lauego
R лауэграмма

-625 **transmission method**
A method of X-ray diffraction analysis in which the diffraction pattern is obtained after the transmission of rays through the specimen.
D Durchstrahlungsmethode
F méthode du diagramme par transmission
P metoda prześwietleniowa, metoda promieni przechodzących
R метод на просвет, метод просвечивания

-630 **back reflection method**
A method of X-ray diffraction analysis in which the diffraction pattern is obtained by reflecting the rays towards the source.
D Rückstrahlverfahren, Rückstrahlmethode
F méthode du diagramme en retour
P metoda promieni zwrotnych
R метод обратной съемки

-635 **(X-ray diffraction) powder pattern, powder (diffraction) photograph, Debye-Scherrer powder pattern**
An X-ray diffraction pattern obtained by means of the powder method.
D Pulveraufnahme, Pulverdiagramm, Debye-Scherrer-Aufnahme, Debye-Scherrer-Diagramm
F diagramme de poudre, diagramme de Debye et Scherrer, cliché de Debye-Scherrer
P debajogram, rentgenogram proszkowy, dyfraktogram proszkowy
R дебаеграмма, порошкограмма, рентгенограмма Дебая, порошковая рентгенограмма

-640 **Debye ring**
Any one of the circular diffraction lines which constitute the Debye-Scherrer powder pattern.
D Debye-Scherrer-Ring, Debye-Scherrer-Kreis
F anneau de Debye
P pierścień debajowski, pierścień Debye'a, linia debajowska
R кольцо Дебая, дебаевское кольцо

-645 **(X-ray) diffraction line**
Any one of the lines occurring in the diffraction pattern obtained by means of X-ray crystal analysis.
D Beugungslinie, Interferenzlinie
F raie de diffraction, ligne de diffraction
P linia dyfrakcyjna, linia interferencyjna
R диффракционная линия, интерференционная линия

-650 **fundamental lines**
Those diffraction lines in the X-ray diffraction pattern of an ordered solid solution which are attributable only to the type of the given crystal lattice, i.e. which do not originate from the phenomenon of ordering, and remain unchanged when the solution passes in the disordered state.
D Grundlinien
F raies principales, raies du réseau principal
P linie podstawowe
R основные линии

-655 **superstructure lines, superlattice reflections**
Additional diffraction lines occurring in the X-ray diffraction pattern of an ordered solid solution, as compared with the diffraction pattern of the same solution in the disordered state.
D Überstrukturlinien
F raies de surstructure
P linie nadstruktury
R сверхструктурные линии

-660 **diffraction spot**
The image of the intersection of a beam diffracted by a crystal with a plane or other surface.
D Beugungsfleck, Interferenzfleck
F tache de diffraction
P plamka dyfrakcyjna
R диффракционное пятно, интерференционное пятно

-665 **rotation pattern, rotation diagram**
An X-ray diffraction pattern obtained by means of the rotating-crystal method.
D Drehkristallaufnahme, Rotationsdiagramm
F diagramme de cristal tournant, cliché de cristal tournant, diagramme de diffraction à cristal tournant
P rentgenogram wykonany metodą kryształu obracanego
R рентгенограмма вращения

-670 **back reflection pattern, back reflection diagram**
An X-ray diffraction pattern obtained by means of the back reflection method.
D Rückstrahlaufnahme, Rückstrahldiagramm
F diagramme en retour
P rentgenogram wykonany metodą promieni zwrotnych
R обратная рентгенограмма, эпиграмма

-675 **transmission pattern, transmission diagram**
An X-ray diffraction pattern obtained by means of the transmission method.
D Durchstrahlungsaufnahme, Röntgendurchstrahlungsaufnahme
F diagramme par transmission
P dyfraktogram wykonany metodą promieni przechodzących, rentgenogram wykonany metodą promieni przechodzących, dyfraktogram prześwietleniowy
R рентгенограмма снятая на просвет, съёмка на просвет

-680 **asterism**
   The occurrence of radial streaks round the circular diffraction spots in X-ray diffraction patterns of plastically deformed metals.
   D Asterismus
   F astérisme
   P asteryzm, gwiazdkowość
   R астеризм

-685 **pole figures**
   A stereographic projection representing the preferred orientation of the normals to specific crystallographic planes in a polycrystalline material.
   D Polfiguren
   F figures de pôle
   P figury biegunowe
   R полюсные фигуры

-690 **orientation of single crystals**
   Determining the position of important sets of crystallographic planes within a given single crystal with respect to a reference system.
   D Orientierungsbestimmung der Kristalle, Bestimmung der Kristallorientierung
   F orientation d'un monocristal
   P orientowanie monokryształu
   R ориентирование монокристалла

-695 **indexing**
   Determining the Miller indices corresponding to the various diffraction lines in a diffraction pattern.
   D Indizierung
   F indexation, determination des indices, assignement des indices
   P wskaźnikowanie
   R индицирование

-700 **fluorescent X-ray spectroscopy, (X-ray) fluorescent analysis, X-ray fluorescence analysis, X-ray (fluorescence) spectroscopy**
   A method of analysis, which can be both qualitative and quantitative, based on the spectrometric observation of the secondary radiation emitted by the substance under examination bombarded with primary X-rays.
   D Röntgenfluoreszenzspektralanalyse, Röntgen(fluoreszenz)-Spektroskopie, Röntgenstrahlspektroskopie, Röntgenspektralanalyse
   F spectrométrie par fluorescence de rayons X, spectrométrie (à rayons) X
   P (rentgenowska) analiza fluorescencyjna, (fluorescencyjna) spektroskopia rentgenowska
   R рентгеновская (флуоресцентная) спектроскопия, рентгеноспектральный анализ

-705 **radiography, radiographic inspection**
   Non-destructive inspection by means of a penetrating radiation, e.g. using X-rays or gamma rays.
   D Durchstrahlungsprüfung, Radiographie
   F radiographie, contrôle radiographique, examen radiographique
   P radiografia, badanie radiograficzne
   R радиографическое исследование, радиография

-710 **radiometallography, radiography of metals**
   Radiography applied to metals.
   D Radiometallographie
   F radiométallographie, métalloradiographie
   P radiografia metali
   R радиография металлов, металлорадиография

-715 **radiograph**
   The picture obtained on a photographic plate or paper by means of a penetrating radiation, e.g. X-rays or gamma rays.
   D Radiogramm(bild)
   F radiogramme, radiographie
   P radiogram
   R радиограмма

-720 **microradiography**
　　The examination of metal structures by passing radiation through a thin specimen in contact with a photographic emulsion and subsequently magnifying the obtained radiograph for observation.
　　D Mikroradiographie
　　F microradiographie
　　P mikroradiografia
　　R микрорадиография

-725 **microradiograph**
　　The radiograph obtained by means of microradiography.
　　D Mikroradiogramm, mikroradiographische Aufnahme
　　F microradiographie
　　P mikroradiogram
　　R микрорадиограмма

-730 **X-ray microscopy**
　　Microradiography in which X-rays are used.
　　D Röntgenmikroskopie
　　F microradiographie par rayons X
　　P mikroskopia rentgenowska, mikrorentgenografia, rentgenomikroskopia
　　R микрорентгенография, рентгеновская микроскопия

-735 **absorption microanalysis**
　　A quantitative method of determining the composition in an area of the microstructure based upon the differences in absorption of different elements in radiation of a given wavelength.
　　D Absorptionsmikroanalyse
　　F microanalyse par absorption
　　P mikroanaliza absorpcyjna
　　R абсорбционный микроанализ

-740 **autoradiography**
　　A technique of determining the distribution of an element in a material by photographically recording the radiation from a sample of this material in which radioactive isotopes of the element in question have been formed by irradiation or added during manufacture.
　　D Autoradiographie
　　F autoradiographie, examen radiographique, étude radiographique
　　P autoradiografia
　　R авторадиография, радиоавтография

-745 **contrast autoradiography**
　　Autoradiography based on the analysis of a distribution pattern consisting of spots of various blackening intensity.
　　D Kontrastautoradiographie
　　F autoradiographie de contraste
　　P autoradiografia kontrastowa
　　R контрастная авторадиография

-750 **track autoradiography**
　　Autoradiography based on the analysis of a distribution pattern consisting of silver grains of different concentration per unit area.
　　D Spurenzählungsautoradiographie
　　F autoradiographie de trace
　　P autoradiografia śladowa
　　R следовая авторадиография

-755 **autoradiograph, radioautograph**
　　The photograph of an object containing radioactive isotopes obtained by developing the photographic film which has been previously placed in contact with this object.
　　D Autoradiogramm, Autoradiographie
　　F autoradiogramme, autoradiographie
　　P autoradiogram
　　R радиоавтограф, авторадиограмма

-760 **radioactive tracer**
　　A radioactive isotope of a given element used for determining the distribution of this element in an alloy.
　　D Radioindikator, radioaktiver Indikator
　　F radiotraceur, traceur radioactif, indicateur radioactif
　　P znacznik promieniotwórczy, radioindykator
　　R радиоактивный индикатор, изотопный индикатор

-765 **(radio)activation analysis**
　　A method of determining small traces of foreign elements in a metal by irradiating the sample in a flux of elementary particles and measuring the radioactivity of the radioelement formed.
　　D Aktivierungsanalyse
　　F analyse par activation
　　P analiza aktywacyjna
　　R радиоактивационный анализ

-770 **electron (diffraction) analysis, electron diffraction study**
The method of investigating of metals based on the diffraction of electrons in the crystal lattice.
D Elektronenbeugungsuntersuchung
F étude par diffraction électronique, analyse (par diffraction) électronique
P elektronografia
R электронография, электронографирование, электронографический анализ

-775 **electron diffraction pattern, electron diffraction photograph, electronogram**
The diffraction pattern obtained as a result of the diffraction of electrons in the crystal lattice.
D Elektronenbeugungsbild, Elektronenbeugungsaufnahme
F diagramme de diffraction électronique, diagramme de diffraction des électrons
P elektronogram, dyfraktogram elektronowy
R электронограмма

-780 **electron probe microanalysis, electron microprobe analysis, X-ray microanalysis**
A technique of chemical analysis based on the spectrographic examination of the characteristic X-radiation emitted by a selected area of the specimen under the influence of an impinging electron beam.
D Elektronenstrahl-Mikroanalyse, Mikrosondenuntersuchung
F microsondage électronique, microanalyse par (micro)sonde électronique, analyse à la microsonde
P mikroanaliza rentgenowska
R рентгеновский (спектральный) анализ с помощью электронного зонда

-785 **non-destructive inspection, non-destructive testing**
Testing of materials without damaging the specimens or objects under examination.
D zerstörungsfreie Prüfung
F contrôle non destructif, examen non destructif, essai non destructif
P badanie nieniszczące
R неразрушающее испытание, испытание без разрушения, неповреждающий контроль

-790 **flaw detection**
Non-destructive inspection of hidden defects in the objects under examination.
D Werkstoffehlerprüfung, Defektoskopie
F défectoscopie, détection des défauts
P defektoskopia
R дефектоскопия

-795 **sonic test(ing), aural test**
Non-destructive testing consisting in striking the part under examination and comparing the ring produced with that of a sound specimen.
D Klangprobe
F essai par le son
P próba dźwiękowa, próba dźwięku; defektoskopia akustyczna
R проба на звучность; акустическая дефектоскопия

-800 **ultrasonic testing, supersonic testing**
Testing of materials by means of ultrasounds.
D Ultraschallprüfung, Ultraschalldefektoskopie, Durchschallung, ultraakustische Prüfung
F contrôle par ultra-sons, inspection ultrasonique, sondage ultra-sonore
P badanie ultradźwiękowe, defektoskopia ultradźwiękowa
R ультразвуковая дефектоскопия, ультраакустическая дефектоскопия, испытание ультразвуком, прозвучивание

-805 **magnetic-particle inspection, magnafluxing, magnetic crack detection**
A non-destructive method of detecting defects in ferromagnetic materials, such as iron or steel, by the application of fine magnetic particles.
D Magnetpulver-Prüfung, Magnetdefektoskopie
F contrôle magnétoscopique, magnétoscopie, examen magnétoscopique, contrôle magnétique
P defektoskopia magnetyczna
R магнитн(о-порошков)ая дефектоскопия

-810 **dry (powder) method**
A method of magnetic particle inspection in which a fine magnetic powder is the indicating medium.
D trockenes Magnetpulververfahren
F méthode à la poudre magnétique
P metoda proszkowa sucha
R метод магнитных порошков

-815 **wet method, suspension method**
  A method of magnetic particle inspection in which a liquid bath, consisting of a suspension of magnetic particles in a light petroleum distillate or a light oil, is the indicating medium.
  D nasses Magnetpulververfahren
  F méthode à la suspension magnétique
  P metoda proszkowa mokra
  R метод магнитных суспензий

-820 **X-ray radiographic inspection**
  Non-destructive inspection of materials by means of X-rays.
  D Röntgenwerkstoffprüfung, Röntgendefektoskopie
  F contrôle radiographique par les rayons X
  P defektoskopia rentgenowska, rentgenodefektoskopia
  R рентгенодефектоскопия

-825 **gamma(-ray) radiography, gamma-ray inspection**
  Non-destructive testing of materials by means of gamma rays.
  D Gamma(-Radio)graphie, Gammadefektoskopie
  F gammagraphie, contrôle gammagraphique, examen gammagraphique, sondage gammagraphique, radiographie par les rayons γ
  P gamma(radio)grafia, radiografia promieniami gamma
  R гаммаграфия, гамма-(лучевая) дефектоскопия

-830 **neutron radiography**
  A technique of non-destructive testing of materials, analogous to X-ray and gamma-ray radiography, but based on the application of a stream of neutrons.
  D Neutronenradiographie
  F neutrographie
  P neutronografia
  R нейтронография, нейтронная дефектоскопия

-835 **fluorescent penetrant inspection, fluorescent particle inspection, fluoroscope test, penetrant flaw detection**
  The use of fluorescent liquids to detect surface cracks, e.g. on castings.
  D Fluoreszenzprüfung
  F contrôle par colorants pénétrants, contrôle par ressuage, examen par fluorescence
  P badanie fluorescencyjne
  R флуоресцентная дефектоскопия

85 MATERIAL DEFECTS
WERKSTOFFEHLER
DÉFAUTS DU MATÉRIAU
WADY MATERIAŁOWE
ДЕФЕКТЫ МАТЕРИАЛА

-005 **material defect**
A defect inherent in a material.
D Werkstoffehler, Materialfehler
F défaut du matériau
P wada materiałowa, wada materiału
R дефект материала, порок материала

-010 **surface defect, surface imperfection**
A defect inherent in the surface of a metal product.
D Oberflächenfehler
F défaut superficiel, défaut de surface
P wada powierzchni(owa)
R поверхностный дефект, поверхностный порок

-015 **cracking**
Formation of cracks or fissures in a material.
D Rißbildung
F fissuration, criquage
P pękanie
R трещинообразование, растрескивание

-020 **residual stresses, internal stresses, locked-in stresses, locked-up stresses**
Those stresses that remain in a material when all externally applied forces as well as changes in temperature are absent.
D Restspannungen, bleibende Spannungen, Eigenspannungen, innere Spannungen, Nachspannungen
F contraintes résiduelles, contraintes internes, tensions internes
P naprężenia własne, naprężenia szczątkowe, naprężenia wewnętrzne
R остаточные напряжения, собственные напряжения, внутренние напряжения

-025 **thermal stresses, temperature stresses**
Stresses that arise in a material owing to a temperature gradient.
D Wärmespannungen, Temperaturspannungen, Abkühlungsspannungen
F contraintes thermiques
P naprężenia cieplne, naprężenia termiczne
R термические напряжения, тепловые напряжения, температурные напряжения

-030 **body stresses, macrostresses, macroscopic residual stresses**
Stresses that act between the different regions of the cross-section or between the different parts of a metal object.
D makroskopische Spannungen, Spannungen 1. Art
F tensions macroscopiques
P naprężenia makroskopowe, naprężenia strefowe, makronaprężenia, naprężenia własne pierwszego rodzaju
R (внутренние) напряжения первого рода, напряжения I рода, объёмные напряжения, макроскопические напряжения

-035 **microstresses, microscopic stresses, constitutional stresses, textural stresses**
Stresses that act within individual grains or between adjacent grains.
D Gefügespannungen, Umwandlungsspannungen, Spannungen 2. Art
F tensions microscopiques
P naprężenia mikroskopowe, naprężenia strukturalne, mikronaprężenia, naprężenia własne drugiego rodzaju
R микронапряжения, (внутренние) напряжения второго рода, структурные напряжения

-040 **fracture, rupture, failure**
A permanent break-up of a piece of material into two or more parts under the influence of external forces.
D Bruch
F rupture
P pęknięcie
R излом, разрыв, разрушение

-045 **microcrack, hairline crack, microfissure**
A crack of microscopic size.
D Mikroriß, Haarriß
F microfissure, microcrique
P mikropęknięcie
R микротрещина, волосовина

-050 **craze, surface crack**
A crack which is restricted to the surface layer of a metal object.
D Oberflächenriß, Anriß
F crique
P pęknięcie powierzchniowe
R поверхностная трещина

-055 **Griffith crack**
Any one of the minute cracks, its existence having been postulated in brittle materials, which produce stress concentration under loading thus raising locally the applied stress to the theoretical strength.
D Griffithscher Riß
F fissure de Griffith
P (mikro)pęknięcie Griffitha
R трещина Гриффи(т)са

-060 **heat crack, thermal crack, fire crack**
A crack produced by thermal stresses.
D Wärmeriß
F tapure, fissure de température
P pęknięcie cieplne
R термическая трещина, тепловая трещина

-065 **stress crack**
A crack produced by residual stress.
D Spannungsriß
F fissure sous tension
P pęknięcie naprężeniowe
R трещина вследствие внутренних напряжений

-070 **strain crack**
A crack occurring when a metal is being plastically deformed.
D Verformungsriß
F fissure de déformation, fissure due à la déformation
P pęknięcie przy odkształcaniu
R деформационная трещина

-075 **cold crack**
A crack which appears at room or subzero temperature.
D Kaltriß, Kaltbruch
F fêlure à froid, cassure à froid
P pęknięcie na zimno
R холодная трещина

-080 **hot crack**
A crack which appears at elevated temperatures.
D Warmriß, Warmbruch
F fêlure à chaud, cassure à chaud
P pęknięcie na gorąco
R горячая трещина

-085 **cleavage (fracture)**
A fracture occurring along a definite crystallographic plane.
D Spaltbruch
F rupture par clivage
P pęknięcie łupliwe, pęknięcie rozdzielcze
R скол, раскалывающий разрыв

-090 **discontinuity of material**
A flaw in the structure of a metal consisting in an abrupt interruption of the metallic matrix caused by the presence of a non-metallic inclusion, pore, or crack.
D Unterbrechung des Zusammenhanges
F solution de continuité du métal
P nieciągłość materiału, przerwa ciągłości materiału
R нарушение сплошности материала

-095 **casting defect**
A defect in a casting due to irregularities in the production process or to the use of improper materials.
D Gußfehler
F défaut de fonderie
P wada odlewnicza
R литейный дефект, литейный порок

-100 **pore, void**
A minute hole or cavity within a solid metal.
D Pore, Hohlraum
F pore, cavité
P por
R пора, пустота

-105 **porosity, sponginess**
The presence of pores in a solid metal.
D Porosität
F porosité
P porowatość
R пористость

-110 **pinholes**
    Extremely fine gas holes in cast metals.
    D Mikroporen
    F micropores
    P mikropory
    R микропоры, микропустоты

-115 **pinhole porosity, microporosity**
    The presence of pinholes in a solid metal.
    D Mikroporosität
    F microporosité
    P mikroporowatość
    R микропористость

-120 **shrinkage porosity**
    A dense concentration of small or even microscopic voids with sharp contours and rough-surface walls within the cast metal.
    D Mikrolunker
    F retassure dispersée, microretassure, porosité interdendritique
    P rzadzizna
    R усадочная пористость, усадочная рыхлота

-125 **shrinkage cavity, contraction cavity, pipe**
    A void, usually of a conical or spindle-like shape, formed in the top of an ingot or the feeder head of a casting by contraction during solidification of the last parts of the liquid metal.
    D Lunker
    F retassure
    P jama skurczowa, jama usadowa
    R усадочная раковина

-130 **blister**
    A defect in cast metal caused by gas bubbles either on the surface or just below the surface, possessing a fairly large volume compared with common blow holes.
    D Luftblase
    F refus
    P bąbel
    R газовая раковина

-135 **blow hole**
    Any one of the spherical or egg-shaped voids formed by the evolution of dissolved gas that fails to escape during the solidification of a metal.
    D Gasblase
    F soufflure
    P pęcherz gazowy
    R внутренняя газовая раковина, закрытая газовая раковина, газовая пустота

-140 **grey spots**
    A material defect occurring in malleable cast iron and characterized by the presence of grey spots of separated graphite, which are clearly visible against the bright background of the fracture.
    D Graphitflecken
    F taches grises
    P zaszarzenie
    R графитная спель

-145 **hard spots**
    A material defect in grey cast iron castings, characterized by the presence of chilled cast-iron spots on the fracture of the casting.
    D harte Stellen
    F points durs
    P zabielenie
    R отбел

-150 **internal chill, inverse chill**
    The appearance of a white or mottled cast iron structure in the interior of grey iron castings.
    D umgekehrter Hartguß
    F trempe inverse
    P odwrotne utwardzenie żeliwa
    R обратный отбел чугуна

-155 **bright fracture**
    A material defect occurring in malleable cast iron, characterized by the bright appearance of the whole fracture of the casting.
    D heller Bruch
    F cassure claire
    P przełom jasny
    R светлый излом

-160 **white fracture**
    A material defect occurring in malleable cast iron, characterized by the appearance of distinct white concentrations of cementite, in addition to bright concentrations of pearlite.
    D weißer Bruch
    F cassure blanche
    P przełom biały
    R белый излом

-165 **tin sweat, tin exudation**
    The formation of frozen droplets of a tin rich low-melting-point phase exuded from the surface of tin bronzes.
    D Zinnschweiß
    F ressuage d'étain, remonte d'étain
    P pot cynowy
    R выпоты олова

-170 **casting stresses**
Stresses arising in castings during the casting process.
D Gußspannungen
F tensions de coulée, contraintes de coulée
P naprężenia odlewnicze
R литейные напряжения

-175 **shrinkage stresses**
Stresses arising in casting due to the prevention of shrinkage by the mould.
D Schwindungsspannungen
F tensions de retrait, contraintes de retrait
P naprężenia skurczowe
R усадочные напряжения

-180 **quenching defect**
A material defect arising during the quenching process, especially during quench hardening.
D Härtefehler
F défaut de trempe
P wada hartownicza
R закалочный порок

-185 **quenching stresses**
Stresses arising within metal objects as a result of quenching, especially during quench hardening.
D Abschreckspannungen, Ablöschspannungen, Härtespannungen
F tensions de trempe, contraintes de trempe
P naprężenia hartownicze
R закалочные напряжения

-190 **quench(ing) crack, hardening crack**
A crack produced by quenching stresses.
D Härteriß
F tapure de trempe, crique de trempe
P pęknięcie hartownicze
R закалочная трещина

-195 **soft spots**
A defect sometimes encountered on the surface of quench-hardened steel products, consisting in the existence of small areas of low hardness scattered within the hardened matrix.
D Weichfleckigkeit, Weichflecke
F plages douces
P miękkie plamy
R мягкие пятна

-200 **warping**
The permanent distortion of a metal object from its true form occurring during quenching as a result of the action of thermal stresses.
D Werfen, Härteverzug
F gauchissement, déjettement
P paczenie się
R коробление

-205 **burning**
Permanent damage caused to a metal by heating it to a high temperature so that oxidation at grain boundaries occurs or melting begins.
D Verbrennen
F brûlure
P przepalenie
R пережог

-210 **surface decarburization**
The loss of carbon from the surface layer of steel as a result of heating in an oxidizing atmosphere.
D Oberflächenentkohlung, Randentkohlung
F décarburation superficielle
P odwęglenie powierzchni(owe)
R поверхностное обезуглероживание, обезуглероживание поверхности

-215 **incipient melting**
A defect sometimes arising during heat treatment when the solidus temperature on heating is surpassed.
D Anschmelzung, örtliches Aufschmelzen
F fusion partielle (au recuit)
P nadtopienie
R оплавление

-220 **temper brittleness, temper embrittlement**
Brittleness of certain low-alloy steels (notably nickel-chromium steels) occurring when they are held within, or cooled slowly through, the temperature range between 600° and 300°C.
D Anlaßsprödigkeit, Krupp-Krankheit
F fragilité de revenu, maladie de Krupp, fragilité Krupp
P kruchość odpuszczania
R отпускная хрупкость, хрупкость отпуска

-225 **abnormality of steel**
    A defect in the structure of carburized steel consisting in the occurrence of massive cementite in the case along with free ferrite.
    D Gefügeanomalie des Stahles
    F anomalie de structure d'acier
    P anormalność stali
    R анормальность стали

-230 **ghost (line), ghost band, ferrite ghost, ferrite band, phosphorus banding, segregation band**
    Any one of the light-coloured streaks, sometimes visible on freshly machined surfaces of steel, representing segregation regions of ferrite containing a high proportion of phosphorus in solution and sulphide inclusions.
    D Seigerungsstreifen, Schattenlinie, Seigerungsbart, Bartseigerung, Stengelseigerung, A-Seigerung, Seigerungszeile
    F bande de ségrégation, bande de ferrite libre, veine sombre
    P (ferrytyczne) pasmo segregacyjne, pasmo ferrytyczne
    R светловина, ликвационная полоса

-235 **carbide banding**
    A defect of some alloy steels, particularly bearing steels, consisting in an arrangement of carbides in the microstructure in the form of elongated parallel conglomerations.
    D Karbidzeiligkeit; Karbidzeilen
    F bandes de carbures
    P pasmowość węglików, pasemkowość węglików; pasma węglików
    R строчечность карбидов

-240 **(snow)flakes, chrome checks, fish-eyes, shatter cracks**
    Minute internal cracks of silvery brightness occurring in some forged or wrought alloy steels when cooling after the working was not sufficiently slow.
    D Flocken, Flockenrisse
    F flocons, œils de poisson
    P płatki
    R флокены

-245 **tin pest, tin disease, tin plague**
    The phenomenon of the crumbling of tin into a powder which accompanies the allotropic transformation of white tin into grey tin.
    D Zinnpest
    F maladie de l'étain, peste de l'étain
    P choroba cynowa, zaraza cynowa, trąd cynowy
    R оловянная чума

-250 **white spots**
    A defect encountered in the microstructure of quench hardened high-carbon steels, mainly bearing steels, as etching-resisting areas on microsections, these areas appearing in the immediate neighbourhood of microcracks caused by contact fatigue.
    D weiße Flecken, weiße Schichten
    F taches blanches, phase blanche
    P białe plamy, biała faza
    R белые пятна

-255 **butterflies**
    White spots, having the contour resembling a butterfly, encountered in the microstructure of bearing steels in the immediate neighbourhood of non-metallic inclusions.
    D Schmetterlingsfehler
    F papillons
    P motyle
    R бабочки

-260 **corrosion**
    The gradual destruction of a metal or alloy by a chemical or electrochemical attack caused by the environment.
    D Korrosion
    F corrosion
    P korozja
    R коррозия

-265 **chemical corrosion**
    Corrosion resulting from a purely chemical action of the environment, i.e. without interference of an electric current.
    D chemische Korrosion
    F corrosion chimique
    P korozja chemiczna
    R химическая коррозия

-270 **electrochemical corrosion, galvanic corrosion**
    Corrosion resulting from the electrolytic action of short-circuited galvanic cells which appear within a heterogeneous material when the latter is placed in an electrolyte.
  D elektrochemische Korrosion, galvanische Korrosion
  F corrosion électrochimique, corrosion galvanique
  P korozja elektrochemiczna
  R электрохимическая коррозия, гальваническая коррозия

-275 **atmospheric corrosion, climatic corrosion**
    Corrosion caused by the action of humid free air.
  D atmosphärische Korrosion
  F corrosion atmosphérique, corrosion à l'air ambiant
  P korozja atmosferyczna
  R атмосферная коррозия

-280 **water corrosion**
    Electrochemical corrosion caused by the action of water.
  D Wasserkorrosion
  F corrosion par l'eau, corrosion à l'eau, corrosion aqueuse
  P korozja wodna
  R водяная коррозия

-285 **soil corrosion**
    Electrochemical corrosion caused by the action of the soil.
  D Bodenkorrosion
  F corrosion souterraine
  P korozja ziemna
  R почвенная коррозия

-290 **sea water corrosion, marine corrosion**
    Corrosion caused by the action of sea water or of an atmosphere containing the fog of sea water.
  D Seewasserkorrosion
  F corrosion par l'eau de mer, corrosion à l'eau de mer, corrosion marine
  P korozja morska
  R морская коррозия

-295 **dry corrosion, gas corrosion**
    Corrosion taking place in a dry gaseous environment.
  D Gaskorrosion
  F corrosion sèche, corrosion par le gaz
  P korozja gazowa
  R сухая коррозия, газовая коррозия

-300 **crevice corrosion**
    Corrosion taking place deep inside tight crevices.
  D Spaltkorrosion, Rißkorrosion
  F corrosion fissurante
  P korozja szczelinowa
  R щелевая коррозия

-305 **selective corrosion**
    Corrosion restricted to some definite components of an alloy.
  D selektive Korrosion
  F corrosion sélective
  P korozja selektywna, korozja wybiorcza
  R избирательная коррозия, селективная коррозия

-310 **stress corrosion**
    Corrosion caused by the simultaneous action of the environment and static stresses.
  D Spannungs(riß)korrosion
  F corrosion sous tension, corrosion sous contrainte
  P korozja naprężeniowa
  R коррозия под напряжением

-315 **fatigue corrosion**
    Corrosion caused by the simultaneous action of the environment and alternating stresses.
  D Ermüdungskorrosion
  F corrosion par fatigue
  P korozja zmęczeniowa
  R усталостная коррозия

-320 **contact corrosion, bimetallic corrosion**
    Electrochemical corrosion taking place when two different metals are in contact in the environment constituting an electrolyte.
  D Kontaktkorrosion; Berührungskorrosion
  F corrosion de contact, corrosion par contact, corrosion bimétallique
  P korozja stykowa, korozja kontaktowa
  R контактная коррозия

-325 **fretting corrosion**
    Corrosion caused by the simultaneous action of the environment and friction.
  D Reibkorrosion
  F corrosion par frottement, fretting
  P korozja cierna
  R коррозия при трении

-330 **surface corrosion**
  Corrosion taking place on the surface of a metal or alloy.
  D Oberflächenkorrosion
  F corrosion superficielle
  P korozja powierzchniowa
  R поверхностная коррозия

-335 **uniform corrosion**
  Surface corrosion which destroys the material uniformly over the whole surface.
  D gleichmäßige Korrosion, ebenmäßige Korrosion
  F corrosion uniforme
  P korozja równomierna
  R равномерная коррозия

-340 **localized corrosion**
  Corrosion restricted to some regions of a metal surface only.
  D örtliche Korrosion
  F corrosion local(isé)e
  P korozja miejscowa, korozja lokalna
  R местная коррозия, локальная коррозия

-345 **point corrosion**
  Localized corrosion restricted to individual points on the surface.
  D Punktkorrosion
  F corrosion ponctuelle
  P korozja punktowa
  R точечная коррозия, булавочная коррозия

-350 **pitting (corrosion)**
  Corrosion restricted to small areas but penetrating deeply into the material.
  D Loch(fraß)korrosion, Lochfraß
  F corrosion par piqûres
  P korozja wżerowa
  R питтинговая коррозия, питтинг, язвенная коррозия, коррозионная язва

-355 **subsurface corrosion**
  Corrosion which takes place within the material below its surface.
  D innere Korrosion
  F corrosion interne
  P korozja podpowierzchniowa, korozja wewnętrzna, korozja ukryta
  R подповерхностная коррозия, внутренняя коррозия

-360 **intercrystalline corrosion, intergranular corrosion, grain boundary corrosion**
  Corrosion propagating itself along grain boundaries.
  D interkristalline Korrosion, Korngrenzenkorrosion
  F corrosion intercristalline, corrosion intergranulaire
  P korozja międzykrystaliczna
  R межкристаллитная коррозия, межзёренная коррозия, коррозия по границам зёрен, интеркристаллитная коррозия

-365 **nuclear corrosion**
  A gradual destruction of materials resulting from the action of nuclear radiation.
  D Kernkorrosion
  F corrosion nucléaire, corrosion radiolytique
  P korozja jądrowa, korozja radiacyjna, korozja popromienna
  R ядерная коррозия, радиационная коррозия

-370 **rust**
  The product of electrochemical corrosion of iron and its alloys.
  D Rost, Eisenrost
  F rouille
  P rdza
  R ржавчина

-375 **patina**
  The product of atmospheric or soil corrosion of copper or copper-base alloys.
  D Patina, Edelrost
  F patine
  P patyna
  R патина

-380 **dezincification**
  A type of corrosion encountered in some brasses, particularly when exposed to warm sea-water, consisting in the apparently preferential removal of zinc.
  D Entzinkung
  F dézincage, dézingage
  P odcynkowanie
  R обесцинкование

-385 **season cracking, stress corrosion cracking**
Cracking of cold worked or stressed brass, occurring in corrosive environments containing ammonia or compounds of ammonia.
D Rißbildung durch Spannungskorrosion
F fissuration saisonnière, fissuration par corrosion sous tension, corrosion fissurante sous tension, season-cracking
P sezonowe pękanie
R сезонное (само)растрескивание, сезонная хрупкость

-390 **hydrogen embrittlement, pickling brittleness, acid embrittlement, acid brittleness, pickle brittleness, hydrogen brittleness, hydrogen fragility**
Reduced ductility of a metal resulting from the absorption of hydrogen gas.
D Wasserstoffbrüchigkeit, Wasserstoffsprödigkeit
F fragilité par l'hydrogène, fragilité due à l'hydrogène
P kruchość wodorowa
R водородная хрупкость

-395 **hydrogen disease, hydrogen unsoundness (of copper)**
Production of cracks in the oxygen-bearing copper occurring when the Cu_2O under the influence of hydrogen absorbed from the environment is reduced with the formation of H_2O, the latter exerting excessive pressure within the metal.
D Wasserstoffkrankheit
F maladie d'hydrogène
P choroba wodorowa
R водородная болезнь

-400 **lap(ping), overlap, cold shut**
A longitudinal surface defect produced in rolling and sometimes in forging by folding a fin of hot metal over the surface of this metal and pressing it into this surface without welding.
D Überlappung, Überwalzung
F repliure
P zawalcowanie
R закатка, закат, губа

-405 **lamination**
A material defect sometimes arising during rolling of sheet metal, consisting in the appearance of a local discontinuity of the material along a plane parallel to the sheet surface.
D Dopplung
F dédoublure
P rozwarstwienie
R расслой, расслоение

-410 **fatigue**
The failure of material subjected to repeated stress cycles, each of which is individually insufficient to cause failure.
D Ermüdung
F fatigue
P zmęczenie
R усталость

-415 **corrosion fatigue**
The reduction of the endurance of a metal or alloy when it is subjected simultaneously to alternating stresses and to corrosion.
D Korrosionsermüdung
F fatigue sous corrosion
P zmęczenie korozyjne
R коррозионная усталость

-420 **contact fatigue**
Fatigue occurring as a result of repeated localized pressure cycles between working surfaces of mating parts (e.g. in roller bearings).
D Kontaktermüdung
F fatigue de contact
P zmęczenie kontaktowe
R контактная усталость

-425 **thermal fatigue**
Surface cracking or deformation of a metal object caused by thermal stresses arising from temperature fluctuations.
D thermische Ermüdung
F fatigue (par cyclage) thermique
P zmęczenie cieplne
R термическая усталость

-430 **(oxide) scale**
A thick layer of oxide formed on a metal at a high temperature.
D Zunder
F écaille, calamine
P zgorzelina
R окалина

DEUTSCHES WÖRTERVERZEICHNIS

A

Abdruck m
Baumann-~ 80-325
Abdruckverfahren n
Baumann-~ 80-320
abgeschreckte Leerstelle f 30-100
Abkühlen n 70-070
~ an ruhender Luft 70-095
kontinuierliches ~ 70-090
langsames ~ 70-080
Abkühlung f 70-070
kontinuierliche ~ 70-090
langsame ~ 70-080
Abkühlungsgeschwindigkeit f 70-120
kritische ~ 70-180
Abkühlungskurve f 80-525
Abkühlungsspannungen fpl 85-025
Abkühlungsvermögen n 70-125
Abkühlungswirkung f 70-125
Ablöschen n 70-345
Ablöschspannungen fpl 85-185
Abnutzfestigkeit f 75-265
Abnutzungsfestigkeit f 75-265
Abriebfestigkeit f 75-265
Abschlußbereich m 40-640
Abschlußbezirk m 40-640
Abschreckalterung f 70-580
Abschreckbad n 70-195
Abschrecken n 70-075, 70-570
Abschreckflüssigkeit f 70-190
Abschreckgeschwindigkeit f 70-120
Abschreckhärten n 70-175
Abschreckintensität f 70-200
Abschreckmartensit m 50-335
Abschreckmittel n 70-185
Abschrecksorbit m 50-385
Abschreckspannungen fpl 85-185
Abschrecktemperatur f 70-140
Abschrecktroostit m 50-400
Abschrecken n 70-075
direktes ~ 70-760
Abschreckung f 70-075
Liquisol-~ 70-085
milde ~ 70-265
schroffe ~ 70-260
absolute Dilatometer-Untersuchung f 80-545
absolute Dilatometrie f 80-545
Absorptionsmikroanalyse f 80-735
Absorptionsquerschnitt m für Neutrone 75-530
Achse f
dreizählige ~ 20-075
kristallographische ~ 20-125
sechszählige ~ 20-085
vierzählige ~ 20-080
zweizählige ~ 20-070
Achsenverhältnis n 20-355
kristallographisches ~ 20-355
Actinium n 05-010
Admiralitätsbronze f 65-205

Admiralitätslegierung f 65-390
Affinität f 75-660
chemische ~ 75-660
Aktinium n 05-010
aktive Atmosphäre f 70-680
Aktivierung f
thermische ~ 45-275
Aktivierungsanalyse f 80-765
Aktivierungsenergie f 45-285
Aktivierungsvolumen n 30-475
Aktivität f 15-280
thermodynamische ~ 15-280
Aktivitätskoeffizient m 15-285
Alclad n 65-560
Aldrey n 65-585
Alitieren n 70-925
alkalibeständiger Stahl m 60-550
alkalibeständiges Gußeisen n 60-1080
Alkalibeständigkeit f 75-700
Alkalimetalle npl 55-030
alkoholische Pikrinsäure f 80-115
allgemeiner Baustahl m 60-665
allotrope Modifikation f 25-390
allotrope Umwandlung f 25-455
Allotropie f 25-385
Alnico n 65-530
AlNiCo-Legierung f 65-530
Alpaka n 65-415
alphagenes Element n 50-455
Alsifer n 65-535
Alsimin n 60-1165
Altern n 70-575
magnetisches ~ 70-620
Alterung f 70-575
beschleunigte ~ 70-590
künstliche ~ 70-590
mechanische ~ 45-225
natürliche ~ 70-585
Alterungsbeständigkeit f 75-845
Alterungshärtung f 70-565
Altmetallegierung f 55-280
Aludur n 65-580
Alumel n 65-460
Aluminieren n 70-925
Aluminium n 05-015
Aluminiumbronze f 65-135
Aluminiumgußeisen n 60-1020
Aluminiummessing n 65-340
Aluminiumstahl m 60-235
Amalgam n 65-850
Americium n 05-020
Amerizium n 05-020
amorph 20-005
Analyse f
chemische ~ 80-470
magnetische ~ 80-560
metallographische ~ 80-035
physikalisch-chemische ~ 80-490
quantitative thermische ~ 80-535

ANALYSE

Analyse
thermische ~ 80-505
thermodynamische ~ 80-535
thermomagnetische ~ 80-565
Anelastizität f 75-050
Anfangspermeabilität f 75-620
angelassener Martensit m 50-345
Anhäufung f von Leerstellen 30-125
Animpfen n von Kristallbildung 25-215
Anisotropie f 20-445
magnetische ~ 75-575
Anisotropieenergie f 75-590
Anisotropiefaktor m 75-595
Anisotropiekoeffizient m 75-595
Anisotropiekonstante f 75-595
Anlaßätzung f 80-185
Anlaßbarkeit f 75-765
anlaßbeständiger Stahl m 60-395
Anlaßbeständigkeit f 75-840
Anlassen n 70-365
~ bei hohen Temperaturen 70-375
~ bei niedrigen Temperaturen 70-370
Anlaßglühen n 70-365
Anlaßmartensit m 50-345
Anlaßsorbit m 50-390
Anlaßsprödigkeit f 85-220
Anlaßtemperatur f 70-380
Anlaßtroostit m 50-405
Anlaufätzung f 80-185
anlaufbeständige Legierung f 55-500
Anlaufbeständigkeit f 75-680
Anlauffarbe f 70-385
Anodenkupfer n 65-040
anodische Ätzung f 80-175
anomaler Stahl m 60-465
anomales Gefüge n 40-135
Anregung f
thermische ~ 45-275
Anriß m 85-050
Anschmelzung f 85-215
Anticorodal n 65-575
Antiferromagnetismus m 75-565
antimagnetischer Stahl m 60-510
Antimon n 05-025
Antimonblei n 65-660
Antimonbronze f 65-165
Antiphasenbereich m 40-625
Antiphasendomäne f 40-625
Antiphasen-Domänengrenze f 40-630
Antiphasengrenze f 40-630
Anzahl f
~ der Freiheiten 35-030
äquikohäsive Temperatur f 30-570
Architekturbronze f 65-270
Argentan n 65-415
Argentanlot n 65-745
Argon n 05-030
Armco-Eisen n 60-010
armiertes Gußeisen n 60-1070
Arsen n 05-035
Arsenkupfer n 65-090
arteigener Keim m 25-110
artfremder Keim m 25-120
A-Seigerung f 85-230
Astat n 05-040
Astatin n 05-040
Asterismus m 80-680

Asymmetrie f
energetische ~ 15-235
athermische Keimbildung f 25-200
athermische Lösung f 15-265
athermische Umwandlung f 25-520
Atom
eingelagertes ~ 30-075
nächstbenachbartes ~ 20-195
substituiertes ~ 30-070
verlagertes ~ 30-510
Atomabstand m 20-230
atomare Fehlstelle f 30-030
Atombindung f 10-330
Atomdurchmesser m 10-080
Atomebene f 20-175
Atomgewicht n 10-075
Atomgrößeneffekt m 10-395
Atomnummer f 75-330
Atomradius m 10-085
Atomrudiment n 10-040
Atomrumpf m 10-040
Atomvolumen n 75-320
Atomwärme f 75-370
Atmosphäre f
aktive ~ 70-680
Cottrell-~ 30-440
geregelte ~ 70-700
inerte ~ 70-675
kontrollierte ~ 70-700
reaktive ~ 70-680
atmosphärische Korrosion f 85-275
Ätzanlassen n 80-185
Ätzdauer f 80-125
Ätzen n 80-120
Ätzfiguren fpl 80-200
Ätzfurche f 80-195
Ätzgraben m 80-195
Ätzgrübchen n 80-205
Ätzmittel n 80-100
Ätzpolieren n 80-085
Ätzung f 80-120
anodische ~ 80-175
chemische ~ 80-170
elektrochemische ~ 80-175
elektrolytische ~ 80-175
thermische ~ 80-190
Ätzzeit f 80-125
Auermetall n 65-875
aufgedampftes Metall n 55-145
aufgefülltes Band n 10-245
aufgekohlte Randzone f 70-780
aufgekohlter Stahl m 60-310
aufgeschwefelter nichtrostender Stahl m 60-525
aufgespaltenes Zwischengitteratom n 30-085
aufgespaltene Versetzung f 30-325
Aufhärtbarkeit f 70-020
Aufkohlen n 70-715
örtliches ~ 70-735
Aufkohlung f 70-715
Aufkohlungsgas n 70-745
Aufkohlungsmedium n 70-740
Aufkohlungsmittel n 70-740
Aufkohlung f
partielle ~ 70-735
Auflegieren n im festen Zustand 70-655
Auflichtmikroskopie f 80-225
Aufnahme f
Debye-Scherrer-~ 80-635

Aufnahme
elektronenmikroskopische ~ 80-365
Laue-~ 80-620
mikroradiographische ~ 80-725
Aufprallabschreckung f 70-085
Aufschmelzen n
örtliches ~ 85-215
Aufschwefeln m 70-895
Aufschweißlegierung f 55-590
Aufspaltung f einer Versetzung 30-320
Aufstauung f von Versetzungen 30-400
Aufsticken f 70-785
Auftraglegierung f 55-590
Aufwachsen n
epitaktisches ~ 25-345
Aufzinken n 70-930
Ausdehnungskoeffizient m
kubischer ~ 75-460
linearer ~ 75-455
thermischer ~ 75-450
Außenelektron n 10-065
äußeres Elektron n 10-065
ausgeprägte Streckgrenze f 75-110
ausgeschiedene Phase f 40-305
Ausglühen n 70-420, 70-465
unvollständiges ~ 70-470
vollständiges ~ 70-465
Aushalten n 70-055
aushärtbare Legierung f 55-385
aushärtbarer Stahl m 60-400
Aushärten n 70-560
aushärtende Legierung f 55-385
Aushärtung f 70-560
Ausheilen n
~ der Defekte 30-135
~ der Gitterbaufehler 30-135
Auslagern n 70-575
Ausscheidung f 40-300
diskontinuierliche ~ 25-380
kontinuierliche ~ 25-375
Ausscheidungen fpl 40-305
inkohärente ~ 40-325
kohärente ~ 40-320
submikroskopische ~ 40-315
ausscheidungshärtbare Legierung f 55-385
Ausscheidungshärten n 70-560
Ausscheidungshärtung f 70-560
Ausscheidungsvorgang m 40-300
Ausschließungsprinzip n
Paulisches ~ 10-170
Ausschlußmetall n 55-640
Austauschstahl m 60-270
Austempering n 70-310
Austenit m 50-025
primärer ~ 50-030
umgewandelter ~ 50-035
unterkühlter ~ 50-265
austenitbildendes Element n 50-450
Austenitbildner m 50-450
Austenitformhärten n 70-960
austenitischer Stahl m 60-420
austenitisches Gußeisen n 60-1050
austenitisch-ferritischer Stahl m 60-455
Austenitisieren n 70-130
Austenitisiertemperatur f 70-135
Austenitisierung f 70-130
Austenitisierungstemperatur f 70-135
Austenitkorn n
ehemaliges ~ 50-040
früheres ~ 50-040

austenitstabilisierendes Element n 50-450
Austenitstabilisierung f 50-425
Austrittsarbeit f 10-060
Automatenlegierung f 55-560
Automatenmessing n 65-370
Automatenstahl m 60-710
Autoradiogramm n 80-755
Autoradiographie f 80-740, 80-755
Avional n 65-610
azikularer Martensit m 50-315
azimutale Quantenzahl f 10-135
Azimutalquantenzahl f 10-135

B

Babbit n 65-645
Babbit-Metall n 65-645
Badabschrecken n 70-245, 70-115
Badabschreckung f 70-115
Badaufkohlen n 70-725
Badeinsetzen n 70-725
Badnitrieren n 70-795
Bahnmetall n 65-685
Bainit m 50-410
oberer ~ 50-420
unterer ~ 50-415
bainitischer Stahl m 60-435
bainitisches Gußeisen n 60-1060
bainitische Umwandlung f 50-290
Bainitisieren n 70-310
Band n
aufgefülltes ~ 10-245
besetztes ~ 10-245
erlaubtes ~ 10-240
verbotenes ~ 10-250
völlig aufgefülltes ~ 10-245
vollständig besetztes ~ 10-245
Bandagenstahl m 60-760
Barium n 05-045
Bartseigerung f 85-230
basischer Stahl m 60-620
Basisebene f 20-360
Baumann-Abdruck m 80-325
Baumann-Abdruckverfahren n 80-320
Baumann-Schwefelabdruckverfahren n 80-320
Bäumchengefüge n 40-090
Bauschinger-Effekt m 75-175
Baustahl m 60-670
allgemeiner ~ 60-665
Bearbeitungsrekristallisation f 45-445
Begleitelement n 55-260
begrenzte Mischbarkeit f 35-320
Behandlung f
chemisch-thermische ~ 70-650
härtesteigernde ~ 70-015
magnetothermische ~ 70-965
thermische ~ 70-006
thermisch-magnetische ~ 70-965
thermochemische ~ 70-650
thermomechanische ~ 70-950
Beilby-Schicht f 80-090
Beimengungen fpl
zufällige ~ 55-200
Bekeimung f
künstliche ~ 25-215
Bereich m
verbotener ~ 10-250
Weißscher ~ 40-635
Bereichsstruktur f 40-620

Bergauf-Diffusion f 45-340
Berggold n 65-765
Berkelium n 05-050
Berthollide npl 35-250
berthollide Verbindungen fpl 35-250
Berthollidverbindungen fpl 35-250
Berührungskorrosion f 85-320
Beryllium n 05-055
Berylliumbronze f 65-150
beruhigter Stahl m 60-655
Beschatten n 80-370
beschleunigte Alterung f 70-590
beschleunigte Diffusion f 45-355
beschleunigtes Kriechen n 45-510
beschränkte Mischbarkeit f 35-320
besetztes Band n 10-245
Bessemerroheisen n 60-845
Bessemerstahl m 60-595
Beständigkeit f
  chemische ~ 75-665
Bestandteil m 35-035
Bestimmung f
  ~ der Kristallorientierung 80-690
Bestrahlungsfehler m 30-485
Bestrahlungsverfestigung f 30-495
betagenes Element n 65-860
Betonstahl m 60-795
Beugungsaufnahme f 80-610
Beugungsbild n 80-610
Beugungsfleck n 80-660
Beugungslinie f 80-645
Bewegung f
  konservative ~ 30-290
Bewehrungsstahl m 60-795
Bezirk m
  magnetischer ~ 40-635
  Weißscher ~ 40-635
Biegefestigkeit f 75-090
Biegungsfestigkeit f 75-090
Bikristall m 20-435
bildsame Formgebung f 45-015
bildsame Formung f 45-015
Bildsamkeit f 75-025
Bildungsenergie f
  freie ~ 15-135
Bildungsenthalpie f 15-160
Bildungsentropie f 15-200
Bimetall n 65-880
binäre Legierung f 55-295
binäre Lösung f 35-405
binärer Stahl m 60-155
binäres Eutektikum n 40-205
binäres Karbid n 40-470
binäres System n 35-120
binär-eutektische Rinne f 35-730
Bindung f
  chemische ~ 10-320
  heteropolare ~ 10-325
  homöopolare ~ 10-325
  kovalente ~ 10-330
  metallische ~ 10-340
  van-der-Waalssche ~ 10-335
Bindungsenergie f 15-090
Bindungsenthalpie f 15-170
bivariantes System n 35-085
Blankglühen n 70-430
Blankhärten n 70-330
Blasenkupfer n 65-015

Blasstahl m 60-590
  Sauerstoff-~ 60-605
Blauanlauf m 70-390
Blaubrüchigkeit f 75-300
Bläuen n 70-400
Blauglühen n 70-445
Blauhitze f 70-395
Blausprödigkeit f 75-300
Blauwärme f 70-395
Blei n 05-240
Blei-Alkali-Lagermetall n 65-685
Bleibadhärten n 70-235
bleibende Formänderung f 45-010
bleibende Spannungen fpl 85-020
bleibende Verformung f 45-010
Bleibronze f 65-140
bleihaltiger Stahl m 60-250
bleihaltiges Messing n 65-360
Bleilot n 65-735
Bleizinnbronze f 65-185
Blindhärten n 70-775
Blindhärteversuch m 70-775
Blindhärtungsversuch m 70-775
Blisterkupfer n 65-015
Bloch-Funktion f 10-195
Bloch-Wand f 40-645
Blockierung f der Versetzungen 30-445
Blockseigerung f 25-415
  umgekehrte ~ 25-440
Bodenkorrosion f 85-285
Bor n 05-065
Borid n 70-905
Borieren n 70-900
borlegierter Stahl m 60-255
Borstahl m 60-255
Braunit m 70-840
Bravaisgitter n 20-275
Bravaissche Indizes mpl 20-170
Bremsvermögen n 75-535
Brennhärten n 70-290
Brillouin-Zone f 10-230
Brom n 05-070
Bronze f 65-095
  hochzinnhaltige ~ 65-115
  niedrigzinnhaltige ~ 65-120
  zinnfreie ~ 65-125
Bruch m 40-670, 85-040
  durchkörniger ~ 40-710
  faseriger ~ 40-720
  feinkörniger ~ 40-700
  glasiger ~ 40-735
  grobkörniger ~ 40-695
  heller ~ 85-155
  innerkristalliner ~ 40-710
  interkristalliner ~ 40-705
  intrakristalliner ~ 40-710
  körniger ~ 40-690
  kristalliner ~ 40-685
  muscheliger ~ 40-730
  samtartiger ~ 40-715
  sehniger ~ 40-720
  spröder ~ 40-675
  transkristalliner ~ 40-710
  verformungsloser ~ 40-675
  verzögerter ~ 40-755
  weißer ~ 85-160
  zäher ~ 40-680
  zwischenkörniger ~ 40-705
  zwischenkristalliner ~ 40-705
Bruchdehnung f 75-120
Brucheinschnürung f 75-125

Bruchfläche f 40-670
Brüchigkeit f 75-270
Bruchoberfläche f 40-670
Bruchprobe f 80-445
Bruchquerschnittsverminderung f 75-125
Buntmetalle npl 55-015
Burgersumlauf m 30-185
Burgersvektor m 30-190
Burgers-Vektor m 30-190

C

Cadmium n 05-075
Cadmiumbronze f 65-190
Caesium n 05-080
Calcium n 05-085
Californium n 05-090
Carbid n 40-430
Carbonitrieren n 70-870
Carbonyleisen n 60-020
Carbonylnickel n 65-425
Cäsium n 05-080
Cer n 05-100
Cerit n 65-865
Cermet m 55-675
charakteristische Temperatur f 10-280
chemisch beständiger Stahl m 60-535
chemische Affinität f 75-660
chemische Analyse f 80-470
chemische Ätzung f 80-170
chemische Beständigkeit f 75-665
chemische Bindung f 10-320
chemische Diffusion f 45-320
chemische Eigenschaften fpl 75-655
chemische Heterogenität f 40-415
chemische Korrosion f 85-265
chemischer Dendrit m 25-320
chemisches Element n 05-005
chemisches Gleichgewicht n 15-330
chemisches Polieren n 80-080
chemisches Potential n 15-175
chemische Verbindung f 35-235
chemische Zusammensetzung f einer Legierung 55-225
chemisch-thermische Behandlung f 70-650
Chlor n 05-105
Chrom n 05-110
Chrombronze f 65-195
Chromel n 65-465
Chromgußeisen n 60-1015
Chromieren n 70-920
Chromkarbid n 40-455
Chromkupfer n 65-200
Chrom-Mangan-Silizium-Stahl m 60-260
Chromnickel n 65-470
Chromnickelstahl m 60-190
Chromnitrid n 70-820
Chromoalitieren n 70-940
Chromstahl m 60-175
Cobalt n 05-115
Cohenit m 50-115
Cottrell-Atmosphäre f 30-440
Cottrellsche Wolke f 30-440
Crowdion n 30-470
C-Übergangszahl f 70-755

Curiepunkt m 25-560
Curietemperatur f 25-560
Curium n 05-125

D

Daltonide npl 35-245
daltonide Verbindungen fpl 35-245
Daltonidverbindungen fpl 35-245
Damaszenerstahl m 60-645
Dämpfungsfähigkeit f 75-205
Dämpfungsvermögen n 75-205
Dauerbruch m 40-745
Dauerfestigkeit f 75-200
Dauermagnetlegierung f 55-450
Dauermagnetstahl m 60-780
Dauerschwingfestigkeit f 75-200
Dauerschwingungsbruch m 40-745
Dauerschwingversuch m 80-420
dauerstandfeste Legierung f 55-510
Dauerstandfestigkeit f 75-225
wahre ~ 75-225
Dauerstandkriechgrenze f 75-225
Dauerversuch m 80-420
Debyesche charakteristische Temperatur f 10-280
Debye-Scherrer-Aufnahme f 80-635
Debye-Scherrer-Diagramm n 80-635
Debye-Scherrer-Kreis m 80-640
Debye-Scherrer-Ring m 80-640
Debye-Scherrer-Verfahren n 80-595
Debyesche Temperatur f 10-280
Debyetemperatur f 10-280
Defekt m
Frenkel-~ 30-060
Defektoskopie f 80-790
Deformation f
elastische ~ 45-005
plastische ~ 45-010
Deformationsband n 45-200
Deformationszwilling m 45-235
Dehnbarkeit f 75-030
Dehngrenze f 75-110
0,2-~ 75-115
Dehnungsmodul m 75-150
Dehnungsvermögen n 75-030
Dehydrierung f 70-555
Dekaleszenz f 25-490
Dekoration f 80-375
Dekorierung f 80-375
Delta-Ferrit m 50-050
Deltaferrit m 50-050
Demodifikationsmittel n 25-235
Demodifizierungsmittel n 25-235
Dendrit m 25-315
chemischer ~ 25-320
thermischer ~ 25-325
Dendritengefüge n 40-090
Dendritenzwischenräume mpl 40-355
dendritischer Kristall m 25-315
dendritische Seigerung f 25-420
dendritisches Gefüge n 40-090
dendritisches Wachstum n 25-310
Denitrieren n 70-850
Dentalgold n 65-785
Dentallegierung f 55-575
Desorientierung f 30-595

Destabilisierung f des Austenits 50-430
Diagramm n
Debye-Scherrer-~ 80-635
isothermes Zeit-Temperatur-Umwandlungs-~ 70-155
isothermes ZTU-~ 70-155
kontinuierliches Zeit-Temperatur-Umwandlungs-~ 70-160
kontinuierliches ZTU-~ 70-160
Schaeffler-~ 60-530
Diamagnetikum n 75-560
diamagnetischer Stoff m 75-560
Diamantgitter n 20-320
Diamantstahl m 60-490
Dichte f 75-345
dichte Kugelpackung f 20-210
dichtest besetzte Gitterebene f 20-215
dichtest besetzte Richtung f 20-220
dichteste Kugelpackung f 20-210
dichtest gepackte Ebene f 20-215
dichtstreifiger Perlit m 50-135
Differentialdilatometrie f 80-550
Differential-Thermoanalyse f 80-515
Diffusion f 45-290, 45-340
beschleunigte ~ 45-355
chemische ~ 45-320
~ über Gitterlücken 45-305
~ über Leerstellen 45-305
einphasige ~ 45-330
gegenseitige ~ 45-380
isotherme ~ 45-350
mehrphasige ~ 45-325
negative ~ 45-340
quasistationäre ~ 45-345
thermische ~ 45-370
diffusionsartige Umwandlung f 25-505
Diffusionsbehandlung f 70-650
Diffusionsfront f 45-410
Diffusionsglühen n 70-485
Diffusionskoeffizient m 45-395
Diffusionskonstante f 45-395
partielle ~ 45-390
Diffusionskriechen n 45-520
diffusionslose Umwandlung f 25-510
Diffusionsmetallisieren n 70-915
Diffusionspaar n 45-375
Diffusionssättigung f 70-655
Diffusionsverchromen n 70-920
Diffusionsvorgang m 25-505
Diffusionszone f 45-415
Digyre f 20-070
Dilatationskurve f 80-555
Dilatometerkurve f 80-555
Dilatometer-Untersuchung f 80-540
Dilatometrie f 80-540
absolute ~ 80-545
Direktabschrecken n 70-760
direktes Abschrecken n 70-760
Direkthärten n 70-760
Disklination f 30-480
diskontinuierliche Ausscheidung f 25-380
diskontinuierliches Eutektikum n 40-235
diskontinuierliche Umwandlung f 25-545
Dislokation f 30-145
Dislokationsanhäufung f 30-400
Dispersat n 40-330
disperse Phase f 40-330
Dispersion f 40-335
dispersionsgehärtete Legierung f 55-390
Dispersionshärtung f 55-220
dispersionsverfestigte Legierung f 55-390

Dispersionsverfestigung f 55-220
Dispersität f 40-335
Dispersitätsgrad m 40-335
Dispersitätsgröße f 40-335
Dispersoid n 40-330
Dissoziationsdruck m 15-345
Dissoziationsspannung f 15-345
Domäne f 40-615
magnetische ~ 40-635
Domänengrenzfläche f 40-645
Domänenstruktur f 40-620
Domänenwand f 40-645
Doppelgleitung f 45-135
Doppelhärten n 70-765
Doppelkarbid n 40-475
Doppelleerstelle f 30-045
Dopplung f 85-405
Drehfestigkeit f 75-085
Drehgrenze f 30-600
Drehimpulsquantenzahl f 10-135
Drehkristallaufnahme f 80-665
Drehkristallmethode f 80-600
Drehungshärten n 70-315
Dreieckschaubild n 35-720
Dreifachleerstelle f 30-050
Dreikomponentensystem n 35-125
dreikomponentiges System n 35-125
Dreiphasensystem n 35-105
dreiphasiges System n 35-105
Dreistofflegierung f 55-300
Dreistoffsystem n 35-125
dreizählige Achse f 20-075
dreizählige Symmetrieachse f 20-075
Druckfestigkeit f 75-075
Druckgußlegierung f 55-540
Druckmodul m 75-160
Drucktextur f 40-590
Dukatengold n 65-795
duktiler Bruch m 40-680
Duktilität f 75-030
Dunkelfeldmikroskopie f 80-230
Dunkelglühen n 70-440
dünne Folie f 80-350
Dünnen n 80-355
Dünnflüssigkeit f 75-785
Dünnschliff m 80-350
Dünnschlifffolie f 80-350
Dural n 65-555
Duralumin n 65-555
durchgreifendes Erwärmen n 70-035
durchgreifendes Härten n 70-280
Durchhärten n 70-280
Durchhärtung f 70-280
Durchhärtungstiefe f 70-350
durchkörniger Bruch m 40-710
Durchlicht-Mikroskopie f 80-220
Durchmesser m
kritischer ~ 70-360
Durchschallung f 80-800
Durchstrahlungsaufnahme f 80-675
Durchstrahlungsmethode f 80-625
Durchstrahlungsmikroskopie f 80-220
Elektronen-~ 80-340
Durchstrahlungsprüfung f 80-705
Durchwärmen n 70-035, 70-055
DV-Legierung f 55-390

dynamische Erholung f 45-425
dynamische Festigkeit f 75-185
dynamische Keimbildung f 25-205
dynamische Rekristallisation f 45-465
Dynamostahl m 60-775
Dysprosium n 05-130

E

Easy-Gleitung f 45-155
easy glide 45-155
Ebene f
   dichtest gepackte ~ 20-215
   kristallographische ~ 20-155
ebenmäßige Korrosion f 85-335
Edelmetall n 55-010
Edelrost m 85-375
Edelstahl m 60-145
Effekt m
   Bauschinger-~ 75-175
   Hall-~ 10-290
   Kirkendall-~ 45-385
   Mößbauer-~ 10-285
   Portevin-Le Chatelier-~ 45-540
   Snoek-~ 30-460
   Suzuki-~ 30-455
effektiver Verteilungskoeffizient m 25-265
ehemaliges Austenitkorn n 50-040
Eigenenergie f der Versetzung 30-230
Eigenfunktion f 10-190
Eigenhalbleiter m 10-380
Eigenkeimbildung f 25-180
Eigenschaften fpl
   chemische ~ 75-655
   elektrische ~ 75-465
   elektrochemische ~ 75-705
   gefügeabhängige ~ 75-860
   gefügeunabhängige ~ 75-865
   kernphysikalische ~ 75-520
   kerntechnische ~ 75-520
   magnetische ~ 75-540
   makroskopische ~ 75-850
   mechanische ~ 75-005, 75-015
   mechanische ~ aus der Zugprobe 75-015
   metallische ~ 10-350
   metallurgische ~ 75-855
   optische ~ 75-500
   physikalische ~ 75-315
   strukturempfindliche ~ 75-800
   strukturunabhängige ~ 75-865
   strukturunempfindliche ~ 75-865
   technologische ~ 75-720
   thermische ~ 75-365
   volumetrische ~ 75-340
Eigenspannungen fpl 85-020
eigentliche Wärmebehandlung f 70-006
einachsige Erstarrung f 25-040
Einbettung f 80-065
eindimensionale Fehlstelle f 30-140
eindimensionaler Gitterbaufehler m 30-140
einfache Gleitung f 45-130
einfache Legierung f 55-295
einfaches Gitter n 20-280
einfaches Karbid n 40-470
einfaches kubisches Gitter n 20-305
Einfachgleitung f 45-130
einfach-hexagonales Gitter n 20-345
Einfachleerstelle f 30-040
einfach legierter Stahl m 60-155
Einfangquerschnitt m
   Neutronen-~ 75-530
Einformung f des Zementits 70-515
eingelagertes Atom n 30-075

eingeschaltete Halbebene f 30-200
Einhärtbarkeit f 75-730
Einhärtetiefe f 70-350
Einhärtungstiefe f 70-350
Einheitsversetzung f 30-215
Einkomponentensystem n 35-115
einkomponentiges System n 35-115
Einkristall m 20-430
Einkristallherstellung f 25-085
   ~ durch Ziehen 25-090
Einkristallzüchtung f 25-085
Einlagerungen fpl
   metallische ~ 40-550
Einlagerungsatom n 30-075
Einlagerungsmischkristalle mpl 35-420
Einlagerungsphase f 35-420
Einphasendiffusion f 45-330
Einphasenfeld n 35-680
Einphasengebiet n 35-680
Einphasensystem n 35-090
einphasige Diffusion f 45-330
einphasige Legierung f 55-330
einphasiges Gefüge n 40-145
einphasiges System n 35-090
Einsatzhärten n 70-670
Einsatzhärtung f 70-670
Einsatzschicht f 70-660
Einsatzstahl m 60-690
Einschlüsse mpl
   endogene ~ 40-525
   exogene ~ 40-520
   metallische ~ 40-550
   nichtmetallische ~ 40-515
   oxidische ~ 40-535
Einschmelzlegierung f 55-600
Einschnürung f 75-125
einseitig-flächenzentriertes rhombisches Gitter n 20-385
Einsetzen n 70-715
Einsteinium n 05-135
Einstoffsystem n 35-115
Einzelkristall m 20-430
Eisen n 05-220
   Armco-~ 60-010
   technisches ~ 60-015
Eisenbronze f 65-175
Eisencarbid n 40-440
Eisen-Graphit-Eutektikum n 50-260
Eisenkarbid n 40-440
Eisen-Kohlenstoffdiagramm n 50-020
Eisen-Kohlenstoff-System n 50-015
Eisen-Kohlenstoff-Zustandsdiagramm n 50-020
Eisenlegierung f 60-070
Eisennitrid n 70-815
Eisenphosphid n 50-250
Eisenrost m 85-370
Eisenschwamm m 60-035
Eisensulfid n 50-245
Eisen-Zementit-Eutektikum n 50-255
E-Kupfer n 65-035
elastische Deformation f 45-005
elastische Formänderung f 45-005
elastische Konstante f 75-145
elastische Nachwirkung f 75-180
elastische Relaxation f 45-525
elastischer Koeffizient m 75-145

elastische Verformung f 45-005
Elastizität f 75-020
Elastizitätsgrenze f 75-105
Elastizitätskonstante f 75-145
Elastizitätsmaß n 75-150
Elastizitätsmodul m 75-150
elektrische Eigenschaften fpl 75-465
elektrische Leitfähigkeit f 75-485
elektrische Überführung f 45-365
Elektroaffinität f 10-115
elektrochemische Ätzung f 80-175
elektrochemische Eigenschaften fpl 75-705
elektrochemische Korrosion f 85-270
elektrochemischer Faktor m 10-400
elektrochemische Spannungsreihe f 75-710
elektrochemisches Potential n 75-715
Elektrodiffusion f 45-365
Elektrokristallisation f 25-080
Elektrolyteisen n 60-025
elektrolytische Ätzung f 80-175
elektrolytische Legierung f 55-290
elektrolytisches Polieren n 80-075
elektrolytisch niedergeschlagenes Metall n 55-130
Elektrolytkupfer n 65-095
Elektrolytmetall n 55-130
Elektron n 65-625
  äußeres ~ 10-065
  gebundenes ~ 10-050
  inneres ~ 10-070
Elektronegativität f 10-095
Elektronenaffinität f 10-115
Elektronenbeugungsaufnahme f 80-775
Elektronenbeugungsbild n 80-775
Elektronenbeugungsuntersuchung f 80-770
Elektronen-Durchstrahlungs-Mikroskopie f 80-340
Elektronenemissionsmikroskopie f 80-355
Elektronenhülle f 10-155
Elektronenkonzentrationsphase f 35-280
Elektronenmetallographie f 80-025
Elektronenmikroskopie f 80-330, 80-335
elektronenmikroskopische Aufnahme f 80-365
Elektronengas n 10-045
Elektronenschale f 10-155
Elektronenstrahl-Mikroanalyse f 80-780
Elektronenstruktur f 10-030
Elektronentheorie f der Metalle 10-025
Elektronen-Transmissionsmikroskopie f 80-340
Elektronenwärme f
  spezifische ~ 10-275
Elektronenwolke f 10-035
elektronische spezifische Wärme f 10-275
Elektropolieren n 80-075
Elektroroheisen n 60-865
Elektrostahl m 60-625
Elektrotransport m 45-365
Elektrotypie-Legierung f 55-635
Elektrum n 65-810
Element n
  alphagenes ~ 50-455
  austenitbildendes ~ 50-450
  austenitstabilisierendes ~ 50-450
  betagenes ~ 65-860
  chemisches ~ 05-005
  ferritbildendes ~ 50-455

Element
  ferritförderndes ~ 50-455
  ferritstabilisierendes ~ 50-455
  gammagenes ~ 50-450
  härtesteigerndes ~ 55-255
  karbidstabilisierendes ~ 50-465
  karbidzerlegendes ~ 50-460
  metallisches ~ 10-360
  nichtmetallisches ~ 10-365
  nitridbildendes ~ 50-470
Elementarbereich m 40-615
Elementarzelle f 20-090
  primitive ~ 20-095
Elinvar n 65-505
Elinvarlegierung f 65-505
Embryo m 25-135
E-Modul m 75-150
Endogas n 70-705
Endogasatmosphäre f 70-705
endogene Einschlüsse mpl 40-525
energetische Asymmetrie f 15-235
Energie f
  gebundene ~ 15-100
  Gibbssche freie ~ 15-140
  Helmholtzsche freie ~ 15-105
  innere ~ 15-095
  partielle freie ~ 15-110
Energieband n 10-235
Energiebändertheorie f 10-020
Energiebereich m 10-235
Energielücke f 10-250
Energieniveau n 10-150
Energiezustand m 10-120
entarteter Perlit m 50-155
entartetes Eutektikum n 40-230
Entfestigung f 45-535
Entgasen n 70-550
Enthalpie f
  freie ~ 15-140
  partielle ~ 15-150
Entmischung f 25-365
  spinodale ~ 25-370
Entordnung f 25-580
Entordnungstemperatur f 25-625
Entropie f
  partielle ~ 15-215
Entspannen n 70-525
Entspannungsglühen n 70-525
Entstickung f 70-850
Entzinkung f 85-380
epitaktisches Aufwachsen n 25-345
epitaktische Schicht f 25-350
Epitaxie f 25-340
Epitaxieschicht f 25-350
epitaxisches Wachsen n 25-345
Erbium n 05-140
Erblichkeit f des Gußeisens 60-985
Erdalkalien pl 55-035
Erdalkalimetalle npl 55-035
Erhitzen n 70-030
Erhitzung f 70-030
Erhitzungskurve f 80-520
Erholung f 30-135, 45-420
  dynamische ~ 45-425
  isochrone ~ 70-495
  statische ~ 45-430
  thermische ~ 45-430
Erichsentiefung f 75-755
erlaubtes Band n 10-240
Ermüdung f 85-410
  thermische ~ 85-425

FERMIUM

Ermüdungsbruch m 40-745
Ermüdungsfestigkeit f 75-200
Ermüdungsgrenze f 75-200
Ermüdungskorrosion f 85-315
Ermüdungskurve f 80-425
Erstarrung f 25-005
 einachsige ~ 25-040
 eutektische ~ 25-025
 gelenkte ~ 25-035
 gerichtete ~ 25-035
 kontrollierte ~ 25-030
 primäre ~ 25-050
Erstarrungsbereich m 35-550
Erstarrungsfront f 25-295
Erstarrungsgefüge n 40-045
Erstarrungsintervall n 35-550
Erstarrungskontraktion f 75-805
Erstarrungspunkt m 75-425
Erstarrungsschrumpfung f 75-805
Erstarrungstemperatur f 75-425
Ermüdungsversuch m 80-420
Erstarrungswärme f 75-385
erste Quantenzahl f 10-130
Erstgefüge n 40-045
erstnächster Nachbar m 20-195
Erwärmen n 70-030
 gepulstes ~ 70-045
Erwärmgeschwindigkeit f 70-040
Erwärmung f 70-030
 durchgreifende ~ 70-035
Erwärmungsgeschwindigkeit f 70-040
Erwärmungskurve f 80-520
E-Stahl m 60-625
Europium n 05-145
Eutektikale f 35-640
Eutektikum n 40-200
 binäres ~ 40-205
 diskontinuierliches ~ 40-235
 Eisen-Graphit-~ 50-260
 Eisen-Zementit-~ 50-255
 entartetes ~ 40-230
 Graphit-~ 50-260
 lamellares ~ 40-220
 quaternäres ~ 40-215
 ternäres ~ 40-210
 Zementit-~ 50-255
eutektische Erstarrung f 25-025
eutektische Horizontale f 35-640
eutektische Karbide npl 50-440
eutektische Kristallisation f 25-025
eutektische Legierung f 55-350
eutektische Linie f 35-640
eutektische Mischung f 40-200
eutektische Reaktion f 35-565
eutektischer Graphit m 50-175
eutektische Rinne f 35-730
eutektischer Punkt m 35-660
eutektischer Zementit m 50-075
eutektisches Gefüge n 40-060
eutektisches Gemisch n 40-200
eutektisches Schmelzen n 35-470
eutektische Struktur f 40-060
eutektisches weißes Gußeisen n 50-505
eutektische Temperatur f 35-605
eutektische Umwandlung f 35-565
Eutektoid n 40-240
Eutektoidale f 35-655
eutektoide Legierung f 55-365
eutektoide Reaktion f 35-590
eutektoider Punkt m 35-675
eutektoider Stahl m 50-490

eutektoides Gefüge n 40-075
eutektoides Gemisch n 40-240
eutektoide Temperatur f 35-630
eutektoide Umwandlung f 35-590
eutektoidische Horizontale f 35-655
eutektoidische Linie f 35-655
exakte Streckgrenze f 75-110
Exogas n 70-710
Exogasatmosphäre f 70-710
exogene Einschlüsse mpl 40-520
Explosivhärten n 45-055
extensive Größe f 15-040
Exzeßfunktion f 15-065

F

Fadenkristall m 25-360
Faktor m
 elektrochemischer ~ 10-400
Farbätzmittel n 80-105
Farbätzung f 80-165
Farbenätzung f 80-165
Farbenmetallographie f 80-030
Faserbruch m 40-720
faseriger Bruch m 40-720
Fasertextur f 40-565
Federstahl m 60-725
fehlerfreier Kristall m 30-005
fehlerhaftes Gitter n 30-020
Fehlordnung f
 Frenkel-~ 30-060
Fehlordnungsgrad m 25-610
Fehlstelle f
 atomare ~ 30-030
 eindimensionale ~ 30-140
 nulldimensionale ~ 30-030
 zweidimensionale ~ 30-535
Feinblei n 65-655
feindendritische Randzone f 40-175
feines Korn n 40-370
Feingefüge n 40-040
Feingehalt m 55-100
Feinglühen n 70-475
Feingold n 65-770
Feinkorn n 30-650, 40-370
Feinkorngrenze f 30-660
feinkörniger Bruch m 40-700
feinkörniges Gefüge n 40-395
Feinkörnigkeit f 40-380
Feinkornstahl m 60-470
Feinkornzusatz m 40-400
feinkristallines Gefüge n 40-395
feinlamellarer Perlit m 50-135
feinnadeliger Martensit m 50-325
Feinsilber n 65-760
feinstreifiger Perlit m 50-135
Feinstruktur f 40-040
Feinstrukturanalyse f
 röntgenographische ~ 80-580
Feinzink n 65-695
Feldionenmikroskopie f 80-380
Fermienergie f 10-210
Fermifläche f 10-215
Fermigrenze f 10-220
Fermikugel f 10-225
Fermioberfläche f 10-215
Fermium n 05-150

Fernordnung f 25-600
Fernordnungsparameter m 25-620
Ferrimagnetismus m 75-580
Ferrit m
Delta-~ 50-050
freier ~ 50-055
perlitischer ~ 50-060
voreutektoider ~ 50-055
ferritbildendes Element n 50-455
Ferritbildner m 50-455
ferritförderndes Element n 50-455
Ferrithof m 50-230
ferritischer nichtrostender Stahl m 60-520
ferritischer Stahl m 60-415
ferritischer Temperguß m 60-1125
ferritisches Gußeisen n 60-1040
ferritisierende Glühbehandlung f 70-645
ferritstabilisierendes Element n 50-455
Ferrobronze f 65-355
Ferrochrom n 60-1145
Ferrolegierungen fpl 60-1135
Ferromagnetikum n 75-550
ferromagnetische Legierung f 55-455
ferromagnetischer Stoff m 75-550
Ferromagnetismus m 75-545
Ferromangan n 60-1150
Ferrosilicomangan n 60-1155
Ferrosilizium n 60-1140
Ferrozer n 65-875
Ferrozerium n 65-875
feste Lösung f 35-375
Festigkeit f 75-040
dynamische ~ 75-185
relative ~ 75-140
Festigkeitseigenschaften fpl 75-010
Festigkeitskenngröße f 75-060
Festigkeitskennwert m 75-060
Festigkeitskennzahl f 75-060
Festigkeitsprüfung f 80-400
Festigkeitssteigerung f 70-025
Festigkeitsversuch m 80-400
Festkontraktion f 75-810
Festkörperphysik f 10-010
Festkörperthermodynamik f 15-005
feuerbeständiges Gußeisen n 60-1085
feuerschweißbarer Stahl m 60-485
flächenzentriertes Gitter n 20-290
Flammenglühen n 70-505
Flammenhärten n 70-290
Flammhärten n 70-290
Flecken mpl
weiße ~ 85-250
Fließen n
plastisches ~ 45-215
viskoses ~ 45-220
Fließfähigkeit f 75-785
Fließgrenze f 75-110
Fließspannung f 45-210
Fließvermögen n 75-785
Flocken fpl 85-240
Flockenfreibehandlung f 70-555
Flockengraphit m 50-185
Flockenrisse mpl 85-240
Flüchtigkeit f 15-290
Fluor n 05-155
Fluoreszenzprüfung f 80-835
Fluß m 15-085
Flußeisen n 60-060

flüssige Lösung f 35-365
flüssiges Härtemittel n 70-190
Flüssigkontraktion f 75-800
Flußstahl m 60-580
fokussierter Stoß m 30-505
Fokusson m 30-505
Folie f
dünne ~ 80-350
Formänderung f
bleibende ~ 45-010
elastische ~ 45-005
plastische ~ 45-010
Formgebung f
bildsame ~ 45-015
Formung f
bildsame ~ 45-015
Fotomikrographie f 80-250
Fraktographie f 80-385
fraktographische Untersuchung f 80-385
Francium n 05-160
Frankium n 05-160
Frank-Read-Quelle f 30-430
Franksche Halbversetzung f 30-335
Frankversetzung f 30-335
freie Bildungsenergie f 15-135
freie Enthalpie f 15-140
freie Grenzflächenenergie f 15-125
freie Mischungsenergie f 15-130
freie Oberflächenenergie f 15-120
freier Ferrit m 50-055
freier Kohlenstoff n 50-005
freier Zementit m 50-095
freies Elektron n 10-055
freie Zusatzenergie f 15-115
Fremddiffusion f 45-335
Fremdhalbleiter m 10-385
Fremdkeim m 25-120
Fremdkern m 25-120
Frenkel-Defekt m 30-060
Frenkel-Fehlordnung f 30-060
Frenkel-Paar n 30-060
Frenkelsche Fehlstelle f 30-060
Frenkelsche Leerstelle f 30-065
Frischeisen n 60-050
Frischfeuereisen n 60-050
früheres Austenitkorn n 50-040
Fugazität f 15-290
Fundamentalzelle f 20-090
Funkenprobe f 80-440
Funkenprüfung f 80-440
Funktion f
integrale thermodynamische ~ 15-055
molare thermodynamische ~ 15-050
partielle thermodynamische ~ 15-060
thermodynamische ~ 15-030

G

Gadolinium n 05-165
Galium n 05-170
Galvanipotential n 75-715
galvanische Korrosion f 85-270
Galvano-Legierung f 55-635
Gammadefektoskopie f 80-825
gammagenes Element n 50-450
Gammagraphie f 80-825
Gamma-Radiographie f 80-825
Garkupfer n 65-025

GITTERBAUFEHLER

Garschaumgraphit m 50-170
Gasätzen n 80-182
Gasaufkohlen n 70-730
Gasaufkohlung f 70-730
Gasblase f 85-135
Gasblasenseigerung f 25-430
Gaseinsetzen n 70-730
Gasgemisch n 35-370
Gaskorrosion f 85-295
Gasnitrieren n 70-790
Gaszyanieren n 70-860
Gebiet n
homogenes ~ 35-680
Gebläselufthärten n 70-220
Gebrauchseigenschaften fpl 75-820
gebrochenes Härten n 70-300
gebundene Energie f 15-100
gebundener Kohlenstoff m 50-010
gebundene Schmelzwärme f 75-380
gebundenes Elektron n 10-050
gebundene Verdampfungswärme f 75-390
gediegenes Metall n 55-110
Gefrierpunkt m 75-425
Gefüge n 40-005
anomales ~ 40-135
dendritisches ~ 40-090
einphasiges ~ 40-145
eutektisches ~ 40-060
eutektoides ~ 40-075
feinkörniges ~ 40-395
feinkristallines ~ 40-395
~ der oberen Zwischenstufe 50-420
~ der unteren Zwischenstufe 50-415
globulares ~ 40-120
globulitisches ~ 40-120
grobkörniges ~ 40-390
grobkristallines ~ 40-390
heterogenes ~ 40-155
homogenes ~ 40-145
körniges ~ 40-120
lamellares ~ 40-105
mehrphasiges ~ 40-160
monophasiges ~ 40-145
nadeliges ~ 40-115
stengeliges ~ 40-125
strahliges ~ 40-115
übereutektisches ~ 40-065
übereutektoides ~ 40-080
untereutektisches ~ 40-070
untereutektoides ~ 40-085
Widmannstättensches ~ 40-130
zweiphasiges ~ 40-150
gefügeabhängige Eigenschaften fpl 75-860
Gefügeanomalie f des Stahles 85-225
Gefügeaufnahme f 80-250
Gefügebestandteil m 40-010
Gefügebild n 80-245
Gefügediagramm n 40-265
Gefügeentwicklung f 80-280
Gefügegleichgewicht n 40-020
Gefügespannungen fpl 85-035
Gefügetextur f 40-555
Gefügeumwandlung f 25-500
gefügeunabhängige Eigenschaften fpl 75-865
Gefügeverfeinerung f 40-385
gegenseitige Diffusion f 45-380
geglühter Stahl m 60-275
gehärteter Stahl m 60-285
geladene Versetzung f 30-465
Gelbguß m 65-300
gelenkte Erstarrung f 25-035

Gelöste n 35-360
Gemenge n
heterogenes ~ der Phasen 40-190
mechanisches ~ der Phasen 40-190
Gemisch n
eutektisches ~ 40-200
eutektoides ~ 40-240
gemischte Versetzung f 30-180
geordnete feste Lösung f 35-430
geordnete Mischphase f 35-200, 35-430
geordnete Phase f 35-200
gepulstes Erwärmen n 70-050
geregelte Atmosphäre f 70-700
gerichtete Erstarrung f 25-035
Germanium n 05-175
gesättigte feste Lösung f 35-395
geschmiedeter Stahl m 60-330
Geschützbronze f 65-245
Gesenkstahl m 60-815
Gesetz n
Henrysches ~ 15-275
Raoultsches ~ 15-270
gesinterte Karbide npl 55-665
gesintertes Hartmetall n 55-665
gesintertes Metall n 55-660
gespeicherte Energie f 45-060
gestuftes Härten n 70-305
Gewicht n
spezifisches ~ 75-350
Gewichtskonzentration f 55-235
gewöhnliche Thermoanalyse f 80-510
Gibbs-Duhem-Gleichung f 15-305
Gibbssche freie Energie f 15-140
Gießbarkeit f 75-780
Gießeigenschaften fpl 75-775
Gießereiroheisen n 60-885
Gießfähigkeit f 75-780
Gitter n
einfaches ~ 20-280
einfaches kubisches ~ 20-305
einfach-hexagonales ~ 20-345
einseitig-flächenzentriertes rhombisches ~ 20-385
fehlerhaftes ~ 30-020
flächenzentriertes ~ 20-290
~ mit hexagonal dichtester Kugelpackung 20-350
hexagonal dichtgepacktes ~ 20-350
hexagonales dichtgepacktes ~ 20-350
hexagonales ~ 20-340
innenzentriertes ~ 20-295
körperzentriertes ~ 20-295
kubisches flächenzentriertes ~ 20-310
kubisches ~ 20-300
kubisches innenzentriertes ~ 20-315
kubisches körperzentriertes ~ 20-315
kubisches raumzentriertes ~ 20-315
kubisch innenzentriertes ~ 20-315
kubisch körperzentriertes ~ 20-315
kubisch raumzentriertes ~ 20-315
monoklines ~ 20-395
raumzentriertes ~ 20-295
reguläres ~ 20-300
reziprokes ~ 20-410
rhombisch-einseitig-flächenzentriertes ~ 20-385
rhombisches ~ 20-380
rhombisch-grundflächenzentriertes ~ 20-385
rhomboedrisches ~ 20-390
tetragonales ~ 20-370
triklines ~ 20-400
würfeliges ~ 20-300
Gitterabstand m 20-120
Gitterbasis f 20-110
Gitterbaufehler m 30-015
eindimensionaler ~ 30-140
punktförmiger ~ 30-030
zweidimensionaler ~ 30-535

Gitterdefekt *m* 30-015
Gitterdiffusion *f* 45-295
Gitterebene *f*
dichtest besetzte ~ 20-215
Gitterfehlanpassung *f* 30-615
Gitterfehler *m* 30-015
linienförmiger ~ 30-140
nulldimensionaler ~ 30-030
punktförmiger ~ 30-030
Gitterfehlstelle *f* 30-015
Gitterkonstante *f* 20-120
Gitterleerstelle *f* 30-035
Gitterlücke *f* 20-105, 30-035
oktaedrische ~ 20-330
tetraedrische ~ 20-325
Gitterparameter *m* 20-120
Gitterplatz *m* 20-140
Gitterpunkt *m* 20-100
Gitterstelle *f* 20-140
Gittertranslation *f* 20-130
Gittervektor *m* 20-135
Gitterverzerrung *f* 30-025
Glanz *m*
metallischer ~ 75-510
glasiger Bruch *m* 40-735
gleichachsige Kristalle *mpl* 40-285
gleichachsige Kristallisation *f* 25-075
gleichachsige Zone *f* 40-185
Gleichartigkeit *f* 40-405
Gleichgewicht *n*
chemisches ~ 15-330
heterogenes ~ 35-015
metastabiles ~ 35-060
stabiles ~ 35-055
thermodynamisches ~ 15-320
Gleichgewichtsdiagramm *n* 35-480
ternäres ~ 35-710
Gleichgewichtserstarrung *f* 25-015
Gleichgewichtskonstante *f*
thermodynamische ~ 15-335
Gleichgewichtskonzentration *f* 35-255
Gleichgewichtsphase *f* 35-180
Gleichgewichtssystem *n* 35-045
Gleichgewichtstemperatur *f* 35-690
Gleichgewichts-Verteilungskoeffizient *m* 25-260
Gleichgewichtszustand *m* 35-050
gleichmäßige Korrosion *f* 85-335
Gleichung *f*
Gibs-Duhem ~ 15-305
Hildebrandsche ~ 15-310
Schrödinger-~ 10-175
Gleitband *n* 45-190
Gleitbewegung *f* 30-290
Gleitbruch *m* 40-660
Gleitebene *f* 45-100
gleitfähige Versetzung *f* 30-265
Gleitfläche *f* 45-100
Gleitlinie *f* 45-185
Gleitmaß *n* 75-155
Gleitmodul *m* 75-155
Gleitrichtung *f* 45-105
Gleitspannung *f* 45-085
kritische ~ 45-090
Gleitsystem *n* 45-110
konjugiertes ~ 45-125
primäres ~ 45-115
sekundäres ~ 45-120
Gleitung *f* 45-095
einfache ~ 45-130
intrakristalline ~ 45-165

Gleitung
prismatische ~ 45-170
zweifache ~ 45-135
Gleitungsbruch *m* 40-660
Gleitversetzung *f* 30-265
Gleitvorgang *m* 45-095
Glimmnitrieren *n* 70-800
globulares Gefüge *n* 40-120
Globulargefüge *n* 40-120
globulitische Kristalle *mpl* 40-285
globulitische Kristallisation *f* 25-075
globulitisches Gefüge *n* 40-120
Glockenbronze *f* 65-250
Glühbehandlung *f*
ferritisierende ~ 70-645
Glühen *n* 70-420
~ auf kugelige Carbide 70-510
~ auf kugeligen Zementit 70-510
isothermisches ~ 70-450
perlitisierendes ~ 70-450
unvollständiges ~ 70-470
Glühfrischen *n* 70-635
Glühtemperatur *f* 70-425
Glühtextur *f* 40-600
Glühzwilling *m* 45-240
Gold *n* 05-180
unechtes ~ 65-815
Gold-Imitation *f* 65-815
Goldlegierung *f*
hochkarätige ~ 65-795
Gosstextur *f* 40-610
Graben *m*
thermischer ~ 80-195
Grabenbildung *f* 80-190
Graphit *m* 50-165
kugeliger ~ 50-190
nestförmiger ~ 50-210
punktförmiger ~ 50-195
sekundärer ~ 50-180
sphärolithischer ~ 50-190
Graphitbildner *m* 50-460
Graphitbronze *f* 65-220
Graphit-Eutektikum *n* 50-260
Graphitflecken *mpl* 85-140
graphitischer Stahl *m* 60-300
Graphitisieren *n* 70-630
graphitisierter Stahl *m* 60-300
Graphitisierung *f* 50-225, 70-630
Graphitkugel *f* 50-215
Graphitsphärolith *m* 50-215
Graphitstahl *m* 60-300
graues Gußeisen *n* 60-950
graues Roheisen *n* 60-915
graues Zinn *n* 65-635
Grauguß *m* 60-950
Grenze *f*
sekundäre ~ 30-660
Tilt-~ 30-605
Twist-~ 30-600
Grenzfläche *f*
kohärente ~ 30-625
nicht-kohärente ~ 30-630
Grenzflächenenergie *f* 30-635
Grenzflächenspannung *f* 30-645
Griffithscher Riß *m* 85-055
grobes Korn *n* 40-365
Grobgefüge *n* 40-025
Grobkorn *n* 40-365
Grobkornglühen *n* 70-480
grobkörniger Bruch *m* 40-695

HÄRTEVERZUG

grobkörniger Martensit m 50-320
grobkörniges Gefüge n 40-390
Grobkörnigkeit f 40-375
Grobkornstahl m 60-475
grobkristallines Gefüge n 40-390
Grobkristallisation f 45-450
grobnadeliger Martensit m 50-320
grobstreifiger Perlit m 50-140
Grobstruktur f 40-025
Größe f
extensive ~ 15-040
integrale thermodynamische ~ 15-055
intensive ~ 15-045
massenproportionale ~ 15-040
massenunabhängige ~ 15-045
molare thermodynamische ~ 15-050
partielle ~ 15-060
partielle thermodynamische ~ 15-060
thermodynamische ~ 15-030
Großwinkelgrenze f 30-585
Grundgefüge n 40-015
Grundlinien fpl 80-650
Grundmasse f 40-015
Grundmetall n 55-185
Grundzelle f 20-090
Guinier-Preston-Zonen fpl 70-605
Guß m 60-945
KG-~ 60-975
Gußbronze f 65-100
Gußeisen n 60-945
alkalibeständiges ~ 60-1080
armiertes ~ 60-1070
austenitisches ~ 60-1050
bainitisches ~ 60-1060
eutektisches weißes ~ 50-505
ferritisches ~ 60-1040
feuerbeständiges ~ 60-1085
graues ~ 60-950
halbgraues ~ 60-960
hochlegiertes ~ 60-1005
KG-~ 60-975
Kugelgraphit-~ 60-975
laugenbeständiges ~ 60-1080
legiertes ~ 60-995
martensitisches ~ 60-1055
meliertes ~ 60-960
modifiziertes ~ 60-970
niedriglegiertes ~ 60-1000
perlitisches ~ 60-1045
säurebeständiges ~ 60-1075
stahlbewehrtes ~ 60-1070
übereutektisches weißes ~ 50-515
unlegiertes ~ 60-990
untereutektisches weißes ~ 50-510
weißes ~ 60-955
zunderbeständiges ~ 60-1085
zunderfestes ~ 60-1085
*Gußeisenerblichkeit f 60-985
Gußfehler m 85-095
Gußgefüge n 40-050
Gußlegierung f 55-535
Messing-~ 65-300
Gußmessing n 65-300
Gußspannungen fpl 85-170
Gußstahl m 60-825
legierter ~ 60-835
unlegierter ~ 60-830
Gußstruktur f 40-050
Gußtextur f 40-580

H

Haarkristall m 25-360
Haarriß m 85-045
Habitus m des Kristalls 20-460

Habitusebene f 20-465
Hadfieldscher Manganhartstahl m 60-785
Hafnium n 05-185
Häggsche Regel f 35-260
halbaustenitischer Stahl m 60-445
halbberuhigter Stahl m 60-660
Halbebene f
eingeschaltete ~ 30-200
halbedles Metall n 55-020
halbferritischer Stahl m 60-450
halbgraues Gußeisen n 60-960
Halb-Hämatit-Roheisen n 60-900
halbiertes Roheisen n 60-920
Halbleiter m 10-375
Halbmetall n 10-370
Halbstahl m 60-1035
Halbversetzung f 30-315
Franksche ~ 30-335
Shockleysche ~ 30-330
Halbwertszeit f 75-525
Hall-Effekt m 10-290
Haltedauer f 70-060
Halten n 70-055
Haltepunkt m
thermischer ~ 80-530
Haltezeit f 70-060
Hämatitroheisen n 60-895
Handelsbaustahl m 60-135
Handelsguß m 60-965
handelsübliche Legierung f 55-520
Hardenit m 50-365
härtbarer Stahl m 60-360
Härtbarkeit f 75-730
Härtbarkeitsstreuband n 80-460
Hartblei n 65-660
Härte f 75-245
Härtebad n 70-195
Härtebereich m 70-145
Härtefähigkeit f 70-020
Härtefehler m 85-180
Härtekurve f
Stirnabschreck-~ 80-455
Härtemittel n 70-185
flüssiges ~ 70-190
Härten n 70-015, 70-175
gebrochenes ~ 70-300
gestuftes ~ 70-305
~ aus dem Einsatz 70-760
~ im Luftstrom 70-220
~ im Salzbad 70-230
~ im Vorschub 70-320
~ in abgekühlten Kokillen 70-255
~ in Salzlösung 70-238
~ in Salzwasser 70-238
örtliches ~ 70-275
übliches ~ 70-270
unterbrochenes ~ 70-300
Härteprüfung f 80-415
Härteriß m 85-190
harter Stahl m 60-350
Härteschicht f 70-355
Härtespannungen fpl 85-185
Härtesteiger m 70-015
härtesteigernde Behandlung f 70-015
härtesteigernde Phase f 40-350
härtesteigerndes Element n 55-255
harte Stellen fpl 85-145
Härtetemperatur f 70-140
Härteversuch m 80-415
Härteverzug m 85-200

Härtezone f 70-355
Hartguß m 60-980
umgekehrter ~ 85-150
Hartlegierung f 55-415
Hartlot n 65-720
hartmagnetische Legierung f 55-465
hartmagnetischer Stahl m 60-505
Hartmannsche Linien fpl 45-195
Hartmetall n 55-415
gesintertes ~ 55-665
Hartmetall-Auftraglegierung f 55-590
Hartschalenguß m 60-980
Hartstahl m 60-350
Härtung f 70-015
Hartzink n 65-700
Hastelloy n 65-455
Hauptlegierungselement n 55-195
Hauptlegierungszusatz m 55-195
Hauptquantenzahl f 10-130
Hauptschale f 10-155
Hebelbeziehung f 35-700
Hebelgesetz n 35-700
Heißdampfbehandlung f 70-880
Heizleiterlegierung f 55-545
Heiztischmikroskopie f 80-240
Helium n 05-190
heller Bruch m 85-155
Hellfeldmikroskopie f 80-235
Helmholtzsche freie Energie f 15-105
Henrysches Gesetz n 15-275
Herdfrischeisen n 60-050
Herring-Nabarro-Kriechen n 45-520
Heterodiffusion f 45-335
heterogene Keimbildung f 25-185
heterogene Legierung f 55-340
heterogener Keim m 25-115
heterogenes Gefüge n 40-155
heterogenes Gemenge n der Phasen 40-190
heterogenes Gleichgewicht n 35-015
heterogenes Marinemessing n 65-385
heterogenes Messing n 65-320
heterogenes System n 35-095
heterogene Umwandlung f 25-535
Heterogenität f 40-410
chemische ~ 40-415
heteropolare Bindung f 10-325
heteropolare Verbindung f 35-230
heterothermisches Schmelzen n 35-455
Heusler-Legierung f 65-420
hexagonal dichtgepacktes Gitter n 20-350
hexagonales dichtgepacktes Gitter n 20-350
hexagonales Gitter n 20-340
hexagonales Kristallsystem n 20-245
Hexagyre f 20-085
Hiduminium n 65-605
Hildebrandsche Gleichung f 15-310
Hipernik n 65-475
hitzebeständiger Stahl m 60-560
Hochbaustahl m 60-685
hochchromhaltiger Stahl m 60-180
Hochdruckphase f 35-165
hochfeste Legierung f 55-420
hochfester Stahl m 60-355
hochgekohlter Stahl m 60-110
hochgekohltes Roheisen n 60-925
Hochglühen n 70-480

hochkarätige Goldlegierung f 65-795
hochlegierte Legierung f 55-325
hochlegierter Stahl m 60-130
hochlegiertes Gußeisen n 60-1005
hochleitfähiges Kupfer n 65-065
Hochofenblei n 65-650
Hochofenroheisen n 60-860
hochpermeable Legierung f 55-475
hochprozentige Legierung f 55-325
hochschmelzbare Legierung f 55-400
hochschmelzende Legierung f 55-400
hochschmelzendes Metall n 55-065
Hochtemperaturanlassen n 70-375
Hochtemperaturfestigkeit f 75-095
Hochtemperaturglühen n 70-480
Hochtemperaturlegierung f 55-510
Hochtemperatur-Metallographie f 80-015
Hochtemperaturmikroskopie f 80-240
Hochtemperaturmodifikation f 25-395
Hochtemperaturphase f 35-160
hochwarmfeste Legierung f 55-515
hochwarmfester Stahl m 60-565
hochwertiges Gußeisen n 60-1065
hochzinnhaltige Bronze f 65-115
Hohlraum m 85-100
Holmium n 05-195
Holzfaserbruch m 40-720
Holzkohlenroheisen n 60-875
homogene Keimbildung f 25-180
homogene Legierung f 55-330
homogenes Gebiet n 35-680
homogenes Gefüge n 40-145
homogenes Marinemessing n 65-390
homogenes Messing n 65-315
homogenes System n 35-090
homogene Umwandlung f 25-530
Homogenglühen n 70-485
Homogenisierungsglühen n 70-485
Homogenität f 40-405
Homogenitätsbereich m 35-680
homologe Temperatur f 45-475
homöopolare Bindung f 10-330
Horizontale f
eutektische ~ 35-640
eutektoide ~ 35-655
eutektoidische ~ 35-655
monotektische ~ 35-650
peritektische ~ 35-645
Hume-Rothery-Phase f 35-280
Hume-Rothersche Regel f 35-325
Hüttenaluminium n 65-545
Hüttenblei n 65-650
Hüttenkupfer n 65-025
Hüttenzink n 65-690
Hydrid n 40-495
Hydronalium n 65-600
Hysterese f
magnetische ~ 75-640

I

ideale Lösung f 15-240
ideale Mischung f 15-240
Idealkristall m 30-005
Identitätsabstand m 20-115
Identitätsperiode f 20-115
idiomorpher Kristall m 25-355

Imitation *f*
Gold-~ 65-815
Impfen *n* von Kristallbildung 25-215
Impfkeim *m* 25-120
Impfkristall *m* 25-130
Impulserwärmen *n* 70-050
inaktive Atmosphäre *f* 70-675
Inchromieren *n* 70-920
Inconel *n* 65-445
Index *m*
kristallographischer ~ 20-160
Indikator *m*
radioaktiver ~ 80-760
Indium *n* 05-205
Indizes *mpl*
Bravaissche ~ 20-170
Millersche ~ 20-165
Indizierung *f* 80-695
Induktion *f*
magnetische ~ 75-610
Induktionshärten *n* 70-295
Induktionshärtung *f* 70-295
Induktionsperiode *f* 70-150
inerte Atmosphäre *f* 70-675
infiltrierte Legierung *f* 55-670
inhomogene Umwandlung *f* 25-535
Inhomogenität *f* 40-410
inkohärente Ausscheidungen *fpl* 40-325
inkongruente intermetallische Verbindung *f* 35-225
inkongruentes Schmelzen *n* 35-465
Inkromieren *n* 70-920
Inkubationszeit *f* 70-150
Innenelektron *n* 10-070
innenzentriertes Gitter *n* 20-295
innere Energie *f* 15-095
innere Korrosion *f* 85-355
innere Oxydation *f* 70-947
innere Reibung *f* 75-210
inneres Elektron *n* 10-070
innere Spannungen *fpl* 85-020
innerkristalliner Bruch *m* 40-710
integrale thermodynamische Funktion *f* 15-055
integrale thermodynamische Größe *f* 15-055
intensive Größe *f* 15-045
Interaktionsparameter *m* 15-300
Interdendritischer Graphit *m* 50-205
interdendritische Seigerung *f* 25-445
Interferenzfleck *n* 80-660
Interferenzlinie *f* 80-645
Interferenzmikroskopie *f* 80-295
Interferogramm *n* 80-300
interkristalline Korrosion *f* 85-360
interkristalliner Bruch *m* 40-705
intermediäre Phase *f* 35-210
intermetallische Phase *f* 35-210
intermetallische Verbindung *f* 35-215
Interstitionsmischkristalle *mpl* 35-420
Intervall *n*
kritisches ~ 35-560
intrakristalline Gleitung *f* 45-165
intrakristalliner Bruch *m* 40-710
Invar *n* 65-500
Invarlegierung *f* 65-500
Invarstahl *m* 65-500

Ionenätzen *n*
katodisches ~ 80-180
Ionenbindung *f* 10-325
Ionenimplantation *f* 70-805
Ionenkristall *m* 20-025
Ionenradius *m* 10-090
Ionisierungsenergie *f* 10-100
Ionisierungspotential *n* 10-105
Ionitrieren *n* 70-800
Iridium *n* 05-215
Iridoosmium *n* 65-845
Isoaktivitätskurve *f* 15-315
Isoaktivitätslinie *f* 15-315
isochrone Erholung *f* 70-495
isochrones Erholungsglühen *n* 70-495
isomorphe Kristalle *mpl* 20-480
Isomorphie *f* 20-475
Isomorphismus *f* 20-475
isotherme Diffusion *f* 45-350
isothermer Schnitt *m* 35-735
isothermes Zeit-Temperatur--Umwandlungs-Diagramm *n* 70-155
isothermes ZTU-Diagramm *n* 70-155
isothermisches Glühen *n* 70-450
isothermisches Schmelzen *n* 35-450
isothermisches Zeit-Temperatur--Umwandlungs-Schaubild *n* 70-155
isothermisches ZTU-Schaubild *n* 70-155
isothermische Umwandlung *f* 25-515
Isotropie *f* 20-450

J

Jod *n* 05-210
Jominy-Probe *f* 80-450
Juweliergold *n* 65-780
Juwelierlegierung *f* 55-580

K

Kadmium *n* 05-075
Kalifornium *n* 05-090
Kalium *n* 05-350
kalorimetrische Untersuchung *f* 80-500
Kalorisieren *n* 70-925
Kaltarbeitsstahl *m* 60-805
Kaltaushärtung *f* 70-585
Kaltauslagern *n* 70-585
Kaltbruch *m* 85-075
Kaltbrüchigkeit *f* 75-285
Kalthärten *n* 45-050
Kaltlagern *n* 70-585
Kaltreckung *f* 45-030
Kaltreckungsgrad *m* 45-040
Kaltriß *m* 85-075
Kaltsprödigkeit *f* 75-285
kaltverfestigter Stahl *m* 60-295
Kaltverfestigung *f* 45-050
Kaltverformung *f* 40-025
Kaltverformungsgrad *m* 45-040
kaltzäher Stahl *m* 60-555
Kalzium *n* 05-085
Kalzium-Silizium *n* 60-1160
Kanalisierung *f* 30-500
Kanonenbronze *f* 65-245
Kantenversetzung *f* 30-150
Kanthal *n* 65-525

KARBID

Karbid *n* 40-430
binäres ~ 40-470
einfaches ~ 40-470
karbidbildendes Element *n* 50-435
Karbidbildner *m* 50-435
Karbide *npl*
eutektische ~ 50-440
gesinterte ~ 55-665
ledeburitische ~ 50-440
Karbidnetz *n* 50-445
karbidstabilisierendes Element *n* 50-465
Karbidzeilen *fpl* 85-235
Karbidzeiligkeit *f* 85-235
karbidzerlegendes Element *n* 50-460
Karbonitrid *n* 40-485
Karbonitrieren *n* 70-870
Karbonyleisen *n* 60-020
Kartuschen-Messing *n* 65-380
Kastenglühen *n* 70-435
Katodenkupfer *n* 65-030
katodisches Ionenätzen *n* 80-180
katodische Vakuumätzung *f* 80-180
Keim *m*
arteigener ~ 25-110
artfremder ~ 25-120
heterogener ~ 25-115
kohärenter ~ 25-155
orientierter ~ 25-160
spontaner ~ 25-110
Keimbildner *m* 25-125
Keimbildung *f* 25-175
athermische ~ 25-200
dynamische ~ 25-205
heterogene ~ 25-185
homogene ~ 25-180
~ durch Reibung 25-210
orientierte ~ 25-195
spontane ~ 25-190
Keimbildungsgeschwindigkeit *f* 25-270
Keimbildungshäufigkeit *f* 25-270
Keimradius *m*
kritischer ~ 25-145
Keimwachstumsgeschwindigkeit *f* 25-275
Keimzentrum *n* 25-135
Kenngröße *f*
plastische ~ 75-065
Kerbempfindlichkeit *f* 75-195
Kerbschlagfestigkeit *f* 75-190
Kerbschlagversuch *m* 80-430
Kerbschlagzähigkeit *f* 75-190
Kern *m* 70-665
~ der Versetzung 30-205
Kernhärten *n* 70-770
Kernkorrosion *f* 85-365
kernphysikalische Eigenschaften *fpl* 75-520
Kernrückfeinen *n* 70-770
kerntechnische Eigenschaften *fpl* 75-520
Kesselbaustahl *m* 60-740
Kesselstahl *m* 60-740
KG-Guß *m* 60-975
KG-Gußeisen *n* 60-975
Kink *m* 30-245
Kippgrenze *f* 30-605
Kirkendall-Effekt *m* 45-385
Kistenglühen *n* 70-435
Klang *m*
metallischer ~ 75-360
Klangprobe *f* 80-795
Klatschkühlung *f* 70-085
Kleingefüge *n* 40-030

Kleinstruktur *f* 40-030
Kleinwinkelgrenze *f* 30-590
Kleinwinkelkorngrenze *f* 30-590
Kletterbewegung *f* 30-285
Klettern *n* von Versetzungen 30-295
Knetbronze *f* 65-105
Knetlegierung *f* 55-530
Messing-~ 65-295
Knick *m* 30-245
Knotengraphit *m* 50-210
Kobalt *n* 05-115
Kobaltstahl *m* 60-240
Koeffizient *m*
elastischer ~ 75-145
Koehler-Quelle *f* 30-435
Koerzitivfeldstärke *f* 75-635
Koerzitivkraft *f* 75-635
Koexistenzkurven *fpl* 35-540
koexistierende Phasen *fpl* 35-190
kohärente Ausscheidungen *fpl* 40-320
kohärente Grenzfläche *f* 30-625
kohärente Phasengrenze *f* 30-625
kohärenter Keim *m* 25-155
Kohärenz *f* 30-620
Kohäsion *f* 10-310
Kohäsionsfestigkeit *f* 75-135
Kohäsionskraft *f* 10-315
Kohlenstoff *n* 05-095
freier ~ 50-005
gebundener ~ 50-010
Kohlenstoffäquivalent *n* 50-480
kohlenstoffarmer Stahl *m* 60-100
Kohlenstoffdiagramm *n*
Eisen-~ 50-020
Kohlenstoffpegel *m* 70-750
Kohlenstoffpotential *n* 70-750
kohlenstoffreicher Stahl *m* 60-110
Kohlenstoffstahl *m* 60-095
Kohlenstoff-Übergangszahl *f* 70-755
Kohlungsatmosphäre *f* 70-695
Kohlungsgas *n* 70-745
Kohlungsmittel *n* 70-740
Koksroheisen *n* 60-870
Kolbenlegierung *f* 55-570
kompaktes Metall *n* 55-155
komplexe Legierung *f* 55-310
Komplexionenzahl *f* 15-230
Komponente *f* 35-035
Kompositwerkstoff *m* 55-680
Kompressibilitätskoeffizient *m* 75-165
Kompressionsmodul *m* 75-160
kondensierte Phase *f* 35-145
konfigurationsbedingte Wärmekapazität *f* 25-635
Konfigurationsentropie *f* 15-210
kongruente intermetallische Verbindung *f* 35-220
kongruentes Schmelzen *n* 35-460
kongruente Umwandlung *f* 25-525
konjugiertes Gleitsystem *n* 45-125
Konnode *f* 35-695
Konode *f* 35-695
Konovalowscher Satz *m* 35-705
konservative Bewegung *f* 30-290
Konstantan *n* 65-450
Konstante *f*
Poisson ~ 75-170

Konstitution f 35-020
konstitutionelle Unterkühlung f 25-255
Konstitutionsdiagramm n 35-480
Konstruktionsstahl m 60-670
Kontaktermüdung f 85-420
Kontakthärten n 70-250
Kontaktkorrosion f 85-320
Kontaktlegierung f 55-585
kontinuierliche Abkühlung f 70-090
kontinuierliche Ausscheidung f 25-375
kontinuierliches Abkühlen n 70-090
kontinuierliches Zeit-Temperatur-
 -Umwandlungs-Diagramm n 70-160
kontinuierliches Zeit-Temperatur-
 -Umwandlungs-Schaubild n 70-160
kontinuierliches ZTU-Diagramm n 70-160
kontinuierliches ZTU-Schaubild n 70-160
kontinuierliche Umwandlung f 25-540
Kontraktion f 75-795
Kontrastautoradiographie f 80-745
kontrollierte Atmosphäre f 70-700
kontrollierte Erstarrung f 25-030
Konverterkupfer n 65-015
Konverterstahl m 60-590
Konzentration f 55-230
 molare ~ 55-240
Konzentrationsdreieck n 35-720
Konzentrationsfluktuationen fpl 25-100
Konzentrationsgefälle n 15-080
Konzentrationsgradient m 15-080
Konzentrationsschwankungen fpl 25-100
Koordinationssphäre f 20-190
Koordinationszahl f 20-185
Korn n 40-360
 feines ~ 40-370
Kornfeinung f 40-385
Kornfelderätzung f 80-150
Kornflächenätzung f 80-150
Korngrenze f 30-565
Korngrenzenätzung f 80-145
Korngrenzenausscheidungen fpl 40-310
Korngrenzenbruch m 40-705
Korngrenzendiffusion f 45-310
Korngrenzenenergie f 30-575
Korngrenzengleitung f 45-160
Korngrenzenkorrosion f 85-360
Korngrenzensegregat n 40-310
Korngrenzenseigerung f 25-450
Korngrenzenversetzung f 30-580
Korngrenzenwanderung f 45-490
Korngröße f 80-285
körniger Bruch m 40-690
körniger Perlit m 50-150
körniger Zementit m 50-100
körniges Gefüge n 40-120
Kornverfeiner m 40-400
Kornverfeinerung f 40-385
Kornvergröberung f 25-290
Kornvergrößerung f 25-290
Kornwachstum n 25-290
körperzentriertes Gitter n 20-295
Korrodierbarkeit f 75-675
Korrosion f 85-260
 atmosphärische ~ 85-275
 chemische ~ 85-265
 ebenmäßige ~ 85-335
 elektrochemische ~ 85-270

Korrosion
 elektrochemische ~ 85-270
 galvanische ~ 85-270
 gleichmäßige ~ 85-335
 innere ~ 85-355
 interkristalline ~ 85-360
 örtliche ~ 85-340
 selektive ~ 85-305
Korrosionsanfälligkeit f 75-675
korrosionsbeständige Legierung f 55-480
korrosionsbeständiger Stahl m 60-535
Korrosionsbeständigkeit f 75-670
Korrosionsempfindlichkeit f 75-675
Korrosionsermüdung f 85-415
Korrosionswiderstand m 75-670
kovalente Bindung f 10-330
Kovar n 65-520
Kraft f
 Peierls-Nabarro-~ 30-300
 treibende ~ 15-070
Krankheit f
 Krupp-~ 85-220
Kreis m
 Debye-Scherrer-~ 80-640
Kriechen n 45-495
 beschleunigtes ~ 45-510
 Herring-Nabarro-~ 45-520
 Nabarro-~ 45-520
 primäres ~ 45-500
 sekundäres ~ 45-505
 stationäres ~ 45-505
 tertiäres ~ 45-510
Kriechfestigkeit f 75-220
Kriechgrenze f 75-220
 wahre ~ 75-225
Kriechkurve f 45-515
Kriechwiderstand m 75-215
Kristall m 20-015
 dendritischer ~ 25-315
 fehlerfreier ~ 30-005
 idiomorpher ~ 25-355
 perfekter ~ 30-005
Kristallachse f 20-125
Kristallaggregat n 20-440
Kristallaufbau m 20-040
Kristallbaufehler m 30-015
Kristalle mpl
 gleichachsige ~ 40-285
 globulitische ~ 40-285
 isomorphe ~ 20-480
 primäre ~ 40-270
 sekundäre ~ 40-275
Kristallebene f 20-155
Kristallfläche f 20-060
Kristallform f 20-460
Kristallgitter n 20-045
Kristallhabitus m 20-460
Kristallhaufwerk n 20-440
kristallin 20-010
kristalliner Bruch m 40-685
kristallinisch 20-010
Kristallisation f 25-010
 eutektische ~ 25-025
 gleichachsige ~ 25-075
 globulitische ~ 25-075
 primäre ~ 25-050
 sekundäre ~ 25-060
 spontane ~ 25-045
Kristallisationsachse f
 primäre ~ 25-300
Kristallisationsfront f 25-295
Kristallisationsgeschwindigkeit f 25-275
 lineare ~ 25-275

Kristallisationskeim *m* 25-105
Kristallit *m* 40-360
Kristallkeim *m* 25-105
  kritischer ~ 25-140
  wachstumsfähiger ~ 25-150
Kristallklasse *f* 20-050
Kristallkorn *n* 40-360
Kristallographie *f* 20-035
kristallographische Achse *f* 20-125
kristallographische Ebene *f* 20-155
kristallographische Richtung *f* 20-150
kristallographischer Index *m* 20-160
kristallographisches Achsenverhältnis *n* 20-355
Kristallorientierung *f* 20-415
Kristallrichtung *f* 20-150
Kristallseigerung *f* 25-420
Kristallstruktur *f* 20-040
Kristallsymmetrie *f* 20-055
Kristallsystem *n* 20-235
  hexagonales ~ 20-245
  kubisches ~ 20-240
  monoklines ~ 20-265
  orthorhombisches ~ 20-255
  quadratisches ~ 20-250
  reguläres ~ 20-240
  rhomboedrisches ~ 20-260
  tetragonales ~ 20-250
  trigonales ~ 20-260
  triklines ~ 20-270
Kristalltracht *f* 20-460
Kristallwachstum *n* 25-285
Kristall-Wachstumsgeschwindigkeit *f* 25-275
Kristallziehen *n* 25-090
Kristallzüchtung *f* 25-085
Kristallzwilling *m* 45-230
kritische Abkühlungsgeschwindigkeit *f* 70-180
kritische Gleitspannung *f* 45-090
kritischer Durchmesser *m* 70-360
kritischer Keimradius *m* 25-145
kritischer Kristallkeim *m* 25-140
kritischer Verformungsgrad *m* 45-075
kritische Schubspannung *f* 45-090
kritisches Intervall *n* 35-560
kritische Temperatur *f* 35-335
kritische Übergangstemperatur *f* 25-570
kritische Verformung *f* 45-075
Krupp-Krankheit *f* 85-220
kryptokristalliner Martensit *m* 50-330
Krypton *n* 05-225
kubischer Ausdehnungskoeffizient *m* 75-460
kubischer Martensit *m* 50-340
kubisches flächenzentriertes Gitter *n* 20-310
kubisches Gitter *n* 20-300
kubisches innenzentriertes Gitter *n* 20-315
kubisches körperzentriertes Gitter *n* 20-315
kubisches Kristallsystem *n* 20-240
kubisches raumzentriertes Gitter *n* 20-315
kubische Textur *f* 40-605
kubisch flächenzentriertes Gitter *n* 20-310
kubisch innenzentriertes Gitter *n* 20-315
kubisch körperzentriertes Gitter *n* 20-315
kubisch raumzentriertes Gitter *n* 20-315
Kugelgraphit *m* 50-190

Kugelgraphitguß *m* 60-975
Kugelgraphit-Gußeisen *n* 60-975
kugeliger Graphit *m* 50-190
kugeliger Zementit *m* 50-100
Kugelkornglühen *n* 70-510
Kugellagerstahl *m* 60-720
Kugelpackung *f*
  dichte ~ 20-210
  dichteste ~ 20-210
Kunstbronze *f* 65-270
künstliche Alterung *f* 70-590
künstliche Bekeimung *f* 25-215
Kupfer *n* 05-120
  E-~ 65-035
  hochleitfähiges ~ 65-065
  OFHC-~ 65-070
  phosphordesoxydiertes ~ 65-055
  sauerstofffreies ~ 65-060
  sauerstofffreies hochleitfähiges ~ 65-070
  sauerstoffhaltiges ~ 65-050
  Silizium-~ 65-080
Kupferlegierung *f* 65-010
Kupfernickel *n* 65-400
Kupferstahl *m* 60-245
Kupfersud *m* 65-045
Kurve *f*
  Spannungs-Dehnungs-~ 80-410
  Spannungs-Verformungs-~ 80-410
  Wöhler-~ 80-425

L

Lagerbronze *f* 65-235
Lagergußeisen *n* 60-1105
Lagerlegierung *f* 55-555
Lagermetall *n*
  Blei-Alkali-~ 65-685
Lagerschalenlegierung *f* 55-555
Lagerstahl *m* 60-715
Lagerweißmetall *n* 65-640
lamellarer Perlit *m* 50-130
lamellares Eutektikum *n* 40-220
lamellares Gefüge *n* 40-105
Lamellenabstand *m* 40-110
Lamellengraphit *m* 50-185
langsame Abkühlung *f* 70-080
langsames Abkühlen *n* 70-080
Lanthan *n* 05-230
latente Schmelzwärme *f* 75-380
latente Verdampfungswärme *f* 75-390
latente Verfestigung *f* 45-180
lattenförmiger Martensit *m* 50-355
Laue-Aufnahme *f* 80-620
Lauediagramm *n* 80-620
Laue-Methode *f* 80-590
laugenbeständige Legierung *f* 55-490
laugenbeständiger Stahl *m* 60-550
laugenbeständiges Gußeisen *n* 60-1080
Laugenbeständigkeit *f* 75-700
Laugenrissigkeit *f* 75-305
laugensicherer Stahl *m* 60-550
Laugensprödigkeit *f* 75-305
Lautal *n* 65-595
Laves-Phasen *fpl* 35-265
Lawrencium *n* 05-235
Lawrentium *n* 05-235
Ledeburit *m* 50-160
ledeburitische Karbide *npl* 50-440
ledeburitischer Stahl *m* 60-440

ledeburitischer Zementit m 50-075
Ledeburitstahl m 60-440
Leerstelle f 30-035
  abgeschreckte ~ 30-095
  strukturelle ~ 30-090
Leerstellenanhäufung f 30-125
Leerstellenkondensation f 30-115
Leerstellenübersättigung f 30-110
Legierbarkeit f 55-175
Legieren n 55-170
legierter Gußstahl m 60-835
legierter Nitrierstahl m 60-700
legierter Stahl m 60-115
legierter Stahlguß n 60-835
legierter Zementit m 50-110
legiertes Gußeisen n 60-995
legiertes Roheisen n 60-935
Legierung f 55-160
  ALNiCo-~ 65-530
  anlaufbeständige ~ 55-500
  aushärtbare ~ 55-385
  ausscheidungshärtbare ~ 55-385
  binäre ~ 55-295
  dauerstandfeste ~ 55-510
  dispersionsgehärtete ~ 55-390
  dispersionsverfestigte ~ 55-390
  einphasige ~ 55-330
  Elektrotypie-~ 55-635
  eutektische ~ 55-350
  eutektoide ~ 55-365
  ferromagnetische ~ 55-455
  Galvano-~ 55-635
  handelsübliche ~ 55-520
  hartmagnetische ~ 55-465
  heterogene ~ 55-340
  Heusler-~ 65-420
  hochfeste ~ 55-420
  hochlegierte ~ 55-325
  hochpermeable ~ 55-475
  hochprozentige ~ 55-325
  hochwarmfeste ~ 55-515
  infiltrierte ~ 55-670
  komplexe ~ 55-310
  korrosionsbeständige ~ 55-430
  laugenbeständige ~ 55-490
  ~ erster Schmelzung 55-275
  ~ mit geringer Wärmedehnung 55-440
  leichtschmelzende ~ 55-395
  Lipowitz-~ 65-680
  magnetische ~ 55-445
  magnetisch harte ~ 55-465
  magnetisch weiche ~ 55-460
  mehrphasige ~ 55-345
  metallkeramische ~ 55-675
  monophasige ~ 55-330
  NE-~ 65-005
  nichtmagnetische ~ 55-470
  niedriglegierte ~ 55-320
  niedrigprozentige ~ 55-320
  niedrigschmelzende ~ 55-395
  Nimonic-~ 65-480
  oxydationsbeständige ~ 55-495
  pyrophore ~ 55-595
  quasibinäre ~ 55-315
  quaternäre ~ 55-305
  Rosesche ~ 65-675
  säurebeständige ~ 55-485
  superplastische ~ 55-430
  synthetische ~ 55-270
  technische ~ 55-520
  ternäre ~ 55-300
  übereutektische ~ 55-360
  übereutektoide ~ 55-375
  übereutektoidische ~ 55-375
  ultrafeste ~ 55-425
  unmagnetische ~ 55-470
  untereutektische ~ 55-355
  untereutektoidische ~ 55-370
  verdünnte ~ 55-320
  verschleißfeste ~ 55-435
  vierkomponentige ~ 55-305
  wärmebehandlungsfähige ~ 55-525

Legierung
  warmfeste ~ 55-510
  weichmagnetische ~ 55-460
  Woodsche ~ 65-670
  Y-~ 65-590
  Ypsylon-~ 65-590
  zahntechnische ~ 55-575
  zunderbeständige ~ 55-505
  zunderfeste ~ 55-505
  zweiphasige ~ 55-335
Legierungsbasis f 55-185
Legierungsbestandteil m 55-180
Legierungselement n 55-190
  stabilisierendes ~ 50-485
Legierungsfähigkeit f 55-175
Legierungskarbid n 40-445
Legierungskomponente f 55-180
Legierungskonzentration f 55-245
Legierungsstahl m 60-115
Legierungssystem n 55-165
Legierungsträger m 55-185
Legierungsverfestigung f 55-210
Legierungszusatz m 55-190
leichte Platinmetalle npl 65-830
leichte Richtung f 75-585
Leichtmetalle npl 55-050
Leichtmetallegierung f 55-405
leichtschmelzende Legierung f 55-395
Leitband n 10-260
Leitbronze f 65-255
Leitfähigkeit f
  elektrische ~ 75-485
  thermische ~ 75-440
Leitfähigkeitsband n 10-260
Leitkupfer n 65-065
Leitlegierung f 55-550
Leitungsband n 10-260
Leitungsbronze f 65-255
Leitungselektron n 10-265
Letternmetall n 55-615
Lichtbogenstahl m 60-625
Lichtmikroskopie f 80-215
lineare Kristallisationsgeschwindigkeit f 25-275
linearer Ausdehnungskoeffizient m 75-455
Linie f
  eutektische ~ 35-640
  eutektoide ~ 35-655
  eutektoidische ~ 35-655
  monotektische ~ 35-650
  peritektische ~ 35-645
Linien fpl
  Hartmannsche ~ 45-195
  Lüderssche ~ 45-195
  Neumannsche ~ 45-205
linienförmiger Gitterfehler m 30-140
linksgängige Schraubenversetzung f 30-175
Linometall n 55-625
Lipowitz-Legierung f 65-680
Lipowitz-Metall n 65-680
Liquidusfläche f 35-510, 35-520
Liquiduskurve f 35-510, 35-515
Liquiduslinie f 35-510, 35-515
Liquidustemperatur f 35-525
Liquisol-Abschreckung f 70-085
Lithium n 05-245
Lochfraß m 85-350
Lochfraßkorrosion f 85-350
Lochkorrosion f 85-350
lockerstreifiger Perlit m 50-140

Lomer-Cottrell-Versetzung f 30-450
Lösemittel n 35-355
Löser m 35-355
Löslichkeit f 35-295
Löslichkeitsgrenze f 35-330
Löslichkeitskurve f 35-530
Löslichkeitslinie f 35-530
Lösung f 35-350
  athermische $\sim$ 15-265
  binäre $\sim$ 35-405
  feste $\sim$ 35-375
  flüssige $\sim$ 35-365
  geordnete feste $\sim$ 35-430
  gesättigte feste $\sim$ 35-395
  ideale $\sim$ 15-240
  metallische $\sim$ 35-385
  nichtideale $\sim$ 15-245
  primäre feste $\sim$ 35-380
  reguläre $\sim$ 15-255
  semireguläre $\sim$ 15-260
  übersättigte feste $\sim$ 35-400
  ungeordnete feste $\sim$ 35-435
  verdünnte $\sim$ 15-250
  wirkliche $\sim$ 15-245
Lösungsglühen n 70-500
  $\sim$ und Abschrecken 70-570
Lösungsmittel n 35-355
Lösungsverfestigung f 55-215
Lösungswärme f 75-405
Lot n 65-710
Lötfähigkeit f 75-740
Lötlegierung f 65-710
Lötmessing n 65-740
Lötmetall n 65-710
Lötzinn f 65-725
lückenlose Mischbarkeit f 35-315
lückenlose Mischkristallreihe f 35-390
Lüderssche Linien fpl 45-195
Luftabkühlen n 70-095
Luftabkühlung f 70-095
Luftabschrecken n 70-095
Luftabschreckung f 70-095
Luftblase f 85-130
Lufthärten n 70-215
Lufthärter m 60-375
Lufthärtestahl m 60-375
Lunker m 85-125
Luppenstahl m 60-570
Lutetium n 05-250

M

Magnalium n 65-620
Magnesium n 05-255
Magnesiumlegierung f 55-410
Magnetdefektoskopie f 80-805
magnetische Analyse f 80-560
magnetische Anisotropie f 75-575
magnetische Domäne f 40-635
magnetische Eigenschaften fpl 75-540
magnetische Hysterese f 75-640
magnetische Induktion f 75-610
magnetische Legierung f 55-445
magnetische Nachwirkung f 75-645
magnetische Permeabilität f 75-615
magnetische Quantenzahl f 10-140
magnetischer Bezirk m 40-635
magnetischer Stahl m 60-495
magnetischer Umwandlungspunkt m 25-560
magnetisches Altern n 70-620
magnetisches Ordnen n 25-640
magnetische Struktur f 40-650
magnetisches Untergitter n 20-485
magnetische Suszeptibilität f 75-625
magnetische Umwandlung f 25-555
magnetische Vorzugsrichtung f 75-585
magnetisch harte Legierung f 55-465
magnetisch harter Stahl m 60-505
magnetisch weiche Legierung f 55-460
magnetisch weicher Stahl m 60-500
Magnetisierung f 75-600
  remanente $\sim$ 75-630
Magnetostriktion f 75-650
magnetothermische Behandlung f 70-965
Magnetpulver-Prüfung f 80-805
Magnetpulververfahren n
  nasses $\sim$ 80-815
  trockenes $\sim$ 80-810
Magnetstahl m 60-780
Makroätzung f 80-135
Makroaufnahme f 80-310
Makrogefüge n 40-025
Makrographie f 80-305
Makrophotographie f 80-315
Makroseigerung f 25-415
makroskopische Eigenschaften fpl 75-850
makroskopische Seigerung f 25-415
makroskopische Spannungen fpl 85-030
makroskopische Untersuchung f 80-045
Makrostruktur f 40-025
Mangan n 05-260
Manganbronze f 65-155
Mangangußeisen n 60-1030
Manganhartstahl m 60-785
Manganin n 65-405
Manganium n 05-260
Mangankupfer n 65-085
Manganmessing n 65-355
Mangansiliziumstahl m 60-205
Manganstahl m 60-195
Mangansulfid n 50-240
Maragingstahl m 60-460
Marinemessing n
  heterogenes $\sim$ 65-385
  homogenes $\sim$ 65-390
Markenbezeichnung f 60-090
Martensit m 50-310
  angelassener $\sim$ 50-345
  azikularer $\sim$ 50-315
  feinnadeliger $\sim$ 50-325
  grobkörniger $\sim$ 50-320
  grobnadeliger $\sim$ 50-320
  kryptokristalliner $\sim$ 50-330
  kubischer $\sim$ 50-340
  lattenförmiger $\sim$ 50-355
  massiver $\sim$ 50-355
  nadeliger $\sim$ 50-315
  plattenförmiger $\sim$ 50-355
  sekundärer $\sim$ 50-350
  strukturloser $\sim$ 50-330
Martensitaltern n 70-615
martensitaushärtbarer Stahl m 60-460
Martensitaushärten n 70-615
martensitaushärtender Stahl m 60-460
Martensitbereich m 50-300
Martensitbildungstemperatur f 70-165
martensitischer Stahl m 60-430
martensitisches Gußeisen n 60-1055
Martensitnadel f 50-375

Martensitplatte f 50-370
Martensitpunkt m
  oberer ~ 70-165
  unterer ~ 70-170
Martensitstufe f 50-300
Martensitumwandlung f 50-280
  sekundäre ~ 50-285
Martinstahl m 60-610
Maschinenbaustahl m 60-680
Maschinenbronze f 65-240
Maschinenmessing n 65-375
Maschinenstahl m 60-680
Massenspektrographie f 80-480
Massenstahl m 60-135
Massenzahl f 75-335
massive Probe f 80-055
massiver Martensit m 50-355
massive Umwandlung f 25-550
Materialfehler m 70-005
Matrix f 35-290, 40-015
Matthiessensche Regel f 75-475
mechanische Alterung f 45-225
mechanische Eigenschaften fpl 75-005
mechanische Eigenschaften fpl aus der Zugprobe 75-015
mechanisches Gemenge n der Phasen 40-190
mechanisches Polieren n 80-070
Medaillenbronze f 65-260
Mehrfachgleitung f 45-140
mehrfach legierter Stahl m 60-170
Mehrphasensystem n 35-110
mehrphasige Diffusion f 45-325
mehrphasige Legierung f 55-345
mehrphasiges Gefüge n 40-160
mehrphasiges System n 35-110
Mehrstofflegierung f 55-310
Mehrstofflösung f 35-410
Mehrstoffsystem n 35-140
meliertes Gußeisen n 60-960
meliertes Roheisen n 60-920
Mendelevium n 05-265
Messerstahl m 60-750
Messing n 65-290
  α-~ 65-325
  α+β-~ 65-330
  β-~ 65-335
  heterogenes ~ 65-320
  homogenes ~ 65-315
  Kartuschen-~ 65-380
  unlegiertes ~ 65-305
  Zinn-~ 65-345
Messing-Gußlegierung f 65-300
Messing-Knetlegierung f 65-295
Messinglot n 65-740
Metall n 10-355, 10-360
  aufgedampftes ~ 55-145
  elektrolytisch niedergeschlagenes ~ 55-130
  gediegenes ~ 55-110
  gesintertes ~ 55-660
  halbedles ~ 55-020
  hochschmelzendes ~ 55-065
  kompaktes ~ 55-155
  Lipowitz-~ 65-680
  niedrigschmelzendes ~ 55-060
  raffiniertes ~ 55-125
  reines ~ 55-090
  Rosesches ~ 65-675
  spaltbares ~ 55-080
  spaltfähiges ~ 55-030
  unedles ~ 55-005
  vakuumgeschmolzenes ~ 55-135

Metall
  Wood-~ 65-670
  Woodsches ~ 65-670
  zonengeschmolzenes ~ 55-140
Metalle npl
  NE-~ 55-015
  T-~ 55-025
Metallegierung f 55-160
Metallehre f 10-005
Metallelektronentheorie f 10-025
Metallgefüge n 40-005
Metallglanz m 75-510
Metallhydrid n 40-495
metallische Bindung f 10-340
metallische Eigenschaften fpl 10-350
metallische Einlagerungen fpl 40-550
metallische Einschlüsse mpl 40-550
metallische Lösung f 35-385
metallische Phase f 35-150
metallischer Glanz m 75-510
metallischer Klang m 75-360
metallischer Zustand m 10-345
metallisches Element n 10-360
metallisches System n 55-165
Metallkarbid n 40-435
Metallkeramik f 55-675
metallkeramische Legierung f 55-675
Metallkristall m 20-020
Metallkunde f 10-005
Metallmikrographie f 80-010
Metallmikroskop n 80-210
Metallmikroskopie f 80-010
Metallogie f 10-005
Metallographie f 80-005
  Hochtemperatur-~ 80-015
  quantitative ~ 80-020
metallographische Analyse f 80-035
metallographische Prüfung f 80-035
metallographische Untersuchung f 80-035
Metalloid n 10-365
Metallphysik f 10-015
Metallschliff m 80-095
Metallschmelze f 55-150
Metallstruktur f 40-005
Metallsystem n 55-165
Metallüberziehen n
  Thermodiffusions-~ 70-915
metallurgische Eigenschaften fpl 75-855
metastabile Phase f 35-175
metastabiler Zustand m 35-070
metastabiles Gleichgewicht n 35-060
Metatektikum m 40-255
metatektische Reaktion f 35-580
metatektische Temperatur f 35-620
metatektische Umwandlung f 35-580
Meteoreisen n 60-005
Methode
  Laue-~ 80-590
  ~ der gemeinsamen Tangente 15-180
  Tangentialschnitt-~ 15-180
Mikroätzung f 80-140
Mikroaufnahme f 80-250
Mikrobild n 80-245
Mikrofotographie f 80-250
Mikrofraktographie f 80-390, 80-395
Mikrogefüge n 40-030
Mikrogefügebestandteil m 40-010
Mikrographie f 80-245

Mikrohärte f 75-250
Mikrokriechen n 45-530
Mikrolunker m 85-120
Mikrophotogramm n 80-250
Mikrophotographie f 80-255
Mikroporen fpl 85-110
Mikroporosität f 85-115
Mikroradiogramm n 80-725
Mikroradiographie f 80-720
mikroradiographische Aufnahme f 80-725
Mikroriß m 85-045
Mikroseigerung f 25-420
Mikroskopie f 80-040
  Durchlicht-~ 80-220
  Reflexions-~ 80-225
  Ultraviolett-~ 80-265
  UV-~ 80-265
mikroskopische Seigerung f 25-420
mikroskopische Untersuchung f 80-040
Mikrosondenuntersuchung f 80-780
Mikrostruktur f 40-030
Mikrostrukturaufnahme f 80-250
milde Abschreckung f 70-265
Mildhärten n 70-265
Millersche Indizes mpl 20-165
Mischbarkeit f 35-300
  begrenzte ~ 35-320
  beschränkte ~ 35-320
  lückenlose ~ 35-315
  ~ im festen Zustand 35-310
  ~ im flüssigen Zustand 35-305
  ~ in der Schmelze 35-305
  völlige ~ 35-315
  vollständige ~ 35-315
Mischkarbid n 40-475
Mischkristall m 40-195
Mischkristallbildung f 25-055
Mischkristalle mpl
  übersättigte ~ 35-400
Mischkristallhärtung f 55-215
Mischkristallisation f 25-055
Mischkristallphase f 35-375
Mischkristallreihe f
  lückenlose ~ 35-390
  ununterbrochene ~ 35-390
Mischkristallverfestigung f 55-215
Mischmetall n 65-865
Mischphase f
  geordnete ~ 35-200, 35-430
  reguläre ~ 15-255
  ungeordnete ~ 35-205, 35-435
Mischung f
  eutektische ~ 40-200
  ideale ~ 15-240
Mischungsenergie f
  freie ~ 15-130
Mischungsenthalpie f 15-155
Mischungsentropie f 15-195
Mischungslücke f 35-345
mittelgekohlter Stahl m 60-105
mittellegierter Stahl m 60-125
mittlerer Teilchenabstand m 40-425
Modifikation f
  allotrope ~ 25-390
  polymorphe ~ 25-290
Modifikator m 25-230
Modifizieren n 25-220
modifiziertes Gußeisen n 60-970
Modifizierungsmittel n 25-230
Modul m

Modul
  E-~ 75-150
  Youngscher ~ 75-150
molare Konzentration f 55-240
molare thermodynamische Funktion f 15-050
molare thermodynamische Größe f 15-050
Molekulargewicht n 75-355
Molgewicht n 75-355
Molybdän n 05-275
  non-sag-~ 65-870
  NS-~ 65-870
Molybdänstahl m 60-220
Molvolumen n 75-325
Molwärme f 75-375
Mond-Nickel n 65-425
Monel n 65-435
Monelmetall n 65-435
monoklines Gitter n 20-395
monoklines Kristallsystem n 20-265
Monokristall m 20-430
Monometall n 55-620
monophasige Legierung f 55-330
monophasiges Gefüge n 40-145
monophasiges System n 50-090
Monotektikale f 35-650
Monotektikum n 40-260
monotektische Horizontale f 35-650
monotektische Linie f 35-650
monotektische Reaktion f 35-575
monotektischer Punkt m 35-670
monotektische Temperatur f 35-615
monotektische Umwandlung f 35-575
monotektoide Reaktion f 35-600
monotektoide Umwandlung f 35-600
Mosaikblock m 30-655
Mosaikstruktur f 40-170
Mößbauer-Effekt m 10-285
Muntzmetall n 65-330
Münzbronze f 65-265
Münzenbronze f 65-265
Münzgold n 65-775
Münzlegierung f 55-565
Münzmetall n 55-565
Münzsilber n 65-755
Muschelbruch m 40-730
muscheliger Bruch m 40-730
Mutterkristall m 40-295
Mutterlauge f 25-095
Mutterphase f 35-290

N

Nachbar m
  erstnächster ~ 20-195
  ~ erster Sphäre 20-195
  ~ zweiter Sphäre 20-200
  nächster ~ 20-195
  zweitnächster ~ 20-200
Nachspannungen fpl 85-020
nächstbenachbartes Atom n 20-195
nächster Nachbar m 20-195
Nachwirkung f
  elastische ~ 75-180
  magnetische ~ 75-645
Nachwirkungseffekt m 75-180
nadeliger Martensit m 50-315
nadeliges Gefüge n 40-115

ÖLHÄRTEND

Nadelkristall m 25-360
Nahordnung f 25-595
Nahordnungsparameter m 25-615
Naphtalinbruch m 40-740
nasses Magnetpulververfahren n 80-815
Natrium n 05-425
naturharter Stahl m 60-350
Naturlegierung f 55-265
natürliche Alterung f 70-585
natürliches Stabilisieren n 70-540
natürliche Streckgrenze f 75-110
Nebenquantenzahl f 10-135
Nebenschale f 10-160
Néel-Punkt m 25-565
Néeltemperatur f 25-565
negative Diffusion f 45-340
negative Stufenversetzung f 30-160
NE-Legierung f 65-005
NE-Metalle npl 55-015
Neodym n 05-280
Neon n 05-285
Neptunium n 05-290
Nest n von Leerstellen 30-125
nestförmiger Graphit m 50-210
Netzebene f 20-155
Netzebenenabstand m 20-225
Netzgefüge n 40-100
Netzstruktur f 40-100
Neulegierung f 55-275
Neumannsche Linien fpl 45-205
Neusilber n 65-415
Neutralgasatmosphäre f 70-675
Neutronen-Einfangsquerschnitt m 75-530
Neutronenradiographie f 80-830
Nichrom n 65-470
nichtaushärtbarer Stahl m 60-405
Nichteisenlegierung f 65-005
Nichteisenmetalle npl 55-015
Nichteisenmetallegierung f 65-005
Nichtgleichgewichtserstarrung f 25-020
nichtgleitfähige Versetzung f 30-280
nichthärtender Stahl m 60-390
nichtideale Lösung f 15-245
nicht-kohärente Grenzfläche f 30-630
nicht-kohärente Phasengrenze f 30-630
nichtkongruente intermetallische Verbindung f 35-225
nichtkongruentes Schmelzen n 35-465
nicht-konservative Versetzungsbewegung f 30-285
nichtmagnetische Legierung f 55-470
nichtmagnetisierbarer Stahl m 60-510
Nichtmetall n 10-365
nichtmetallische Einschlüsse mpl 40-515
nichtmetallische Phase f 35-155
nichtmetallisches Element n 10-365
nichtrostender Stahl m 60-515
Nickel n 05-295
Mond-∼ 65-425
Nickelbasislegierung f 65-430
Nickelbronze f 65-160
Nickelgußeisen n 60-1025
Nickelin n 65-410
Nickellegierung f 65-430
Nickelmessing n 65-350
Nickelstahl m 60-185

Nicrosilal n 60-1095
niedriggekohlter Stahl m 60-100
niedriglegierte Legierung f 55-320
niedriglegierter Stahl m 60-120
niedriglegiertes Gußeisen n 60-1000
niedrigprozentige Legierung f 55-320
niedrigschmelzende Legierung f 55-395
niedrigschmelzendes Metall n 55-060
niedrigzinnhaltige Bronze f 65-120
Nietstahl m 60-745
Nimonic-Legierung f 65-480
Niob n 05-300
Niobid n 40-505
Niobkarbid n 40-465
Niresist n 60-1090
Nital n 80-110
Nitrid n 70-810
nitridbildendes Element n 50-470
Nitridbildner m 50-470
Nitrideinschlüsse mpl 40-545
nitrierbarer Stahl m 60-695
Nitrieren n 70-785
Nitrierschicht f 70-845
Nitrierstahl m 60-695
legierter ∼ 60-700
nitrierter Stahl m 60-315
Nitroaustenit m 70-835
Nitrokarburieren n 70-875
Nitrooxidieren n 70-885
Nobelium n 05-310
non-sag-Molybdän n 65-870
nonvariantes System n 35-075
normale Seigerung f 25-435
normalgeglühter Stahl m 60-280
Normalglühen n 70-475
Normalisieren n 70-475
Normstahl m 60-265
NS-Molybdän n 65-870
nulldimensionale Fehlstelle f 30-030
nulldimensionaler Gitterfehler m 30-030

O

oberer Bainit m 50-420
oberer Martensitpunkt m 70-165
Oberflächenabdruck m 80-345
Oberflächenbehandlung f 70-010
Oberflächendiffusion f 45-300
Oberflächenenergie f 30-640
freie ∼ 15-120
Oberflächenentkohlung f 85-210
Oberflächenfehler m 85-010
Oberflächenhärten n 70-285
Oberflächenkorrosion f 85-330
Oberflächenriß m 85-050
Oberflächenspannung f 30-645
Ofenabkühlen n 70-110
Ofenabkühlung f 70-110
OFHC-Kupfer n 65-070
Oktaederebene f 20-335
Oktaederlücke f 20-330
oktaedrische Gitterlücke f 20-330
Ölabschrecken n 70-105
Ölabschreckung f 70-105
Ölhärten n 70-210
ölhärtender Stahl m 60-370

Ölhärter *m* 60-370
Ölhärtestahl *m* 60-370
optische Eigenschaften *fpl* 75-500
Ordnen *n*
magnetisches ~ 25-640
Ordnung *f* 25-590
Ordnungsenergie *f* 25-630
Ordnungsgrad *m* 25-605
Ordnungsphase *f* 35-200
Ordnungsvorgang *m* 25-575
Ordnungszahl *f* 75-330
Ordnungszustand *m* 25-590
Ordnung-Unordnung-Umwandlung *f* 25-580
orientierte Keimbildung *f* 25-195
orientierter Keim *m* 25-160
Orientierung *f*
regellose ~ 20-420
Orientierungsbestimmung *f* der Kristalle 80-690
Orientierungskontrast *m* 80-155
Orientierungsquantenzahl *f* 10-140
Orientierungsunterschied *m* 30-595
Orientierungszwilling *m* 45-240
orthorhombisches Kristallsystem *n* 20-255
örtliche Korrosion *f* 85-340
örtliches Aufkohlen *n* 70-735
örtliches Aufschmelzen *n* 85-215
örtliches Härten *n* 70-275
Osmiridium *n* 65-845
Osmium *n* 05-315
Osteoplastiklegierung *f* 55-645
Oxideinschlüsse *mpl* 40-535
oxidische Einschlüsse *mpl* 40-535
Oxydation *f*
innere ~ 70-947
oxydationsbeständige Legierung *f* 55-495
Oxydationsbeständigkeit *f* 75-685
Oxydieren *n* 70-400
Oxygenstahl *m* 60-605

P

Paar *n*
Frenkel-~ 30-060
Palladium *n* 05-325
Pantal *n* 65-615
Panzerplattenstahl *m* 60-765
Panzerstahl *m* 60-765
Paramagnetikum *n* 75-555
paramagnetischer Stoff *m* 75-555
partielle Aufkohlung *f* 70-735
partielle Diffusionskonstante *f* 45-390
partielle Enthalpie *f* 15-150
partielle Entropie *f* 15-215
partielle freie Energie *f* 15-110
partielle Größe *f* 15-060
partielle thermodynamische Funktion *f* 15-060
partielle thermodynamische Größe *f* 15-060
Patentieren *n* 70-415
Patina *f* 85-375
Patronenmessing *n* 65-380
Pauli-Prinzip *n* 10-170
Paulisches Ausschließungsprinzip *n* 10-170
Pauli-Verbot *n* 10-170
Peierls-Nabarro-Kraft *f* 30-300

Pendelglühen *n* 70-520
perfekter Kristall *m* 30-005
Peritektikale *f* 35-645
Peritektikum *n* 40-245
peritektische Horizontale *f* 35-645
peritektische Linie *f* 35-645
peritektische Reaktion *f* 35-570
peritektischer Punkt *m* 35-665
peritektisches Schmelzen *n* 35-475
peritektische Temperatur *f* 35-610
peritektische Umwandlung *f* 35-570
Peritektoid *n* 40-250
peritektoide Reaktion *f* 35-595
peritektoide Temperatur *f* 35-635
peritektoide Umwandlung *f* 35-595
Perlit *m* 50-125
dichtstreifiger ~ 50-135
entarteter ~ 50-155
feinlamellarer ~ 50-135
feinstreifiger ~ 50-135
grobstreifiger ~ 50-140
körniger ~ 50-150
lamellarer ~ 50-130
streifiger ~ 50-130
Perlitglühen *n* 70-450
perlitischer Ferrit *m* 50-060
perlitischer Stahl *m* 60-425
perlitischer Temperguß *m* 60-1130
perlitischer Zementit *m* 50-085
perlitisches Gußeisen *n* 60-1045
Perlitisieren *n* 70-540
perlitisierendes Glühen *n* 70-450
Perlitstufe *f* 50-295
Perlitumwandlung *f* 50-275
perlmutterartiger Bruch *m* 40-740
Permalloy *n* 65-440
Permeabilität *f*
magnetische ~ 75-615
Permendur *m* 65-490
Perminvar *n* 65-510
Phase *f* 35-025
ausgeschiedene ~ 40-305
disperse ~ 40-330
geordnete ~ 35-200
härtesteigernde ~ 40-350
Hume-Rothery-~ 35-280
intermediäre ~ 35-210
intermetallische ~ 35-210
kondensierte ~ 35-140
metallische ~ 35-150
metastabile ~ 35-175
nichtmetallische ~ 35-155
Sigma-~ 35-275
stabile ~ 35-170
ungeordnete ~ 35-205
voreutektische ~ 40-340
voreutektoidische ~ 40-345
Phasen *fpl*
koexistierende ~ 35-190
Laves-~ 35-265
Zintl-~ 35-270
Phasenanalyse *f*
röntgenographische ~ 80-585
Phasendiagramm *n* 35-480
ternäres ~ 35-710
Phasengemisch *n* 40-190
Phasengesetz *n* 35-010
Phasengleichgewicht *n* 35-015
Phasengrenze *f* 35-535
kohärente ~ 30-625
nicht-kohärente ~ 30-630
Phasengrenzfläche *f* 30-610
Phasengrenzlinie *f* 35-535

Phasenkontrast *m* 80-275
Phasenkontrastmikroskopie *f* 80-270
Phasenregel *f* 35-010
Phasenumsetzung *f* 25-460
Phasenumwandlung *f* 25-460
Phasenverfestigung *f* 45-080
Phonon *n* 10-305
Phosphideutektikum *n* 50-235
ternäres ~ 50-235
Phosphor *m* 05-330
Phosphorbronze *f* 65-215
phosphordesoxydiertes Kupfer *n* 65-055
Phosphorkupfer *n* 65-075
phosphorreiches Roheisen *n* 60-890
Phosphorroheisen *n* 60-890
Phosphor-Zinnbronze *f* 65-210
Photomikrographie *f* 80-250, 80-255
physikalisch-chemische Analyse *f* 80-490
physikalische Eigenschaften *fpl* 75-315
physikalische Streckgrenze *f* 75-110
Picral *n* 80-115
Pikrinsäure *f*
alkoholische ~ 80-115
Plasmanitrieren *n* 70-800
plastische Deformation *f* 45-010
plastische Formänderung *f* 45-010
plastische Kenngröße *f* 75-065
plastisches Fließen *n* 45-215
plastische Verformbarkeit *f* 75-025
plastische Verformung *f* 45-010, 45-015
Plastizität *f* 75-025
Platin *n* 05-335
Platiniridium *n* 65-835
Platinit *n* 65-515
Platinmetalle *npl* 65-820
leichte ~ 65-830
schwere ~ 65-825
Platinrhodium *n* 65-840
plattenförmiger Martensit *m* 50-355
plattierter Stahl *m* 60-335
Plutonium *n* 05-340
Poisson-Konstante *f* 75-170
Poisson-Zahl *f* 75-170
Poissonsche Zahl *f* 75-170
Polarisationsmikroskopie *f* 80-260
Polfiguren *fpl* 80-685
Polierelysieren *n* 80-075
Polieren *n*
chemisches ~ 80-080
elektrolytisches ~ 80-075
mechanisches ~ 80-070
Polonium *n* 05-345
Polversetzung *f* 30-365
Polygonisation *f* 45-435
Polygonisierung *f* 45-435
Polykristall *m* 20-440
polymorphe Modifikation *f* 25-390
polymorphe Umwandlung *f* 25-455
Polymorphie *f* 25-385
Polymorphismus *m* 25-385
Pore *f* 85-100
Porosität *f* 80-105
Portevin-Le Chatelier-Effekt *m* 45-540
positive Stufenversetzung *f* 30-155
Potential *n*
chemisches ~ 15-175
elektrochemisches ~ 75-715
~ des Metalles 75-715

Potential
thermodynamisches ~ bei konstantem Druck 15-140
thermodynamisches ~ bei konstantem Volumen 15-105
Potentialberg *m* 10-110
Potentialschwelle *f* 10-110
Potentialwall *m* 10-110
Prägeabdruck *m* 80-345
Praseodym *n* 05-355
Preßlufthärten *n* 70-220
Preßmessing *n* 65-295
preßschweißbarer Stahl *m* 60-485
Primäraustenit *m* 50-030
primäre Erstarrung *f* 25-050
primäre feste Lösung *f* 35-380
primäre Kristalle *mpl* 40-270
primäre Kristallisation *f* 25-050
primäre Kristallisationsachse *f* 25-300
primärer Austenit *m* 50-030
primäre Rekristallisation *f* 45-445
primärer Zementit *m* 50-070
primäre Seigerung *f* 25-405
primäres Gleitsystem *n* 45-115
primäres Kriechen *n* 45-500
primäre Versetzungen *fpl* 30-250
Primärgefüge *n* 40-045
Primärgraphit *m* 50-170
Primärkristalle *mpl* 40-270
Primärkristallisation *f* 25-050
Primärmetall *n* 55-115
Primärrekristallisation *f* 45-445
Primärseigerung *f* 25-405
Primärstruktur *f* 40-045
Primärversetzungen *fpl* 30-250
Primärzementit *m* 50-070
primitive Elementarzelle *f* 20-095
primitive Zelle *f* 20-095
Prinzip *n*
~ der gemeinsamen Tangente 15-180
Pauli-~ 10-170
prismatische Gleitung *f* 45-170
prismatischer Versetzungsring *m* 30-345
prismatische Versetzung *f* 30-345
Prismenebene *f* 20-365
Probe *f*
Jominy-~ 80-450
massive ~ 80-055
Probeentnahme *f* 80-060
Probekörper *m* 80-050
Probenahme *f* 80-060
Probestab *m* 80-435
Probestück *n* 80-050
Promethium *n* 05-360
Proportionalitätsgrenze *f* 70-100
Proportionsgrenze *f* 75-100
Protactinium *n* 05-365
Protaktinium *n* 05-365
Prozeß *m*
thermisch aktivierter ~ 45-280
Prüfung *f*
Magnetpulver-~ 80-805
metallographische ~ 80-035
ultraakustische ~ 80-800
zerstörungsfreie ~ 80-785
pseudobinäres System *n* 35-130
Pseudolegierung *f* 55-650
Puddeleisen *n* 60-055

Puddelstahl m 60-575
Pulveraufkohlen n 70-720
Pulveraufkohlung f 70-720
Pulveraufnahme f 80-635
Pulverdiagramm n 80-635
Punkt m
eutektischer ~ 35-660
eutektoider ~ 35-675
monotektischer ~ 35-670
peritektischer ~ 35-665
spinodaler ~ 15-185
Punktdefekt m 30-030
Punktfehler m 30-030
Punktfehlstelle f 30-030
punktförmiger Gitterbaufehler m 30-030
punktförmiger Gitterfehler m 30-030
punktförmiger Graphit m 50-195
Punktkorrosion f 85-345
pyrophore Legierung f 55-595

Q

quadratisches Kristallsystem n 20-250
Qualitätsgröße f 15-045
Qualitätsstahl m 60-140
Quantenzahl f 10-125
azimutale ~ 10-135
erste ~ 10-130
räumliche ~ 10-140
zweite ~ 10-135
Quantenzustand m 10-120
quantitative Metallographie f 80-020
quantitative thermische Analyse f 80-535
Quantitätsgröße f 15-040
quasibinäre Legierung f 55-315
quasibinärer Schnitt m 35-745
quasibinäres System n 35-130
Quasieutektikum n 40-225
Quasiisotropie f 20-455
quasistationäre Diffusion f 45-345
quaternäre Legierung f 55-305
quaternärer Stahl m 60-165
quaternäres Eutektikum n 40-215
quaternäres System n 35-135
Quecksilber n 05-270
Quecksilbernitratlösung f 80-485
Quecksilbernitratprobe f 80-485
Quecksilberprobe f 80-485
Quelle f
Frank-Read-~ 30-430
Koehler-~ 30-435
Quergleitebene f 45-150
Quergleitung f 45-145
Quettenhärten n 70-255

R

radioaktiver Indikator m 80-760
radioaktives Metall n 55-075
Radiogramm n 80-715
Radiogrammbild n 80-715
Radiographie f 80-705
Radioindikator m 80-760
Radiometallographie f 80-710
Radium n 05-370
Radon n 05-375
Raffinadeblei n 65-655
Raffinadekupfer n 65-020
Raffinal n 65-550

Raffinatkupfer n 65-020
Raffinierstahl m 60-635
raffiniertes Metall n 55-125
Randentkohlung f 85-210
Randschichthärten n 70-285
Randsystem n 35-715
Randzone f
aufgekohlte ~ 70-780
feindendritische ~ 40-175
Raoultsches Gesetz n 15-270
Rasterelektronenmikroskopie f 80-360
Rastermikroskopie f 80-360
Rastlinien fpl 40-750
Raum m
k-~ 10-205
Raumerfüllung f 20-205
Raumerfüllungsgrad m 20-205
Raumerfüllungszahl f 20-205
Raumgitter n 20-045
Raumgitterfehler m 30-015
räumliche Quantenzahl f 10-140
Raummodul m 75-160
raumzentriertes Gitter n 20-295
Raumzustandsdiagramm n 35-750
ternäres ~ 35-750
Raumzustandsschaubild n 35-750
ternäres ~ 35-750
Reaktion f
eutektische ~ 35-565
eutektoide ~ 35-590
metatektische ~ 35-580
monotektische ~ 35-575
monotektoide ~ 35-600
peritektische ~ 35-570
peritektoide ~ 35-595
syntektische ~ 35-585
reaktionsfreudiges Metall n 55-070
reaktive Atmosphäre f 70-680
reaktives Metall n 55-070
Reaktormetall n 55-085
Realkristall m 30-010
rechtsgängige Schraubenversetzung f 30-170
Reckalterung f 45-225
Reflexionsfaktor m 75-515
Reflexionsgrad m 75-515
Reflexionskoeffizient m 75-515
Reflexions-Mikroskopie f 80-225
Reflexionszahl f 75-515
Regel f
Häggsche ~ 35-260
Hume-Rotherysche ~ 35-325
Matthiessensche ~ 75-475
Vegard-~ 35-425
Vegardsche ~ 35-425
regellose Orientierung f 20-420
Reglettenmetall n 55-640
reguläre Lösung f 15-255
reguläre Mischphase f 15-255
reguläres Gitter n 20-300
reguläres Kristallsystem n 20-240
Regulus m 65-665
Reibkorrosion f 85-325
Reibung f
innere ~ 75-210
Reineisen n
technisches ~ 60-015
reines Metall n 55-090
reine Wärmebehandlung f 70-006
Reinheitsgrad m 55-095

Reinmetall n 55-090
Reinstmetall n 55-105
Reißfestigkeit f 75-130
Rekaleszenz f 25-485
Rekristallisation f 45-440
dynamische ~ 45-465
primäre ~ 45-445
sekundäre ~ 45-450
spontane ~ 45-455
tertiäre ~ 45-460
Rekristallisationsdiagramm n 45-485
Rekristallisationsglühen n 70-490
Rekristallisationsgrenze f 45-470
Rekristallisationskeim m 25-170
Rekristallisationsschaubild n 45-485
Rekristallisationstemperatur f 45-470
Rekristallisationstextur f 40-600
Rekristallisationszwilling m 45-240
relative Atommasse f 10-075
relative Festigkeit f 45-140
Relaxation f
elastische ~ 45-525
~ der Spannungen 45-525
Reliefabdruck m 80-345
remanente Magnetisierung f 75-630
Remanenz f 75-630
Renneisen n 60-045
Rennfeuereisen n 60-045
Restaustenit m 50-270
Restenthalpie f 15-145
Restentropie f 15-220
Restgröße f 15-065
Restkarbide npl 40-480
Restmagnetismus m 75-630
Restschmelze f 25-280
Restspannungen fpl 85-020
reversible Umwandlung f 25-495
reziprokes Gitter n 20-410
Rhenium n 05-380
rheotrope Sprödigkeit f 75-310
Rhodium n 05-385
rhombisch-einseitig-flächenzentriertes
   Gitter n 20-385
rhombisches Gitter n 20-380
rhombisch-grundflächenzentriertes Gitter
   n 20-385
rhomboedrisches Gitter n 20-390
rhomboedrisches Kristallsystem n 20-260
Richtreihennummer f 80-290
Richtung f
dichtest besetzte ~ 20-220
kristallographische ~ 20-150
leichte ~ 75-585
~ leichtester Magnetisierbarkeit 75-585
Richtzahl f 80-290
Riffelstahl m 60-490
Ring m
Debye-Scherrer-~ 80-640
Rinne f
binär-eutektische ~ 35-730
eutektische ~ 35-730
Riß m
Griffithscher ~ 85-055
Rißbildung f 85-015
~ durch Spannungskorrosion 85-385
Rißkorrosion f 85-300
Rohaluminium n 65-545
Rohblei n 65-650
Roheisen n 60-840

Roheisen
graues ~ 60-915
Halb-Hämatit-~ 60-900
halbiertes ~ 60-920
hochgekohltes ~ 60-925
legiertes ~ 60-935
meliertes ~ 60-920
phosphorreiches ~ 60-890
Siemens-Martin-~ 60-855
synthetisches ~ 60-940
weißes ~ 60-910
Rohkupfer n 65-015
Röhrendiffusion f 45-360
Röntgenbeugungsaufnahme f 80-615
Röntgendefektoskopie f 80-820
Röntgendurchstrahlungsaufnahme f 80-675
Röntgenfeinstrukturanalyse f 80-580
Röntgenfeinstrukturdiagramm n 80-615
Röntgenfluoreszenzspektralanalyse f
   80-700
Röntgenfluoreszenz-Spektroskopie f
   80-700
Röntgenkristallstrukturanalyse f 80-580
Röntgenmetallographie f 80-575
Röntgenmikroskopie f 80-730
röntgenographische Feinstrukturanalyse
   f 80-580
röntgenographische Phasenanalyse f 80-585
röntgenographische Strukturanalyse f
   80-580
röntgenographische Untersuchung f 80-570
Röntgenspektralanalyse f 80-700
Röntgen-Spektroskopie f 80-700
Röntgenstrahlspektroskopie f 80-700
Röntgenstrukturanalyse f 80-580
Röntgenwerkstoffprüfung f 80-820
Rosesche Legierung f 65-675
Rosesches Metall n 65-675
Rosettengraphit m 50-200
Rost m 85-370
rostbeständiger Stahl m 60-515
rostsicherer Stahl m 60-515
Rotationsdiagramm n 80-665
Rotbrüchigkeit f 75-290
Rotglühhärte f 75-260
Rotgluthärte f 75-260
Rotgold n 65-800
Rotguß m 65-205
Rotmessing n 65-395
Rotwarmhärte f 75-260
Rubidium n 05-390
Rückbildung f 70-610
rückläufige Soliduskurve f 35-505
Rückstrahlaufnahme f 80-670
Rückstrahldiagramm n 80-670
Rückstrahlmethode f 80-630
Rückstrahlverfahren n 80-630
Ruthen n 05-395
Ruthenium n 05-395

S

Salzbadaufkohlung f 70-725
Salzbadeinsatzhärtung f 70-725
Salzbadhärten n 70-230
Salzbadnitrieren n 70-795
Samarium n 05-400
Sammelrekristallisation f 45-450
samtartiger Bruch m 40-715

SÄTTIGUNGSKURVE

Sättigungskurve f 35-530
Sättigungsmagnetisierung f 75-605
Satz m
  Konovalowscher $\sim$ 35-705
Sauerstoff m 05-320
Sauerstoff-Blasstahl m 60-605
sauerstofffreies hochleitfähiges Kupfer n 65-070
sauerstofffreies Kupfer n 65-060
sauerstoffhaltiges Kupfer n 65-050
Sauerstoffpotential n 15-350
säurebeständige Legierung f 55-485
säurebeständiger Stahl m 60-545
säurebeständiges Gußeisen n 60-1075
Säurebeständigkeit f 75-695
saurer Stahl m 60-615
Scandium n 05-405
Schaeffler-Diagramm n 60-530
Schalenguß m 60-980
Schalenhartguß m 60-980
Schattenlinie f 85-230
Schaubild n
  isothermisches Zeit-Temperatur-
    -Umwandlungs-$\sim$ 70-155
  isothermisches ZTU-$\sim$ 70-155
  kontinuierliches Zeit-Temperatur-
    -Umwandlungs-$\sim$ 70-160
  kontinuierliches ZTU-$\sim$ 70-160
Schaumgraphit m 50-170
Scherfestigkeit f 75-080
Schermodul m 75-155
Schicht f
  Beilby-$\sim$ 80-090
  epitaktische $\sim$ 25-350
  weiße $\sim$ 85-250
Schiebungsmartensit m 50-355
Schiebungsumwandlung f 25-510
Schieferbruch m 40-725
Schienenstahl m 60-730
Schiffbaustahl m 60-755
Schlackeneinschlüsse mpl 40-540
Schlagfestigkeit f 75-185
Schlaglot n 65-720
Schlagversuch m 80-430
Schleiffunkenanalyse f 80-440
Schleifmartensit m 50-360
Schliff m 80-095
Schlifffläche f 80-095
Schmelzbarkeit f 75-790
Schmelzbereich m 35-545
Schmelze f 25-095, 55-150
Schmelzen n 35-445
  eutektisches $\sim$ 35-470
  heterothermisches $\sim$ 35-455
  inkongruentes $\sim$ 35-465
  isothermisches $\sim$ 35-450
  kongruentes $\sim$ 35-460
  nichtkongruentes $\sim$ 35-465
  peritektisches $\sim$ 35-475
Schmelzenthalpie f 15-165
Schmelzentropie f 15-225
Schmelzintervall n 35-545
Schmelzpunkt m 75-420
schmelzschweißbarer Stahl m 60-485
Schmelztemperatur f 75-420
Schmelzwärme f 75-380
  gebundene $\sim$ 75-380
  latente $\sim$ 75-380
Schmetterlingsfehler m 85-255
Schmiedbarkeit f 75-725

Schmiedelegierung f 55-530
Schmiedestahl m 60-330
Schmuckgold n 65-780
Schmucklegierung f 55-580
Schnellarbeitsstahl m 60-800
Schnellot n 65-715
Schnellstahl m 60-800
Schnitt m
  isothermer $\sim$ 35-735
  quasibinärer $\sim$ 35-745
  vertikaler $\sim$ 35-740
Schraubenmessing n 65-370
Schraubenversetzung f 30-165
  linksgängige $\sim$ 30-175
  rechtsgängige $\sim$ 30-170
Schriftmetall n 55-615
Schrödinger-Gleichung f 10-175
schroffe Abschreckung f 70-260
Schrumpfen n 75-795, 75-800
Schrumpfung f 75-795, 75-800
Schubfestigkeit f 75-080
Schubmodul m 75-155
Schubspannung f 45-085
  kritische $\sim$ 45-090
Schutzgasatmosphäre f 70-690
schwach einhärtender Stahl m 60-380
schwachlegierter Stahl m 60-120
Schwammeisen n 60-035
Schwärzen n 70-400
schwarzer Temperguß m 60-1115
Schwarzglühen n 70-440
Schwarzkernguß m 60-1115
Schwarzkerntemperguß m 60-1115
Schwarzkupfer n 65-015
Schwarzmetall n 60-070
schwedisches Eisen n 60-030
Schwefel m 05-435
Schwefelabdruck m 80-325
Schweißbarkeit f 75-735
Schweißeisen n 60-040
Schweißstahl m 60-570
schwere Platinmetalle npl 65-825
Schwereseigerung f 25-425
Schwerkraftseigerung f 25-425
Schwermetalle npl 55-055
schwerrostender Stahl m 60-540
Schwinden n 75-795, 75-810
Schwindung f 75-795, 75-810
Schwindungsspannungen fpl 85-175
Schwingkristallmethode f 80-605
Schwingungen fpl
  thermische $\sim$ 10-270
Schwingungsentropie f 15-205
Schwingungsfestigkeit f 75-200
sechszählige Achse f 20-085
sechszählige Symmetrieachse f 20-085
Seewasserkorrosion f 85-290
Segregat n 40-305
Segregationskoeffizient m 40-420
Segregatphase f 40-305
sehniger Bruch m 40-720
sehr feinstreifiger Perlit m 50-145
sehr weicher Stahl m 60-345
Seigerung f 25-400
  A-$\sim$ 85-230
  dendritische $\sim$ 25-420
  interdendritische $\sim$ 25-445
  makroskopische $\sim$ 25-415

SPEZIFISCH

Seigerung
mikroskopische ~ 25-420
normale ~ 25-435
primäre ~ 25-405
~ in Dendritenzwischenräumen 25-445
sekundäre ~ 25-410
Seigerungsbart m 85-230
Seigerungsgrad m 40-420
Seigerungskennzahl f 40-420
Seigerungskoeffizient m 40-420
Seigerungsstreifen m 85-230
Seigerungszeile f 85-230
sekundäre Grenze f 30-660
sekundäre Kristalle mpl 40-275
sekundäre Kristallisation f 25-060
sekundäre Martensitumwandlung f 50-285
sekundäre Rekristallisation f 45-450
sekundärer Graphit m 50-180
sekundärer Martensit m 50-350
sekundärer Zementit m 50-080
sekundäre Seigerung f 25-410
sekundäres Gleitsystem n 45-120
sekundäres Kriechen n 45-505
sekundäre Umwandlung f 25-475
sekundäre Versetzungen fpl 30-255
Sekundärgefüge n 40-055
Sekundärgraphit m 50-180
Sekundärhärtung f 70-405
Sekundärkristalle mpl 40-275
Sekundärkristallisation f 25-060
Sekundärmartensit m 50-350
Sekundärmetall n 55-120
Sekundärrekristallisation f 45-450
Sekundärseigerung f 25-410
Sekundärstruktur f 40-055
Sekundärversetzungen fpl 30-255
Sekundärzementit m 50-080
Selbstdiffusion f 45-400
Selbstdiffusionskoeffizient m 45-405
Selbstenergie f der Versetzung 30-230
selbsthärtender Stahl m 60-375
selektive Korrosion f 85-305
Selen n 05-410
seltene Erdmetalle npl 55-040
seltene Metalle npl 55-045
Seltenerdmetalle npl 55-040
semireguläre Lösung f 15-260
Sendust n 65-535
Senke f 30-130
Sensibilisieren n 70-545
Sensibilisierungsglühen f 70-545
seßhafte Versetzung f 30-280
sessile Versetzung f 30-280
Sherardisieren n 70-930
Shockleysche Halbversetzung f 30-330
Shockley-Versetzung f 30-330
Sichtbarmachtung f des Gefüges 80-280
Siedepunkt m 75-430
Siedetemperatur f 75-430
Siemens-Martin-Roheisen n 60-855
Siemens-Martin-Stahl m 60-610
Sigma-Phase f 35-275
Silal n 60-1100
Silber n 05-420
Silberlot n 65-750
Silberstahl m 60-820
Silicieren n 70-910

Silicium n 05-415
Silicospiegel m 60-1155
Silizid n 40-490
Silizieren n 70-910
Silizium n 05-415
Siliziumbronze f 65-145
Siliziumeisen n 60-210
Siliziumferrit m 50-475
Siliziumgußeisen n 60-1010
Silizium-Kupfer n 60-080
Siliziummessing n 65-365
Siliziumstahl m 60-200
Silumin n 65-565
veredeltes ~ 65-570
singulärer Punkt m 80-495
Sinterbronze f 65-225
Sintereisen n 60-065
Sinterhartmetall n 55-665
Sinterkarbid n 55-665
Sinterlegierung f 55-655
Sintermetall n 55-660
Skandium n 05-405
SM-Stahl m 60-610
Snoek-Effekt m 30-460
Solidusfläche f 35-485, 35-495
Soliduskurve f 35-485, 35-490
rückläufige ~ 35-505
Soliduslinie f 35-485, 35-490
Solidustemperatur f 35-500
Sonderbronze f 65-130
Sonderkarbid n 40-445
Sondermessing n 65-310
Sonderstahl m 60-115, 60-150
Sorbit m 50-380
spaltbares Metall n 55-080
Spaltbruch m 40-655, 85-085
spaltfähiges Metall n 55-080
Spaltfläche f 40-665
Spaltkorrosion f 85-300
Spannungen fpl
bleibende ~ 85-020
innere ~ 85-020
makroskopische ~ 85-030
~ 1. Art 85-030
~ 2. Art 85-035
Spannungsarmglühen n 70-525
Spannungsausgleichsglühung f 70-525
Spannungs-Dehnungs-Kurve f 80-410
Spannungsfeld n 30-225
Spannungsfreiglühen n 70-525
Spannungskorrosion f 85-310
Spannungsreihe f
elektrochemische ~ 75-710
Spannungsrelaxation f 45-525
Spannungsriß m 85-065
Spannungsrißkorrosion f 85-310
Spannungs-Verformungs-Kurve f 80-410
Spektralanalyse f 80-475
Spektroskopie f
Röntgen-~ 80-700
Röntgenfluoreszenz-~ 80-700
Spezialstahl m 80-150
spezifische Elektronenwärme f 10-275
spezifischer elektrischer Widerstand m 75-470
spezifischer Widerstand m 75-470
spezifisches Gewicht n 75-350
spezifische Wärme f 75-415

SPEZIFISCH

spezifische Wärmeleitfähigkeit f 75-440
Sphärolith m 40-290
sphärolithischer Graphit m 50-190
Spiegelbronze f 65-280
Spiegeleisen n 60-930
Spinodale f 15-190
spinodale Entmischung f 25-370
spinodaler Punkt m 15-185
Spinodalkurve f 15-190
Spinquantenzahl f 10-145
spontane Keimbildung f 25-190
spontane Kristallisation f 25-045
spontane Rekristallisation f 45-455
spontaner Keim m 25-110
Spritzhärten n 70-240
Sprödbruch m 40-675
Sprödbruchtemperatur f 75-280
spröder Bruch m 40-675
Sprödigkeit f 75-270
rheotrope ~ 75-310
Sprühharten n 70-240
Sprung m 30-240
Sprungpunkt m 25-570
Sprungtemperatur f 25-570
Spurenbestandteil m 55-250
Spurenelement n 55-250
Spurenmetalle npl 55-045
Spurenzählungsautoradiographie f 80-750
SS-Stahl m 60-800
Stäbchengleitung f 45-175
stabile Phase f 35-170
stabiler Zustand m 35-065
stabiles Gleichgewicht n 35-055
Stabilglühen n 70-530
Stabilisation f des Austenits 50-425
Stabilisieren n
  natürliches ~ 70-540
stabilisierendes Legierungselement n 50-485
stabilisierter Stahl m 60-305
Stabilisierung f 70-535
Stabilisierungselement n 50-485
Stabilisierungsglühen n 70-535
Stabilität f
  thermische ~ 75-835
Stahl m 60-075
  alkalibeständiger ~ 60-550
  anlaßbeständiger ~ 60-395
  anomaler ~ 60-465
  antimagnetischer ~ 60-510
  aufgekohlter ~ 60-310
  aufgeschwefelter nichtrostender ~ 60-525
  aushärtbarer ~ 60-400
  austenitischer ~ 60-420
  austenitisch-ferritischer ~ 60-455
  bainitischer ~ 60-435
  basischer ~ 60-620
  beruhigter ~ 60-655
  bleihaltiger ~ 60-250
  borlegierter ~ 60-255
  chemisch beständiger ~ 60-535
  Chrom-Mangan-Silizium-~ 60-260
  E-~ 60-625
  einfach legierter ~ 60-155
  eutektoider ~ 50-490
  ferritischer ~ 60-415
  ferritischer nichtrostender ~ 60-520
  feuerschweißbarer ~ 60-485
  geglühter ~ 60-275
  gehärteter ~ 60-285
  geschmiedeter ~ 60-330
  graphitischer ~ 60-300
  graphitisierter ~ 60-300

Stahl
  Hadfieldscher ~ 60-785
  halbaustenitischer ~ 60-445
  halbberuhigter ~ 60-660
  halbferritischer ~ 60-450
  härtbarer ~ 60-360
  harter ~ 60-350
  hartmagnetischer ~ 60-505
  hitzebeständiger ~ 60-560
  hochchromhaltiger ~ 60-180
  hochfester ~ 60-355
  hochgekohlter ~ 60-110
  hochlegierter ~ 60-130
  hochwarmfester ~ 60-565
  kaltverfestigter ~ 60-295
  kaltzäher ~ 60-555
  kohlenstoffarmer ~ 60-100
  kohlenstoffreicher ~ 60-110
  korrosionsbeständiger ~ 60-535
  laugenbeständiger ~ 60-550
  laugensicherer ~ 60-550
  ledeburitischer ~ 60-440
  legierter ~ 60-115
  magnetischer ~ 60-495
  magnetisch harter ~ 60-505
  magnetisch weicher ~ 60-500
  martensitaushärtbarer ~ 60-460
  martensitaushärtender ~ 60-460
  martensitischer ~ 60-430
  mehrfach legierter ~ 60-170
  mittelgekohlter ~ 60-105
  mittellegierter ~ 60-125
  naturharter ~ 60-350
  nichtaushärtbarer ~ 60-405
  nichthärtender ~ 60-390
  nichtmagnetisierbarer ~ 60-510
  nichtrostender ~ 60-515
  niedriggekohlter ~ 60-100
  niedriglegierter ~ 60-120
  nitrierbarer ~ 60-695
  nitrierter ~ 60-315
  normalgeglühter ~ 60-280
  ölhärtender ~ 60-370
  perlitischer ~ 60-425
  plattierter ~ 60-335
  preßschweißbarer ~ 60-485
  quaternärer ~ 60-165
  rostbeständiger ~ 60-515
  rostsicherer ~ 60-515
  säurebeständiger ~ 60-545
  saurer ~ 60-615
  schmelzschweißbarer ~ 60-485
  schwach einhärtender ~ 60-380
  schwachlegierter ~ 60-120
  schwerrostender ~ 60-540
  sehr weicher ~ 60-345
  selbsthärtender ~ 60-375
  Siemens-Martin-~ 60-610
  SM-~ 60-610
  SS-~ 60-800
  stabilisierter ~ 60-305
  starkhärtender ~ 60-385
  stehenbleibender ~ 60-410
  ternärer ~ 60-160
  tief einhärtender ~ 60-385
  übereutektoider ~ 50-500
  übereutektoidischer ~ 50-500
  überhitzter ~ 60-320
  überhitzungsempfindlicher ~ 60-475
  überhitzungsunempfindlicher ~ 60-470
  unberuhigter ~ 60-650
  unlegierter ~ 60-095
  unmagnetischer ~ 60-510
  unmagnetisierbarer ~ 60-510
  untereutektoider ~ 50-495
  untereutektoidischer ~ 50-495
  vakuumbehandelter ~ 60-640
  verbrannter ~ 60-325
  vergüteter ~ 60-290
  verschleißfester ~ 60-480
  verzugsfreier ~ 60-410
  warmfester ~ 60-565
  wasserhärtender ~ 60-365
  weicher ~ 60-340
  weichmagnetischer ~ 60-500
  witterungsbeständiger ~ 60-540
  zweifach legierter ~ 60-160

Stahl
zunderbeständiger ~ 60-560
zunderfester ~ 60-560
Stahlart f 60-080
stahlbewehrtes Gußeisen n 60-1070
Stahlbezeichnung f 60-090
Stahlbronze f 65-355
Stahleisen n 60-880
Stahlguß m 60-825
legierter ~ 60-835
unlegierter ~ 60-830
Stahllegierung f 60-115
Stahllot n 65-745
Stahlmarke f 60-085
Stahlroheisen n 60-880
Stahlsorte f 60-085
Stahltyp m 60-080
Stapelfehler m 30-540
~ I. Art 30-545
~ II. Art 30-550
Stapelfehlerenergie f 30-560
Stapelfolge f 20-180
Stärke f der Versetzung 30-210
starkhärtender Stahl m 60-385
stationäres Kriechen n 45-505
statische Erholung f 45-430
Statuenbronze f 65-270
Steadit m 50-235
stehenbleibender Stahl m 60-410
Stellen fpl
harte ~ 85-145
Stellit n 65-495
stengeliges Gefüge n 40-125
Stengelkristalle mpl 40-280
Stengelkristallisation f 25-065
Stengelseigerung f 85-230
Stereometall n 55-630
Sterlinggold n 65-790
Stickstoff m 05-305
Stickstoffaustenit m 70-835
Stirnabschreck-Härtekurve f 80-455
Stirnabschreckprobe f 80-450
Stirnabschreckversuch m 80-450
stöchiometrische Zusammensetzung f 35-240
Stoff m
diamagnetischer ~ 75-560
ferromagnetischer ~ 75-550
paramagnetischer ~ 75-555
Störungsbereich m
thermischer ~ 30-530
Stoß m
fokussierter ~ 30-505
Stoßerwärmen n 70-050
Stoßkaskade f 30-515
Stoßlawine f 30-515
Stoßwellenbehandlung f 45-055
Strahlenschädigung f 30-490
strahliges Gefüge n 40-115
Strahlungsschädigung f 30-490
Streckgrenze f 75-110
ausgeprägte ~ 75-110
exakte ~ 75-110
natürliche ~ 75-110
physikalische ~ 75-110
streifiger Perlit m 50-130
Strenglot n 65-720
Streudiagramm n 80-610
Strontium n 05-430

Struktur f
eutektische ~ 40-060
magnetische ~ 40-650
Strukturanalyse f
röntgenographische ~ 80-580
Strukturbestandteil m 40-010
strukturelle Leerstelle f 30-090
strukturempfindliche Eigenschaften fpl 75-860
Strukturgleichgewicht n 40-020
strukturloser Martensit m 50-330
Strukturumwandlung f 25-500
strukturunabhängige Eigenschaften fpl 75-865
strukturunempfindliche Eigenschaften fpl 75-865
Stufenglühen n 70-455
Stufenhärten n 70-305
Stufenversetzung f 30-150
negative ~ 30-160
positive ~ 30-155
Subgefüge n 40-035
Subkorn n 30-650
Subkorngrenze f 30-660
Sublimationspunkt m 75-435
Sublimationstemperatur f 75-435
Sublimationswärme f 75-395
Submikrogefüge n 40-040
submikroskopische Ausscheidungen fpl 40-315
Submikrostruktur f 40-040
substituiertes Atom n 30-070
Substitutionsatom n 30-070
Substitutionsmischkristalle mpl 35-415
Substrat n 25-165
Substruktur f 40-035
Sulfidieren n 70-895
Sulf-Inuzieren n 70-890
Sulfocarbonitrieren n 70-892
Sulfokarbonitrieren n 70-892
Sulfonitrieren n 70-890
Sulphideinschlüsse mpl 40-530
Superlegierung f 55-515
Superparamagnetismus m 75-570
superplastische Legierung f 55-430
Superplastizität f 75-055
Superversetzung f 30-220
Supraleiter m 10-390
Supraleitfähigkeit f 75-490
Supraleitung f 75-490
Suszeptibilität f
magnetische ~ 75-625
Suzuki-Effekt m 30-455
Symmetrieachse f 20-065
dreizählige ~ 20-075
sechszählige ~ 20-085
vierzählige ~ 20-080
zweizählige ~ 20-070
Symmetrieklasse f 20-050
Syngonie f 20-235
syntektische Reaktion f 35-585
syntektische Temperatur f 35-625
syntektische Umwandlung f 35-585
synthetische Legierung f 55-270
synthetisches Roheisen n 60-940
System n 35-005
binäres ~ 35-120
bivariantes ~ 35-085
dreikomponentiges ~ 35-125

SYSTEM

System
- einkomponentiges ~ 35-115
- einphasiges ~ 35-090
- heterogenes ~ 35-095
- homogenes ~ 35-090
- mehrphasiges ~ 35-110
- metallisches ~ 55-165
- monophasiges ~ 50-090
- nonvariantes ~ 35-075
- pseudobinäres ~ 35-130
- quasibinäres ~ 35-130
- quaternäres ~ 35-135
- Eisen-Kohlenstoff-~ 50-015
- ternäres ~ 35-125
- unäres ~ 35-115
- unitäres ~ 35-115
- univariantes ~ 35-080
- vierkomponentiges ~ 35-135
- zweikomponentiges ~ 35-120
- zweiphasiges ~ 35-100

T

Tangentialschnitt-Methode f 15-180
Tannenbaumgefüge n 40-090
Tantal n 05-440
tatsächlicher Verteilungskoeffizient m 25-265
Tauchhärten n 70-245
Technetium n 05-445
technische Legierung f 55-520
technisches Eisen n 60-015
technisches Reineisen n 60-015
technologische Eigenschaften fpl 75-720
Teilchenabstand m
- mittlerer ~ 40-425

Teilchenhärtung f 55-220
Teilchenverfestigung f 55-220
Teilgitter n 20-285
Teilversetzung f 30-315
Tellur n 05-450
Temperatur f
- äquikohäsive ~ 30-570
- charakteristische ~ 10-280
- Debyesche ~ 10-280
- Debyesche charakteristische ~ 10-280
- eutektische ~ 35-605
- eutektoide ~ 35-630
- homologe ~ 45-475
- kritische ~ 35-335
- Mf-~ 70-170
- Ms-~ 70-165
- metatektische ~ 35-620
- monotektische ~ 35-615
- peritektische ~ 35-610
- peritektoide ~ 35-635
- syntektische ~ 35-625
- ~ vollständiger Martensitumwandlung 70-170

Temperaturbeständigkeit f 75-835
Temperaturgefälle n 15-075
Temperaturgradient m 15-075
Temperaturkoeffizient m
- ~ des elektrischen Widerstandes 75-480
- ~ des Widerstandes 75-480

Temperaturleitfähigkeit f 75-445
Temperaturleitungsvermögen n 75-445
Temperaturleitzahl f 75-445
Temperaturspannungen fpl 85-025
Temperaturwechselbeständigkeit f 75-830
Temperatur-Zeit-Folge f 70-970
Temperguß m 60-1110
- ferritischer ~ 60-1125
- perlitischer ~ 60-1130
- schwarzer ~ 60-1115
- weißer ~ 60-1120

Temperkohle f 50-220

Tempern n 70-625
Temperroheisen n 60-905
Terbium n 05-455
ternäre Legierung f 55-300
ternärer Stahl m 60-160
ternäres Eutektikum n 40-210
ternäres Gleichgewichtsdiagramm n 35-710
ternäres Phasendiagramm n 35-710
ternäres Phosphideutektikum n 50-235
ternäres Raumzustandsdiagramm n 35-750
ternäres Raumzustandsschaubild n 35-750
ternäres System n 35-125
ternäres Zustandsdiagramm n 35-710
tertiäre Rekristallisation f 45-460
tertiärer Zementit m 50-090
tertiäres Kreichen n 45-510
Tertiärzementit m 50-090
tetraedrische Gitterlücke f 20-325
tetragonaler Martensit m 50-335
tetragonales Gitter n 20-370
tetragonales Kristallsystem n 20-250
tetragonale Verzerrung f 20-375
Tetragonalität f 20-375
Tetragyre f 20-080
Textur f 40-555
- Goss-~ 40-610
- kubische ~ 40-605

texturbehaftete Legierung f 55-380
Thallium n 05-460
Thermalloy n 65-540
thermisch aktivierter Prozeß m 45-280
thermische Aktivierung f 45-275
thermische Analyse f 80-505
thermische Anregung f 45-275
thermische Ätzung f 80-190
thermische Behandlung f 70-006
thermische Diffusion f 45-370
thermische Eigenschaften fpl 75-365
thermische Erholung f 45-430
thermische Ermüdung f 85-425
thermische Leerstelle f 30-095
thermische Leitfähigkeit f 75-440
thermischer Ausdehnungskoeffizient m 75-450
thermischer Dendrit m 25-325
thermischer Graben m 80-195
thermischer Haltepunkt m 80-530
thermischer Störungsbereich m 30-530
thermische Schwingungen fpl 10-270
thermische Stabilität f 75-835
thermisch-magnetische Behandlung f 70-965
Thermoanalyse f 80-505
- Differential-~ 80-515
- gewöhnliche ~ 80-510

thermochemische Behandlung f 70-650
Thermodiffusion f 45-370
Thermodiffusions-Metallüberziehen n 70-915
Thermodynamik f der Legierungen 15-010
thermodynamische Aktivität f 15-280
thermodynamische Analyse f 80-535
thermodynamische Funktion f 15-030
thermodynamische Gleichgewichtskonstante f 15-335
thermodynamische Größe f 15-030
thermodynamischer Zustand m 15-015

thermodynamisches Gleichgewicht *n* 15-320
thermodynamisches Potential *n* bei konstantem Druck 15-140
thermodynamisches Potential *n* bei konstantem Volumen 15-105
thermodynamische Variable *f* 15-020
thermodynamische Wahrscheinlichkeit *f* 15-230
thermodynamische Zusatzfunktion *f* 15-065
Thermogramm *n* 80-527
Thermokraft *f* 75-495
thermomagnetische Analyse *f* 80-565
thermomechanische Behandlung *f* 70-950
thermomechanische Behandlung *f* bei hohen Temperaturen 70-955
Thermotransport *m* 45-370
Thomasroheisen *n* 60-850
Thomasstahl *m* 60-600
Thorium *n* 05-465
Th-Stahl *m* 60-600
Thulium *n* 05-470
Tiefätzung *f* 80-160
tief einhärtender Stahl *m* 60-385
Tiefkühlen *n* 70-340
Tieftemperaturbehandeln *n* 70-340
Tieftemperaturhärten *n* 70-340
Tieftemperatursprödigkeit *f* 75-295
Tieftemperaturstahl *m* 60-555
Tiefungswert *m* 75-755
Tiefziehfähigkeit *f* 75-750
Tiefziehkenngröße *f* 75-755
Tiefziehstahl *m* 60-790
Tiegelstahl *m* 60-630
Tilt-Grenze *f* 30-605
Titan *n* 05-480
Titanid *n* 40-510
Titanieren *n* 70-945
Titannitrid *n* 70-825
Titanstahl *m* 60-230
T-Metalle *npl* 55-025
Tombak *m* 65-395
Topfglühen *n* 70-435
Torsionsfestigkeit *f* 75-085
Torsionsmodul *m* 75-155
Torsionstextur *f* 40-595
Totglühen *n* 70-465
Trafostahl *m* 60-770
Trägergas *n* 70-685
Tränklegierung *f* 55-670
Transformatorenstahl *m* 60-770
transkristalliner Bruch *m* 40-710
Transkristallisation *f* 25-070
Translation *f* 20-130
Translationsgitter *n* 20-275
Translationsperiode *f* 20-115
Translationsvektor *m* 20-135
Transmissionsmikroskopie *f* 80-220
  Elektronen-∼ 80-340
treibende Kraft *f* 15-070
Trennbruch *m* 40-655
Trennungsbruch *m* 40-655
Trennungswiderstand *m* 75-135
Triangulation *f* des Systems 35-725
Tribonukleation *f* 25-210
Triebkraft *f* 15-070

trigonales Kristallsystem *n* 20-260
Trigyre *f* 20-075
triklines Gitter *n* 20-400
triklines Kristallsystem *n* 20-270
Tripelpunkt *m* 35-040
Trockenätzung *f* 80-185
trockenes Magnetpulververfahren *n* 80-810
Trockenhärten *n* 70-250
Troostit *m* 50-395
Tüpfelanalyse *f* 80-465
Tüpfelprobe *f* 80-465
Twist-Grenze *f* 30-600
Typometall *n* 55-610

U

Überaltern *n* 70-595
Überalterung *f* 70-595
Überätzung *f* 80-130
übereutektische Legierung *f* 55-360
übereutektisches Gefüge *m* 40-065
übereutektisches weißes Gußeisen *n* 50-515
übereutektoide Legierung *f* 55-375
übereutektoider Stahl *m* 50-500
übereutektoider Zementit *m* 50-080
übereutektoides Gefüge *n* 40-080
übereutektoidische Legierung *f* 55-375
übereutektoidischer Stahl *m* 55-500
Überführung *f*
  elektrische ∼ 45-365
Übergangselemente *mpl* 55-025
Übergangskriechen *n* 45-500
Übergangsmetalle *npl* 55-025
Übergangsphase *f* 35-195
Übergangstemperatur *f* 25-570, 75-280
  kritische ∼ 25-570
Übergangszahl *f*
  C-∼ 70-755
  Kohlenstoff-∼ 70-755
Überhitzen *n* 70-065
überhitzter Stahl *m* 60-320
Überhitzung *f* 70-065
überhitzungsempfindlicher Stahl *m* 60-475
Überhitzungsempfindlichkeit *f* 75-760
überhitzungsunempfindlicher Stahl *m* 60-470
Überlappung *f* 85-400
Übermikroskopie *f* 80-330
übersättigte feste Lösung *f* 35-400
übersättigte Mischkristalle *mpl* 35-400
Überschießen *n* 30-260
Überschuß *n* 30-260
Überschußenergie *f* 15-115
  freie ∼ 15-115
Überschußenthalpie *f* 15-145
Überschußentropie *f* 15-220
Überschußfunktion *f* 15-065
Überschußleerstelle *f* 30-105
Überschwingen *n* 30-260
Übersichtsgefüge *n* 40-025
Überstruktur *f* 35-440
Überstrukturlinien *fpl* 80-655
Überstrukturphase *f* 35-200
Überversetzung *f* 30-220
Überwalzung *f* 85-400
übliches Härten *n* 70-270
U-Legierung *f* 55-280

ULTRAAKUSTISCH

ultraakustische Prüfung f 80-800
ultrafeste Legierung f 55-425
Ultrafestigkeit f 75-045
Ultraschalldefektoskopie f 80-800
Ultraschallprüfung f 80-800
Ultraviolett-Mikroskopie f 80-265
umgekehrte Blockseigerung f 25-440
umgekehrter Hartguß m 85-150
umgewandelter Austenit m 50-035
Umklappmartensit m 50-315
Umklappumwandlung f 25-510
Umkörnen n 70-475
Umlaufhärten n 70-315
Umlaufmantelhärten n 70-315
Umlaufvorschubhärten n 70-325
Umschmelzbronze f 65-285
Umschmelzlegierung f 55-280
Umschmelzmetall n 55-120
Umwandlung f
  allotrope ~ 25-455
  athermische ~ 25-520
  diffusionsartige ~ 25-505
  diffusionslose ~ 25-510
  diskontinuierliche ~ 25-545
  eutektische ~ 35-565
  eutektoide ~ 35-590
  heterogene ~ 25-535
  homogene ~ 25-530
  inhomogene ~ 25-535
  isothermische ~ 25-515
  kongruente ~ 25-525
  konunuierliche ~ 25-540
  magnetische ~ 25-555
  massive ~ 25-550
  metatektische ~ 35-580
  monotektische ~ 35-575
  monotektoide ~ 35-600
  Ordnung-Unordnung-~ 25-580
  peritektische ~ 35-570
  peritektoide ~ 35-595
  polymorphe ~ 25-455
  reversible ~ 25-495
  sekundäre ~ 25-475
  syntektische ~ 35-585
  ~ erster Art 25-465
  ~ erster Ordnung 25-465
  ~ im festen Zustand 25-475
  ~ in der Bainitstufe 50-290
  Unordnung-Ordnung-~ 25-575
  ~ zweiter Art 25-470
  ~ zweiter Ordnung 25-470
  Zwischenstufen-~ 50-290
Umwandlungsbereich m 35-560
Umwandlungshärten n 70-175
Umwandlungsintervall m 35-560
Umwandlungspunkt m 35-555
  magnetischer ~ 25-560
Umwandlungsspannungen fpl 85-035
Umwandlungstemperatur f 35-555
Umwandlungsverzögerung f 25-480
Umwandlungswärme f 75-400
unäres System n 35-115
unberuhigter Stahl m 60-650
Unbestimmtheitsprinzip n 10-165
Unbestimmtheitsrelation f 10-165
Undurchsichtigkeit f 75-505
unechtes Gold n 65-815
Unedelmetall n 55-005
unedles Metall n 55-005
ungeordnete feste Lösung f 35-435
ungeordnete Mischphase f 35-205, 35-435
ungeordnete Phase f 35-205
Ungleichgewichtsphase f 35-185
unitäres System n 35-115

univariantes System n 35-080
unlegierter Gußstahl m 60-830
unlegierter Stahl m 60-095
unlegierter Stahlguß m 60-830
unlegiertes Gußeisen n 60-990
unlegiertes Messing n 65-305
unmagnetische Legierung f 55-470
unmagnetischer Stahl m 60-510
unmagnetisierbarer Stahl m 60-510
Unmischbarkeit f 35-340
Unordnung f 25-585
Unordnung-Ordnung-Umwandlung f 25-575
Unordnungsgrad m 25-610
Unordnungszustand m 25-585
Unterbrechung f des Zusammenhanges 85-090
unterbrochenes Härten n 70-300
unterer Bainit m 50-415
unterer Martensitpunkt m 70-170
untereutektische Legierung f 55-355
untereutektisches Gefüge n 40-070
untereutektisches weißes Gußeisen n 50-510
untereutektoide Legierung f 55-370
untereutektoider Stahl m 50-495
untereutektoides Gefüge n 40-085
untereutektoidische Legierung f 55-370
untereutektoidischer Stahl m 55-495
Untergitter n 20-285
  magnetisches ~ 20-485
Unterhärten n 70-335
Unterkeim m 25-135
Unterkorn n 30-650
Unterkorngrenze f 30-660
unterkühlter Austenit m 50-265
Unterkühlung f 25-240, 25-245
  konstitutionelle ~ 25-255
Unterkühlungsgrad m 25-250
Unterschale f 10-160
Unterstruktur f 40-035
Untersuchung f
  absolute Dilatometer-~ 80-545
  Dilatometer-~ 80-540
  fraktographische ~ 80-385
  kalorimetrische ~ 80-500
  makroskopische ~ 80-045
  metallographische ~ 80-035
  mikroskopische ~ 80-040
  röntgenographische ~ 80-570
ununterbrochene Mischkristallreihe f 35-390
unvollständiges Ausglühen n 70-470
unvollständiges Glühen n 70-470
unvollständige Versetzung f 30-310
Uran n 05-490
UV-Mikroskopie f 80-265

V

Vakuumätzung f
  katodische ~ 80-180
vakuumbehandelter Stahl m 60-640
vakuumgeschmolzenes Metall n 55-135
Vakuumstahl m 60-640
Valenzband n 10-255
Valenzelektron n 10-065
Valenzelektronenkonzentration f 35-285

VERTIKAL

Valenzfaktor *m* 10-405
Valenzkristall *m* 20-030
Vanadin *n* 05-495
Vanadinkarbid *n* 40-460
Vanadinmessing *n* 65-180
Vanadinnitrid *n* 70-830
Vanadinstahl *m* 60-225
Vanadium *n* 05-495
Vanadiumstahl *m* 60-225
van-der-Waalssche Bindung *f* 10-335
Variable *f*
  thermodynamische ~ 15-020
Vegard-Regel *f* 35-425
Vegardsche Regel *f* 35-425
Ventilbronze *f* 65-275
Ventilstahl *m* 60-735
Veraluminieren *n* 70-925
verankerte Versetzung *f* 30-270
Verankerungspunkt *m* 30-275
Verarbeitungseigenschaften *fpl* 75-720
Verarmungszone *f* 30-120
Verbindung *f*
  chemische ~ 35-235
  heteropolare ~ 35-230
  inkongruente intermetallische ~ 35-225
  intermetallische ~ 35-215
  kongruente intermetallische ~ 35-220
  nichtkongruente intermetallische ~ 35-225
Verbindungen *fpl*
  berthollide ~ 35-250
  daltonide ~ 35-245
Verbot *m*
  Pauli-~ 10-170
verbotener Bereich *m* 10-250
verbotenes Band *n* 10-250
verbrannter Stahl *m* 60-325
Verbrennen *n* 85-205
Verbundstahl *m* 60-335
Verbundwerkstoff *m* 55-680
Verdampfungswärme *f* 75-390
  gebundene ~ 75-390
  latente ~ 75-390
Verdrehungsfestigkeit *f* 75-085
Verdünnen *n* 80-355
verdünnte Legierung *f* 55-320
verdünnte Lösung *f* 15-250
verdünnte Zone *f* 30-120
veredeltes Silumin *n* 65-570
Veredelung *f* des Silumins 25-225
Verfahren *n*
  Debye-Scherrer-~ 80-595
Verfestigbarkeit *f* 75-770
Verfestigung *f* 45-045, 70-025
  latente ~ 45-180
Verfestigungsbehandlung *f* 70-025
Verfestigungsfähigkeit *f* 75-770
Verfestigungskoeffizient *m* 45-065
Verfestigungskurve *f* 45-070
Verformbarkeit *f*
  plastische ~ 75-025
Verformung *f*
  bleibende ~ 45-010
  elastische ~ 45-005
  kritische ~ 45-075
  plastische ~ 45-010, 45-015
Verformungsband *n* 45-200
Verformungsbruch *m* 40-680
Verformungsgrad *m* 45-035
  kritischer ~ 45-075

verformungsloser Bruch *m* 40-675
Verformungsmartensit *m* 50-360
Verformungsriß *m* 85-070
Verformungstextur *f* 40-560
Verformungszwilling *m* 45-235
Vergießbarkeit *f* 75-780
Vergrößerung *f* 80-282
Vergüten *n* 70-410
vergüteter Stahl *m* 60-290
Vergütung *f* 70-410
Vergütungsstahl *m* 60-705
verlagertes Atom *n* 30-510
Verlagerungskaskade *f* 30-520
Verlagerungslawine *f* 30-520
Verlagerungsspitze *f* 30-525
Verschiebungsbruch *m* 40-660
verschleißfeste Legierung *f* 55-435
verschleißfester Stahl *m* 60-480
Verschleißfestigkeit *f* 75-265
versetztes Atom *n* 30-510
Versetzung *f* 30-145
  aufgespaltene ~ 30-325
  geladene ~ 30-465
  gemischte ~ 30-180
  gleitfähige ~ 30-265
  Lomer-Cottrell-~ 30-450
  prismatische ~ 30-345
  seßhafte ~ 30-280
  sessile ~ 30-280
  Shockley-~ 30-330
  unvollständige ~ 30-310
  verankerte ~ 30-270
  vollständige ~ 30-305
Versetzungen *fpl*
  primäre ~ 30-250
  sekundäre ~ 30-255
  Vorläufer-~ 30-370
Versetzungsansammlung *f* 30-385
Versetzungsaufstauung *f* 30-400
Versetzungsblockierung *f* 30-445
Versetzungsdichte *f* 30-235
Versetzungsdiffusion *f* 45-360
Versetzungsdipol *m* 30-380
Versetzungskern *m* 30-205
Versetzungsknäuel *m* 30-385
Versetzungsknoten *m* 30-350
Versetzungslinie *f* 30-195
Versetzungsmultiplikation *f* 30-420
Versetzungsnetz *n* 30-390
Versetzungsnetzwerk *n* 30-390
Versetzungsquelle *f* 30-425
Versetzungsring *m* 30-340
  prismatischer ~ 30-345
Versetzungsschlauchdiffusion *f* 45-360
Versetzungsschleife *f* 30-340
Versetzungsspirale *f* 30-355
Versetzungssprung *m* 30-240
Versetzungsstärke *f* 30-210
Versetzungsstruktur *f* 30-415
Versetzungswald *m* 30-405
Versetzungswand *f* 30-395
Versetzungswendel *m* 30-355
Versprödung *f* 75-275
Versprödungstemperatur *f* 75-280
Vertauschungsentropie *f* 15-210
Verteilungskoeffizient *m*
  effektiver ~ 25-265
  Gleichgewichts-~ 25-260
  tatsächlicher ~ 25-265
vertikaler Schnitt *m* 35-740

VERUNREINIGUNGEN

Verunreinigungen *fpl* 55-205
Verunreinigungsatmosphäre *f* 30-440
Vervielfachung *f* von Versetzungen 30-420
Vervielfältigung *f* von Versetzungen 30-420
Verwachsungsebene *f* 20-470
Verzerrung *f*
tetragonale ~ 20-375
verzögerter Bruch *m* 40-755
verzugsfreier Stahl *m* 60-410
Verzweigungsstruktur *f* 40-165
Vielfachgleitung *f* 45-140
Vielgestaltigkeit *f* 25-385
Vielkomponentensystem *n* 35-140
Vielkristall *m* 20-440
Vielstoffsystem *n* 35-140
Vierfachleerstelle *f* 30-055
Vierkomponentensystem *n* 35-135
vierkomponentige Legierung *f* 55-305
vierkomponentiges System *n* 35-135
Vierstofflegierung *f* 55-305
Vierstoffsystem *n* 35-135
vierzählige Achse *f* 20-080
vierzählige Symmetrieachse *f* 20-080
viskoses Fließen *n* 45-220
Vitallium *n* 65-485
völlig aufgefülltes Band *n* 10-245
völlige Mischbarkeit *f* 35-315
vollständig besetztes Band *n* 10-245
vollständige Mischbarkeit *f* 35-315
vollständiges Ausglühen *n* 70-465
vollständige Versetzung *f* 30-305
Volumendiffusion *f* 45-295
Volumenhärten *n* 70-287
volumetrische Eigenschaften *fpl* 75-340
Vorausscheidung *f* 70-600
voreutektische Phase *f* 40-340
voreutektoider Ferrit *m* 50-055
voreutektoider Zementit *m* 50-080
voreutektoidische Phase *f* 40-345
Vorläufer-Versetzungen *fpl* 30-370
Vorlegierung *f* 55-285
Vorschubhärten *n* 70-320
Vorwärmen *n* 70-045
Vorwärmung *f* 70-045
Vorzugsorientierung *f* 20-425
Vorzugsrichtung *f*
magnetische ~ 75-585

W

Wachsen *n*
epitaxisches ~ 25-345
~ des Gußeisens 70-640
Wachstum *n*
dendritisches ~ 25-310
wachstumsfähiger Kristallkeim *m* 25-150
Wachstumsfront *f* 45-480
Wachstumsgeschwindigkeit *f* 25-275
Wachstumshemmer *m* 40-400
Wachstumsspirale *f* 25-335
Wachstumsstufe *f* 25-330
Wachstumszwilling *m* 45-240
wahre Dauerstandfestigkeit *f* 75-225
wahre Kriechgrenze *f* 75-225
wahre Zerreißfestigkeit *f* 75-130
Waldversetzung *f* 30-410
Walzbronze *f* 65-105

Wälzlagerstahl *m* 60-715
Walzmessing *n* 65-295
Walztextur *f* 40-570
Wand *f*
Bloch-~ 40-645
Wanddickenempfindlichkeit *f* 75-815
Warmarbeitsstahl *m* 60-810
Warmaushärtung *f* 70-590
Warmauslagern *n* 70-590
Warmbadhärten *n* 70-225, 70-305
Warmbeständigkeit *f* 75-835
Warmbruch *m* 85-080
Warmbrüchigkeit *f* 75-290
Wärme *f*
elektronische spezifische ~ 10-275
spezifische ~ 75-415
Wärmeausdehnungskoeffizient *m* 75-450
Wärmeausdehnungszahl *f* 75-450
Wärmebehandlung *f* 70-005
eigentliche ~ 70-006
reine ~ 70-006
wärmebehandlungsfähige Legierung *f* 55-525
Wärmebeständigkeit *f* 75-835
Wärmegeschwindigkeit *f* 70-040
Wärmegleichgewicht *n* 15-325
Wärmekapazität *f* 75-410
konfigurationsbedingte ~ 25-635
Wärmeleitfähigkeit *f* 75-440
spezifische ~ 75-440
Wärmeleitvermögen *n* 75-440
Wärmeleitzahl *f* 75-440
Wärmen *n* 70-030
Wärmeriß *m* 85-060
Wärmespannungen *fpl* 85-025
Wärmestoßbeständigkeit *f* 75-825
warmfeste Legierung *f* 55-510
warmfester Stahl *m* 60-565
Warmfestigkeit *f* 75-240
Warmhärte *f* 75-255
Warmlagern *n* 70-590
Warmriß *m* 85-080
Warmsprödigkeit *f* 75-290
Warmverformung *f* 45-020
Warmversprödung *f* 75-290
Wasserabschrecken *n* 70-100
Wasserabschreckung *f* 70-100
Wasserhärten *n* 70-205
wasserhärtender Stahl *m* 60-365
Wasserhärter *m* 60-365
Wasserhärtestahl *m* 60-365
Wasserkorrosion *f* 85-280
Wasserstoff *m* 05-200
Wasserstoffbrüchigkeit *f* 85-390
Wasserstoffentzug *m* 70-555
Wasserstofffreiglühen *n* 70-555
Wasserstoffkrankheit *f* 85-395
Wasserstoffsprödigkeit *f* 85-390
Wechselwirkungsenergie *f* 10-295
Wechselwirkungskoeffizient *m* 15-295
Wechselwirkungsparameter *m* 15-300
Wechselwirkungspotential *n* 10-300
Weichblei *n* 65-655
Weicheisen *n* 60-015
weicher Stahl *m* 60-340
Weichflecke *mpl* 85-195
Weichfleckigkeit *f* 85-195

Weichglühen n 70-460, 70-510
Weichlot n 65-715
weichmagnetische Legierung f 55-460
weichmagnetischer Stahl m 60-500
Weichstahl m 60-340
Weißbronze f 65-230
weiße Flecken mpl 85-250
weißer Bruch m 85-160
weißer Temperguß m 60-1120
weiße Schichten fpl 85-250
weißes Gußeisen n 60-955
weißes Roheisen n 60-910
weißes Zinn n 65-630
Weißgold n 65-805
Weißguß m 60-955, 60-1120
Weißkernguß m 60-1120
Weißlot n 65-715
Weißmetall n 65-640
Weißscher Bereich m 40-635
Weißscher Bezirk m 40-635
Wellenfunktion f 10-180
Wellenvektor m 10-185
Werfen n 85-200
Werkblei n 65-650
Werkstoffehler m 85-005
Werkstoffehlerprüfung f 80-790
Werkzeugstahl m 60-675
Whisker m 25-360
Wichte f 75-350
Widerstand m
  spezifischer elektrischer ~ 75-470
  spezifischer ~ 75-470
Widerstandslegierung f 55-545
Widmannstättensches Gefüge n 40-130
Winkelversetzung f 30-360
wirkliche Lösung f 15-245
wirksamer Verteilungskoeffizient m 25-265
Wirkungskoeffizient m 15-295
Wirkungsparameter m 15-300
Wirtskristall m 40-295
Wismut n 05-060
witterungsbeständiger Stahl m 60-540
Wöhler-Kurve f 80-425
Wolfram m 05-485
Wolframbronze f 65-170
Wolframid n 40-500
Wolframieren n 70-935
Wolframkarbid m 40-450
Wolframstahl m 60-215
Wolke f
  Cottrellsche ~ 30-440
Wood-Metall n 65-670
Woodsche Legierung f 65-670
Woodsches Metall n 65-670
würfeliges Gitter n 20-300
Würfeltextur f 40-605

X

Xenon n 05-500

Y

Y-Legierung f 65-590
Youngscher Modul m 75-150
Ypsylon-Legierung f 65-590

Ytterbium n 05-505
Yttrium n 05-510

Z

zäher Bruch m 40-680
Zähfestigkeit f 75-035
Zähigkeit f 75-035
Zahl f
  Poissonsche ~ 75-170
  Poisson-~ 75-170
  ~ der Freiheitsgrade 35-030
zahntechnische Legierung f 55-575
Zamak n 65-705
Zäsium n 05-080
Zeilengefüge n 40-140
Zeilenstruktur f 40-140
Zeitbruchgrenze f 75-235
Zeitdehngrenze f 75-230
Zeitstandfestigkeit f 75-235
Zeitstandkriechgrenze f 75-235
Zelle f
  primitive ~ 20-095
Zellgefüge n 40-095
Zellstruktur f 40-095
Zementationsgas n 70-745
Zementationsmittel n 70-740
Zementieren n 70-715
Zementit m 50-065
  eutektischer ~ 50-075
  freier ~ 50-095
  körniger ~ 50-100
  kugeliger ~ 50-100
  ledeburitischer ~ 50-075
  legierter ~ 50-110
  perlitischer ~ 50-085
  primärer ~ 50-070
  sekundärer ~ 50-080
  tertiärer ~ 50-090
  übereutektoider ~ 50-080
  voreutektoider ~ 50-080
Zementit-Eutektikum n 50-255
Zementitlamelle f 50-120
Zementitnetz n 50-105
Zementkupfer n 65-045
Zementstahl m 60-585
Zentrum n der Versetzung 30-205
Zer n 05-100
Zereisen n 65-875
Zerreißfestigkeit f
  wahre ~ 75-130
Zersetzungspunkt m 15-340
Zersetzungstemperatur f 15-340
Zerspanbarkeit f 75-745
zerstörungsfreie Prüfung f 80-785
Ziehen n von Kristallen 25-090
Ziehtextur f 40-575
Zieratlegierung f 55-605
Zink n 05-515
Zinklot n 65-730
Zinn n 05-475
  graues ~ 65-635
  weißes ~ 65-630
Zinnbronze f 65-110
  Phosphor-~ 65-210
zinnfreie Bronze f 65-125
Zinngeschrei n 45-270
Zinnlot n 65-725
Zinn-Messing n 65-345
Zinnpest f 85-245

Zinnschweiß m 85-165
Zinn-Weißmetall n 65-645
Zintl-Phasen fpl 35-270
Zircaloy n 65-855
Zirkon n 05-520
Zirkonium n 05-520
Zone f
  gleichachsige ~ 40-185
  verdünnte ~ 30-120
  ~ des gerichteten Dendritenwachstums 40-180
Zonen fpl
  Guinier-Preston-~ 70-605
zonengeschmolzenes Metall n 55-140
Zonenseigerung f 25-415
Züchtung f von Einkristallen 25-085
zufällige Beimengungen fpl 55-200
Zugfestigkeit f 75-070
Zugprobe f 80-405
Zugtextur f 40-585
Zugversuch m 80-405
Zunder m 85-430
zunderbeständige Legierung f 55-505
zunderbeständiger Stahl m 60-560
zunderbeständiges Gußeisen n 60-1085
Zunderbeständigkeit f 75-690
zunderfeste Legierung f 55-505
zunderfester Stahl m 60-560
zunderfestes Gußeisen n 60-1085
Zunderfestigkeit f 75-690
Zusammenballung f des Zementits 70-515
Zusammenlegieren n 55-170
Zusammensetzung f
  chemische ~ einer Legierung 55-225
  stöchiometrische ~ 35-240
Zusatzenergie f
  freie ~ 15-115
Zusatzenthalpie f 15-145
Zusatzentropie f 15-220
Zusatzfunktion f
  thermodynamische ~ 15-065
Zusatzgröße f 15-065
Zustand m
  metallischer ~ 10-345
  metastabiler ~ 35-070
  stabiler ~ 35-065
  thermodynamischer ~ 15-015
Zustandsdiagramm n 35-480
  Eisen-Kohlenstoff-~ 50-020
  ternäres ~ 35-710
Zustandsdichte f 10-200
Zustandsfunktion f 15-035
Zustandsgleichung f 15-025
Zustandsgröße f 15-035
Zustandsschaubild n 35-480
Zustandsveränderliche f 15-020
Zwangslösung f 35-400
zweidimensionale Fehlstelle f 30-535

zweidimensionaler Gitterbaufehler m 30-535
zweifache Gleitung f 45-135
zweifach legierter Stahl m 60-160
Zweikomponentensystem n 35-120
zweikomponentiges System n 35-120
Zweikristall m 20-435
Zweiphasenfeld n 35-685
Zweiphasengebiet n 35-685
Zweiphasensystem n 35-100
zweiphasige Legierung f 55-335
zweiphasiges Gefüge n 40-150
zweiphasiges System n 35-100
Zweistofflegierung f 55-295
Zweistoffsystem n 35-120
zweite Quantenzahl f 10-135
zweitnächster Nachbar m 20-200
zweizählige Achse f 20-070
zweizählige Symmetrieachse f 20-070
Zwilling m 45-230
Zwillingsachse f 45-260
Zwillingsband n 45-255
Zwillingsbildung f 45-245
Zwillingsebene f 45-250
Zwillingsgrenze f 45-265
Zwillingslamelle f 45-255
Zwillingsstapelfehler m 30-555
Zwillingsstreifen m 45-255
Zwillingsversetzung f 30-375
zwischendendritischer Graphit m 50-205
Zwischengitter n 20-405
Zwischengitteratom n 30-075
  aufgespaltenes ~ 30-085
Zwischengitterdiffusion f 45-315
Zwischengitterhantel m 30-085
Zwischengitterlage f 20-145
Zwischengitterlücke f 20-105
Zwischengittermischkristalle mpl 35-420
Zwischengitterplatz m 20-145
Zwischengitterraum m 20-105
Zwischengitterverunreinigung f 30-080
zwischenkörniger Bruch m 40-705
zwischenkristalliner Bruch m 40-705
Zwischenphase f 35-195
Zwischenstufe f 50-305
Zwischenstufengefüge n 50-410
Zwischenstufenglühen n 70-450
Zwischenstufenhärten n 70-310
Zwischenstufen-Umwandlung f 50-290
Zwischenstufenvergüten n 70-310
Zwischenwerkstoff m 55-680
Zyanbadhärten n 70-865
Zyanhärten n 70-855
Zyanieren n 70-855

ENGLISH INDEX

A

abnormal steel 60-465
abnormal structure 40-135
abnormality of steel 85-225
abrasion-resisting steel 60-480
absorption microanalysis 80-735
accelerated creep 45-510
accelerated diffusion 45-355
accelerating creep 45-510
acicular cast iron 60-1060
acicular grey cast iron 60-1060
acicular martensite 50-315
acicular structure 40-115
acid brittleness 85-390
acid embrittlement 85-390
acid resistance 75-695
acid-resisting alloy 55-485
acid-resisting cast iron 60-1075
acid-resisting steel 60-545
acid steel 60-615
actinium 05-010
activation
 thermal ~ 45-275
activation analysis 80-765
activation energy 45-285
activation volume 30-475
active atmosphere 70-680
activity 15-280
 thermodynamic ~ 15-280
activity coefficient 15-285
activity factor 15-295
actual solution 15-245
addition
 alloying ~ 55-190
Admiralty brass 65-390
Admiralty bronze 65-205
Admiralty gun metal 65-205
Admiralty metal 65-390
affinity 75-660
 chemical ~ 75-660
 electron ~ 10-115
aftereffect
 magnetic ~ 75-645
age-hardening alloy 55-385
age-hardenable alloy 55-385
age hardening 70-565
ageing 70-575
 artificial ~ 70-590
 magnetic ~ 70-620
 natural ~ 70-585
 quench ~ 70-580
 strain ~ 45-225
ageing resistance 75-845
ageing-resisting steel 60-405
ageing steel 60-400
agent
 nucleating ~ 25-125
 quenching ~ 70-185
aggregate
 polycrystalline ~ 20-440

aging 70-575
aging-resisting steel 60-405
aging steel 60-400
agitation
 thermal ~ 10-270
air-blast quenching 70-220
air cooling 70-095
air hardening 70-215
air-hardening steel 60-375
air quenching 70-215
alclad 65-560
aldrey 65-585
aldural 65-560
alkali metals pl 55-030
alkaline earth bearing alloy 65-685
alkaline earth metals pl 55-035
alkali resistance 75-700
alkali-resisting alloy 55-490
alkali-resisting cast iron 60-1080
alkali-resisting steel 60-550
allotriomorphic crystal 40-360
allotrope 25-390
 high-temperature ~ 25-395
allotropic change 25-455
allotropic form 25-390
allotropic transformation 25-455
allotropy 25-385
allowed band 10-240
alloy 55-160
 acid-resisting ~ 55-485
 age-hardening ~ 55-385
 alkaline earth bearing ~ 65-685
 alkali-resisting ~ 55-490
 Alnico ~ 65-530
 alpax ~ 65-565
 anatomical ~ 55-645
 antifriction ~ 55-555
 bearing ~ 55-555
 brazing ~ 65-710
 bronze ~ 65-130
 cast ~ 55-535
 castable ~ 55-535
 casting ~ 55-535
 cerium standard ~ 65-865
 coinage ~ 55-565
 commercial ~ 55-520
 complex ~ 55-310
 concentrated ~ 55-325
 constant-modulus ~ 65-505
 contact ~ 55-585
 copper ~ 65-010
 copper-base ~ 65-010
 corrosion-resistant ~ 55-480
 corrosion-resisting ~ 55-480
 creep-resistant ~ 55-510
 creep-resisting ~ 55-510
 decorative ~ 55-605
 dental ~ 55-575
 die-casting ~ 55-540
 dilute ~ 55-320
 dispersion ~ 55-390
 dispersion-hardened ~ 55-390
 dispersion-strengthened ~ 55-390
 electrical contact ~ 55-585

— 263 —

ALLOY

alloy
electrical resistance ~ 55-545
electrodeposited ~ 55-290
electron ~ 65-625
eutectic ~ 55-350
eutectoid ~ 55-365
ferromagnetic ~ 55-455
ferrous ~ 60-070
forging ~ 55-530
foundry ~ 55-285
four-component ~ 55-305
free cutting ~ 55-560
fusible ~ 55-395
hard ~ 55-415
hard-facing ~ 55-590
hard magnetic ~ 55-465
hard-surfacing ~ 55-590
heat-resistant ~ 55-505
heat-resisting ~ 55-505
heat-treatable ~ 55-525
heterogeneous ~ 55-340
Heusler ~ 65-420
Heusler's ~ 65-420
high ~ 55-325
high-carat gold ~ 65-795
high coercive ~ 55-465
high-conductivity ~ 55-550
high melting-point ~ 55-400
high permeability ~ 55-475
high-resistance ~ 55-545
high-strength ~ 55-420
high-temperature ~ 55-510
homogeneous ~ 55-330
hypereutectic ~ 55-360
hypereutectoid ~ 55-375
hypoeutectic ~ 55-355
hypoeutectoid ~ 55-370
ignition ~ 55-595
imitation gold ~ 65-815
industrial ~ 55-520
infiltration ~ 55-670
iron-base ~ 60-070
jewellery ~ 55-580
kanthal ~ 65-525
light ~ 55-405
light metal ~ 55-405
Lipowitz ~ 65-680
low ~ 55-320
low expansion ~ 55-440
low melting ~ 55-395
low melting-point ~ 55-395
magnetic ~ 55-445
magnetically hard ~ 55-465
magnetically soft ~ 55-460
master ~ 55-285
metallic ~ 55-160
mirror ~ 65-280
multicomponent ~ 55-310
native ~ 55-265
natural ~ 55-265
new ~ 55-275
nickel ~ 65-430
nickel-base ~ 65-430
nimonic ~ 65-480
non-corrosive ~ 55-480
non-ferrous ~ 65-005
non-magnetic ~ 55-470
nonmagnetic ~ 55-470
non-tarnishing ~ 55-500
ornamental ~ 55-605
oxidation-resistant ~ 55-495
oxidation-resisting ~ 55-495
permanent magnet ~ 55-450
permeability ~ 55-475
piston ~ 55-570
polyphase ~ 55-345
precipitation-hardenable ~ 55-385
pseudo- ~ 55-650
pseudo-binary ~ 55-315
pyrophoric ~ 55-595
quasibinary ~ 55-315
refractory ~ 55-400
re-melted ~ 55-280
resistance ~ 55-545
rich ~ 55-285
Rose's ~ 65-675
seal-in ~ 55-600
secondary ~ 55-280

alloy
silver brazing ~ 65-750
sintered ~ 55-655
sintered hard ~ 55-665
solder ~ 65-710
soldering ~ 65-710
sparking ~ 55-595
speculum ~ 65-280
strong ~ 55-420
superplastic ~ 55-430
super-strength ~ 55-425
synthetic ~ 55-270
synthetized ~ 55-270
tarnish-resistant ~ 55-500
tarnish-resisting ~ 55-500
technical ~ 55-520
textured ~ 55-380
three-component ~ 55-300
tombac ~ 65-395
two-component ~ 55-295
two-phase ~ 55-335
ultra-high-strength ~ 55-425
ultra-light ~ 55-410
wear-resisting ~ 55-435
Wood's ~ 65-670
workable ~ 55-530
wrought ~ 55-530
Y-~ 65-590
zamak ~ 65-705
alloy carbide 40-445
alloy cast iron 60-995
alloy cast steel 60-835
alloy cementite 50-110
alloyed cementite 50-110
alloy hardening 55-210
alloying 55-170
diffusion ~ 70-655
alloying addition 55-190
alloying component 55-180
alloying element 55-190
alloying power 55-175
alloy pig iron 60-935
alloys pl
ferro- ~ 60-1135
alloy steel 60-115
alloy system 55-165
alnico 65-530
Alnico alloy 65-530
alpax 65-565
modified ~ 65-570
alpax alloy 65-565
alpha-beta brass 65-330
alpha brass 65-325
alpha martensite 50-335
alpha-producing element 50-455
alsifer 60-1165, 65-535
alternate steel 60-270
aludur 65-580
alumel 65-460
aluminium 05-015
primary ~ 65-545
refined ~ 65-550
aluminium brass 65-340
aluminium bronze 65-135
aluminium cast iron 60-1020
aluminium impregnation 70-925
aluminium solder 65-730
aluminium steel 60-235
aluminizing 70-925
amalgam 65-850
americium 05-020
ammonia carburizing 70-860
ammonia nitriding 70-790

ATOMIC

amorphous 20-005
analysis
activation ~ 80-765
chemical ~ 80-470
differential dilatometric ~ 80-550
differential thermal ~ 80-515
dilatometric ~ 80-540
electron ~ 80-770
electron diffraction ~ 80-770
electron microprobe ~ 80-780
fluorescent ~ 80-700
magnetic ~ 80-560
microscopic ~ 80-040
physicochemical ~ 80-490
quantitative thermal ~ 80-535
radioactivation ~ 80-765
simple thermal ~ 80-510
spectral ~ 80-475
spectrum ~ 80-475
spot ~ 80-465
thermal ~ 80-505
thermodynamical ~ by calorimetry 80-535
thermomagnetic ~ 80-565
X-ray crystal ~ 80-580
X-ray diffraction ~ 80-580
X-ray fluorescence ~ 80-700
X-ray fluorescent ~ 80-700
X-ray phase ~ 80-585
X-ray structure ~ 80-580
anatomical alloy 55-645
anatomical metal 55-645
anchored dislocation 30-270
anchoring point 30-275
anelasticity 75-050
angle
misorientation ~ 30-595
anisotropy 20-445
magnetic ~ 75-575
anisotropy coefficient 75-595
anisotropy constant 75-595
anisotropy energy 75-590
anneal
magnetic ~ 70-965
annealing 70-420
~ of defects 30-135
black ~ 70-440
blue ~ 70-445
box ~ 70-435
bright ~ 70-430
clean ~ 70-430
close ~ 70-435
cyclic ~ 70-520
dead ~ 70-465
decarburizing ~ 70-635
diffusion ~ 70-485
ferritizing ~ 70-645
flame ~ 70-505
full ~ 70-465
grain growth ~ 70-480
high-temperature ~ 70-480
isochronal ~ 70-495
isochronous ~ 70-495
isothermal ~ 70-450
magnetic ~ 70-965
pack ~ 70-435
partial ~ 70-470
pot ~ 70-435
quench ~ 70-500
recrystallization ~ 70-490
self- ~ 45-455
soft ~ 70-460
solution ~ 70-500
spheroidizing ~ 70-510
spontaneous ~ 45-455
stabilizing ~ 70-530
stepped ~ 70-455
stress relief ~ 70-525
under ~ 70-470
annealing carbon 50-220
annealing temperature 70-425
annealing texture 40-600

annealing twin 45-240
anode copper 65-040
anorthic system 20-270
anticorodal 65-575
antiferromagnetism 75-565
antifriction alloy 55-555
antifriction metal 55-555
antimonial bronze 65-165
antimonial copper 65-165
antimonial lead 65-660
antimony 05-025
star ~ 65-665
antimony bronze 65-165
antiphase boundary 40-630
antiphase domain 40-625
antiphase domain boundary 40-630
arborescent crystal 25-315
architectural bronze 65-270
area
one-phase ~ 35-680
argentan 65-415
argon 05-030
Armco iron 60-010
armoured cast iron 60-1070
armour plate 60-765
arrest
thermal ~ 80-530
arrest lines 40-750
arsenic 05-035
arsenical copper 65-090
artificial ageing 70-590
as cast structure 40-050
astatine 05-040
asterism 80-680
asymmetric system 20-270
asymmetry
energetic ~ 15-235
athermal nucleation 25-200
athermal solution 15-265
athermal transformation 25-520
atmosphere
active ~ 70-680
carburizing ~ 70-695
controlled ~ 70-700
Cottrell ~ 30-440
endothermic ~ 70-705
exothermic ~ 70-710
impurity ~ 30-440
inert ~ 70-675
neutral ~ 70-675
protective ~ 70-690
reactive ~ 70-680
atmospheric corrosion 85-275
atom 30-075
displaced ~ 30-510
interstitial ~ 30-075
nearest neighbour ~ 20-195
substitutional ~ 30-070
atom core 10-040
atomic bond 10-330
atomic core 10-040
atomic diameter 10-080
atomic heat 75-370
atomic mass 10-075
atomic number 75-330
atomic plane 20-175
atomic radius 10-085
atomic size effect 10-395
atomic size factor 10-395

ATOMIC

atomic trunk 10-040
atomic volume 75-320
atomic weight 10-075
attack
  polish ~ 80-085
aural test 80-795
ausageing 70-560
ausforming 70-960
austempering 70-310
austenite 50-025
  nitrogen ~ 70-835
  primary ~ 50-030
  retained ~ 50-270
  transformed ~ 50-035
  undercooled ~ 50-265
  unstable ~ 50-265
austenite former 50-450
austenite forming element 50-450
austenite stabilization 50-425
austenite stabilizer 50-450
austenitic cast iron 60-1050
austenitic-ferritic steel 60-455
austenitic steel 60-420
austenitizer 50-450
austenitizing 70-130
austenitizing temperature 70-135
automatic steel 60-710
autoradiograph 80-755
autoradiography 80-740
  contrast ~ 80-745
  track ~ 80-750
avional 65-610
axial length 20-115
axial ratio 20-355
axis
  ~ of easy magnetization 75-585
  ~ of four-fold symmetry 20-080
  ~ of six-fold symmetry 20-085
  ~ of symmetry 20-065
  ~ of three-fold symmetry 20-075
  ~ of two-fold symmetry 20-070
  binary ~ 20-070
  crystal ~ 20-125
  crystallographic ~ 20-125
  diad ~ 20-070
  four-fold rotation ~ 20-080
  hexad ~ 20-085
  preferred magnetic ~ 75-585
  principal crystallization ~ 25-300
  quaternary ~ 20-080
  rotation ~ 20-065
  rotatory reflection ~ 20-065
  secondary crystallization ~ 25-305
  senary ~ 20-085
  six-fold rotation ~ 20-085
  ternary ~ 20-075
  tetrad ~ 20-080
  three-fold rotation ~ 20-075
  triad ~ 20-075
  twin ~ 45-260
  two-fold rotation ~ 20-070
azimuthal quantum number 10-135

B

babbit 65-645
Babbitt metal 65-645
back reflection diagram 80-670
back reflection method 80-630
back reflection pattern 80-670
Bahn metal 65-685
bahnmetall 65-685
bainite 50-410
  lower ~ 50-415

bainite
  upper ~ 50-420
bainite range 50-305
bainite reaction 50-290
bainite transformation 50-290
bainitic steel 60-435
bainitic transformation 50-290
baking 70-550
balanced steel 60-660
ball-bearing steel 60-720
ball iron 60-050
band
  allowed ~ 10-240
  conduction ~ 10-260
  deformation ~ 45-200
  energy ~ 10-235
  ferrite ~ 85-230
  filled ~ 10-245
  forbidden energy ~ 10-250
  ghost ~ 85-230
  glide ~ 45-190
  hardenability ~ 80-460
  segregation ~ 85-230
  slip ~ 45-190
  twin ~ 45-255
  twinning ~ 45-255
  valence ~ 10-255
  valency ~ 10-255
banded pearlite 50-130
banded structure 40-140
banding
  carbide ~ 85-235
  phosphorus ~ 85-230
bands pl
  Neumann ~ 45-205
band theory 10-020
bar
  test ~ 80-435
barium 05-045
barrier
  Cottrell-Lomer ~ 30-450
  energy ~ 10-110
  potential ~ 10-110
bar specimen 80-435
basal plane 20-360
base
  ~ of an alloy 55-185
  lattice ~ 20-110
base centred orthorhombic lattice 20-385
base diameter 70-360
base metal 55-005
basic steel 60-620
basis metal 55-185
bath
  hardening ~ 70-195
  quench ~ 70-195
  quenching ~ 70-195
bath carburizing 70-725
bath cooling 70-115
bath nitriding 70-795
Baumann printing 80-320
Bauschinger effect 75-175
beach marks 40-750
bearing alloy 55-555
bearing bronze 65-235
bearing cast iron 60-1105
bearing metal 55-555
bearing steel 60-715
Beilby layer 80-090
bell bronze 65-250
bell metal 65-250
bending strength 75-090

berkelium 05-050
berthollides *pl* 35-250
beryllium 05-055
beryllium bronze 65-150
beryllium copper 65-150
Bessemer pig iron 60-845
Bessemer steel 60-595
beta brass 65-335
beta martensite 50-340
beta-producing element 65-860
bicrystal 20-435
bimetal, bi-metal 65-880
bimetallic corrosion 85-320
binary alloy 55-295
binary axis 20-070
binary eutectic 40-205
binary solution 35-405
binary steel 60-155
binary sub-system 35-715
binary system 35-120
binding
  chemical ∼ 10-320
binding energy 15-090
binding enthalpy 15-170
biphasic system 35-100
bismuth 05-060
bivariant system 35-085
black annealing 70-440
black copper 65-015
black heart malleable cast iron 60-1115
black heart process 70-630
blank hardening 70-775
blast furnace pig iron 60-860
blister 85-130
blister copper 65-015
blister steel 60-585
Bloch function 10-195
Bloch wall 40-645
block
  mosaic ∼ 30-655
block structure 40-170
bloomary iron 60-045
bloomery iron 60-045
blow hole 85-135
blowhole segregation 25-430
blue annealing 70-445
blue brittleness 75-300
blue-heat range 70-395
blue shortness 75-300
blue temper 70-390
blueing 70-400
bluing 70-400
  ∼ in steam 70-880
body centred cubic lattice 20-315
body-centred lattice 20-295
body stresses *pl* 85-030
boiler embrittlement 75-305
boiler steel 60-740
boiling point 75-430
boiling temperature 75-430
bond
  atomic ∼ 10-330
  chemical ∼ 10-320
  covalent ∼ 10-330
  electrovalent ∼ 10-325
  heteropolar ∼ 10-325
  homopolar ∼ 10-330
  ionic ∼ 10-325

bond
  metallic ∼ 10-340
  molecular ∼ 10-335
  polar ∼ 10-325
  van der Waals ∼ 10-335
bond energy 15-090
bonding energy 15-090
bonding enthalpy 15-170
boride 70-905
boriding 70-900
boron 05-065
boron steel 60-255
boronizing 70-900
boundary
  antiphase ∼ 40-630
  antiphase domain ∼ 40-630
  coherent ∼ 30-625
  grain ∼ 30-565
  high-angle ∼ 30-585
  high-angle grain ∼ 30-585
  incoherent ∼ 30-630
  interphase ∼ 30-610
  large angle ∼ 30-585
  low-angle ∼ 30-590
  low-angle grain ∼ 30-590
  non-coherent ∼ 30-630
  small angle ∼ 30-590
  sub-grain ∼ 30-660
  subgrain ∼ 30-660
  tilt ∼ 30-605
  twin ∼ 45-265
  twinned ∼ 45-265
  twist ∼ 30-600
bound electron 10-050
bound energy 15-100
box annealing 70-435
box carburizing 70-720
Bragg method 80-600
brass 65-290
  α-∼ 65-325
  α+β-∼ 65-330
  β-∼ 65-335
  Admiralty ∼ 65-390
  alpha ∼ 65-325
  alpha-beta ∼ 65-330
  aluminium ∼ 65-340
  beta ∼ 65-335
  brazing ∼ 65-740
  cartridge ∼ 65-380
  casting ∼ 65-300
  common ∼ 65-305
  complex ∼ 65-310
  free-cutting ∼ 65-370
  high-tensile ∼ 65-355
  leaded ∼ 65-360
  machinery ∼ 65-375
  malleable ∼ 65-295
  manganese ∼ 65-355
  naval ∼ 65-385
  nickel ∼ 65-350
  red ∼ 65-395
  shrapnel ∼ 65-380
  silicon ∼ 65-365
  single-phase ∼ 65-315
  special ∼ 65-310
  straight ∼ 65-305
  wrought ∼ 65-295
  tin ∼ 65-345
  two-phase ∼ 65-320
  yellow ∼ 65-330
braunite 70-840
Bravais lattice 20-275
Bravais space lattice 20-275
brazing alloy 65-710
brazing brass 65-740
brazing solder 65-720
breakdown temperature 15-340
bright annealing 70-430
bright-field microscopy 80-235

BRIGHT

bright fracture 85-155
bright hardening 70-330
Brillouin zone 10-230
brine hardening 70-238
brine quenching 70-238
brittle fracture 40-675
brittle fracture transition temperature 75-280
brittleness 75-270
  acid ~ 85-390
  blue ~ 75-300
  hot ~ 75-290
  hydrogen ~ 85-390
  low-temperature ~ 75-295
  pickle ~ 85-390
  pickling ~ 85-390
  rheotropic ~ 75-310
  temper ~ 85-220
bromine 05-070
bronze 65-095
  Admiralty ~ 65-205
  aluminium ~ 65-135
  antimonial ~ 65-165
  antimony ~ 65-165
  architectural ~ 65-270
  bearing ~ 65-235
  bell ~ 65-250
  beryllium ~ 65-150
  cadmium ~ 65-190
  casting ~ 65-100
  chromium ~ 65-195
  coinage ~ 65-265
  conductivity ~ 65-255
  free-cutting ~ 65-370
  ferro ~ 65-175
  high-conductivity ~ 65-255
  high-tin ~ 65-115
  Government ~ 65-205
  graphite ~ 65-220
  iron ~ 65-175
  lead ~ 65-140
  leaded ~ 65-185
  leaded tin ~ 65-185
  low-tin ~ 65-120
  machine ~ 65-240
  manganese ~ 65-155, 65-355
  medal ~ 65-260
  nickel ~ 65-160
  ordnance ~ 65-245
  phosphor ~ 65-215
  phosphor tin ~ 65-210
  plastic ~ 65-185
  porous ~ 65-225
  red ~ 75-290
  scrap ~ 65-285
  silicon ~ 65-145
  sintered ~ 65-225
  special ~ 65-130
  statuary ~ 65-270
  statue ~ 65-270
  steel ~ 65-355
  valve ~ 65-275
  vanadium ~ 65-180
  white ~ 65-230
  wolfram ~ 65-170
  wrought ~ 65-105
  tin ~ 65-110
  tin-free ~ 65-125
  tungsten ~ 65-170
  zinc ~ 65-205
bronze alloy 65-130
bulk concentration 55-245
bulk diffusion 45-295
bulk elastic modulus 75-160
bulk hardening 70-287
bulk modulus of elasticity 75-160
bulk quenching 70-287
bulk specimen 80-055
Burgers circuit 30-185
Burgers dislocation 30-165

Burgers vector 30-190
burning 85-205
burnt steel 60-325
butterflies pl 85-255

C

cadmium 05-075
cadmium bronze 65-190
cadmium copper 65-190
caesium 05-080
calcium 05-085
calcium-silicon 60-1160
californium 05-090
calorimetric investigation 80-500
calorimetric study 80-500
calorizing 70-925
capability
  deep drawing ~ 75-750
capacitive properties pl 75-340
capacity
  damping ~ 75-205
  hardening ~ 75-020
  heat ~ 75-410
  magnetic inductive ~ 75-615
  thermal ~ 75-410
  work-hardening ~ 75-770
capacity property 15-040
carbide 40-430
  alloy ~ 40-445
  chromium ~ 40-455
  complex ~ 40-475
  double ~ 40-475
  iron ~ 40-440
  metal ~ 40-435
  niobium ~ 40-465
  simple ~ 40-470
  tungsten ~ 40-450
  vanadium ~ 40-460
carbide banding 85-235
carbide former 50-435
carbide-forming element 50-435
carbide network 50-445
carbide stabilizer 50-465
carbides pl
  cemented ~ 55-665
  eutectic ~ 50-440
  ledeburitic ~ 50-440
  sintered ~ 55-665
  undissolved ~ 40-480
carbon 05-095
  annealing ~ 50-220
  combined ~ 50-010
  free ~ 50-005
  graphitic ~ 50-005
  temper ~ 50-220
carbon cast steel 60-830
carbon cementation 70-715
carbon equivalent 50-480
carbon level 70-750
carbon potential 70-750
carbon steel 60-095
carbon transfer coefficient 70-755
carbonyl iron 60-020
carbonyl nickel 65-425
carbonitride 40-485
carbonitriding 70-870
carburization 70-715
carburized case 70-780
carburized steel 60-310
carburizer 70-740

carburizing 70-715
 ammonia ~ 70-860
 bath ~ 70-725
 box ~ 70-720
 gas ~ 70-730
 liquid ~ 70-725
 pack ~ 70-720
 powder ~ 70-720
 salt bath ~ 70-725
 selective ~ 70-735
 solid ~ 70-720
carburizing atmosphere 70-695
carburizing gas 70-745
carburizing steel 60-690
carrier gas 70-685
cartridge brass 65-380
cascade
 ~ of displacements 30-520
 collision ~ 30-515
 displacement ~ 30-520
case 70-660
 carburized ~ 70-780
 hardened ~ 70-355
 nitride ~ 70-845
 nitrided ~ 70-845
case hardening 70-670
case-hardening steel 60-690
castability 75-780
castable alloy 55-535
cast alloy 55-535
casting brass 65-300
casting bronze 65-100
casting defect 85-095
casting properties pl 75-775
casting stresses pl 85-170
casting texture 40-580
cast iron 60-945
cast iron heredity 60-985
cast steel 60-825
cast structure 40-050
catalyst
 nucleation ~ 25-125
category of steel 60-080
cathode copper 65-030
cathodic etching 80-180
cathodic vacuum etching 80-180
caustic embrittlement 75-305
cavity
 contraction ~ 85-125
 shrinkage ~ 85-125
cell
 primitive ~ 20-095
 primitive unit ~ 20-095
 simple ~ 20-095
 unit ~ 20-090
cell structure 40-095
cellular structure 40-095
cement copper 65-045
cementation 70-655
 carbon ~ 70-715
 metallic ~ 70-915
cementation steel 60-690
cemented carbides pl 55-665
cementite 50-065
 alloy ~ 50-110
 alloyed ~ 50-110
 divorced ~ 50-100
 eutectic ~ 50-075
 excess ~ 50-095
 free ~ 50-095
 globular ~ 50-100
 hypereutectic ~ 50-070
 hypereutectoid ~ 50-080
 nodular ~ 50-100

cementite
 pearlitic ~ 50-085
 primary ~ 50-070
 pro-eutectoid ~ 50-080
 spheroidal ~ 50-100
 spheroidized ~ 50-100
 tertiary ~ 50-090
cementite eutectic 50-255
cementite lamella 50-120
cementite network 50-105
cementitic steel 50-500
ceramal 55-675
cerium 05-100
cerium standard alloy 65-865
cermet 55-675
cerrobend 65-680
cesium (USA) 05-080
change
 allotropic ~ 25-455
 eutectic ~ 35-565
 eutectoid ~ 35-590
 martensitic ~ 50-280
 metatectic ~ 35-580
 monotectic ~ 35-575
 monotectoid ~ 35-600
 peritectic ~ 35-570
 peritectoid ~ 35-595
 phase ~ 25-460
 polymorphic ~ 25-455
 structural ~ 25-500
 synthetic ~ 35-585
change point 35-555
channeling 30-500
chaotic orientation 20-420
characteristic
 deep drawing ~ 75-755
 plasticity ~ 75-065
 strength ~ 75-060
charcoal-hearth iron 60-050
charcoal iron 60-050
charcoal pig iron 60-875
charged dislocation 30-465
checks pl
 chrome ~ 85-240
chemical affinity 75-660
chemical analysis 80-470
chemical binding 10-320
chemical bond 10-320
chemical composition of an alloy 55-225
chemical compound 35-235
chemical corrosion 85-265
chemical dendrite 25-320
chemical diffusion 45-320
chemical element 05-005
chemical equilibrium 15-330
chemical etching 80-170
chemical inhomogeneity 40-415
chemical non-uniformity 40-415
chemical polishing 80-080
chemical potential 15-175
chemical properties pl 75-655
chemical resistance 75-665
chemico-thermal treatment 70-650
chill
 internal ~ 85-150
 inverse ~ 85-150
chilled cast iron 60-980
chill hardening 70-255
chill zone 40-175
chlorine 05-105
chromansil 60-260

CHROME

chrome checks *pl* 85-240
chromel 65-465
chrome-nickel 65-470
chrome steel 60-175
chromium 05-110
chromium bronze 65-195
chromium carbide 40-455
chromium cast iron 60-1015
chromium copper 65-200
chromium impregnation 70-920
chromium-nickel steel 60-190
chromium nitride 70-820
chromium steel 60-175
chromizing 70-920
chromoaluminizing 70-940
circuit
Burgers $\sim$ 30-185
clad steel 60-335
clamshell markings *pl* 40-750
clamshell marks *pl* 40-750
class
crystal $\sim$ 20-050
clean annealing 70-430
clean hardening 70-330
cleavage fracture 85-085
cleavage plane 40-665
climatic corrosion 85-275
climb
dislocation $\sim$ 30-295
close annealing 70-435
close-packed hexagonal lattice 20-350
close packing 20-210
closest packing 20-210
closure domain 40-640
cloud
electron $\sim$ 10-035
impurity $\sim$ 30-440
cluster
vacancy $\sim$ 30-125
coalescence of cementite 70-515
coarse grain 40-365
coarse-grained fracture 40-695
coarse grained martensite 50-320
coarse grainedness 40-375
coarse-grained steel 60-475
coarse-grained structure 40-390
coarse lamellar pearlite 50-140
coarsely lamellar pearlite 50-140
coarseness
grain $\sim$ 40-375
coarsening 25-290
grain $\sim$ 25-290
coarse pearlite 50-140
coarse structure 40-390
cobalt 05-115
cobalt steel 60-240
concentration gradient 15-080
coefficient
activity $\sim$ 15-285
anisotropy $\sim$ 75-595
carbon transfer $\sim$ 70-755
$\sim$ of compressibility 75-165
$\sim$ of cubical expansion 75-460
$\sim$ of linear expansion 75-455
$\sim$ of reflection 75-515
$\sim$ of thermal conduction 75-440
$\sim$ of thermal expansion 75-450
diffusion $\sim$ 45-395
distribution $\sim$ 25-260
effective distribution $\sim$ 25-265

coefficient
elastic $\sim$ 75-145
equilibrium distribution $\sim$ 25-260
interaction $\sim$ 15-295
intrinsic diffusion $\sim$ 45-390
linear $\sim$ of thermal expansion 75-455
partial diffusion $\sim$ 45-390
self-diffusion $\sim$ 45-405
strain-hardening $\sim$ 45-065
temperature $\sim$ of electrical resistivity 75-480
temperature $\sim$ of resistivity 75-480
thermal $\sim$ of expansion 75-450
thermal expansion $\sim$ 75-450
volume $\sim$ of thermal expansion 75-460
work-hardening $\sim$ 45-065
coercive force 75-635
coercivity 75-635
coexisting phases *pl* 35-190
cohenite 50-115
coherency 30-620
coherent boundary 30-625
coherent interface 30-625
coherent nucleus 25-155
coherent precipitate 40-320
cohesion 10-310
cohesive force 10-315
cohesive strength 75-135
coinage alloy 55-565
coinage bronze 65-265
coinage copper 65-265
coinage metal 55-565
coin gold 65-775
coin metal 55-565
coin silver 65-755
coke pig iron 60-870
cold crack 85-075
cold cracking 85-075
cold-forming tool steel 60-805
cold plastic working 45-025
cold shortness 75-285
cold shut 85-400
cold work 45-030
cold working 45-025
cold-working tool steel 60-805
cold-work steel 60-805
cold-work tool steel 60-805
cold treatment 70-340
collision
focused $\sim$ 30-505
collision cascade 30-515
colour
temper $\sim$ 70-385
colour etching 80-165
colour metallography 80-030
columbium 05-300
columnar crystal zone 40-180
columnar crystallization 25-065
columnar crystals *pl* 40-280
columnar structure 40-125
columnar zone 40-180
combined carbon 50-010
commercial alloy 55-520
commercial iron 60-015
commercially pure iron 60-015
commercial steel 60-135
common brass 65-305
common metal 55-005
common tangent construction 15-180
common tangent principle 15-180

common tangent rule 15-180
compact metal 55-155
complete dislocation 30-305
complete miscibility 35-315
complete solubility 35-315
complex alloy 55-310
complex brass 65-310
complex carbide 40-475
complex solution 35-410
complex steel 60-170
compliance
  elastic ~ 75-145
component 35-035
  alloying ~ 55-180
  ~ of an alloy 55-180
composite material 55-680
composition
  chemical ~ of an alloy 55-225
  stoichiometric ~ 35-240
compositional fluctuations pl 25-100
composition fluctuations pl 25-100
composition plane 20-470
composition triangle 35-720
compound 35-235
  chemical ~ 35-235
  congruently melting ~ 35-220
  electron ~ 35-280
  incongruently melting ~ 35-225
  intermetallic ~ 35-215
  ionic ~ 35-230
  peritectic ~ 35-225
  polar ~ 35-230
compound dislocation 30-180
compounds pl
  Laves ~ 35-265
compression modulus of elasticity 75-160
compressive texture 40-590
concentrated alloy 55-325
concentration 55-230
  bulk ~ 55-245
  electron ~ 35-285
  equilibrium ~ 35-255
  mass ~ 55-235
  molar ~ 55-240
concentration fluctuations pl 25-100
concentration triangle 35-720
conchoidal fracture 40-730
conchoidal markings pl 40-750
conchoidal marks pl 40-750
condensation
  vacancy ~ 30-115
condensed phase 35-145
condition
  metastable ~ 35-070
  stable ~ 35-065
conductance
  specific ~ 75-485
conduction band 10-260
conduction electron 10-265
conductivity
  electrical ~ 75-485
  heat ~ 75-440
  thermal ~ 75-440
conductivity bronze 65-255
configurational entropy 15-210
configurational specific heat 25-635
congruent intermediate phase 35-220
congruently melting compound 35-220
congruently-melting intermediate phase 35-220
congruent melting 35-460

congruent transformation 25-525
conjugate curves pl 35-540
conjugate lines pl 35-540
conjugate phases pl 35-190
conjugate slip system 45-125
conode 35-695
conservative motion 30-290
constant
  anisotropy ~ 75-595
  grating ~ 20-225
  elastic ~ 75-145
  lattice ~ 20-120
  thermodynamic equilibrium ~ 15-335
constantan 65-450
constant-modulus alloy 65-505
constant-rate creep 45-505
constituent
  hardening ~ 55-255
  major ~ 55-185
  microstructural ~ 40-010
  oligo ~ 55-250
  structural ~ 40-010
constitution
  phase ~ 35-020
constitutional diagram 35-480
constitutional stresses pl 85-035
constitutional supercooling 25-255
constitutional undercooling 25-255
constitutional vacancy 30-090
constitution diagram 35-480
constitution phase 35-020
construction
  common tangent ~ 15-180
constructional steel 60-670
contact alloy 55-585
contact corrosion 85-320
contact fatigue 85-420
contact hardening 70-250
continuous cooling 70-090
continuous cooling transformation diagram 70-160
continuous precipitation 25-375
continuous series of solid solutions 35-390
continuous solid solutions pl 35-390
continuous TTT diagram 70-160
continuous transformation 25-540
contraction 75-795
  ~ of area 75-125
contraction cavity 85-125
contrast
  orientation ~ 80-155
  phase ~ 80-275
contrast autoradiography 80-745
controlled atmosphere 70-700
controlled solidification 25-030
conventional quenching 70-270
converter copper 65-015
converter steel 60-590
cooling 70-070
  air ~ 70-095
  bath ~ 70-115
  continuous ~ 70-090
  ~ diagram 80-525
  ~ in air 70-095
  ~ in still air 70-095
  furnace ~ 70-110
  oil ~ 70-105
  slow ~ 70-080
  splat ~ 70-085
  water ~ 70-100
cooling curve 80-525

COOLING

cooling rate 70-120
coordination number 20-185
coordination shell 20-190
coordination zone 20-190
copper 05-120
  anode ~ 65-040
  antimonial ~ 65-165
  arsenical ~ 65-090
  beryllium ~ 65-150
  black ~ 65-015
  blister ~ 65-015
  cadmium ~ 65-190
  cathode ~ 65-030
  cement ~ 65-065
  coinage ~ 65-265
  chromium ~ 65-200
  converter ~ 65-015
  dry ~ 65-050
  electrolytic ~ 65-035
  fire-refined ~ 65-025
  H.C. ~ 65-065
  high conductivity ~ 65-065
  lake ~ 65-025
  manganese ~ 65-155
  OFHC ~ 65-070
  oxygen-free ~ 65-060
  oxygen-free high conductivity ~ 65-070
  phosphor ~ 65-075
  phosphorized ~ 65-055
  refined ~ 65-020
  set ~ 65-050
  white ~ 65-415
copper alloy 55-010
copper-base alloy 65-010
copper-bearing steel 60-245
copper-manganese 65-085
copper-silicon 65-080
copper steel 60-245
core 70-665
  atom ~ 10-040
  atomic ~ 10-040
  dislocation ~ 30-205
core refining 70-770
coring 25-420
corrodibility 75-675
corrosion 85-260
  atmospheric ~ 85-275
  bimetallic ~ 85-320
  chemical ~ 85-265
  climatic ~ 85-275
  contact ~ 85-320
  crevice ~ 85-300
  dry ~ 85-295
  electrochemical ~ 85-270
  fatigue ~ 85-315
  fretting ~ 85-325
  galvanic ~ 85-270
  gas ~ 85-295
  grain boundary ~ 85-360
  intercrystalline ~ 85-360
  intergranular ~ 85-360
  localized ~ 85-340
  marine ~ 85-290
  nuclear ~ 85-365
  pitting ~ 85-350
  point ~ 85-345
  sea water ~ 85-290
  selective ~ 85-305
  soil ~ 85-285
  stress ~ 85-310
  subsurface ~ 85-355
  surface ~ 85-330
  uniform ~ 85-335
  water ~ 85-280
corrosion fatigue 85-415
corrosion resistance 75-670
corrosion-resistant alloy 55-480
corrosion-resisting alloy 55-480
corrosion-resisting steel 60-535
Cottrell atmosphere 30-440

Cottrell-Lomer barrier 30-450
Cottrell-Lomer lock 30-450
couple
  diffusion ~ 45-375
covalent bond 10-330
crack
  cold ~ 85-075
  fire ~ 85-060
  Griffith ~ 85-055
  hairline ~ 85-045
  hardening ~ 85-190
  heat ~ 85-060
  quench ~ 85-190
  quenching ~ 85-190
  strain ~ 85-070
  stress ~ 85-065
  surface ~ 85-050
  thermal ~ 85-060
cracking 85-015
  cold ~ 85-075
  hot ~ 85-080
  season ~ 85-385
  stress corrosion ~ 85-385
cracks *pl*
  shatter ~ 85-240
crap bronze 65-285
craze 85-050
creep 45-495
  accelerated ~ 45-510
  accelerating ~ 45-510
  constant-rate ~ 45-505
  diffusional ~ 45-520
  Herring-Nabarro ~ 45-520
  Herring-Nabarro diffusion ~ 45-520
  increasing-rate ~ 45-510
  initial ~ 45-500
  Nabarro ~ 45-520
  primary ~ 45-500
  secondary ~ 45-505
  steady-rate ~ 45-505
  steady-state ~ 45-505
  tertiary ~ 45-510
  transient ~ 45-500
  unsteady ~ 45-500
creep curve 45-515
creep limit 75-220
creep resistance 75-215
creep-resistant alloy 55-510
creep-resistant steel 60-565
creep-resisting alloy 55-510
creep-resisting steel 60-565
creep rupture strength 75-235
creep strength 75-220
crescent
  fatigue ~ 40-750
crevice corrosion 85-300
critical cooling rate 70-180
critical degree of deformation 45-075
critical diameter 70-360
critical heat 75-400
critical nucleus 25-140
critical point 35-555
critical quenching rate 70-180
critical radius of a nucleus 25-145
critical range 35-560
critical resolved shear stress 45-090
critical size nucleus 25-140
critical solution temperature 35-335
critical strain 45-075
critical temperature 35-335, 35-555
critical temperature range 35-560
cross slip 45-145
cross-slip plane 45-150
crowdion 30-470

crucible steel 60-630
crude lead 65-650
cry
  tin ~ 45-270
cryptocrystalline martensite 50-330
crystal 20-015
  allotriomorphic ~ 40-360
  arborescent ~ 25-315
  dendritic ~ 25-315
  fir-tree ~ 25-315
  idiomorphic ~ 25-355
  imperfect ~ 30-010
  ionic ~ 20-025
  metallic ~ 20-020
  mixed ~ 40-195
  original ~ 40-295
  parent ~ 40-295
  perfect ~ 30-005
  pine-tree ~ 25-315
  polar ~ 20-025
  real ~ 30-010
  single ~ 20-430
  twin ~ 45-230
  twinned ~ 45-230
  valence ~ 20-030
  xenomorphic ~ 40-360
crystal axis 20-125
crystal class 20-050
crystal face 20-060
crystal grain 40-360
crystal growing 25-085
crystal growth 25-285
crystal habit 20-460
crystal imperfection 30-015
crystal lattice 20-045
crystalline 20-010
crystalline direction 20-150
crystalline fracture 40-685
crystalline lattice 20-045
crystalline symmetry 20-055
crystallite 40-360
crystallization 25-010
  columnar ~ 25-065
  equal-axed ~ 25-075
  fully-columnar ~ 25-070
  mixed ~ 25-055
  primary ~ 25-050
  secondary ~ 25-060
  spontaneous ~ 25-045
crystallization interval 35-550
crystallographic axis 20-125
crystallographic direction 20-150
crystallographic imperfection 30-015
crystallographic index 20-160
crystallographic orientation 20-415
crystallographic plane 20-155
crystallographic system 20-235
crystallographic structure 20-040
crystallography 20-035
crystal nucleus 25-105
crystal plane 20-155
crystal pulling 25-090
crystals pl
  columnar ~ 40-280
  directional ~ 40-280
  equi-axed ~ 40-285
  equiaxial ~ 40-285
  fringe ~ 40-280
  growing single ~ 25-085
  isomorphous ~ 20-480
  primary ~ 40-270
  secondary ~ 40-275
crystal seed 25-130
crystal structure 20-040

crystal symmetry 20-055
crystal system 20-235
crystal twin 40-230
cube-on-edge texture 40-610
cube texture 40-605
cubic lattice 20-300
cubic system 20-240
cupro-nickel 65-400
Curie point 25-560
Curie temperature 25-560
curium 05-125
curly pearlite 50-145
curve
  cooling ~ 80-525
  creep ~ 45-515
  dilatometer ~ 80-555
  dilatometric ~ 80-555
  end-quench hardenability ~ 80-455
  fatique ~ 80-425
  freezing-point ~ 35-515
  hardenability ~ 80-455
  heating ~ 80-520
  isoactivity ~ 15-315
  liquidus ~ 35-515
  melting-point ~ 35-490
  retrograde solidus ~ 35-505
  S-N ~ 80-425
  solidus ~ 35-490
  solubility ~ 35-530
  solvus ~ 35-530
  spinodal ~ 15-190
  strain-hardening ~ 45-070
  stress-strain ~ 80-410
  Wöhler ~ 80-425
  work-hardening ~ 45-070
curves pl
  conjugate ~ 35-540
customary quenching 70-270
cutlery steel 60-750
cyanide case hardening 70-855
cyanide hardening 70-855
cyaniding 70-855
  dry ~ 70-860
  gas ~ 70-860
  liquid ~ 70-865
cycle
  heat treatment ~ 70-970
  thermal ~ 70-970
cyclic annealing 70-520
cyclic fracture 40-745
cyclic strength 75-200
cycling
  thermal ~ 70-520

D

daltonides pl 35-245
damage
  irradiation ~ 30-490
  radiation ~ 30-490
damascene steel 60-645
damping capacity 75-205
dark-field microscopy 80-230
dead annealing 70-465
dead soft steel 60-345
Debye characteristic temperature 10-280
Debye ring 80-640
Debye-Scherrer method 80-595
Debye-Scherrer powder pattern 80-635
Debye temperature 10-280
decalescence 25-490
decarburization
  surface ~ 85-210

decarburizing annealing 70-635
decomposition
 eutectoid ~ 35-590
 eutectoidal ~ 35-590
 spinodal ~ 25-370
decomposition point 15-340
decorating 80-375
decoration 80-375
decorative alloy 55-605
deep-drawability 75-750
deepdrawability 75-750
deep drawing capability 75-750
deep drawing characteristic 75-755
deep-drawing steel 60-790
deep etching 80-160
deep-hardening steel 60-385
defect
 casting ~ 85-095
 Frenkel ~ 30-060
 irradiation-induced ~ 30-485
 irradiation-produced ~ 30-485
 lattice ~ 30-015
 line ~ 30-140
 material ~ 85-005
 one-dimensional ~ 30-140
 planar ~ 30-535
 plane ~ 30-535
 point ~ 30-030
 quenching ~ 85-180
 radiation ~ 30-485
 surface ~ 30-535, 85-010
 two-dimensional ~ 30-535
 zero-dimensional ~ 30-030
defect lattice 30-020
deformation
 elastic ~ 45-005
 permanent ~ 45-010
 plastic ~ 45-010
deformation band 45-200
deformation martensite 50-360
deformation texture 40-560
deformation twin 45-235
degenerate pearlite 50-155
degree
 ~ of cold work 45-040
 ~ of deformation 45-035
 ~ of dendritic segregation 40-420
 ~ of disorder 25-610
 ~ of disordering 25-610
 ~ of dispersion 40-335
 ~ of order 25-605
 ~ of ordering 25-605
 ~ of supercooling 25-250
 ~ of undercooling 25-250
degrees pl
 ~ of freedom 35-030
dehydrogenation 70-555
delayed elasticity 75-180
delayed fracture 40-755
delta ferrite 50-050
demodifier 25-235
dendrite 25-315
 chemical ~ 25-320
 thermal ~ 25-325
dendritic crystal 25-315
dendritic growth 25-310
dendritic segregation 25-420
dendritic structure 40-090
denitriding 70-850
denitrogenizing 70-850
dense lamellar pearlite 50-135
density
 ~ of states 10-200
 dislocation ~ 30-235

density
 magnetic flux ~ 75-610
 specific ~ 75-345
 weight ~ 75-350
dental alloy 55-575
dental gold 65-785
denuded zone 30-120
depleted zone 30-120
depth
 ~ of hardening 70-350
 hardening ~ 70-350
 hardness penetration ~ 70-350
designation
 ~ of steel 60-090
 steel ~ 60-090
destabilization of austenite 50-430
detection
 flaw ~ 80-790
 magnetic crack ~ 80-805
 penetrant flaw ~ 80-835
dezincification 85-380
diad axis 20-070
diagram
 back reflection ~ 80-670
 constitution ~ 35-480
 constitutional ~ 35-480
 continuous cooling transformation ~ 70-160
 continuous TTT ~ 70-160
 cooling ~ 80-525
 heating ~ 80-520
 iron-carbon ~ 50-020
 iron-carbon phase ~ 50-020
 isothermal time-temperature-transformation ~ 70-155
 isothermal transformation ~ 70-155
 Laue ~ 80-620
 phase ~ 35-480
 phase equilibrium ~ 35-480
 recrystallization ~ 45-485
 rotation ~ 80-665
 Sauveur's ~ 40-265
 stress-strain ~ 80-410
 ternary equilibrium ~ 35-710
 ternary equilibrium space ~ 35-750
 ternary phase equilibrium ~ 35-710
 ternary phase equilibrium spatial ~ 35-750
 ternary space ~ 35-750
 ternary spatial ~ 35-750
 thermal equilibrium ~ 35-480
 transmission ~ 80-675
diamagnetic 75-560
diamagnetic material 75-560
diamagnetic substance 75-560
diameter
 atomic ~ 10-080
 base ~ 70-360
 critical ~ 70-360
 isothermal TTT ~ 70-155
 Schaefler ~ 60-530
 Schaeffler's ~ 60-530
diamond cubic lattice 20-320
die-casting alloy 55-540
die steel 60-815
differential dilatometric analysis 80-550
differential dilatometry 80-550
differential hardening 70-275
differential thermal analysis 80-515
diffraction line 80-645
diffraction pattern 80-610
diffraction spot 80-660
diffusion 45-290
 accelerated ~ 45-355
 bulk ~ 45-295
 chemical ~ 45-320
 dislocation pipe ~ 45-360
 enhanced ~ 45-355
 grain boundary ~ 45-310
 interstitial ~ 45-315

DISPLACEMENT

diffusion
isothermal ~ 45-350
lattice ~ 45-295
pipe ~ 45-360
polyphase ~ 45-325
self-~ 45-400
single-phase ~ 45-330
stationary ~ 45-345
steady-state ~ 45-345
surafce ~ 45-300
thermal ~ 45-370
up-hill ~ 45-340
vacancy ~ 45-305
volume ~ 45-295
diffusional creep 45-520
diffusion alloying 70-655
diffusional process 25-505
diffusional transformation 25-505
diffusion annealing 70-485
diffusion coefficient 45-395
diffusion couple 45-375
diffusion-free process 25-510
diffusion front 45-410
diffusionless transformation 25-510
diffusion metallization 70-915
diffusion zone 45-415
diffusivity 45-395
intrisic ~ 45-390
thermal ~ 75-445
dilatometer curve 80-555
dilatometric analysis 80-540
dilatometric curve 80-555
dilatometry 80-540
differential ~ 80-550
simple ~ 80-545
dilute alloy 55-320
diluted solution 15-250
dilute solution 15-250
dipole
dislocation ~ 30-380
directed solidification 25-035
direct hardening 70-760
direction
crystalline ~ 20-150
crystallographic ~ 20-150
easy ~ of magnetization 75-585
slip ~ 45-105
strong ~ 20-220
directional crystals pl 40-280
directional solidification 25-035
direct quenching 70-760
disclination 30-480
discontinuity of material 85-090
discontinuous eutectic 40-235
discontinuous precipitation 25-380
discontinuous transformation 25-545
disease
hydrogen ~ 85-395
tin ~ 85-245
dislocation 30-145
anchored ~ 30-270
Burgers ~ 30-165
charged ~ 30-465
complete ~ 30-305
compound ~ 30-180
edge ~ 30-150
extended ~ 30-325
Frank ~ 30-335
Frank partial ~ 30-335
Frank sessile ~ 30-335
glide ~ 30-265
glissile ~ 30-265
grain boundary ~ 30-580
half-~ 30-315
helical ~ 30-355

dislocation
helicoidal ~ 30-355
imperfect ~ 30-310
left-handed screw ~ 30-175
mixed ~ 30-180
negative edge ~ 30-160
partial ~ 30-315
penny-shaped ~ 30-345
perfect ~ 30-305
pinned ~ 30-270
pole ~ 30-365
positive edge ~ 30-155
prismatic ~ 30-345
right-handed screw ~ 30-170
Schockley ~ 30-330
Schockley partial ~ 30-330
screw ~ 30-165
sessile ~ 30-280
slip ~ 30-265
stair-rod ~ 30-360
Taylor ~ 30-150
Taylor-Orowan ~ 30-150
twinning ~ 30-375
unit ~ 30-215
dislocation climb 30-295
dislocation core 30-205
dislocation density 30-235
dislocation dipole 30-380
dislocation forest 30-405
dislocation generator 30-425
dislocation jog 30-240
dislocation kernel 30-205
dislocation line 30-195
dislocation locking 30-445
dislocation loop 30-340
dislocation multiplication 30-420
dislocation network 30-390
dislocation node 30-350
dislocation pile-up 30-400
dislocation pipe diffusion 45-360
dislocation ring 30-340
dislocations pl
emissary ~ 30-370
primary ~ 30-250
secondary ~ 30-255
dislocation source 30-425
dislocation structure 30-415
dislocation tangle 30-385
dislocation tree 30-410
dislocation wall 30-395
disorder 25-585
Frenkel ~ 30-060
disordered phase 35-205
disordered solid solution 35-435
disordered state 25-585
disordering 25-580
disordering temperature 25-625
disorientation 30-595
dispersed phase 40-330
dispersed phase hardening 55-220
disperse phase 40-330
dispersion 40-335
dispersion alloy 55-390
dispersion-hardened alloy 55-390
dispersion hardening 55-220
dispersion-strengthened alloy 55-390
dispersion strengthening 55-220
dispersoid 40-330
displaced atom 30-510
displacement cascade 30-520
displacement series 75-710
displacement spike 30-525

DISREGISTRY

disregistry
lattice ~ 30-615
disruptive strength 75-135
dissociation of a dislocation 30-320
dissociation pressure 15-345
distance
interatomic ~ 20-230
interplanar ~ 20-225
distortion
lattice ~ 30-025
distribution coefficient 25-260
divacancy 30-045
divorced cementite 50-100
divorced eutectic 40-230
divorced pearlite 50-150
domain 40-615
antiphase ~ 40-625
closure ~ 40-640
magnetic ~ 40-635
domain structure 40-620
domain wall 40-645
double carbide 40-475
double hardening treatment 70-765
double quenching 70-765
double slip 45-135
drastic quenching 70-260
drawability
deep- ~ 75-750
drawing 70-365
drawing texture 40-575
driving force 15-070
droplet segregation 25-430
dry copper 65-050
dry corrosion 85-295
dry cyaniding 70-860
dry method 80-810
dry powder method 80-810
dry quenching 70-220
ductile cast iron 60-975
ductile fracture 40-680
ductility 75-030
ductility transition temperature 75-280
dumbbell interstitial 30-085
duplex metal 65-880
duplex slip 45-135
duplex structure 40-150
duralumin 65-555
dynamic nucleation 25-205
dynamic recovery 45-425
dynamic recrystallization 45-465
dynamic strength 75-185
dynamo steel 60-775
dysprosium 05-130

E

easy-cutting steel 60-710
easy direction of magnetization 75-585
easy glide 45-155
edge dislocation 30-150
effect
atomic size ~ 10-395
Bauschinger ~ 75-175
elastic after-~ 75-180
electronegative valency ~ 10-400
Hall ~ 10-290
Kirkendall ~ 45-385
Mössbauer ~ 10-285
Portevin-Le Chatelier ~ 45-540

effect
relative valency ~ 10-405
Snoek ~ 30-460
Soret ~ 45-370
Suzuki ~ 30-455
effective distribution coefficient 25-265
efficiency of space filling 20-205
eigenfunction 10-190
einsteinium 05-135
elastic after-effect 75-180
elastic coefficient 75-145
elastic compliance 75-145
elastic constant 75-145
elastic deformation 45-005
elasticity 75-020
delayed ~ 75-180
elastic limit 75-105
elastic modulus 75-150
elastic relaxation 45-525
elastic strain 45-005
electrical conductivity 75-485
electrical contact alloy 55-585
electrical properties pl 75-465
electrical resistance alloy 55-545
electrical resistivity 75-470
electric furnace pig iron 60-865
electric resistivity 75-470
electric steel 60-625
electrochemical corrosion 85-270
electrochemical factor 10-400
electrochemical potential 75-715
electrochemical properties pl 75-705
electrochemical series pl 75-710
electrocopper 65-035
electrocrystallization 25-080
electrodeposited alloy 55-290
electrodeposited metal 55-130
electrodiffusion 45-365
electrolytic copper 65-035
electrolytic etching 80-175
electrolytic iron 60-025
electrolytic metal 55-130
electrolytic polishing 80-075
electromigration 45-365
electromotive series pl 75-710
electron 65-625
bound ~ 10-050
conduction ~ 10-265
external ~ 10-065
free ~ 10-055
inner-shell ~ 10-070
outer ~ 10-065
outer-shell ~ 10-065
valency ~ 10-065
electron affinity 10-115
electron alloy 65-625
electron analysis 80-770
electron cloud 10-035
electron compound 35-280
electron concentration 35-285
electron diffraction analysis 80-770
electron diffraction pattern 80-775
electron diffraction photograph 80-775
electron diffraction study 80-770
electronegative valency effect 10-400
electronegativity 10-095
electron emission microscopy 80-335
electron energy gap 10-250

EQUILIBRIUM

electron gas 10-045
electronic shell 10-155
electronic specific heat 10-275
electronic work function 10-060
electronic structure 10-030
electron metallography 80-025
electron micrograph 80-365
electron microprobe analysis 80-780
electron microscopy 80-330
electronogram 80-775
electron phase 35-280
electron photomicrograph 80-365
electron probe microanalysis 80-780
electron shell 10-155
electron state 10-115
electron theory of metals 10-025
electrotransport 45-365
electrotype metal 55-635
electrovalent bond 10-325
electrum 65-810
element 05-005
  alloying ~ 55-190
  alpha-producing ~ 50-455
  austenite forming ~ 50-450
  beta-producing ~ 65-860
  carbide-forming ~ 50-435
  chemical ~ 05-005
  hardening alloy ~ 55-255
  ferrite-forming ~ 50-455
  gamma-producing ~ 50-450
  graphite-forming ~ 50-460
  major alloying ~ 55-195
  metallic ~ 10-360
  nitride-forming ~ 50-470
  non-metal ~ 10-365
  non-metallic ~ 10-365
  principal alloying ~ 55-195
  residual ~ 55-260
  trace ~ 55-250
elements pl
  tramp ~ 55-200
  transition ~ 55-025
elinvar 65-505
elongation 75-120
  ~ at fracture 75-120
  percentage ~ 75-120
embrittlement 75-275
  acid ~ 85-390
  boiler ~ 75-305
  caustic ~ 75-305
  hydrogen ~ 85-390
  temper ~ 85-220
embryo 25-135
emissary dislocations pl 30-370
end-quench hardenability curve 80-455
end-quench hardenability test 80-450
endogas 70-705
endothermic atmosphere 70-705
endothermic gas 70-705
endurance
  thermal ~ 75-830
endurance limit 75-200
endurance strength 75-200
endurance test 80-420
energetic asymmetry 15-235
energy
  activation ~ 45-285
  anisotropy ~ 75-590
  binding ~ 15-090
  bond ~ 15-090
  bonding ~ 15-090
  bound ~ 15-100
  excess free ~ 15-115
  Fermi ~ 10-210

energy
  free ~ 15-105
  Gibbs free ~ 15-140
  grain boundary ~ 30-575
  Helmholtz free ~ 15-105
  interaction ~ 10-295
  interface ~ 30-635
  interfacial ~ 30-635
  interfacial free ~ 15-125
  internal ~ 15-095
  intrinsic ~ 15-095
  ionization ~ 10-100
  latent ~ 15-100
  ordering ~ 25-630
  partial free ~ 15-110
  self- ~ of a dislocation line 30-230
  stacking fault ~ 30-560
  stored ~ 45-060
  surface ~ 30-640
  surface free ~ 15-120
energy band 10-235
energy barrier 10-110
energy gap 10-250
energy level 10-150
energy state 10-120
enhanced diffusion 45-355
enthalpy
  binding ~ 15-170
  bonding ~ 15-170
  ~ of formation 15-160
  ~ of fusion 15-165
  ~ of melting 15-165
  ~ of mixing 15-155
  excess ~ 15-145
  free ~ 15-140
  partial ~ 15-150
entropy
  configurational ~ 15-210
  ~ of formation 15-200
  ~ of fusion 15-225
  ~ of melting 15-225
  ~ of mixing 15-195
  excess ~ 15-220
  partial ~ 15-215
  positional ~ 15-210
  thermal ~ 15-205
  vibrational ~ 15-205
envelope structure 40-100
epitaxial film 25-350
epitaxial growth 25-345
epitaxial layer 25-350
epitaxy 25-340
equalization
  temperature ~ 70-035
equation
  ~ of state 15-025
  Gibbs-Duhem ~ 15-305
  Hildebrand's ~ 15-310
  Schrödinger ~ 10-175
equi-axed crystallization 25-075
equi-axed crystals pl 40-285
equiaxed grains pl 40-285
equiaxed zone 40-185
equiaxial crystals pl 40-285
equicohesive temperature 30-570
equilibrium
  chemical ~ 15-330
  metastable ~ 35-060
  microstructural ~ 40-020
  phase ~ 35-015
  stable ~ 35-055
  thermal ~ 15-325
  thermodynamic ~ 15-320
equilibrium concentration 35-255
equilibrium diagram 35-480
equilibrium distribution coefficient 25-260
equilibrium freezing 25-015

EQUILIBRIUM

equilibrium phase 35-180
equilibrium solidification 25-015
equilibrium state 35-050
equilibrium system 35-045
equilibrium temperature 35-690
equilibrium vacancy 30-090
equivalent
carbon ~ 50-480
erbium 05-140
etch
grain-contrast ~ 80-150
grain-contrasting ~ 80-150
grain surface ~ 80-150
polish ~ 80-085
etch figures pl 80-200
etch pattern 80-200
etch pit 80-205
etchant 80-100
etching 80-120
cathodic ~ 80-180
cathodic vacuum ~ 80-180
chemical ~ 80-170
colour ~ 80-165
deep ~ 80-160
electrolytic ~ 80-175
gas ~ 80-182
grain-boundary ~ 80-145
ion bombardment ~ 80-180
macro- ~ 80-135
micro- ~ 80-140
thermal ~ 80-190
etching pattern 80-200
etching period 80-125
etching pit 80-205
etching reagent 80-100
etching time 80-125
europium 05-145
eutectic
binary ~ 40-205
cementite ~ 50-255
discontinuous ~ 40-235
divorced ~ 40-230
graphite ~ 50-260
lamellar ~ 40-220
phosphide ~ 50-235
pseudo- ~ 40-225
quaternary ~ 40-215
ternary ~ 40-210
eutectic alloy 55-350
eutectic carbides pl 50-440
eutectic cementite 50-075
eutectic change 35-565
eutectic graphite 50-175
eutectic horizontal 35-640
eutectic isothermal 35-640
eutectic line 35-640
eutectic melting 35-470
eutectic mixture 40-200
eutectic point 35-660
eutectic reaction 35-565
eutectic solidification 25-025
eutectic structure 40-060
eutectic temperature 35-605
eutectic transformation 35-565
eutectic trough 35-730
eutectic valley 35-730
eutectic white cast iron 50-505
eutectoid 40-240
eutectoidal decomposition 35-590
eutectoid alloy 55-365
eutectoidal reaction 35-590
eutectoidal transformation 35-590

eutectoid change 35-590
eutectoid decomposition 35-590
eutectoid horizontal 35-655
eutectoid isothermal 35-655
eutectoid mixture 40-240
eutectoid line 35-655
eutectoid point 35-675
eutectoid reaction 35-590
eutectoid steel 50-490
eutectoid temperature 35-630
eutectoid transformation 35-590
even fracture 40-700
examination
macrographic ~ 80-045
macroscopic ~ 80-045
microscopic ~ 80-040
metallographic ~ 80-035
X-ray ~ 80-570
excess cementite 50-095
excess enthalpy 15-145
excess entropy 15-220
excess free energy 15-115
excess function 15-065
excess quantity 15-065
excess thermodynamic function 15-065
excess thermodynamic quantity 15-065
excess vacancy 30-105
exogas 70-710
exogenous inclusions pl 40-520
exothermic atmosphere 70-710
exothermic gas 70-710
explosive hardening 45-055
extended dislocation 30-325
extensive property 15-040
extensive quantity 15-040
external electron 10-065
extra half plane 30-200
extra-hard steel 60-490
extra-soft steel 60-345
extrinsic semiconductor 10-385
extrinsic stacking fault 30-550
exudation
tin ~ 85-165
eyes pl
fish- ~ 85-240

F

face
crystal ~ 20-060
face-centred cubic lattice 20-310
face-centred lattice 20-290
factor
activity ~ 15-295
atomic size ~ 10-395
electrochemical ~ 10-400
failure 80-040
fast quench 70-260
fatigue 85-410
contact ~ 85-420
corrosion ~ 85-415
thermal ~ 85-425
fatigue corrosion 85-315
fatigue crescent 40-750
fatigue curve 80-425
fatigue fracture 40-745
fatigue limit 75-200
fatigue strength 75-200
fatigue test 80-420

fault
extrinsic stacking ~ 30-550
intrinsic stacking ~ 30-545
stacking ~ 30-540
twin ~ 30-555
Fermi energy 10-210
Fermi level 10-220
Fermi sphere 10-225
Fermi surface 10-215
fermium 05-150
fernico 65-520
ferrimagnetism 75-580
ferrite 50-045
delta ~ 50-050
free ~ 50-055
high-temperature ~ 50-050
hypoeutectoid ~ 50-055
pearlitic ~ 50-060
pro-eutectoid ~ 50-055
silicon ~ 50-475
ferrite band 85-230
ferrite former 50-455
ferrite-forming element 50-455
ferrite ghost 85-230
ferrite halo 50-230
ferrite stabilizer 50-455
ferritic cast iron 60-1040
ferritic malleable cast iron 60-1125
ferritic malleable iron 60-1125
ferritic stainless steel 60-520
ferritic steel 60-415
ferritizer 50-455
ferritizing annealing 70-645
ferro-alloys pl 60-1135
ferro-bronze 65-175
ferro-cerium 65-875
ferro-chromium 60-1145
ferromagnetic 75-550
ferromagnetic alloy 55-455
ferromagnetic material 75-550
ferromagnetic substance 75-550
ferromagnetism 75-545
ferro-manganese 60-1150
ferro-manganese-silicon 60-1155
ferro-silico-aluminium 60-1165
ferro-silicon 60-1140
ferro-steel 60-1035
ferrous alloy 60-070
fibre texture 40-565
fibrous fracture 40-720
fibrous texture 40-565
field-ion microscopy 80-380
figures pl
etch ~ 80-200
flow ~ 45-195
pole ~ 80-685
filamentary microcrystal 25-360
filled band 10-245
film
epitaxial ~ 25-350
thin ~ 80-350
fine acicular martensite 50-325
fined iron 60-050
fine gold 65-770
fine grain 40-370
fine-grained fracture 40-700
fine-grained steel 60-470
fine-grained structure 40-395
fine lamellar pearlite 50-135

fineness 55-100
grain ~ 40-380
fine pearlite 50-135, 50-385
finery-fire iron 60-050
finery iron 60-050
fine silver 65-760
fine structure 40-040
fire crack 85-060
fire-refined copper 65-025
first-degree transformation 25-465
first-degree transition 25-465
first-nearest neighbour 20-195
first-order transformation 25-465
first-order transition 25-465
fir-tree crystal 25-315
fish-eyes pl 85-240
fish-scale fracture 40-740
fissile metal 55-080
fissionable metal 55-080
flake graphite 50-185
flakes pl 85-240
flame annealing 70-505
flame hardening 70-290
flaw detection 80-790
flow
plastic ~ 45-215
viscous ~ 45-220
flow figures pl 45-195
flowing power 75-785
flow lines pl 45-195
flow stress 45-210
fluctuations pl
compositional ~ 25-100
composition ~ 25-100
concentration ~ 25-100
fluid
quenching ~ 70-190
fluidity 75-785
fluorescent particle inspection 80-835
fluorescent penetrant inspection 80-835
fluorescent X-ray spectroscopy 80-700
fluorine 05-155
fluoroscope test 80-835
flux 15-085
focused collision 30-505
focuson 30-505
foil
thin ~ 80-350
forbidden energy band 10-250
force
coercive ~ 75-635
cohesive ~ 10-315
driving ~ 15-070
Peierls ~ 30-300
Peierls-Nabarro ~ 30-300
thermal electromotive ~ 75-495
thermodynamic ~ 15-070
thermoelectromotive ~ 75-495
foreign nucleus 25-120
forest
dislocation ~ 30-405
forgeability 75-725
forged steel 60-330
forging alloy 55-530
form
allotropic ~ 25-390
high-temperature ~ 25-395
polymorphic ~ 25-390
former
austenite ~ 50-450

former
  carbide ~ 50-435
  ferrite ~ 50-455
  nitride ~ 50-470
former austenite grain 50-040
founding properties pl 75-775
foundry alloy 55-285
foundry pig iron 60-885
foundry type metal 55-615
four-component alloy 55-305
four-component system 35-135
four-fold rotation axis 20-080
fourth quantum number 10-145
fractography 80-385
fracture 85-040
  bright ~ 85-155
  brittle ~ 40-675
  cleavage ~ 85-085
  coarse-grained ~ 40-695
  conchoidal ~ 40-730
  crystalline ~ 40-685
  cyclic ~ 40-745
  delayed ~ 40-755
  ductile ~ 40-680
  even ~ 40-700
  fatigue ~ 40-745
  fibrous ~ 40-720
  fine-grained ~ 40-700
  fish-scale ~ 40-740
  glassy ~ 40-735
  granular ~ 40-690
  intercrystalline ~ 40-705
  intergranular ~ 40-705
  intracrystalline ~ 40-710
  laminated ~ 40-725
  parting ~ 40-655
  shear ~ 40-660
  shearing ~ 40-660
  silky ~ 40-715
  smooth ~ 40-700
  woody ~ 40-720
  transcrystalline ~ 40-710
  transgranular ~ 40-710
  vitreous ~ 40-735
  white ~ 85-160
fracture stress 75-130
fracture surface 40-670
fracture test 80-445
fragility
  hydrogen ~ 85-390
francium 05-160
Frank dislocation 30-335
Frank partial dislocation 30-335
Frank-Read dislocation generator 30-430
Frank-Read source 30-430
Frank sessile dislocation 30-335
free carbon 50-005
free cementite 50-095
free cutting alloy 55-560
free-cutting brass 65-370
free-cutting bronze 65-370
free-cutting steel 60-710
free electron 10-055
free energy 15-105
free energy of formation 15-135
free energy of mixing 15-130
free enthalpy 15-140
free ferrite 50-055
free-machining steel 60-710
freezing 25-005
  equilibrium ~ 25-015
  non-equilibrium ~ 25-020
freezing point 75-425
freezing-point curve 35-515

freezing point surface 35-520
freezing range 35-550
freezing schrinkage 75-805
freezing temperature 75-425
Frenkel defect 30-060
Frenkel disorder 30-060
frenkel vacancy 30-065
fretting corrosion 85-325
friction
  internal ~ 75-210
fringe crystals pl 40-280
front
  diffusion ~ 45-410
  growth ~ 45-480
  solidification ~ 25-295
fugacity 15-290
fugitiveness 15-290
full annealing 70-465
full hardening 70-280
fully annealed steel 60-275
fully-columnar crystallization 25-070
function
  Bloch ~ 10-195
  electronic work ~ 10-060
  excess ~ 15-065
  excess thermodynamic ~ 15-065
  Gibbs ~ 15-140
  integral thermodynamic ~ 15-055
  partial thermodynamic ~ 15-060
  state ~ 15-035
  thermodynamic ~ 15-030
  wave ~ 10-180
  work ~ 10-060, 15-105
fundamental lines pl 80-650
furnace cooling 70-110
fused salt quenching 70-230
fusibility 75-790
fusible alloy 55-395
fusing point 75-420
fusion 35-445
fusion point 75-420
fusion range 35-545

G

gadolinium 05-165
gallium 05-170
galvanic corrosion 85-270
galvanic series pl 75-710
gamma-producing element 50-450
gamma-ray inspection 80-825
gamma-ray radiography 80-825
gap
  electron energy ~ 10-250
  energy ~ 10-250
  miscibility ~ 35-345
  solubility ~ 35-345
gas
  carburizing ~ 70-745
  carrier ~ 70-685
  electron ~ 10-045
  endothermic ~ 70-705
  exothermic ~ 70-710
gas carburizing 70-730
gas corrosion 85-295
gas cyaniding 70-860
gaseous solution 35-370
gas etching 80-182
gas mixture 35-370
gas quenching 70-220
general-purpose steel 60-665

GROWTH

generator
 dislocation ~ 30-425
 Frank-Read dislocation ~ 30-430
germanium 05-175
German silver 65-415
ghost 85-230
 ferrite ~ 85-230
ghost band 85-230
ghost line 85-230
Gibbs-Duhem equation 15-305
Gibbs free energy 15-140
Gibbs function 15-140
Gibbs triangle 35-720
glassy fracture 40-735
glide 45-095
 easy ~ 45-155
 multiple ~ 45-140
 pencil ~ 45-175
 simple ~ 45-130
glide band 45-190
glide dislocation 30-265
glide plane 45-100
gliding
 pencil ~ 45-175
 prismatic ~ 45-170
gliding plane 45-100
glissile dislocation 30-265
globular cementite 50-100
globular pearlite 50-150
globular structure 40-120
glow discharge nitriding 70-800
glow nitriding 70-800
gold 05-180
 coin ~ 65-775
 dental ~ 65-785
 fine ~ 65-770
 jewellery ~ 65-780
 mock ~ 65-815
 native ~ 65-765
 red ~ 65-800
 standard ~ 65-790
 sterling ~ 65-790
 white ~ 65-805
Goss texture 40-610
Government bronze 65-205
grade
 ~ of purity 55-095
 ~ of steel 60-085
gradient
 concentration ~ 15-080
 temperature ~ 15-075
 thermal ~ 15-075
graduated hardening 70-305
grain 40-360
 coarse ~ 40-365
 crystal ~ 40-360
 fine ~ 40-370
 former austenite ~ 50-040
 prior austenite ~ 50-040
 prior austenitic ~ 50-040
grain boundary 30-565
grain boundary corrosion 85-360
grain boundary diffusion 45-310
grain boundary dislocation 30-580
grain boundary energy 30-575
grain boundary etching 80-145
grain-boundary migration 45-490
grain boundary precipitate 40-310
grain boundary segregation 25-450
grain boundary sliding 45-160
grain boundary slip 45-160
grain coarseness 40-375

grain coarsening 25-290
grain-contrast etch 80-150
grain-contrasting etch 80-150
grainedness
 coarse ~ 40-375
grain fineness 40-380
grain growth 25-290
grain growth annealing 70-480
grain growth inhibitor 40-400
grain refinement 40-385
grain refiner 40-400
grain refining 40-385
grains pl
 equiaxed ~ 40-285
grain size 80-285
grain-size index 80-290
grain-size number 80-290
grain surface etch 80-150
granular fracture 40-690
granular pearlite 50-150
granular structure 40-120
graphite 50-165
 eutectic ~ 50-175
 flake ~ 50-185
 interdendritic ~ 50-205
 nodular ~ 50-210
 point ~ 50-195
 primary ~ 50-170
 rosette ~ 50-200
 secondary ~ 50-180
 spheroidal ~ 50-190
 spherolitic ~ 50-190
 spherulitic ~ 50-190
 temper ~ 50-220
graphite bronze 65-220
graphite eutectic 50-260
graphite-forming element 50-460
graphite spherule 50-215
graphitic carbon 50-005
graphitic steel 60-300
graphitization 50-225
graphitized steel 60-300
graphitizer 50-460
graphitizing 70-630
grating constant 20-225
gravitational segregation 25-425
gravity
 specific ~ 75-350
gravity segregation 25-425
grey cast iron 60-950
grey pig iron 60-915
grey spots pl 85-140
grey tin 65-635
Griffith crack 85-055
groove
 thermal ~ 80-195
grooving 80-190
 thermal ~ 80-190
gross properties pl 75-850
groundmass 40-015
growing
 crystal ~ 25-085
 single crystal ~ 25-085
growing single crystals pl 25-085
growth
 crystal ~ 25-285
 dendritic ~ 25-310
 epitaxial ~ 25-345
 grain ~ 25-290
growth front 45-480

GROWTH

growth of cast iron 70-640
growth rate 25-275
growth spiral 25-335
growth step 25-330
growth twin 45-240
Guinier-Preston zones *pl* 70-605
gun iron 60-1035
gun metal 65-205

H

habit
  crystal ~ 20-460
  ~ of a crystal 20-460
habit plane 20-465
Hadfield steel 60-785
Hadfield's manganese steel 60-785
haematite pig iron 60-895
hafnium 05-185
Hägg's rule 35-260
hairline crack 85-045
half life 75-525
half-life time 75-525
Hall effect 10-290
hamatite pig iron 60-895
hammered steel 60-330
halo
  ferrite ~ 50-230
hard alloy 55-415
hard lead 65-660
hard magnetic alloy 55-465
hard magnetic steel 60-505
hard solder 65-720
hard spots *pl* 85-145
hard steel 60-350
hard-surfacing alloy 55-590
hard zinc 65-700
hardenable steel 60-360
hardenability 75-730
hardenability band 80-460
hardenability curve 80-455
hardenability line 80-455
hardened case 70-355
hardened steel 60-285
hardened zone 70-355
hardener 55-255
hardener phase 40-350
hardening 70-015
hardening
  age ~ 70-565
  air ~ 70-215
  alloy ~ 55-210
  blank ~ 70-775
  bright ~ 70-330
  brine ~ 70-238
  bulk ~ 70-287
  case ~ 70-670
  chill ~ 70-255
  clean ~ 70-330
  contact ~ 70-250
  cyanide ~ 70-855
  cyanide case ~ 70-855
  differential ~ 70-275
  direct ~ 70-760
  dispersed phase ~ 55-220
  dispersion ~ 55-220
  explosive ~ 45-055
  flame ~ 70-290
  full ~ 70-280
  graduated ~ 70-305
  ~ by quenching 70-175

hardening
  high frequency ~ 70-295
  induction ~ 70-295
  internal work ~ 45-080
  interrupted ~ 70-300
  irradiation ~ 30-495
  latent ~ 45-180
  lead bath ~ 70-235
  local ~ 70-275
  nitrogen ~ 70-785
  nitrogen case ~ 70-785
  non-oxydizing ~ 70-330
  oil ~ 70-210
  precipitation ~ 70-560
  progressive ~ 70-320
  progressive spin ~ 70-325
  quench ~ 70-175
  radiation ~ 30-495
  salt bath ~ 70-230
  scanning ~ 70-320
  secondary ~ 70-405
  selective ~ 70-275
  self ~ 70-215
  shallow ~ 70-285
  skin ~ 70-285
  solution ~ 55-215
  spin ~ 70-315
  spray ~ 70-240
  strain ~ 45-045
  surface ~ 70-285
  through ~ 70-280
  torch ~ 70-290
  water ~ 70-205
  work ~ 45-050
hardening alloy element 55-255
hardening bath 70-195
hardening by quenching 70-175
hardening capacity 70-020
hardening constituent 55-255
hardening crack 85-190
hardening depth 70-350
hardening liquid 70-190
hardening medium 70-185
hardening phase 40-350
hardening power 70-020
hardening temperature 70-140
hardening temperature range 70-145
hardenite 50-365
hardness 75-245
  hot ~ 75-255
  micro- ~ 75-250
  red ~ 75-260
hardness penetration depth 70-350
hardness test 80-415
Hartmann lines *pl* 45-195
hastelloy 65-455
H.C.-copper 65-065
heart 70-665
heat
  atomic ~ 75-370
  configurational specific ~ 25-635
  critical ~ 75-400
  electronic specific ~ 10-275
  ~ of fusion latent 75-380
  ~ of melting latent 75-380
  ~ of solidification 75-385
  ~ of solution 75-405
  ~ of sublimation 75-395
  ~ of transformation 75-400
  ~ of transition 75-400
  molecular ~ 75-375
  specific ~ 75-415
heat capacity 75-410
heat conductivity 75-440
heat crack 85-060
heating 70-030
  pulse ~ 70-050
heating curve 80-520

HOT

heating diagram 80-520
heating rate 70-040
heating through 70-035
heating throughout 70-035
heat refining 70-475
heat resistance 75-690
heat-resistant alloy 55-505
heat-resistant steel 60-560
heat-resisting alloy 55-505
heat-resisting cast iron 60-1085
heat-resisting steel 60-560
heat stability 75-835
heat tint 70-385
heat tinting 80-185
heat-treatable alloy 55-525
heat treatable steel 60-705
heat treatment 70-005
heat treatment cycle 70-970
heavy metals pl 55-055
heavy plantinum metals pl 65-825
Heisenberg uncertainty principle 10-165
helical dislocation 30-355
helicoidal dislocation 30-355
helium 05-190
Helmholtz free energy 15-105
Henry's law 15-275
heredity
cast iron ~ 60-985
Herring-Nabarro creep 45-520
Herring-Nabarro diffusion creep 45-520
heterodiffusion 45-335
heterogeneity 40-410
heterogeneous alloy 55-340
heterogeneous equilibrium 35-015
heterogeneous nucleation 25-185
heterogeneous nucleus 25-115
heterogeneous structure 40-155
heterogeneous system 35-095
heterogeneous transformation 25-535
heteropolar bond 10-325
heterothermal melting 35-455
Heusler alloy 65-420
Heusler's alloy 65-420
hexad axis 20-085
hexagonal close-packed lattice 20-350
hexagonal indices pl 20-170
hexagonal lattice 20-340
hexagonal system 20-245
hiduminium 65-605
high alloy 55-325
high-alloy cast iron 60-1005
high alloy steel 60-130
high-angle boundary 30-585
high-carat gold alloy 65-795
high-carbon pig iron 60-925
high-carbon steel 60-110
high-chromium steel 60-180
high coercive alloy 55-465
high-conductivity alloy 55-550
high-conductivity bronze 65-255
high conductivity copper 65-065
high-density metals pl 55-055
high-duty cast iron 60-1065
high frequency hardening 70-295
high-grade cast iron 60-1065
high grade steel 60-145

high-hardenability steel 60-385
highly-alloy steel 60-130
highly alloyed cast iron 60-1005
highly-alloyed steel 60-130
high melting-point alloy 55-400
high melting-point metal 55-065
high permeability alloy 55-475
high-pressure phase 35-165
high-quality steel 60-145
high-resistance alloy 55-545
high-speed steel 60-800
high-strength alloy 55-420
high-strength steel 60-355
high-temperature allotrope 25-395
high-temperature alloy 55-510
high-temperature annealing 70-480
high-temperature creep resistance 75-240
high-temperature creep strength 75-240
high-temperature ferrite 50-050
high-temperature form 25-395
high-temperature metallography 80-015
high-temperature phase 35-160
high-temperature strength 75-095, 75-240
high-temperature tempering 70-375
high-temperature thermo-mechanical
  treatment 70-955
high temperature tool steel 60-810
high-tensile brass 65-355
high-tensile steel 60-355
high-test cast iron 60-1065
high-tin bronze 65-115
Hildebrand's equation 15-310
hipernik 65-475
holding 70-055
holding time 70-060
hole
blow ~ 85-135
holmium 05-195
homogeneity 40-405
homogeneity range 35-680
homogeneous alloy 55-330
homogeneous nucleation 25-180
homogeneous nucleus 25-110
homogeneous region 35-680
homogeneous structure 40-145
homogeneous system 35-090
homogeneous transformation 25-530
homogenizing 70-485
homologous temperature 45-475
homopolar bond 10-330
horizontal
  eutectic ~ 34-640
  eutectoid ~ 35-655
  monotectic ~ 35-650
  peritectic ~ 35-645
horizontal section 35-735
hot brittleness 75-290
hot cracking 85-080
hot hardness 75-255
hot quenching 70-225
hot plastic working 45-020
hot shortness 75-290
hot-stage microscopy 80-240
hot strength 75-095
hot-work die steel 60-810
hot working 45-020
hot-working die steel 60-810

— 283 —

HOT

hot-working steel 60-810
hot-work tool steel 60-810
Hume-Rothery phase 35-280
Hume-Rothery rule 35-325
hydride 40-495
hydrogen 05-200
hydrogen brittleness 85-390
hydrogen disease 85-395
hydrogen embrittlement 85-390
hydrogen fragility 85-390
hydrogen unsoundness of copper 85-395
hydronalium 65-600
hypereutectic alloy 55-360
hyper-pure metal 55-105
hypereutectic cementite 50-070
hypereutectic structure 40-065
hypereutectic white cast iron 50-515
hypereutectoid alloy 55-375
hypereutectoid cementite 50-080
hypereutectoid steel 50-500
hypereutectoid structure 40-080
hypoeutectic alloy 55-355
hypoeutectic structure 40-070
hypoeutectic white cast iron 50-510
hypoeutectoid alloy 55-370
hypoeutectoid ferrite 50-055
hypoeutectoid steel 50-495
hypoeutectoid structure 40-085
hysteresis
  magnetic $\sim$ 75-640
  temperature $\sim$ 25-480
  thermal $\sim$ 25-480

I

ideal solution 15-240
identity period 20-115
idiomorphic crystal 25-355
ignition alloy 55-595
imitation gold alloy 65-815
immersion quenching 70-245
immiscibility 35-340
impact strength 75-190
impact test 80-430
imperfect crystal 30-010
imperfect dislocation 30-310
imperfection
  crystal $\sim$ 30-015
  crystallographic $\sim$ 30-015
  point $\sim$ 30-030
  surface $\sim$ 85-010
implantation
  ion $\sim$ 70-805
impregnation
  aluminium $\sim$ 70-925
  chromium $\sim$ 70-920
  metal $\sim$ 70-915
  silicon $\sim$ 70-910
  tungsten $\sim$ 70-935
  zinc $\sim$ 70-930
impurities pl 55-205
impurity
  interstitial $\sim$ 30-080
impurity atmosphere 30-440
impurity cloud 30-440
impurity semiconductor 10-385
incipient melting 85-215
inclusions pl
  exogenous $\sim$ 40-520
  indigenous $\sim$ 40-525

inclusions
  metallic $\sim$ 40-550
  nitride $\sim$ 40-545
  non-metallic $\sim$ 40-515
  oxide $\sim$ 40-535
  slag $\sim$ 40-540
  sulfide $\sim$ 40-530
incoherent boundary 30-630
incoherent interface 30-630
incoherent precipitate 40-325
inconel 65-445
incongruent intermediate phase 35-225
incongruently melting compound 35-225
incongruently-melting intermediate phase 35-225
incongruent melting 35-465
increasing-rate creep 45-510
incubation period 70-150
incubation time 70-150
index
  crystallographic $\sim$ 20-160
  grain-size $\sim$ 80-290
  $\sim$ of segregation 40-420
  strain-hardening $\sim$ 45-065
indexing 80-695
index number 80-290
indices pl
  hexagonal $\sim$ 20-170
  Miller-Bravais $\sim$ 20-170
  Miller $\sim$ 20-165
indigenous inclusions pl 40-525
indium 05-205
induction
  magnetic $\sim$ 75-610
induction hardening 70-295
industrial alloy 55-520
industrial iron 60-015
inert atmosphere 70-675
infiltration alloy 55-670
ingot iron 60-060
ingot steel 60-580
inhibitor
  grain growth $\sim$ 40-400
inhomogeneity 40-410
  chemical $\sim$ 40-415
initial creep 45-500
initial permeability 75-620
inner electron 10-070
inner-shell electron 10-070
inoculant 25-230
inoculated cast iron 60-970
inoculation 25-220
  $\sim$ of crystallization 25-215
inspection
  gamma-ray $\sim$ 80-825
  fluorescent particle $\sim$ 80-835
  fluorescent penetrant $\sim$ 80-835
  magnetic-particle $\sim$ 80-805
  non-destructive $\sim$ 80-785
  radiographic $\sim$ 80-705
  X-ray radiographic $\sim$ 80-820
integral quantity 15-055
integral thermodynamic function 15-055
integral thermodynamic quantity 15-055
intensity
  $\sim$ of magnetization 75-600
  quenching $\sim$ 70-200
  saturation magnetic $\sim$ 75-605
intensive property 15-045
intensive quantity 15-045
interaction coefficient 15-295

IRON

interaction energy 10-295
interaction parameter 15-300
interaction potential 10-300
interatomic distance 20-230
interatomic spacing 20-230
intercept ratio 20-355
intercrystalline corrosion 85-360
intercrystalline fracture 40-705
intercrystalline segregation 25-450
interdendritic graphite 50-205
interdendritic segregation 25-445
interdendritic spaces pl 40-355
interdiffusion 45-380
interface 30-610
  coherent ~ 30-625
  incoherent ~ 30-630
  interphase ~ 30-610
interface energy 30-635
interfacial energy 30-635
interfacial free energy 15-125
interfacial tension 30-645
interference microscopy 80-295
interferogram 80-300
intergranular corrosion 85-360
intergranular fracture 40-705
interlamellar spacing 40-110
intermediate phase 35-210
intermetallic compound 35-215
intermetallic phase 35-210
internal chill 85-150
internal energy 15-095
internal friction 75-210
internal oxidation 70-947
internal stresses pl 85-020
internal structure 40-035
internal work hardening 45-080
interparticle spacing 40-425
interphase boundary 30-610
interphase interface 30-610
interplanar distance 20-225
interplanar spacing 20-225
interrupted hardening 70-300
interrupted quenching 70-300
intersolubility 35-300
  mutual ~ 35-300
interstice 20-105
  octahedral ~ 20-330
  tetrahedral ~ 20-325
intersticialcy 30-075
interstitial
  dumbbell ~ 30-085
  split ~ 30-085
interstitial atom 30-075
interstitial diffusion 45-315
interstitial impurity 30-080
interstitial position 20-145
interstitial site 20-145
interstitial solid solution 35-420
interstitial void 20-105
interval
  crystallization ~ 35-550
intracrystalline fracture 40-710
intracrystalline slip 45-165
intrinsic diffusion coefficient 45-390
intrinsic diffusivity 45-390
intrinsic energy 15-095
intrinsic semiconductor 10-380

intrinsic stacking fault 30-545
invar 65-500
invariant system 35-075
inverse chill 85-150
inverse segregation 25-440
investigation
  calorimetric ~ 80-500
iodin (USA) 05-210
iodine 05-210
ion bombardment etching 80-180
ionic bond 10-325
ionic compound 35-230
ionic crystal 20-025
ionic radius 10-090
ion implantation 70-805
ionitriding 70-800
ionization energy 10-100
ionization potential 10-105
ionizing potential 10-105
ion nitriding 70-800
iridium 05-215
iridosmine 65-845
iridosmium 65-845
iron 05-220
  acicular cast ~ 60-1060
  acicular grey cast ~ 60-1060
  acid-resisting cast ~ 60-1075
  alkali-resisting cast ~ 60-1080
  alloy cast ~ 60-995
  alloy pig ~ 60-935
  aluminium cast ~ 60-1020
  Armco ~ 60-010
  armoured cast ~ 60-1070
  austenitic cast ~ 60-1050
  ball ~ 60-050
  bearing cast ~ 60-1105
  Bessemer pig ~ 60-845
  black heart malleable cast ~ 60-1115
  blast furnace pig ~ 60-860
  bloomary ~ 60-045
  bloomery ~ 60-045
  carbonyl ~ 60-020
  cast ~ 60-945
  charcoal ~ 60-050
  charcoal-hearth ~ 60-050
  charcoal pig ~ 60-875
  chilled cast ~ 60-980
  chromium cast ~ 60-1015
  coke pig ~ 60-870
  commercial ~ 60-015
  commercially pure ~ 60-015
  ductile cast ~ 60-975
  electric furnace pig ~ 60-865
  electrolytic ~ 60-025
  eutectic white cast ~ 50-505
  ferritic cast ~ 60-1040
  ferritic malleable ~ 60-1125
  ferritic malleable cast ~ 60-1125
  fined ~ 60-050
  finery ~ 60-050
  finery-fire ~ 60-050
  foundry pig ~ 60-885
  grey cast ~ 60-950
  grey pig ~ 60-915
  gun ~ 60-1035
  haematite pig ~ 60-895
  heat-resisting cast ~ 60-1085
  hematite pig ~ 60-895
  high-alloy cast ~ 60-1005
  high-carbon pig ~ 60-925
  high-duty cast ~ 60-1065
  high-grade cast ~ 60-1065
  highly alloyed cast ~ 60-1005
  high-test cast ~ 60-1065
  hypereutectic white cast ~ 50-515
  hypoeutectic white cast ~ 50-510
  industrial ~ 60-015
  ingot ~ 60-060
  inoculated cast ~ 60-970
  low-alloy cast ~ 60-1000

IRON

iron
  malleable cast ~ 60-1110
  malleable ~ 60-1110
  malleable pig ~ 60-905
  manganese cast ~ 60-1030
  martensitic cast ~ 60-1055
  medium-phosphoric pig ~ 60-900
  meteoric ~ 60-005
  mottled cast ~ 60-960
  mottled pig ~ 60-920
  nickel cast ~ 60-1025
  nodular cast ~ 60-975
  open-hearth pig ~ 60-855
  ordinary cast ~ 60-965
  plain cast ~ 60-990
  pearlitic cast ~ 60-1045
  pearlitic malleable ~ 60-1130
  pearlitic malleable cast ~ 60-1130
  phosphoric pig ~ 60-890
  pig ~ 60-840
  puddled ~ 60-055
  rustless ~ 60-520
  silicon ~ 60-210
  silicon cast ~ 60-1010
  sintered ~ 60-065
  specular pig ~ 60-930
  spheroidal cast ~ 60-975
  spheroidal graphite cast ~ 60-975
  sponge ~ 60-035
  spongy ~ 60-035
  Swedish ~ 60-030
  stainless ~ 60-520
  steeled cast ~ 60-1035
  steelmaking pig ~ 60-880
  synthetic pig ~ 60-940
  Thomas pig ~ 60-850
  white cast ~ 60-955
  white heart malleable cast ~ 60-1120
  white pig ~ 60-910
  wrought ~ 60-040
iron-base alloy 60-070
iron bronze 65-175
iron carbide 40-440
iron-carbon diagram 50-020
iron-carbon phase diagram 50-020
iron-carbon system 50-015
iron nitride 70-815
iron phosphide 50-250
iron sponge 60-035
iron sulfide 50-245
irradiation damage 30-490
irradiation hardening 30-495
irradiation-induced defect 30-485
irradiation-produced defect 30-485
isoactivity curve 15-315
isoactivity line 15-315
isochronal annealing 70-495
isochronous annealing 70-495
isometric system 20-240
isomorphism 20-475
isomorphous crystals pl 20-480
isopleth section 35-740
isothermal
  eutectic ~ 35-640
  eutectoid ~ 35-655
  monotectic ~ 35-650
  peritectic ~ 35-645
isothermal annealing 70-450
isothermal diffusion 45-350
isothermal melting 35-450
isothermal section 35-735
isothermal time-temperature-transformation diagram 70-155
isothermal transformation 25-515
isothermal transformation diagram 70-155

isothermal TTT diagram 70-155
isotropy 20-450
  quasi- ~ 20-455

J

jewellery alloy 55-580
jewellery gold 65-780
jog 30-240
  dislocation ~ 30-240
Jominy end quench test 80-450
Jominy test 80-450

K

kanthal 65-525
kanthal alloy 65-525
kernel
  dislocation ~ 30-205
killed steel 60-655
kink 30-245
Kirkendall effect 45-385
kish 50-170
Koehler source 30-435
Konovalov's rule 35-705
kovar 65-520
krypton 05-225
k space 10-205

L

lag
  thermal ~ 25-480
lake copper 65-025
lamella
  cementite ~ 50-120
lamellae
  Neumann ~ 45-205
lamellar eutectic 40-220
lamellar microstructure 40-105
lamellar pearlite 50-130
lamellar spacing 40-110
lamellar structure 40-105
laminated fracture 40-725
laminated pearlite 50-130
lamination 85-405
lanthanum 05-230
lap 85-400
lapping 85-400
large angle boundary 30-585
latent energy 15-100
latent hardening 45-180
latent heat of fusion 75-380
latent heat of melting 75-380
latent heat of vaporization 75-390
lath martensite 50-355
lattice
  base-centred orthorhombic ~ 20-385
  body-centred ~ 20-295
  body-centred cubic ~ 20-315
  Bravais ~ 20-275
  Bravais space ~ 20-275
  close-packed hexagonal ~ 20-350
  crystal ~ 20-045
  crystalline ~ 20-045
  cubic ~ 20-300
  defect ~ 30-020
  diamond cubic ~ 20-320
  face-centred ~ 20-290

lattice
  face-centred cubic ~ 20-310
  hexagonal ~ 20-340
  hexagonal close-packed ~ 30-350
  monoclinic ~ 20-395
  orthorhombic ~ 20-380
  point ~ 20-275
  reciprocal ~ 20-410
  rhombohedral ~ 20-390
  simple ~ 20-280
  simple cubic ~ 20-305
  simple hexagonal ~ 20-345
  space ~ 20-045
  tetrahedral cubic ~ 20-320
  tetragonal ~ 20-370
  transition ~ 20-405
  translation ~ 20-275
  triclinic ~ 20-400
  trigonal ~ 20-390
lattice base 20-110
lattice constant 20-120
lattice defect 30-015
lattice diffusion 45-295
lattice disregistry 30-615
lattice distortion 30-025
lattice misfit 30-615
lattice parameter 20-120
lattice plane 20-155
lattice point 20-100
lattice position 20-140
lattice site 20-140
lattice spacing 20-120
lattice vektor 20-135
Laue diagram 80-620
Laue method 80-590
Laue pattern 80-620
Laue photogram 80-620
Laue photograph 80-620
Laue picture 80-620
lantal 65-595
Laves compounds pl 35-265
Laves phases pl 35-265
law
  Henry's ~ 15-275
  lever ~ 35-700
  Raoult's ~ 15-270
  Vegard's ~ 35-425
lawrencium 05-235
layer
  Beilby ~ 80-090
  epitaxial ~ 25-350
lead 05-240
  antimonial ~ 65-660
  crude ~ 65-650
  hard ~ 65-660
  refined ~ 65-655
lead bath hardening 70-235
lead bath quenching 70-235
lead-bearing steel 60-250
lead bronze 65-140
lead solder 65-735
ledeburite 50-160
ledeburite carbides pl 50-440
ledeburitic steel 60-440
leaded brass 65-360
leaded bronze 65-185
leaded Muntz metal 65-360
leaded tin-bronze 65-185
ledloy 60-250
left-handed screw dislocation 30-175
length
  axial ~ 20-115
  carbon ~ 70-750

length
  energy ~ 10-150
  Fermi ~ 10-220
letting down 70-365
lever law 35-700
lever rule 35-700
life
  half ~ 75-525
light alloy 55-405
light metal alloy 55-405
light metals pl 55-050
light microscopy 80-215
light platinum metals pl 65-830
limit
  creep ~ 75-220
  fatigue ~ 75-200
  elastic ~ 75-105
  endurance ~ 75-200
  proportional ~ 75-100
  proportionality ~ 75-100
  solubility ~ 35-330
limited solubility 35-320
limiting creep stress 75-230
limit of proportionality 75-100
limit of solubility 35-330
line
  dislocation ~ 30-195
  eutectic ~ 35-640
  eutectoid ~ 35-655
  ghost ~ 85-230
  hardenability ~ 80-455
  isoactivity ~ 15-315
  monotectic ~ 35-650
  peritectic ~ 35-645
  phase boundary ~ 35-535
  slip ~ 45-185
  solvus ~ 35-530
  spinodal ~ 15-190
  tie ~ 35-695
  X-ray diffraction ~ 80-645
lineage structure 40-165
linear coefficient of thermal expansion 75-455
line defect 30-140
lines pl
  arrest ~ 40-750
  conjugate ~ 35-540
  flow ~ 45-195
  fundamental ~ 80-650
  Hartmann ~ 45-195
  Lüders ~ 45-195
  Neumann ~ 45-205
  superstructure ~ 80-655
linotype metal 55-625
Lipowitz alloy 65-680
Lipowitz's metal 65-680
liquid
  hardening ~ 70-190
  quenching ~ 70-190
  remaining ~ 25-280
liquid carburizing 70-725
liquid cyaniding 70-865
liquidity 75-785
liquid miscibility 35-305
liquid quenching 70-245
liquid shrinkage 75-800
liquid solubility 35-305
liquid solution 35-365
liquidus 35-510
liquidus curve 35-515
liquidus surface 35-520
liquidus temperature 35-525
liquisol quenching 70-085

LIQUOR

liquor
 mother ~ 25-095
lithium 05-245
local hardening 70-275
localized corrosion 85-340
lock
 Cottrell-Lomer ~ 30-450
locked-in stresses pl 85-020
locked-up stresses pl 85-020
locking
 dislocation ~ 30-445
longitudinal modulus of elasticity 75-150
long-range order 25-600
long-range order parameter 25-620
long-range segregation 25-415
loop
 dislocation ~ 30-340
 prismatic ~ 30-345
 prismatic dislocation ~ 30-345
low alloy 55-320
low-alloy iron 60-1000
low-alloy steel 60-120
low-angle boundary 30-590
low-angle grain boundary 30-590
low-carbon martensite 50-340
low-carbon steel 60-100
lower bainite 50-415
low expansion alloy 55-440
low-hardenability steel 60-380
low melting alloy 55-395
low-melting-point alloy 55-395
low-melting-point metal 55-060
low-temperature brittleness 75-295
low-temperature steel 60-555
low-temperature tempering 70-370
low-tin bronze 65-120
Lüders lines pl 45-195
lustre
 metallic ~ 75-510
lutetium 05-250

M

machinability 75-745
machine bronze 65-240
machinery brass 65-375
machinery steel 60-680
machine steel 60-680
macro-etching 80-135
macroexamination 80-045
macrograph 80-305
macrographic examination 80-045
macrography 80-045
macroscopic examination 80-045
macroscopic properties pl 75-850
macroscopic residual stresses pl 85-030
macrosegregation 25-415
macrostresses pl 85-030
macrostructure 40-025
magnafluxing 80-805
magnalium 65-620
magnesium 05-255
magnetic after-effect 75-645
magnetic after effect 75-645
magnetic ageing 70-620
magnetic alloy 55-445

magnetically hard alloy 55-465
magnetically hard steel 60-505
magnetically soft alloy 55-460
magnetically soft steel 60-500
magnetic analysis 80-560
magnetic anisotropy 75-575
magnetic anneal 70-965
magnetic annealing 70-965
magnetic change point 25-560
magnetic crack detection 80-805
magnetic domain 40-635
magnetic flux density 75-610
magnetic hysteresis 75-640
magnetic induction 75-610
magnetic inductive capacity 75-615
magnetic ordering 25-640
magnetic-particle inspection 80-805
magnetic permeability 75-615
magnetic properties pl 75-540
magnetic quantum number 10-140
magnetic steel 60-495
magnetic structure 40-650
magnetic sublattice 20-485
magnetic susceptibility 75-625
magnetic transformation 25-555
magnetic transformation point 25-560
magnetism
 residual ~ 75-630
magnetization 75-600
 saturation ~ 75-605
magnetoelasticity 75-650
magnetostriction 75-650
magnification 80-282
major alloying element 55-195
major constituent 55-185
major segregation 25-415
malleability 75-025, 75-725
malleable brass 65-295
malleable cast iron 60-1110
malleable iron 60-1110
malleable pig iron 60-905
malleablizing 70-625
manganese 05-260
manganese brass 65-355
manganese bronze 65-155, 65-355
manganese cast iron 60-1030
manganese copper 65-155
manganese-silicon steel 60-205
manganese steel 60-195
manganese sulfide 50-240
manganin 65-405
maraging 70-615
maraging steel 60-460
marine corrosion 85-290
markings pl
 clamshell ~ 40-750
 conchoidal ~ 40-750
marks pl
 beach ~ 40-750
 clamshell ~ 40-750
 conchoidal ~ 40-750
marquenching 70-305
martempering 70-305
martensite 50-310
 acicular ~ 50-315
 alpha ~ 50-335
 beta ~ 50-340
 coarse grained ~ 50-320

METALLIC

martensite
cryptocrystalline ~ 50-330
deformation ~ 50-360
fine acicular ~ 50-325
low-carbon ~ 50-340
massive ~ 50-355
secondary ~ 50-350
strain induced ~ 50-360
tempered ~ 50-345
tetragonal ~ 50-335
troostite ~ 50-410
martensite finish temperature 70-170
martensite needle 50-375
martensite plate 50-370
martensite range 50-300
martensite start temperature 70-165
martensite transformation 50-280
martensitic cast iron 60-1055
martensitic change 50-280
martensitic steel 60-430
martensitic transformation 50-280
mass
atomic ~ 10-075
molecular ~ 75-355
relative atomic ~ 10-075
mass concentration 55-235
massive martensite 50-355
massive transformation 25-550
mass number 75-335
mass spectrography 80-480
master alloy 55-285
material
composite ~ 55-680
diamagnetic ~ 75-560
ferromagnetic ~ 75-550
paramagnetic ~ 75-555
material defect 85-005
matrix 40-015
Matthiessen's rule 75-475
mazak 65-705
mechanical mixture of phases 40-190
mechanical polishing 80-070
mechanical properties *pl* 75-005
mechanical strength 75-040
mechanical testing 80-400
mechanical twin 45-235
mechanical working 45-015
medal bronze 65-260
medal metal 65-260
medium
hardening ~ 70-185
quenching ~ 70-185
medium-alloy steel 60-125
medium-carbon steel 60-105
medium-phosphoric pig iron 60-900
melt 25-095, 55-150
melting 35-445
congruent ~ 35-460
eutectic ~ 35-470
heterothermal ~ 35-455
incipient ~ 85-215
incongruent ~ 35-465
isothermal ~ 35-450
peritectic ~ 35-475
melting point 75-420
melting-point curve 35-490
melting range 35-545
mendelevium 05-265
mercurous nitrate test 80-485
mercury 05-270
metal 10-355, 10-360
Admiralty ~ 65-390

metal
Admiralty gun ~ 65-204
anatomical ~ 55-645
antifriction ~ 55-555
Babbitt ~ 65-645
Bahn ~ 65-685
base ~ 55-005
basis ~ 55-185
bearing ~ 55-555
bell ~ 65-250
coin ~ 55-565
coinage ~ 55-565
common ~ 55-005
compact ~ 55-155
duplex ~ 65-880
electrodeposited ~ 55-130
electrolytic ~ 55-130
electrotype ~ 55-635
evaporated ~ 55-145
fissile ~ 55-080
fissionable ~ 55-080
foundry type ~ 55-615
high melting-point ~ 55-065
hyper-pure ~ 55-105
leaded Muntz ~ 65-360
linotype ~ 55-625
Lipowitz's ~ 65-680
low-melting-point ~ 55-060
medal ~ 65-260
Monel ~ 65-435
money ~ 55-565
monotype ~ 55-620
Muntz ~ 65-330
native ~ 55-110
new ~ 55-115
noble ~ 55-010
nuclear ~ 55-085
ordnance ~ 65-245
parent ~ 55-185
precious ~ 55-010
primary ~ 55-115
printer's ~ 55-615
printing ~ 55-610
pure ~ 55-090
radioactive ~ 55-075
reactive ~ 55-070
refined ~ 55-125
refractory ~ 55-065
regulus ~ 65-660
reused ~ 55-120
Rose's ~ 65-675
secondary ~ 55-120
semi- ~ 10-370
semi-noble ~ 55-020
sintered ~ 55-660
slug-casting ~ 55-625
spacing ~ 55-640
star ~ 65-665
stereotype ~ 55-630
super-pure ~ 55-105
tempering ~ 55-285
type ~ 55-610
ultra-pure ~ 55-105
vacuum ~ 55-135
vacuum-evaporated ~ 55-145
vacuum-melted ~ 55-135
virgin ~ 55-115
white ~ 65-640
white bearing ~ 65-640
Wood's ~ 65-670
zone-refined ~ 55-140
metal carbide 40-435
metal crystal 20-020
metal impregnation 70-915
metallic alloy 55-160
metallic bond 10-340
metallic cementation 70-915
metallic crystal 20-020
metallic element 10-360
metallic inclusions *pl* 40-550
metallic lustre 75-510
metallic phase 35-150
metallic properties *pl* 10-350

METALLIC

metallic ring 75-360
metallic solution 35-385
metallic state 10-345
metallic system 55-165
metallization
diffusion ~ 70-915
metallographic examination 80-035
metallographic microscope 80-210
metallographic microscopy 80-010
metallographic study 80-035
metallograph 80-210
metallography 80-005
colour ~ 80-030
electron ~ 80-025
high-temperature ~ 80-015
quantitative ~ 80-020
X-ray ~ 80-575
metalloid 10-365
metallurgical microscope 80-210
metallurgical properties pl 75-855
metallurgical thermodynamics 15-010
metallurgy
physical ~ 10-005
structural ~ 10-005
metal phase 35-150
metal physics 10-015
metals pl
alkali ~ 55-030
alkaline earth ~ 55-035
heavy ~ 55-055
heavy platinum ~ 65-825
high-density ~ 55-055
light ~ 55-050
light platinum ~ 65-830
non-ferrous ~ 55-015
platinum ~ 65-820
platinum group ~ 65-820
rare ~ 55-045
rare earth ~ 55-040
transition ~ 55-025
metal structure 40-005
metal system 55-165
metastable condition 35-070
metastable equilibrium 35-060
metastable phase 35-175
metastable state 35-070
metatectic 40-255
metatectic change 35-580
metatectic reaction 35-580
metatectic temperature 35-620
metatectic transformation 35-580
meteoric iron 60-005
method
back reflection ~ 80-630
Bragg ~ 80-600
dry ~ 80-810
dry powder ~ 80-810
Debye-Scherrer ~ 80-595
Laue ~ 80-590
oscillating crystal ~ 80-605
powder ~ 80-595
revolving crystal ~ 80-600
rotating-crystal ~ 80-600
suspension ~ 80-815
transmission ~ 80-625
wet ~ 80-815
microanalysis
absorption ~ 80-735
electron probe ~ 80-780
X-ray ~ 80-780
microconstituent 40-010
microcrack 85-045
microcreep 45-530
micro-etching 80-140

microcrystal
filamentary ~ 25-360
microfissure 85-045
microfractograph 80-395
microfractography 80-390
micrograph 80-245
electron ~ 80-365
micro-hardness 75-250
microporosity 85-115
microradiograph 80-725
microradiography 80-720
microscope
metallographic ~ 80-210
metallurgical ~ 80-210
microscopic analysis 80-040
microscopic examination 80-040
microscopic stresses pl 85-035
microscopy 80-040
bright-field ~ 80-235
dark-field ~ 80-230
electron ~ 80-330
electron emission ~ 80-335
field-ion ~ 80-380
flying-spot ~ 80-360
hot-stage ~ 80-240
interference ~ 80-295
light ~ 80-215
optical ~ 80-215
phase-contrast ~ 80-270
polarized light ~ 80-260
reflection ~ 80-225
scanning ~ 80-360
scanning electron ~ 80-360
transmission ~ 80-220
transmission electron ~ 80-340
ultra-violet ~ 80-265
X-ray ~ 80-730
microsection 80-095
microsegregation 25-420
microstresses pl 85-035
microstructural constituent 40-010
microstructural equilibrium 40-020
microstructure 40-030
lamellar ~ 40-105
migration
grain-boundary ~ 45-490
mild-alloy steel 60-120
mild quenching 70-265
mild steel 60-100, 60-340
Miller-Bravais indices pl 20-170
Miller indices pl 20-165
minor segregation 25-420
mirror alloy 65-280
mischmetal 65-865
miscibility 35-300
complete ~ 35-315
liquid ~ 35-305
solid ~ 35-310
miscibility gap 35-345
misfit
lattice ~ 30-615
misorientation angle 30-595
mixed crystal 40-195
mixed crystallization 25-055
mixed dislocation 30-180
mixture
eutectic ~ 40-200
eutectoid ~ 40-240
gas ~ 35-370
mechanical ~ of phases 40-190
phase ~ 40-190
mock gold 65-815
moderately alloyed steel 60-125

NITRIDE

moderating power 75-535
modification 25-220, 25-225
modified alpax 65-570
modified silumin 65-570
modifier 25-230
modulus
bulk ~ of elasticity 75-160
bulk elastic ~ 75-160
elastic ~ 75-150
longitudinal ~ of elasticity 75-150
~ of rigidity 75-155
~ of strain hardening 45-065
rigidity ~ 75-155
shear ~ of elasticity 75-155
torsion ~ 75-155
Voight's ~ 75-160
volumetric ~ of elasticity 75-160
Young's ~ 75-150
molar concentration 55-240
molar property 15-050
molar quantity 15-050
molar volume 75-325
molecular bond 10-335
molecular heat 75-375
molecular mass 75-355
molecular weight 75-355
molybdenum 05-275
non-sag ~ 65-870
molybdenum steel 60-220
Monel 65-435
Monel metal 65-435
money metal 55-565
monoclinic lattice 20-395
monoclinic system 20-265
monocrystal 20-430
monophasic system 35-090
monotectic 40-260
monotectic change 35-575
monotectic horizontal 35-650
monotectic isothermal 35-650
monotectic line 35-650
monotectic point 35-670
monotectic reaction 35-575
monotectic temperature 35-615
monotectic transformation 35-575
monotectoid change 35-600
monotectoid reaction 35-600
monotectoid transformation 35-600
monotype metal 55-620
monovariant system 35-080
mosaic block 30-655
mosaic structure 40-170
Mössbauer effect 10-285
mother liquor 25-095
motion
conservative ~ 30-290
non-conservative ~ 30-285
mottled cast iron 60-960
mottled pig iron 60-920
mounting 80-065
multicomponent alloy 55-310
multicomponent solution 35-140
multicomponent system 35-140
multiple glide 45-140
multiple slip 45-140
multiplication
dislocation ~ 30-420
multiplicity
thermodynamic ~ 15-230

multiphase system 35-110
Muntz metal 65-330
mutual intersolubility 35-300

N

Nabarro creep 45-520
native alloy 55-265
native gold 65-765
native metal 55-110
natural alloy 55-265
natural ageing 70-585
natural stabilizing treatment 70-540
naval brass 65-385
nearest neighbour 20-195
nearest neighbour atom 20-195
needle
martensite ~ 50-375
Néel point 25-565
Néel temperature 25-565
negative edge dislocation 30-160
negative segregation 25-440
neodymium 05-280
neon 05-285
neptunium 05-290
network
carbide ~ 50-445
cementite ~ 50-105
dislocation ~ 30-390
network structure 40-100
Neumann bands pl 45-205
Neumann lamellae pl 45-205
Neumann lines pl 45-205
neutral atmosphere 70-675
neutron absorption cross-section 75-530
neutron capture cross-section 75-530
neutron radiography 80-830
new alloy 55-275
new metal 55-115
next-nearest neighbour 20-200
ni-carbing 70-875
Ni-resist 60-1090
nichrome 65-470
nickel 05-295
carbonyl ~ 65-425
nickel alloy 65-430
nickel-base alloy 65-430
nickel brass 65-350
nickel bronze 65-160
nickel cast iron 60-1025
nickelin 65-410
nickel silver 65-415
nickel steel 60-185
nicrosilal 60-1095
nil-ductility temperature 75-280
nilvar 65-500
nimonic alloy 65-480
niobide 40-505
niobium 05-300
niobium carbide 40-465
nital 80-810
nitralloy steel 60-700
nitride 70-810
chromium ~ 70-820
iron ~ 70-815
titanium ~ 70-825
vanadium ~ 70-830

NITRIDE

nitride case 70-845
nitride former 50-470
nitride-forming element 50-470
nitride-hardened steel 60-315
nitride inclusions pl 40-545
nitride steel 60-315
nitrided case 70-845
nitrided steel 60-315
nitriding 70-785
  ammonia ~ 70-790
  bath ~ 70-795
  glow ~ 70-800
  glow discharge ~ 70-800
  ion ~ 70-800
  wet ~ 70-795
nitriding steel 60-695
nitrocarburizing 70-875
nitrogen 05-305
nitrogen austenite 70-835
nitrogen case hardening 70-785
nitrogen hardening 70-785
nitrooxidizing 70-885
nobelium 05-310
noble metal 55-010
node
  dislocation ~ 30-350
nodular cast iron 60-975
nodular cementite 50-100
nodular graphite 50-210
nodular troostite 50-400
nodule 40-290
non-ageing steel 60-405
non-aging steel 60-405
non-coherent boundary 30-630
non-conservative motion 30-285
non-corrosive alloy 55-480
non-deforming steel 60-410
non-destructive inspection 80-785
non-destructive testing 80-785
non-equilibrium freezing 25-020
nonequilibrium phase 35-185
non-equilibrium solidification 25-020
non-ferrous alloy 65-005
non-ferrous metals pl 55-015
non-hardenable steel 60-390
nonideal solution 15-245
nonmagnetic alloy 55-470
non-magnetic alloy 55-470
non-magnetic steel 60-510
nonmetal 10-365
non-metal 10-365
non-metallic element 10-365
non-metallic inclusions pl 40-515
nonmetallic phase 35-155
non-oxydizing hardening 70-330
non-sag molybdenum 65-870
non-tarnishing alloy 55-500
normalized steel 60-280
normalizing 70-475
normal segregation 25-435
notch impact toughness 75-190
notch sensitivity 75-195
notch toughness 75-190
nucleant 25-125
nuclear corrosion 85-365
nuclear metal 55-085
nuclear properties pl 75-520

nucleating agent 25-125
nucleation 25-175
  athermal ~ 25-200
  dynamic ~ 25-205
  heterogenous ~ 25-185
  homogeneous ~ 25-180
  oriented ~ 25-195
  spontaneous ~ 25-190
nucleation catalyst 25-125
nucleation rate 25-270
nucleus 25-105
  coherent ~ 25-155
  critical ~ 25-140
  critical size ~ 25-140
  crystal ~ 25-105
  foreign ~ 25-120
  heterogeneous ~ 25-115
  homogeneous ~ 25-110
  ~ of crystallization 25-105
  oriented ~ 25-160
  recrystallization ~ 25-170
  spontaneous ~ 25-110
  stable ~ 25-150
number
  atomic ~ 75-330
  azimuthal quantum ~ 10-135
  coordination ~ 20-185
  fourth quantum ~ 10-145
  grain-size ~ 80-290
  index ~ 80-290
  magnetic quantum ~ 10-140
  mass ~ 75-335
  ~ of complexions 15-230
  ~ of degrees of freedom 35-030
  ~ of elementary complexions 15-230
  ~ of microscopic complexions 15-230
  principal quantum ~ 10-130
  quantum ~ 10-125
  second quantum ~ 10-135
  secondary quantum ~ 10-135
  shell quantum ~ 10-130
  spin quantum ~ 10-145
  subsidiary quantum ~ 10-135
  third quantum ~ 10-140
  total quantum ~ 10-130

O

oblique system 20-265
octahedral interstice 20-330
octahedral plane 20-335
octahedral void 20-330
offset yield stress 75-115
OFHC copper 65-070
oil cooling 70-105
oil hardening 70-210
oil-hardening steel 60-370
oil quenching 70-210
oligo-constituent 55-250
one-component system 35-115
one-dimensional defect 30-140
one-phase area 35-680
one-phase region 35-680
opacity 75-505
open-hearth pig iron 60-855
open-hearth steel 60-610
optical miscroscopy 80-215
optical properties pl 75-500
optical reflectivity 75-515
order 25-590
  long-range ~ 25-600
  short-range ~ 25-595
order-disorder transformation 25-580
order-disorder transition 25-580
ordered phase 35-200
ordered solid solution 35-430

ordered state 25-590
ordering 25-575
  magnetic ~ 25-640
ordering energy 25-630
ordinary cast iron 60-965
ordinary steel 60-095
ordnance bronze 65-245
ordnance metal 65-245
orientation
  chaotic ~ 20-420
  crystal ~ 20-415
  crystallographic ~ 20-415
  ~ of single crystals 80-690
  preferred ~ 20-425
  random ~ 20-420
orientation contrast 80-155
oriented nucleation 25-195
oriented nucleus 25-160
oriented solidification 25-035
original crystal 40-295
ornamental alloy 55-605
orthorhombic lattice 20-380
orthorhombic system 20-255
oscillating crystal method 80-605
osmiridium 65-845
osmium 05-315
outer electron 10-065
outer-shell electron 10-065
overageing 70-595
overetching 80-130
overheated steel 60-320
overheating 70-065
overheating sensitivity 75-760
overlap 85-400
overshoot 30-260
overshooting 30-260
oxidation
  internal ~ 70-947
oxidation resistance 75-685
oxidation-resistant alloy 55-495
oxidation-resistant steel 60-560
oxidation-resisting alloy 55-495
oxide inclusions pl 40-535
oxide scale 85-430
oxygen 05-320
oxygen-free copper 65-060
oxygen-free high conductivity copper 65-070
oxygen potential 15-350
oxygen steel 60-605
oxynitriding 70-885

P

pack annealing 70-435
pack carburizing 70-720
packing
  close ~ 20-210
  closest ~ 20-210
palladium 05-325
pantal 65-615
paramagnetic 75-555
paramagnetic material 75-555
paramagnetic substance 75-555
parameter
  interaction ~ 15-300
  lattice ~ 20-120
  long-range order ~ 25-620
  plasticity ~ 75-065

parameter
  short-range ~ 25-615
  strength ~ 75-060
  thermodynamic ~ 15-020
parent crystal 40-295
parent metal 55-185
parent phase 35-290
partial annealing 70-450
partial dislocation 30-315
partial diffusion coefficient 45-390
partial enthalpy 15-150
partial entropy 15-215
partial free energy 15-110
partial quantity 15-060
partial solubility 35-320
partial thermodynamic function 15-060
partial thermodynamic quantity 15-060
particle reinforcement 55-220
parting fracture 40-655
parting rupture 40-655
patina 85-375
patenting 70-415
pattern
  back reflection ~ 80-670
  Debye-Scherrer powder ~ 80-635
  diffraction ~ 80-610
  electron diffraction ~ 80-775
  etch ~ 80-200
  etching ~ 80-200
  Laue ~ 80-620
  powder ~ 80-635
  rotation ~ 80-665
  transmission ~ 80-675
  X-ray diffraction ~ 80-615
  X-ray diffraction powder ~ 80-635
Pauli exclusion principle 10-170
Pauli principle 10-170
pearlite 50-125
  banded ~ 50-130
  coarse ~ 50-140
  coarse lamellar ~ 50-140
  coarsely lamellar ~ 50-140
  curly ~ 50-145
  degenerate ~ 50-155
  dense lamellar ~ 50-135
  divorced ~ 50-150
  fine ~ 50-135
  fine lamellar ~ 50-135
  globular ~ 50-150
  granular ~ 50-150
  incubation ~ 70-150
  lamellar ~ 50-130
  laminated ~ 50-130
  sorbitic ~ 50-145
  spheroidal ~ 50-150
  very fine ~ 50-400
pearlite range 50-295
pearlite reaction 50-275
pearlitic cast iron 60-1045
pearlitic cementite 50-085
pearlitic ferrite 50-060
pearlitic malleable cast iron 60-1130
pearlitic malleable iron 60-1130
pearlitic steel 60-425
pearlitic transformation 50-275
Peierls force 30-300
Peierls-Nabarro force 30-300
pencil glide 45-175
pencil gliding 45-175
penetrant flaw detection 80-835
penny-shaped dislocation 30-345
percentage elongation 75-120
percentage reduction of area 75-125
perfect crystal 30-005

PERFECT

perfect dislocation 30-305
perfect solution 15-240
permalloy 65-440
permanent deformation 45-010
permanent magnet alloy 55-450
permanent magnet steel 60-780
permeability 75-615
  initial ~ 75-620
  magnetic ~ 75-615
permendur 65-490
perminvar 65-510
period
  etching ~ 80-125
  identity ~ 20-115
  translation ~ 20-115
peritectic 40-245
peritectic change 35-570
peritectic compound 35-225
peritectic horizontal 35-645
peritectic isothermal 35-645
peritectic line 35-645
peritectic melting 35-475
peritectic point 35-665
peritectic reaction 35-570
peritectic temperature 35-610
peritectic transformation 35-570
peritectoid 40-250
peritectoid change 35-595
peritectoid reaction 35-595
peritectoid temperature 35-635
peritectoid transformation 35-595
pest
  tin ~ 85-245
phase 35-025
  σ ~ 35-275
  condensed ~ 35-145
  congruent intermediate ~ 35-220
  congruently-melting intermediate ~ 35-220
  disordered ~ 35-205
  disperse ~ 40-330
  dispersed ~ 40-330
  electron ~ 35-280
  equilibrium ~ 35-180
  hardening ~ 40-350
  high-pressure ~ 35-165
  high-temperature ~ 35-160
  Hume-Rothery ~ 35-280
  incongruent intermediate ~ 35-225
  incongruently-melting intermediate ~ 35-225
  intermediate ~ 35-210
  intermetallic ~ 35-210
  metal ~ 35-150
  metallic ~ 35-150
  metastable ~ 35-175
  nonequilibrium ~ 35-185
  nonmetallic ~ 35-155
  ordered ~ 35-200
  parent ~ 35-290
  precipitated ~ 40-305
  proeutectic ~ 40-340
  pro-eutectic ~ 40-340
  proeutectoid ~ 40-345
  pro-eutectoid ~ 40-345
  sigma ~ 35-275
  stable ~ 35-170
  superlattice ~ 35-200, 35-430
  terminal ~ 35-380
  transient ~ 35-195
  transition ~ 35-195
σ phase 35-275
phase boundary line 35-535
phase change 25-460
phase constitution 35-020
phase contrast 80-275
phase-contrast microscopy 80-270

phase diagram 35-480
phase equilibrium 35-015
phase equilibrium diagram 35-480
phase mixture 40-190
phase rule 35-010
phase transformation 25-460
phase transition 25-460
phases pl
  coexisting ~ 35-190
  conjugate ~ 35-190
  Laves ~ 35-265
  Zintl ~ 35-270
phonon 10-305
phosphide
  iron ~ 50-250
phosphide eutectic 50-235
phosphor bronze 65-215
phosphor copper 65-075
phosphoric pig iron 60-890
phosphorized copper 65-055
phosphor tin-bronze 65-210
phosphorus 05-330
phosphorus banding 85-230
photograph
  electron diffraction ~ 80-775
  Laue ~ 80-620
  powder ~ 80-635
  powder diffraction ~ 80-635
  X-ray diffraction ~ 80-615
photogram
  Laue ~ 80-620
photomacrograph 80-310
photomacrography 80-315
photomicrograph 80-250
  electron ~ 80-365
photomicrography 80-255
physical metallurgy 10-005
physical properties pl 75-315
physics
  metal ~ 10-015
  ~ of metals 10-015
  ~ of solids 10-010
  solid state ~ 10-010
pickle brittleness 85-390
pickling brittleness 85-390
picral 80-115
picture
  Laue ~ 80-620
pig iron 60-840
pile-up of dislocations 30-400
pine-tree crystal 25-315
pinhole porosity 85-115
pinholes pl 85-110
pinned dislocation 30-270
pinning point 30-275
pipe 85-125
pipe diffusion dislocation 45-360
piston alloy 55-570
pit
  etch ~ 80-205
  etching ~ 80-205
pitting corrosion 85-350
plague
  tin ~ 85-245
plain cast iron 60-990
plain carbon steel 60-095
planar defect 30-535
plane
  atomic ~ 20-175
  basal ~ 20-360

— 294 —

plane
 cleavage ~ 40-665
 composition ~ 20-470
 cross-slip ~ 45-150
 crystal ~ 20-155
 crystallographic ~ 20-155
 extra half ~ 30-200
 glid ~ 45-100
 gliding ~ 45-100
 habit ~ 20-465
 lattice ~ 20-155
 octahedral ~ 20-335
 prism ~ 20-365
 prismatic ~ 20-365
 slip ~ 45-100
 strong ~ 20-215
 twin ~ 45-250
 twinning ~ 45-250
plane defect 30-535
plane spacing 20-225
plastic bronze 65-185
plastic deformation 45-010
plastic flow 45-215
plasticity 75-025
plasticity characteristic 75-065
plasticity parameter 75-065
plastic strain 45-010
plastic working 45-015
plate
 armour ~ 60-767
 martensite ~ 50-370
 martensitic ~ 50-370
platin iridium 65-835
platinite 65-515
platinum 05-335
platinum group metals pl 65-820
platinum-iridium 65-835
platinum metals pl 65-820
platinum-rhodium 65-840
plutonium 05-340
point
 anchoring ~ 30-275
 boiling ~ 75-430
 change ~ 35-555
 critical ~ 35-555
 Curie ~ 25-560
 decomposition ~ 15-340
 eutectic ~ 35-660
 eutectoid ~ 35-675
 freezing ~ 75-425
 fusing ~ 75-420
 fusion ~ 75-420
 lattice ~ 20-100
 magnetic change ~ 25-560
 magnetic transformation ~ 25-560
 melting ~ 75-420
 monotectic ~ 35-670
 Néel ~ 25-565
 pinning ~ 30-275
 peritectic ~ 35-665
 saturation ~ 35-330
 sharp yield ~ 75-110
 singular ~ 80-495
 spinodal ~ 15-185
 solidifying ~ 75-425
 sublimation ~ 75-435
 transformation ~ 35-555
 transition ~ 25-570, 35-555
 triple ~ 35-040
 yield ~ 75-110
point corrosion 85-345
point defect 30-030
point graphite 50-195
point imperfection 30-030
point lattice 20-275
Poisson's ratio 75-170
polar bond 10-325
polar compound 35-230

polar crystal 20-025
polarized light microscopy 80-260
pole dislocation 30-365
pole figures pl 80-685
polish attack 80-085
polish etch 80-085
polishing
 chemical ~ 80-080
 electrolytic ~ 80-075
 mechanical ~ 80-070
polonium 05-345
polycomponent system 35-140
polycrystal 20-440
polycrystalline aggregate 20-440
polygonization 45-435
polyhedral structure 40-095
polymorph 25-390
polymorphic change 25-455
polymorphic form 25-390
polymorphic transformation 25-455
polymorphism 25-385
polymorphy 25-385
polynary system 35-140
polyphase alloy 55-345
polyphase diffusion 45-325
polyphase structure 40-160
polyphase system 35-110
polythermal section 35-740
pore 85-100
porosity 85-105
 pinhole ~ 85-115
 shrinkage ~ 85-120
porous bronze 65-225
Portevin-Le Chatelier effect 45-540
position
 interstitial ~ 20-145
 lattice ~ 20-140
positional entropy 15-210
positive edge dislocation 30-155
pot annealing 70-435
potassium 05-350
potential
 carbon ~ 70-750
 chemical ~ 15-175
 electrochemical ~ 75-715
 interaction ~ 10-300
 ionization ~ 10-105
 ionizing ~ 10-105
 oxygen ~ 15-350
 thermodynamic ~ at constant volume 15-105
potential barrier 10-110
pot quenching 70-760
powder carburizing 70-720
powder diffraction photograph 80-635
powder method 80-595
powder pattern 80-635
powder photograph 80-635
power
 alloying ~ 55-175
 flowing ~ 75-785
 hardening ~ 70-020
 moderating ~ 75-535
 quenching ~ 70-125
 slowing-down ~ 75-535
 thermoelectric ~ 75-495
praseodymium 05-355
precious metal 55-010
precipitate 40-305
 coherent ~ 40-320
 grain boundary ~ 40-310
 incoherent ~ 40-325
 submicroscopic ~ 40-315

precipitated phase 40-305
precipitation 40-300
 continuous ~ 25-375
 discontinuous ~ 25-380
precipitation-hardenable alloy 55-385
precipitation hardening 70-560
preferred magnetic axis 75-585
preferred orientation 20-425
preheating 70-045
pre-precipitation 70-600
pressure
 dissociation ~ 15-345
pressure-spray quenching 70-240
pressure vessel steel 60-740
primary aluminium 65-545
primary austenite 50-030
primary cementite 50-070
primary creep 45-500
primary crystallization 25-050
primary crystals pl 40-270
primary dislocations pl 30-250
primary glide system 45-115
primary graphite 50-170
primary metal 55-115
primary recrystallization 45-445
primary segregation 25-405
primary slip system 45-115
primary solidification 25-050
primary solid solution 35-380
primary structure 40-045
primitive cell 20-095
primitive unit cell 20-095
principal alloying element 55-195
principal crystallization axis 25-300
principal quantum number 10-130
principal shell 10-155
principle
 common tangent ~ 15-180
 Heisenberg uncertainty ~ 10-165
 Pauli ~ 10-170
 Pauli exclusion ~ 10-170
 uncertainty ~ 10-165
print
 sulfur ~ 80-325
printer's metal 55-615
printing
 Baumann ~ 80-320
printing metal 55-610
prior austenite grain 50-040
prior austenitic grain 50-040
prismatic dislocation 30-345
prismatic dislocation loop 30-345
prismatic gliding 45-170
prismatic loop 30-345
prismatic plane 20-365
prismatic slip 45-170
prism plane 20-365
process
 black heart ~ 70-630
 diffusional ~ 25-505
 diffusion-free ~ 25-510
 thermally activated ~ 45-280
 white heart ~ 70-635
processing properties pl 75-720
proeutectic 40-340
pro-eutectic phase 40-340
proeutectic phase 40-340
pro-eutectoid 40-345
pro-eutectoid cementite 50-080

pro-eutectoid ferrite 50-055
pro-eutectoid phase 40-345
proeutectoid phase 40-345
progressive hardening 70-320
progressive spin hardening 70-325
promethium 05-360
proof stress 75-115
properties pl
 capacitive ~ 75-340
 casting ~ 75-775
 chemical ~ 75-655
 electrical ~ 75-465
 electrochemical ~ 75-705
 founding ~ 75-775
 gross ~ 75-850
 macroscopic ~ 75-850
 magnetic ~ 75-540
 mechanical ~ 75-005
 metallic ~ 10-350
 metallurgical ~ 75-855
 nuclear ~ 75-520
 optical ~ 75-500
 physical ~ 75-315
 processing ~ 75-720
 strength ~ 75-010
 structural ~ 75-860
 structure-insensitive ~ 75-865
 structure-sensitive ~ 75-860
 technological ~ 75-720
 tensile ~ 75-015
 thermal ~ 75-365
 volumetric ~ 75-340
 working ~ 75-820
property
 capacity ~ 15-040
 extensive ~ 15-040
 intensive ~ 15-045
 molar ~ 15-050
 state ~ 15-035
proportionality limit 75-100
proportional limit 75-100
protactinium 05-365
protective atmosphere 70-690
pseudo-alloy 55-650
pseudo-binary alloy 55-315
pseudo-binary section 35-745
pseudo-binary system 35-130
pseudo-eutectic 40-225
puddled iron 60-055
puddled steel 60-575
pulling
 crystal ~ 25-090
 ~ of crystals 25-090
pulse heating 70-050
pure metal 55-090
pyramidal system 20-250
pyrophoric alloy 55-595

Q

quadric system 20-250
quality steel 60-140
quantitative metallography 80-020
quantitative thermal analysis 80-535
quantity
 excess ~ 15-065
 excess thermodynamic ~ 15-065
 extensive ~ 15-040
 integral ~ 15-055
 integral thermodynamic ~ 15-055
 intensive ~ 15-045
 molar ~ 15-050
 partial ~ 15-060
 partial thermodynamic ~ 15-060
 state ~ 15-035
 thermodynamic ~ 15-030

REACTION

quantum number 10-125
quantum state 10-120
quasi-binary alloy 55-315
quasi-binary section 35-745
quasi-binary system 35-130
quasi-isotropy 20-455
quaternary alloy 55-305
quaternary axis 20-080
quaternary eutectic 40-215
quaternary steel 60-165
quaternary system 35-135
quench
  fast ~ 70-260
  slow ~ 70-265
quench ageing 70-580
quench annealing 70-500
quenchant 70-185
quench bath 70-195
quench crack 85-190
quenched-in vacancy 30-100
quenched steel 60-285
quench-hardened steel 60-285
quench hardening 70-175
quenching
  air ~ 70-215
  air-blast ~ 70-220
  brine ~ 70-238
  bulk ~ 70-287
  conventional ~ 70-270
  customary ~ 70-270
  direct ~ 70-760
  drastic ~ 70-260
  dry ~ 70-220
  fused salt ~ 70-230
  gas ~ 70-220
  hot ~ 70-225
  immersion ~ 70-245
  interrupted ~ 70-300
  lead bath ~ 70-235
  liquid ~ 70-245
  liquisol ~ 70-085
  mild ~ 70-265
  oil ~ 70-210
  pot ~ 70-760
  pressure-spray ~ 70-240
  ~ from high temperatures 70-345
  ~ of vacancies 30-110
  rapid ~ 70-260
  regenerative ~ 70-765
  salt bath ~ 70-230
  splat ~ 70-085
  spray ~ 70-240
  step ~ 70-305
  time ~ 70-300
  water ~ 70-100, 70-205
quenching agent 70-185
quenching bath 70-195
quenching crack 85-190
quenching defect 85-180
quenching fluid 70-190
quenching intensity 70-200
quenching liquid 70-190
quenching medium 70-185
quenching power 70-125
quenching rate 70-120
quenching sorbite 50-385
quenching stresses pl 85-185
quenching temperature 70-140
quenching troostite 50-400
quenching velocity 70-120

R

radiation damage 30-490
radiation defect 30-485
radiation hardening 30-495
radiation-induced strengthening 30-495
radioactivation analysis 80-765
radioactive metal 55-075
radioactive tracer 80-760
radioautograph 80-755
radiograph 80-715
radiographic inspection 80-705
radiography 80-705
  gamma ~ 80-825
  gamma-ray ~ 80-825
  neutron ~ 80-830
  ~ of metals 80-710
radiometallography 80-710
radium 05-370
radius
  atomic ~ 10-085
  critical ~ of a nucleus 25-145
  ionic ~ 10-090
radon 05-375
rail steel 60-730
random orientation 20-420
random solid solution 35-435
range
  bainite ~ 50-305
  blue-heat ~ 70-395
  critical ~ 35-560
  critical temperature ~ 35-560
  freezing ~ 35-550
  fusion ~ 35-545
  hardening temperature ~ 70-145
  homogeneity ~ 35-680
  martensite ~ 50-300
  melting ~ 35-545
  pearlite ~ 50-295
  solidification ~ 35-550
  transformation ~ 35-560
Raoultian solution 15-240
Raoult's law 15-270
rapid quenching 70-260
rapid steel 60-800
rare earth metals pl 55-040
rare metals pl 55-045
rate
  cooling ~ 70-120
  critical cooling ~ 70-180
  critical quenching ~ 70-180
  growth ~ 25-275
  heating ~ 70-040
  nucleation ~ 25-270
  quenching ~ 70-120
  ~ of cooling 70-120
  ~ of crystallization 25-275
  ~ of growth 25-275
  ~ of heating 70-040
  ~ of nucleation 25-270
ratio
  axial ~ 20-355
  intercept ~ 20-355
  Poisson's ~ 75-170
  segregation ~ 40-420
reaction
  bainite ~ 50-290
  eutectic ~ 35-565
  eutectoid ~ 35-590
  eutectoidal ~ 35-590
  metatectic ~ 35-580
  monotectic ~ 35-575
  monotectoid ~ 35-600
  pearlite ~ 50-275
  peritectic ~ 35-570
  peritectoid ~ 35-595
  syntetic ~ 35-585

REACTIVE

reactive atmosphere 70-680
reactive metal 55-070
reagent
  etching ~ 80-100
  staining metallographic ~ 80-105
  staining metallographic etching ~ 80-105
real crystal 30-010
real solution 15-245
recalescence 25-485
reciprocal lattice 20-410
recovery 45-420
  dynamic ~ 45-425
  static ~ 45-430
  thermal ~ 45-430
recrystallization 45-440
  dynamic ~ 45-465
  primary ~ 45-445
  secondary ~ 45-450
  tertiary ~ 45-460
recrystallization annealing 70-490
recrystallization diagram 45-485
recrystallization nucleus 25-170
recrystallization temperature 45-470
recrystallization texture 40-600
recrystallization twin 45-240
rectified zinc 65-695
red brass 65-395
red brittleness 75-290
red gold 65-800
red hardness 75-260
red shortness 75-290
reduction of area at fracture 75-125
refined aluminium 65-550
refined copper 65-020
refined lead 65-655
refined metal 55-125
refined steel 60-635
refinement
  grain ~ 40-385
refiner
  grain ~ 40-400
refining
  core ~ 70-770
  grain ~ 40-385
  heat ~ 70-475
reflection microscopy 80-225
reflectivity 75-515
  optical ~ 75-515
reflections pl
  superlattice ~ 80-655
refractory alloy 55-400
refractory metal 55-065
regenerative quenching 70-765
region
  homogeneous ~ 35-680
  one-phase ~ 35-680
  single-phase ~ 35-680
  two-phase ~ 35-685
regular solution 15-255
regular system 20-240
regulus 65-665
regulus metal 65-660
reinforcement
  particle ~ 55-220
reinforcing steel 60-795
relative atomic mass 10-075
relative valency effect 10-405
relaxation
  elastic ~ 45-525
  stress ~ 45-525

relieving
  stress ~ 70-525
remaining liquid 25-280
remanence 75-630
re-melted alloy 55-280
residual element 55-260
residual magnetism 75-630
residual stresses pl 85-020
resistance
  abrasion ~ 75-265
  acid ~ 75-695
  ageing ~ 75-845
  alkali ~ 75-700
  chemical ~ 75-665
  corrosion ~ 75-670
  creep ~ 75-215
  heat ~ 75-690
  high-temperature creep ~ 75-240
  oxidation ~ 75-685
  ~ to oxidation 75-685
  ~ to scaling 75-690
  ~ to softening 75-840
  ~ to thermal cycling 75-830
  scale ~ 75-690
  shock ~ 75-185
  specific ~ 75-470
  tarnish ~ 75-680
  tempering ~ 75-840
  thermal ~ 75-835
  thermal shock ~ 75-825
  wear ~ 75-265
resistance alloy 55-545
resistivity 75-470
  electric ~ 75-470
  electrical ~ 75-470
restricted solubility 35-320
resulfurized stainless steel 60-525
resulfurized steel 60-710
retained austenite 50-270
retention of hardness 75-840
reticular structure 40-100
retrograde solidus curve 35-505
retrogression 70-610
reused metal 55-120
revealing
  ~ of the microstructure 80-280
  ~ the microstructure 80-280
reversible transformation 25-495
reversible transition 25-495
reversion 70-610
revolving crystal method 80-600
rhenium 05-380
rheotropic brittleness 75-310
rhodio-platinum 65-840
rhodium 05-385
rhombohedral lattice 20-390
rhombohedral system 20-260
rich alloy 55-285
right-handed screw dislocation 30-170
rigidity modulus 75-155
rimming steel 60-650
ring
  Debye ~ 80-640
  dislocation ~ 30-340
  metallic ~ 75-360
rivet steel 60-745
rolling texture 40-570
Rose's alloy 65-675
Rose's metal 65-675
rosette graphite 50-200
rotating-crystal method 80-600
rotation axis 20-065
rotation diagram 80-665

rotation pattern 80-665
rotatory reflection axis 20-065
rubidium 05-390
rule
  common tangent ~ 15-180
  Hägg's ~ 35-260
  Hume-Rothery ~ 35-325
  Konovalov's ~ 35-705
  lever ~ 35-700
  Matthiessen's ~ 75-475
  parting ~ 40-655
  phase ~ 35-010
runnability 75-780
rupture 85-040
  parting ~ 40-655
rust 85-370
rustless iron 60-520
rustless steel 60-515
rust-resisting steel 60-515
ruthenium 05-395

S

salt bath carburizing 70-725
salt bath hardening 70-230
salt bath quenching 70-230
samarium 05-400
sample 80-050
sampling 80-060
saturated solid solution 35-395
saturation magnetic intensity 75-605
saturation magnetization 75-605
saturation point 35-330
Sauveur's diagram 40-265
scale 85-430
  oxide ~ 85-430
scale resistance 75-690
scandium 05-405
scanning electron microscopy 80-360
scanning hardening 70-320
scanning microscopy 80-360
Schaeffler's diagram 60-530
Schrödinger equation 10-175
screw dislocation 30-165
sea water corrosion 85-290
seal-in alloy 55-600
season cracking 85-385
secondary alloy 55-280
secondary creep 45-505
secondary crystallization 25-060
secondary crystallization axis 25-305
secondary crystals pl 40-275
secondary dislocations pl 30-255
secondary graphite 50-180
secondary hardening 70-405
secondary martensite 50-350
secondary martensitic transformation 50-285
secondary metal 55-120
secondary quantum number 10-135
secondary recrystallization 45-450
secondary segregation 25-410
secondary slip system 45-120
secondary structure 40-055
second-degree transformation 25-470
second-degree transition 25-470
second-nearest neighbour 20-200

second-order transformation 25-470
second-order transition 25-470
second quantum number 10-135
section
  horizontal ~ 35-735
  isopleth ~ 35-740
  isothermal ~ 35-735
  quasibinary ~ 35-745
  polythermal ~ 35-740
  pseudobinary ~ 35-745
  vertical ~ 35-740
sectional sensitivity 75-815
seed
  crystal ~ 25-130
seeding of crystallization 25-215
segregate 40-305
segregation 25-400
  blowhole ~ 25-430
  dentritic ~ 25-420
  droplet ~ 25-430
  grain boundary ~ 25-450
  gravitational ~ 25-425
  gravity ~ 25-425
  intercrystalline ~ 25-450
  interdentritic ~ 25-445
  inverse ~ 25-440
  long-range ~ 25-415
  major ~ 25-415
  minor ~ 25-420
  negative ~ 25-440
  normal ~ 25-435
  primary ~ 25-405
  secondary ~ 25-410
  short-range ~ 25-420
segregation band 85-230
segregation ratio 40-420
selective carburizing 70-735
selective corrosion 85-305
selective hardening 70-275
selenium 05-410
self-annealing 45-455
self-diffusion 45-400
self-diffusion coefficient 45-405
self-energy of a dislocation line 30-230
self hardening 70-215
self-hardening steel 60-375
semi-austenitic steel 60-445
semiconductor 10-375
  extrinsic ~ 10-385
  impurity ~ 10-385
  intrinsic ~ 10-380
semi-ferritic steel 60-450
semi-killed steel 60-660
semi-metal 10-370
semimetal 10-370
semi-noble metal 55-020
semi-regular solution 15-260
semi-rimming steel 60-660
semi-steel 60-1035
senary axis 20-085
sendust 65-535
sensitivity
  notch ~ 75-195
  overheating ~ 75-760
  sectional ~ 75-815
sensitizing 70-545
sensitization 70-545
sequence
  stacking ~ 20-180
series pl
  displacement ~ 75-710
  electrochemical ~ 75-710
  electromotive ~ 75-710
  galvanic ~ 75-710

SESSILE

sessile dislocation 30-280
set copper 65-050
shadowing 80-370
shallow hardening 70-285
shallow-hardening steel 60-380
sharp yield point 75-110
shatter cracks pl 85-240
shear fracture 40-660
shear modulus of elasticity 75-155
shear strength 75-080
shear stress 45-085
shear transformation 25-510
shearing fracture 40-640
shell
  coordination ~ 20-190
  electron ~ 10-155
  electronic ~ 10-155
  principal ~ 10-155
shell quantum number 10-130
sherardizing 70-930
ship steel 60-755
Shockley dislocation 30-330
Shockley partial dislocation 30-330
shock resistance 75-185
shock-wave hardening 45-055
shortness 75-270
  blue ~ 75-300
  cold ~ 75-285
  hot ~ 75-290
  red ~ 75-290
short-range order 25-595
short-range order parameter 25-615
short-range segregation 25-420
shrapnel brass 65-380
shrinkage 75-795
  freezing ~ 75-805
  liquid ~ 75-800
  solid ~ 75-810
  solidification ~ 75-805
  volume ~ 75-795
shrinkage cavity 85-125
shrinkage porosity 85-120
shrinkage stresses pl 85-175
shut
  cold ~ 85-400
Siemens-Martin steel 60-610
sigma phase 35-275
silal 60-1100
silicide 40-490
siliciding 70-910
silicoferrite 50-475
silico-manganese 60-1155
silicon 05-415
silicon brass 65-365
silicon bronze 65-145
silicon cast iron 60-1010
silicon ferrite 50-475
silicon impregnation 70-910
silicon iron 60-210
siliconizing 70-910
silicon steel 60-200
silky fracture 40-715
silumin 65-565
  modified ~ 65-570
silver 05-420
  coin ~ 65-755
  fine ~ 65-760
  German ~ 65-415
  modified ~ 65-570
  nickel ~ 65-415

silver brazing alloy 65-750
silver solder 65-750
silver steel 60-820
simple carbide 40-470
simple cell 20-095
simple cubic lattice 20-305
simple dilatometry 80-545
simple glide 45-130
simple heat treatment 70-006
simple hexagonal lattice 20-345
simple lattice 20-280
simple slip 45-130
simple thermal analysis 80-510
single crystal 20-430
single crystal growing 25-085
single-phase alloy 55-330
single-phase brass 65-315
single-phase diffusion 45-330
single-phase region 35-680
single-phase structure 40-145
single-phase system 35-090
single slip 45-130
single vacancy 30-040
singular point 80-495
sink 30-130
sintered alloy 55-655
sintered carbides pl 55-665
sintered bronze 65-225
sintered hard alloy 55-665
sintered iron 60-065
sintered metal 55-660
site
  interstitial ~ 20-145
  lattice ~ 20-140
six-fold rotation axis 20-085
size
  grain ~ 80-285
skin hardening 70-285
slag inclusions pl 40-540
slip 45-095
  cross ~ 45-145
  double ~ 45-135
  duplex ~ 45-135
  grain boundary ~ 45-160
  intracrystalline ~ 45-165
  multiple ~ 45-140
  prismatic ~ 45-170
  simple ~ 45-130
  single ~ 45-130
slip band 45-190
slip direction 45-105
slip dislocation 30-265
slip line 45-185
slip plane 45-100
slip system 45-110
slow cooling 70-080
slowing-down power 75-535
slow quench 70-265
slug-casting metal 55-625
small angle boundary 30-590
smooth fracture 40-700
Snoek effect 30-460
snowflakes pl 85-240
soaking 70-055
soaking time 70-060
sodium 05-425
soft annealing 70-460

softening 70-460
work ~ 45-535
soft magnetic steel 60-500
soft solder 65-715
soft spots pl 85-195
soft steel 60-340
soil corrosion 85-285
solder 65-710
  aluminium ~ 65-730
  brazing ~ 65-720
  hard ~ 65-720
  lead ~ 65-735
  silver ~ 65-750
  soft ~ 65-715
  spelter ~ 65-740
  tin ~ 65-725
  white ~ 65-745
solderability 75-740
soldering alloy 65-710
solid carburizing 70-720
solidification 25-005
  controlled ~ 25-030
  directed ~ 25-035
  directional ~ 25-035
  eutectic ~ 25-025
  equilibrium ~ 25-015
  oriented ~ 25-035
  primary ~ 25-050
  uniaxial ~ 25-040
  unidirectional ~ 25-040
solidification front 25-295
solidification range 35-550
solidification shrinkage 75-805
solidification temperature 75-425
solidifying point 75-425
solid miscibility 35-310
solid shrinkage 75-810
solid solubility 35-310
solid solution 35-375
solid state physics 10-010
solid state transformation 25-475
solid state transition 25-475
solid transformation 25-475
solidus 35-485
solidus curve 35-490
solidus surface 35-495
solidus temperature 35-500
solubility 35-295
  complete ~ 35-315
  limited ~ 35-320
  liquid ~ 35-305
  partial ~ 35-320
  restricted ~ 35-320
  solid ~ 35-310
solubility curve 35-530
solubility gap 35-345
solubility limit 35-330
solute 35-360
solution 35-350
  actual ~ 15-245
  athermal ~ 15-265
  binary ~ 35-405
  complex ~ 35-410
  dilute ~ 15-250
  diluted ~ 15-250
  disordered solid ~ 35-435
  gaseous ~ 35-370
  ideal ~ 15-240
  interstitial solid ~ 35-420
  liquid ~ 35-365
  metallic ~ 35-385
  multicomponent ~ 35-410
  non-equilibrium ~ 25-020
  nonideal ~ 15-245
  ordered solid ~ 35-430
  perfect ~ 15-240

solution
  primary solid ~ 35-380
  random solid ~ 35-435
  Raoultian ~ 15-240
  real ~ 15-245
  regular ~ 15-255
  saturated solid ~ 35-395
  semi-regular ~ 15-260
  solid ~ 35-375
  substitutional solid ~ 35-415
  supersaturated solid ~ 35-400
  terminal solid ~ 35-380
solution annealing 70-500
solution hardening 55-215
solution heat treatment 70-570
solution strengthening 55-215
solution treatment 70-500
solvent 35-355
solvus 35-530
solvus curve 35-530
solvus line 35-530
sonic test 80-795
sonic testing 80-795
sonims pl 40-515
sorbite 50-380
  quenching ~ 50-385
  temper ~ 50-390
sorbitic pearlite 50-145
Soret effect 45-370
source
  dislocation ~ 30-425
  Frank-Read ~ 30-430
  Koehler ~ 30-435
space
  k ~ 10-205
  wave ~ 10-205
  wave vector ~ 10-205
space filling 20-205
space lattice 20-045
spaces pl
  interdendritic ~ 40-355
spacing
  interatomic ~ 20-230
  interlamellar ~ 40-110
  interparticle ~ 40-425
  interplanar ~ 20-225
  lamellar ~ 40-110
  lattice ~ 20-120
  plane ~ 20-225
spacing metal 55-640
sparking alloy 55-595
spark test 80-440
spark testing 80-440
special brass 65-310
special bronze 65-130
special steel 60-115, 60-150
specific conductance 75-485
specific density 75-345
specific gravity 75-350
specific heat 75-415
specific resistance 75-470
specific strength 75-140
specific tenacity 75-140
specimen 80-050
  bar ~ 80-435
  bulk ~ 80-055
spectral analysis 80-475
spectrography
  mass ~ 80-480
spectroscopy
  fluorescent X-ray ~ 80-700
  X-ray ~ 80-700
  X-ray fluorescence ~ 80-700

spectrum analysis 80-475
specular pig iron 60-930
speculum 65-280
speculum alloy 65-280
speed of growth 25-275
spelter 65-690
spelter solder 65-740
sphere
  Fermi ~ 10-225
spheroidal cast iron 60-975
spheroidal cementite 50-100
spheroidal graphite 50-190
spheroidal graphite cast iron 60-975
spheroidal pearlite 50-150
spheroidite 50-150
spheroidization of cementite 70-515
spheroidized cementite 50-100
spheroidizing annealing 70-510
spherolitic graphite 50-190
spherule
  graphite ~ 50-215
spherulite 40-290
spherulitic graphite 50-190
spiegel 60-930
spiegeleisen 60-930
spike
  displacement ~ 30-525
  thermal ~ 30-530
spin hardening 70-315
spinodal 15-190
spinodal curve 15-190
spinodal decomposition 25-370
spinodal line 15-190
spinodal point 15-185
spinode 15-185
spin quantum number 10-145
spiral
  growth ~ 25-335
splat cooling 70-085
splat quenching 70-085
split interstitial 30-085
split of a dislocation 30-320
split transformation of austenite 50-275
splitting of a dislocation 30-320
sponge iron 60-035
sponginess 85-105
spongy iron 60-035
spontaneous annealing 45-455
spontaneous crystallization 25-045
spontaneous nucleation 25-190
spontaneous nucleus 25-110
spot
  diffraction ~ 80-660
spot analysis 80-465
spots pl
  grey ~ 85-140
  hard ~ 85-145
  soft ~ 85-195
  white ~ 85-250
spot test 80-465
spray hardening 70-240
spring steel 60-725
stabilization
  austenite ~ 50-425
  ~ of austenite 50-425
stabilizer 50-485
  austenite ~ 50-450
  carbide ~ 50-465
  ferrite ~ 50-455

stabilized steel 60-305
stabilizing annealing 70-530
stabilizing treatment 70-535
stable condition 35-065
stable equilibrium 35-055
stable nucleus 25-150
stable phase 35-170
stable state 35-065
stacking fault 30-540
stacking fault energy 30-560
stacking sequence 20-180
staining metallographic etching reagent 80-105
staining metallographic reagent 80-105
stainless iron 60-520
stainless steel 60-515
stair-rod dislocation 30-360
standard gold 65-790
standard steel 60-265
star antimony 65-665
star metal 65-665
state
  disordered ~ 25-585
  electron ~ 10-115
  energy ~ 10-120
  equilibrium ~ 35-050
  quantum ~ 10-120
  metallic ~ 10-345
  metastable ~ 35-070
  order ~ 25-590
  ordered ~ 25-590
  stable ~ 35-065
  thermodynamic ~ 15-015
state function 15-035
state property 15-035
state quantity 15-035
state variable 15-020
static recovery 45-430
stationary diffusion 45-345
statuary bronze 65-270
statue bronze 65-270
steadite 50-235
steady-rate creep 45-505
steady-state creep 45-505
steady-state diffusion 45-345
steady-state diffusion 45-345
steam treating 70-880
steam treatment 70-880
steel 60-075
  abnormal ~ 60-465
  abrasion-resisting ~ 60-480
  acid ~ 60-615
  acid-resisting ~ 60-545
  ageing ~ 60-400
  ageing-resisting ~ 60-405
  aging ~ 60-400
  aging-resisting ~ 60-405
  air-hardening ~ 60-375
  alkali-resisting ~ 60-550
  alloy ~ 60-115
  alloy cast ~ 60-835
  alternate ~ 60-270
  aluminium ~ 60-235
  annealed ~ 60-275
  austenitic ~ 60-420
  austenitic-ferritic ~ 60-455
  automatic ~ 60-710
  bainitic ~ 60-435
  balanced ~ 60-660
  ball-bearing ~ 60-720
  basic ~ 60-620
  bearing ~ 60-715
  Bessemer ~ 60-595
  binary ~ 60-155
  blister ~ 60-585

steel
 boiler ~ 60-740
 boron ~ 65-255
 burnt ~ 60-325
 carbon ~ 60-095
 carbon-cast ~ 60-830
 carburized ~ 60-310
 carburizing ~ 60-690
 case-hardening ~ 60-690
 cast ~ 60-825
 cementation ~ 60-690
 cementitic ~ 50-500
 chrome ~ 60-175
 chromium ~ 60-175
 chromium-nickel ~ 60-190
 clad ~ 60-335
 coarse-grained ~ 60-475
 cobalt ~ 60-240
 cold-forming tool ~ 60-805
 cold-working tool ~ 60-805
 cold-work ~ 60-805
 cold-work tool ~ 60-805
 commercial ~ 60-135
 complex 60-170
 constructional ~ 60-670
 converter ~ 60-590
 copper ~ 60-245
 cooper-bearing ~ 60-245
 corrosion-resisting ~ 60-535
 creep-resistant ~ 60-565
 creep-resisting ~ 60-565
 crucible ~ 60-630
 cutlery ~ 60-750
 damascene ~ 60-645
 dead soft ~ 60-345
 deep-drawing ~ 60-790
 deep-hardening ~ 60-385
 die ~ 60-815
 dynamo ~ 60-775
 easy-cutting ~ 60-710
 electric ~ 60-625
 eutectoid ~ 50-490
 extra-hard ~ 60-490
 extra-soft ~ 60-345
 ferritic ~ 60-415
 ferritic stainless ~ 60-520
 ferro- ~ 60-1035
 fine-grained ~ 60-470
 forged ~ 60-330
 free-cutting ~ 60-710
 free-machining ~ 60-710
 fully annealed ~ 60-275
 general-purpose ~ 60-665
 graphitic ~ 60-300
 graphitized ~ 60-300
 Hadfield ~ 60-785
 Hadfield's manganese ~ 60-785
 hammered ~ 60-330
 hard ~ 60-350
 hardenable ~ 60-360
 hardened ~ 60-285
 hard magnetic ~ 60-505
 heat-resistant ~ 60-560
 heat-resisting ~ 60-560
 heat treatable ~ 60-705
 high alloy ~ 60-130
 high-carbon ~ 60-110
 high-chromium ~ 60-180
 high-grade ~ 60-145
 high-hardenability ~ 60-385
 highly-alloyed ~ 60-130
 high-quality ~ 60-145
 high-speed ~ 60-800
 high-strength 60-355
 high temperature tool ~ 60-810
 high-tensile ~ 60-355
 hot-work die ~ 60-810
 hot-working ~ 60-810
 hot-working die ~ 60-810
 hot-work tool ~ 60-810
 hypoeutectoid ~ 50-495
 hypereutectoid ~ 50-500
 ingot ~ 60-580
 killed ~ 60-655
 lead-bearing ~ 60-250
 ledeburitic ~ 60-440
 low-alloy ~ 60-120
 low-carbon ~ 60-100

steel
 low-hardenability ~ 60-330
 low-temperature ~ 60-555
 machine ~ 60-680
 machinery ~ 60-680
 magnet ~ 60-780
 magnetic ~ 60-495
 magnetically hard ~ 60-505
 magnetically soft ~ 60-500
 manganese ~ 60-195
 manganese-silicon ~ 60-205
 maraging ~ 60-460
 martensitic ~ 60-430
 medium-alloy ~ 60-125
 medium-carbon ~ 60-105
 mild ~ 60-100, 60-340
 mild-alloy ~ 60-120
 moderately alloyed ~ 60-125
 molybdenum ~ 60-220
 nickel ~ 60-185
 nitralloy ~ 60-700
 nitrided ~ 60-315
 nitride-hardened ~ 60-315
 nitride ~ 60-315
 nitriding ~ 60-695
 non-ageing ~ 60-405
 non-aging 60-405
 non-deforming ~ 60-410
 non-hardenable ~ 60-390
 non-magnetic ~ 60-510
 normalized ~ 60-280
 oil-hardening ~ 60-370
 open-hearth ~ 60-610
 ordinary ~ 60-095
 overheated ~ 60-320
 oxidation-resistant ~ 60-560
 oxygen ~ 60-605
 pearlitic ~ 60-425
 permanent magnet ~ 60-780
 plain carbon ~ 60-095
 pressure vessel ~ 60-740
 puddled ~ 60-575
 quality ~ 60-140
 quaternary ~ 60-165
 quenched ~ 60-285
 quench-hardened ~ 60-285
 rail ~ 60-730
 rapid ~ 60-800
 refined ~ 60-635
 reinforcing ~ 60-795
 resulfurized ~ 60-710
 resulfurized stainless ~ 60-525
 rimming ~ 60-650
 rivet ~ 60-745
 rustless ~ 60-515
 rust-resisting ~ 60-515
 self-hardening ~ 60-375
 semi-~ 60-1035
 semi-austenitic ~ 60-445
 semi-ferritic ~ 60-450
 semi-killed ~ 60-660
 semi-rimming ~ 60-660
 shallow-hardening ~ 60-380
 ship ~ 60-755
 Siemens-Martin ~ 60-610
 silicon ~ 60-200
 silver ~ 60-820
 soft ~ 60-340
 soft magnetic ~ 60-500
 special ~ 60-115, 60-150
 specialty ~ 60-150
 spring ~ 60-725
 stabilized ~ 60-305
 stainless ~ 60-515
 standard ~ 60-265
 ~ of commercial grade 60-135
 ~ tough at subzero 60-555
 straight carbon ~ 60-095
 structural ~ 60-685
 surface-hardening ~ 60-380
 temper resistant ~ 60-395
 ternary ~ 60-160
 Thomas ~ 60-600
 titanium ~ 60-230
 tonnage ~ 60-135
 tool ~ 60-675
 toughened ~ 60-290
 toughening ~ 60-705

STEEL

steel
 transformer ~ 60-770
 tungsten ~ 60-215
 tyre ~ 60-760
 unalloyed ~ 60-095
 unkilled ~ 60-650
 vacuum-melted ~ 60-640
 vacuum-treated ~ 60-640
 valve ~ 60-735
 vanadium ~ 60-225
 water-hardening ~ 60-365
 wear-resistant ~ 60-480
 wear-resisting ~ 60-480
 weather-resisting ~ 60-540
 weldable ~ 60-485
 work-hardened ~ 60-295
 wrought ~ 60-570
steel bronze 65-355
steel designation 60-090
steeled cast iron 60-1035
steelmaking pig iron 60-880
stellite 65-495
step
 growth ~ 25-330
stepped annealing 70-455
step quenching 70-305
stereotype metal 55-630
sterling gold 65-790
stoichiometric composition 35-240
stored energy 45-060
straight brass 65-305
straight carbon steel 60-095
strain
 critical ~ 45-075
 elastic ~ 45-005
 plastic ~ 45-010
 threshold ~ 45-075
strain ageing 45-225
strain crack 85-070
strain hardening 45-045
strain-hardening coefficient 45-065
strain-hardening curve 45-070
strain-hardening index 45-065
strain induced martensite 50-360
strains pl
 stretcher ~ 45-195
strength 75-040
 bending ~ 75-090
 cohesive ~ 75-135
 creep ~ 75-220
 creep rupture ~ 75-235
 cyclic ~ 75-200
 disruptive ~ 75-135
 dynamic ~ 75-185
 endurance ~ 75-200
 fatigue ~ 75-200
 high-temperature ~ 75-095, 75-240
 high-temperature creep ~ 75-240
 hot ~ 75-095
 impact ~ 75-190
 mechanical ~ 75-040
 shear ~ 75-080
 specific ~ 75-140
 ~ of dislocation 30-210
 stress rupture ~ 75-235
 tensile ~ 75-070
 torsional ~ 75-085
 ultimate ~ 75-040
 ultimate compressive ~ 75-075
 ultimate tensile ~ 75-070
 ultrahigh ~ 75-045
 yield ~ 75-115
strength characteristic 75-060
strengthening 70-025
 dispersion ~ 55-220
 radiation-induced ~ 30-495
 solution ~ 55-215

strengthening treatment 70-025
strength parameter 75-060
strength properties pl 75-010
stress
 critical resolved shear ~ 45-090
 critical shear ~ 45-090
 flow ~ 45-210
 fracture ~ 75-130
 limiting creep ~ 75-230
 offset yield ~ 75-115
 proof ~ 75-115
 shear ~ 45-085
 true limiting creep ~ 75-225
 ultimate tensile ~ 75-070
 yield ~ 75-110
stress crack 85-065
stress corrosion 85-310
stress corrosion cracking 85-385
stresses pl
 body ~ 85-030
 casting ~ 85-170
 constitutional ~ 85-035
 internal ~ 85-020
 locked-in ~ 85-020
 locked-up ~ 85-020
 macroscopic residual ~ 85-030
 microscopic ~ 80-035
 quenching ~ 85-185
 residual ~ 85-020
 shrinkage ~ 85-175
 temperature ~ 85-025
 textural ~ 85-035
 thermal ~ 85-025
stress field of a dislocation 30-225
stress relaxation 45-525
stress relief annealing 70-525
stress relieving 70-525
stress rupture strength 75-235
stress-strain curve 80-410
stress-strain diagram 80-410
stretcher strains pl 45-195
strong alloy 55-420
strong direction 20-220
strong plane 20-215
strontium 05-430
structural change 25-500
structural constituent 40-010
structural metallurgy 10-005
structural properties pl 75-860
structural steel 60-685
structural vacancy 30-090
structure 40-005
 abnormal ~ 40-135
 acicular ~ 40-115
 as cast ~ 40-050
 banded ~ 40-140
 block ~ 40-170
 cast ~ 40-050
 cell ~ 40-095
 cellular ~ 40-095
 coarse ~ 40-390
 coarse-grained ~ 40-390
 columnar ~ 40-125
 crystal ~ 20-040
 crystallographic ~ 20-040
 dendritic ~ 40-090
 dislocation ~ 30-415
 domain ~ 40-620
 duplex 40-150
 electronic ~ 10-030
 envelope ~ 40-100
 eutectic ~ 40-060
 eutectoid ~ 40-075
 fine ~ 40-040
 fine-grained ~ 40-395
 globular ~ 40-120
 granular ~ 40-120
 heterogeneous ~ 40-155
 homogeneous ~ 40-145

— 304 —

SYSTEM

structure
  hypereutectic ~ 40-065
  hypereutectoid ~ 40-080
  hypoeutectic ~ 40-070
  hypoeutectoid ~ 40-085
  internal ~ 40-035
  lamellar ~ 40-105
  lineage ~ 40-165
  magnetic ~ 40-650
  metal ~ 40-005
  mosaic ~ 40-170
  network ~ 40-100
  polyhedral ~ 40-095
  polyphase ~ 40-160
  primary ~ 40-045
  reticular ~ 40-100
  secondary ~ 40-055
  single-phase ~ 40-145
  sub-boundary ~ 40-035
  subgrain ~ 40-035
  Widmannstätten ~ 40-130
  two-phase ~ 40-150
structure-insensitive properties pl 75-865
structure metal 40-005
structure-sensitive properties pl 75-860
study
  calorimetric ~ 80-500
  electron diffraction ~ 80-770
  metallographic ~ 80-035
  X-ray ~ 80-570
subboundary 30-660
sub-boundary structure 40-035
sub-grain 30-650
subgrain 30-650
sub-grain boundary 30-660
subgrain boundary 30-660
subgrain structure 40-035
sublattice 20-285
  magnetic ~ 20-485
sublimation point 75-435
sublimation temperature 75-435
submicroscopic precipitate 40-315
subquenching 70-340
sub-shell 10-160
subsidiary quantum number 10-135
substance
  diamagnetic ~ 75-560
  ferromagnetic ~ 75-550
  paramagnetic ~ 75-555
substitutional atom 30-070
substitutional solid solution 35-415
substrate 25-165
substructure 40-035
subsurface corrosion 85-355
sub-zero treatment 70-340
sulfide
  iron ~ 50-245
  manganese ~ 50-240
sulfide inclusions pl 40-530
sulfiding 70-895
sulfinuzing 70-890
sulfocarbonitriding 70-892
sulfonitriding 70-890
sulfur 05-435
sulfurizing 70-895
sulfur print 80-325
sulfur print test 80-320
sulphur 05-435
superalloy 55-515
superconducting critical temperature 25-570
superconducting transition temperature 25-570
superconductivity 75-490

superconductor 10-390
supercooling 25-240
  constitutional ~ 25-255
superdislocation 30-220
superfusion 25-240
superlattice 35-440
superlattice phase 35-200, 35-430
superlattice reflections pl 80-655
superparamagnetism 75-570
superplastic alloy 55-430
superplasticity 75-055
super-pure metal 55-105
supersaturated solid solution 35-400
supersonic testing 80-800
superstrength 75-045
super-strength alloy 55-425
superstructure 35-440
superstructure lines pl 80-655
surface
  Fermi ~ 10-215
  fracture ~ 40-670
  freezing point ~ 35-520
  liquidus ~ 35-520
  solidus ~ 35-495
surface corrosion 85-330
surface crack 85-050
surface decarburization 85-210
surface defect 85-010, 30-535
surface diffusion 45-300
surface energy 30-640
surface free energy 15-120
surface hardening 70-285
surface-hardening steel 60-380
surface imperfection 85-010
surface tension 30-645
surface treatment 70-010
surfusion 25-240
susceptibility
  magnetic ~ 75-625
  ~ to grain growth 75-760
  ~ to overheating 75-760
suspension method 80-815
Suzuki effect 30-455
sweat
  tin ~ 85-165
sweating out 25-430
Swedish iron 60-030
symmetry
  crystal ~ 20-055
  crystalline ~ 20-055
synthetic alloy 55-270
synthetic change 35-585
synthetic pig iron 60-940
synthetic reaction 35-585
synthetic temperature 35-625
synthetic transformation 35-585
synthetized alloy 55-270
system 35-005
  alloy ~ 55-165
  anorthic ~ 20-270
  asymmetric ~ 20-270
  binary ~ 35-120
  biphasic ~ 35-100
  bivariant ~ 35-085
  conjugate slip ~ 45-125
  crystal ~ 20-235
  crystallographic ~ 20-235
  cubic ~ 20-240
  equilibrium ~ 35-045
  four-component ~ 35-135
  heterogeneous ~ 35-095
  hexagonal ~ 20-245

system
 invariant ~ 35-075
 iron-carbon ~ 50-015
 isometric ~ 20-240
 metal ~ 55-165
 metallic ~ 55-165
 monoclinic ~ 20-265
 monophasic ~ 35-090
 monovariant ~ 35-080
 multicomponent ~ 35-140
 multiphase ~ 35-110
 oblique ~ 20-265
 one-component ~ 35-115
 orthorhombic ~ 20-255
 polycomponent ~ 35-140
 polynary ~ 35-140
 polyphase ~ 35-110
 primary glide ~ 45-115
 primary slip ~ 45-115
 pseudo-binary ~ 35-130
 pseudobinary ~ 35-130
 pyramidal ~ 20-250
 quadratic ~ 20-250
 quasi-binary ~ 35-130
 quaternary ~ 35-135
 regular ~ 20-240
 rhombohedral ~ 20-260
 secondary slip ~ 45-120
 single-phase ~ 35-090
 slip ~ 45-110
 terminal ~ 35-715
 ternary ~ 35-125
 tesseral ~ 20-240
 tetragonal ~ 20-250
 three-component ~ 35-125
 three-phase ~ 35-105
 triclinic ~ 20-270
 trigonal ~ 20-260
 two-component ~ 35-120
 two-phase ~ 35-100
 unary ~ 35-115
 unicomponent ~ 35-115
 unitary ~ 35-115

T

tangle
 dislocation ~ 30-385
tantalum 05-440
tarnish resistance 75-680
tarnish-resistance alloy 55-500
tarnish-resisting alloy 55-500
Taylor dislocation 30-150
Taylor-Orowan dislocation 30-150
technetium 05-445
technical alloy 55-520
technological properties pl 75-720
tellurium 05-450
temper
 blue ~ 70-390
temperability 75-765
temperature
 annealing ~ 70-425
 austenitizing ~ 70-135
 boiling ~ 75-430
 breakdown ~ 15-340
 brittle fracture transition ~ 75-280
 critical ~ 35-335, 35-555
 critical solution ~ 35-335
 Curie ~ 25-560
 Debye ~ 10-280
 Debye characteristic ~ 10-280
 disordering ~ 25-625
 ductility transition ~ 75-280
 equicohesive ~ 30-570
 equilibrium 35-690
 eutectic ~ 35-605
 eutectoid ~ 35-630
 freezing ~ 75-425
 hardening ~ 70-140
 homologous ~ 45-475
 liquidus ~ 35-525

temperature
 M_f ~ 70-170
 M_s ~ 70-165
 martensite finish ~ 70-170
 martensite start ~ 70-165
 metatectic ~ 35-620
 monotectic ~ 35-615
 Néel ~ 25-565
 nil-ductility ~ 75-280
 peritectic ~ 35-610
 peritectoid ~ 35-635
 quenching ~ 70-140
 recrystallization ~ 45-470
 solidification ~ 75-425
 solidus ~ 35-500
 sublimation ~ 75-435
 superconducting critical ~ 25-570
 superconducting transition ~ 25-570
 syntetic ~ 35-625
 tempering ~ 70-380
 transformation ~ 35-555
 transition ~ 25-570, 35-555, 75-280
temperature coefficient of electrical resistivity 75-480
temperature coefficient of resistance 75-480
temperature coefficient of resistivity 75-480
temperature equalization 70-035
temperature gradient 15-075
temperature hysteresis 25-480
temperature stresses pl 85-025
temper carbon 50-220
temper colour 70-385
temper brittleness 85-220
tempered martensite 50-345
temper embrittlement 85-220
temper graphite 50-220
tempering 70-365
 high-temperature ~ 70-375
 low-temperature ~ 70-370
tempering metal 55-285
tempering resistance 75-840
tempering temperature 70-380
temper resistant steel 60-395
temper sorbite 50-390
temper troostite 50-405
tenacity 75-070
 specific ~ 75-140
tensile properties pl 75-015
tensile strength 75-070
tensile test 80-405
tensile texture 40-585
tension
 interfacial ~ 30-645
 surface ~ 30-645
tension test 80-405
terbium 05-455
terminal phase 35-380
terminal solid solution 35-380
terminal system 35-715
ternary alloy 55-300
ternary axis 20-075
ternary equilibrium diagram 35-710
ternary equilibrium space diagram 35-750
ternary eutectic 40-210
ternary phase equilibrium diagram 35-710
ternary phase equilibrium spatial diagram 35-750
ternary space diagram 35-750
ternary spatial diagram 35-750
ternary steel 60-160

TIME

ternary system 35-125
tertiary cementite 50-090
tertiary creep 45-510
tertiary recrystallization 45-460
tesseral system 20-240
test
  aural ∼ 80-795
  end quench hardenability ∼ 80-450
  endurance ∼ 80-420
  fatigue ∼ 80-420
  fluoroscope ∼ 80-835
  fracture ∼ 80-445
  hardness ∼ 80-415
  impact ∼ 80-430
  Jominy ∼ 80-450
  Jominy end quench ∼ 80-450
  mercurous nitrate ∼ 80-485
  sonic ∼ 80-795
  spark ∼ 80-440
  spot ∼ 80-465
  sulfur print ∼ 80-320
  tensile ∼ 80-405
  tension ∼ 80-405
test bar 80-435
testing
  mechanical ∼ 80-400
  non-destructive ∼ 80-785
  sonic ∼ 80-795
  spark ∼ 80-440
  supersonic ∼ 80-800
  ultrasonic ∼ 80-800
tetrad axis 20-080
tetragonality 20-375
tetragonal lattice 20-370
tetragonal martensite 50-335
tetragonal system 20-250
tetrahedral cubic lattice 20-320
tetrahedral interstice 20-325
tetrahedral void 20-325
tetravacancy 30-055
textural stresses pl 85-035
texture 40-555
  annealing ∼ 40-600
  casting ∼ 40-580
  compressive ∼ 40-590
  cube ∼ 40-605
  cube-on-edge ∼ 40-610
  deformation ∼ 40-560
  drawing ∼ 40-575
  fibre ∼ 40-565
  fibrous ∼ 40-565
  Goss ∼ 40-610
  recrystallization ∼ 40-600
  rolling ∼ 40-570
  tensile ∼ 40-585
  torsional ∼ 40-595
textured alloy 55-380
thalium 05-460
theory
  band ∼ 10-020
thermal activation 45-275
thermal agitation 10-270
thermal analysis 80-505
thermal arrest 80-530
thermal capacity 75-410
thermal coefficient of expansion 75-450
thermal conductivity 75-440
thermal crack 85-060
thermal cycle 70-970
thermal cycling 70-520
thermal dendrite 25-325
thermal diffusion 45-370
thermal diffusivity 75-445
thermal electromotive force 75-495
thermal endurance 75-830

thermal entropy 15-205
thermal equilibrium 15-325
thermal equilibrium diagram 35-480
thermal etching 80-190
thermal expansion coefficient 75-450
thermal fatigue 85-425
thermal gradient 15-075
thermal groove 80-195
thermal grooving 80-190
thermal hysteresis 25-480
thermal lag 25-480
thermalloy 65-540
thermally activated process 45-280
thermal properties pl 75-365
thermal recovery 45-430
thermal resistance 75-835
thermal shock resistance 75-825
thermal spike 30-530
thermal stresses pl 85-025
thermal treatment 70-006
thermal vacancy 30-095
thermal vibration 10-270
thermo-chemical treatment 70-650
thermodynamic activity 15-280
thermodynamical analysis by calorimetry 80-535
thermodynamic equilibrium 15-320
thermodynamic equilibrium constant 15-335
thermodynamic force 15-070
thermodynamic function 15-030
thermodynamic multiplicity 15-230
thermodynamic parameter 15-020
thermodynamic potential at constant volume 15-105
thermodynamic quantity 15-030
thermodynamics
  metallurgical ∼ 15-010
  ∼ of alloys 15-010
  ∼ of solids 15-005
thermodynamic state 15-015
thermodynamic variable 15-020
thermoelectric power 75-495
thermoelectromotive force 75-495
thermogram 80-527
thermomagnetic analysis 80-565
thermo-magnetic treatment 70-965
thermo-mechanical treatment 70-950
thermomigration 45-370
thermopower 75-495
thin foil 80-350
third quantum number 10-140
Thomas pig iron 60-850
Thomas steel 60-600
thorium 05-465
three-component alloy 55-300
three-component system 35-125
three-fold rotation axis 20-075
three-phase system 35-105
threshold strain 45-075
through-hardening 70-280
thulium 05-470
tie line 35-695
tilt boundary 30-605
time
  etching ∼ 80-125
  half-life ∼ 75-525
  holding ∼ 70-060

TIME

time
incubation ~ 70-150
soaking ~ 70-060
time quenching 70-300
tin 05-475
grey ~ 65-635
white ~ 65-630
tin brass 65-345
tin bronze 65-110
tin cry 45-270
tin disease 85-245
tin exudation 85-165
tin-free bronze 65-125
tin pest 85-245
tin plague 85-245
tin solder 65-725
tin sweat 85-165
tint
heat ~ 70-385
tinting
heat ~ 80-185
titanide 40-510
titanium 05-480
titanium nitride 70-825
titanium steel 60-230
titanzing 70-945
tombac 65-395
tombac alloy 65-395
tonnage steel 60-135
tool steel 60-675
torch hardening 70-290
torsional strength 75-085
torsional texture 40-595
torsion modulus 75-155
toughened steel 60-290
toughening 70-410
toughening steel 60-705
toughness 75-035
notch ~ 75-190
notch impact ~ 75-190
total quantum number 10-130
trace element 55-250
tracer
radioactive ~ 80-760
track autoradiography 80-750
tramp elements pl 55-200
transcrystalline fracture 40-710
transcrystallization 25-070
transformation
allotropic ~ 25-455
athermal ~ 25-520
bainite ~ 50-290
bainitic ~ 50-290
congruent ~ 25-525
continuous ~ 25-540
diffusional ~ 25-505
diffusionless ~ 25-510
discontinuous ~ 25-545
eutectic ~ 35-565
eutectoid ~ 35-590
eutectoidal ~ 35-590
first-degree ~ 25-465
first-order ~ 25-465
heterogeneous ~ 25-535
homogeneous ~ 25-530
isothermal ~ 25-515
magnetic ~ 25-555
martensite ~ 50-280
martensitic ~ 50-280
massive ~ 25-550
metatectic ~ 35-580
monotectic ~ 35-575
monotectoid ~ 35-600
order-disorder ~ 25-580

transformation
pearlitic ~ 50-275
peritectic ~ 35-570
peritectoid ~ 35-595
phase ~ 25-460
polymorphic ~ 25-455
reversible ~ 25-495
secondary martensitic ~ 50-285
second-degree ~ 25-470
second-order ~ 25-470
shear ~ 25-510
solid state ~ 25-475
solid ~ 25-475
split ~ of austenite 50-275
syntetic ~ 35-585
transformation point 35-555
transformation range 35-560
transformation temperature 35-555
transformed austenite 50-035
transformer steel 60-770
transgranular fracture 40-710
transient creep 45-500
transient phase 35-195
transition
first-degree ~ 25-465
first-order ~ 25-465
order-disorder ~ 25-580
phase ~ 25-460
reversible ~ 25-495
second-degree ~ 25-470
second-order ~ 25-470
solid state ~ 25-475
transition elements pl 55-025
transition lattice 20-405
transition metals pl 55-025
transition phase 35-195
transition point 25-570, 35-555
transition temperature 35-555, 25-570
translation 20-130
translation lattice 20-275
translation period 20-115
transmission diagram 80-675
transmission electron microscopy 80-340
transmission method 80-625
transmission microscopy 80-220
transmission pattern 80-675
transition temperature brittle fracture 75-280
treating
steam ~ 70-880
treatment
chemico-thermal ~ 70-650
cold ~ 70-340
double hardening ~ 70-765
heat ~ 70-005
high-temperature thermo-mechanical ~ 70-955
natural stabilizing ~ 70-540
solution ~ 70-500
solution heat ~ 70-570
stabilizing ~ 70-535
steam ~ 70-880
strengthening ~ 70-025
sub-zero ~ 70-340
surface ~ 70-010
thermal ~ 70-005
thermo-chemical ~ 70-650
thermo-magnetic ~ 70-965
thermo-mechanical ~ 70-950
tree 30-410
tree dislocation 30-410
triad axis 20-075
triangle
composition ~ 35-720
concentration ~ 35-720
Gibbs ~ 35-720
triangulation of the system 35-725

— 308 —

tribonucleation 25-210
triclinic lattice 20-400
triclinic system 20-270
trigonal lattice 20-390
trigonal system 20-260
triple point 35-040
trivacancy 30-050
troostite 50-395
  nodular ~ 50-400
  quenching ~ 50-400
  temper ~ 50-405
troostite-martensite 50-410
troosto-martensite 50-410
trough
  eutectic ~ 35-730
true limiting creep stress 75-225
trunk
  atomic ~ 10-040
tungsten 05-485
tungsten bronze 65-170
tungsten carbide 40-450
tungsten impregnation 70-935
tungsten steel 60-215
tungstide 40-500
twin 45-230
  annealing ~ 45-240
  crystal ~ 45-230
  deformation ~ 45-235
  growth ~ 45-240
  mechanical ~ 45-235
  recrystallization ~ 45-240
twin axis 45-260
twin band 45-255
twin boundary 45-265
twin crystal 45-230
twin fault 30-555
twinned boundary 45-265
twinned crystal 45-230
twinning 45-245
twinning band 45-255
twinning dislocation 30-375
twinning plane 45-250
twin plane 45-250
twist boundary 30-600
two-component alloy 55-295
two-component system 35-120
two-dimensional defect 30-535
two-fold rotation axis 20-070
two-phase alloy 55-335
two-phase brass 65-320
two-phase region 35-685
two-phase structure 40-150
two-phase system 35-100
type metal 55-610
type of steel 60-080
tyre steel 60-760

U

ulminium 65-555
ultimate compressive strength 75-075
ultimate strength 75-040
ultimate tensile strength 75-070
ultimate tensile stress 75-070
ultrahigh strength 75-045
ultra-high-strength alloy 55-425
ultra-light alloy 55-410
ultra-pure metal 55-105

ultrasonic testing 80-800
ultra-violet microscopy 80-265
unalloyed steel 60-095
unary system 35-115
uncertainty principle 10-165
under-annealing 70-470
undercooled austenite 50-265
undercooling 25-240, 25-245
  constitutional ~ 25-255
underhardening 70-335
undissolved carbides pl 40-480
uniaxial solidification 25-040
unicomponent system 35-115
unidirectional solidification 25-040
uniform corrosion 85-335
unitary system 35-115
unit cell 20-090
unit cell vector 20-135
unit dislocation 30-215
unit lattice vector 20-135
unkilled steel 60-650
unmixing 25-365
unsoundness of copper 85-395
unstable austenite 50-265
unsteady creep 45-500
up-hill diffusion 45-340
upper bainite 50-420
uranium 05-490

V

vacancy 30-035
  constitutional ~ 30-090
  equilibrium ~ 30-090
  excess ~ 30-105
  Frenkel ~ 30-065
  quenched-in ~ 30-100
  single ~ 30-040
  structural ~ 30-090
  thermal ~ 30-095
vacancy cluster 30-125
vacancy condensation 30-115
vacancy diffusion 45-305
vacuum-evaporated metal 55-145
vacuum-melted metal 55-135
vacuum-melted steel 60-640
vacuum metal 55-135
vacuum-treated steel 60-640
valence band 10-255
valence crystal 20-030
valency band 10-255
valency electron 10-065
valley
  eutectic ~ 35-730
valve bronze 65-275
valve steel 60-735
vanadium 05-495
vanadium bronze 65-180
vanadium carbide 40-460
vanadium nitride 70-830
vanadium steel 60-225
van der Waals bond 10-335
variable
  state ~ 15-020
  thermodynamic ~ 15-020
variance 35-030
vector
  Burgers ~ 30-190
  lattice ~ 20-135

vector
  unit cell ~ 20-135
  unit lattice ~ 20-135
  wave ~ 10-185
Vegard's law 35-425
very fine pearlite 50-400
vibration
  thermal ~ 10-270
vibrational entropy 15-205
virgin metal 55-115
viscous flow 45-220
vitallium 65-485
vitreous fracture 40-735
velocity
  quenching ~ 70-120
vertical section 35-740
void 85-100
  interstitial ~ 20-105
  octahedral ~ 20-330
  tetrahedral ~ 20-325
Voight's modulus 75-160
volume
  activation ~ 30-475
  atomic ~ 75-320
  molar ~ 75-325
volume coefficient of thermal expansion 75-460
volume diffusion 45-295
volume shrinkage 75-795
volumetric modulus of elasticity 75-160
volumetric properties pl 75-340

W

wall
  Bloch ~ 40-645
  dislocation ~ 30-395
  domain ~ 40-645
warping 85-200
water cooling 70-100
water corrosion 85-280
water hardening 70-205
water-hardening steel 60-365
water quenching 70-100, 70-205
wave function 10-180
wave space 10-205
wave vector 10-185
wave vector space 10-205
wear resistance 75-265
wear-resistant steel 60-480
wear-resisting alloy 55-435
wear-resisting steel 60-480
weather-resisting steel 60-540
weight
  atomic ~ 10-075
  molecular ~ 75-355
weight density 75-350
weldability 75-735
weldable steel 60-485
wet method 80-815
wet nitriding 70-795
whisker 25-360
white bearing metal 65-640
white bronze 65-230
white cast iron 60-955
white copper 65-415
white fracture 85-160
white gold 65-805
white heart malleable cast iron 60-1120

white heart process 70-635
white metal 65-640
white pig iron 60-910
white solder 65-745
white spots pl 85-250
white tin 65-630
Widmannstätten structure 40-130
Wöhler curve 80-425
wolfram bronze 65-170
Wood's alloy 65-670
Wood's metal 65-670
woody fracture 40-720
work
  cold ~ 45-030
workable alloy 55-530
work function 10-060, 15-105
work-hardened steel 60-295
work hardening 45-045
work-hardening capacity 75-770
work-hardening coefficient 45-065
work-hardening curve 45-070
working
  cold plastic ~ 45-025
  cold ~ 45-025
  hot ~ 45-020
  hot plastic ~ 45-020
  mechanical ~ 45-015
  plastic ~ 45-015
working properties pl 75-820
work softening 45-535
wrought alloy 55-530
wrought brass 65-295
wrought bronze 65-105
wrought iron 60-040
wrought steel 60-570

X

xenomorphic crystal 40-360
xenon 05-500
X-ray crystal analysis 80-580
X-ray diffraction analysis 80-580
X-ray diffraction line 80-645
X-ray diffraction pattern 80-615
X-ray diffraction photograph 80-615
X-ray diffraction powder pattern 80-635
X-ray examination 80-570
X-ray fluorescence analysis 80-700
X-ray fluorescence spectroscopy 80-700
X-ray fluorescent analysis 80-700
X-ray metallography 80-575
X-ray microanalysis 80-780
X-ray microscopy 80-730
X-ray phase analysis 80-585
X-ray radiographic inspection 80-820
X-ray spectroscopy 80-700
X-ray study 80-570
X-ray structure analysis 80-580

Y

Y-alloy 65-590
yellow brass 65-330
yield point 75-110
yield strength 75-115
yield stress 75-110

Young's modulus 75-150
ytterbium 05-505
yttrium 05-510

Z

zamak 65-705
zamak alloy 65-705
zero-dimensional defect 30-030
zinc 05-515
hard ~ 65-700
rectified ~ 65-695
zinc bronze 65-205
zinc impregnation 70-930
Zintl phases *pl* 35-270

zircaloy 65-855
zirconium 05-520
zone
Brillouin ~ 10-230
chill ~ 40-175
columnar ~ 40-180
columnar crystal ~ 40-180
coordination ~ 20-190
denuded ~ 30-120
depleted ~ 30-120
diffusion ~ 45-415
equiaxed ~ 40-185
hardened ~ 70-355
~ of equiaxed grains 40-185
zone-refined metal 55-140
zones *pl*
Guinier-Preston ~ 70-605

INDEX FRANÇAIS

A

acier *m* 60-075
- ~ à aimants 60-780
- ~ à bandages 60-760
- ~ à bas carbone 60-100
- ~ à bonne trempabilité 60-385
- ~ à carbone élevé 60-110
- ~ à carbone moyen 60-105
- ~ à chaudières 60-740
- ~ à coupe rapide 60-800
- ~ à grain fin 60-470
- ~ à gros grain 60-475
- ~ à haute carbone 60-110
- ~ à haute résistance mécanique 60-355
- ~ à haute tenacité à froid 60-555
- ~ à haute teneur en carbone 60-110
- ~ à haute teneur en chrome 60-180
- ~ à l'aluminium 60-235
- ~ à l'oxygène 60-605
- ~ à matrices 60-815
- ~ à moyen carbone 60-105
- ~ à outils 60-675
- ~ à rail 60-730
- ~ à résistance élevée 60-355
- ~ à ressort 60-725
- ~ à rivets 60-745
- ~ à roulements 60-715
- ~ à soufflures 60-585
- ~ à soupapes 60-735
- ~ acide 60-615
- ~ allié 60-115
- ~ ~ de nitruration 60-700
- ~ amagnétique 60-510
- ~ améliorable par traitement 60-705
- ~ amélioré 60-290
- ~ anormal 60-465
- ~ argenté 60-820
- ~ au bore 60-255
- ~ au carbone 60-095
- ~ au chrome 60-175
- ~ au chrome-manganèse-silicium 60-260
- ~ au chrome-nickel 60-190
- ~ au cobalt 60-240
- ~ au convertisseur 60-590
- ~ au creuset 60-630
- ~ au cuivre 60-245
- ~ au manganèse 60-195
- ~ au molybdène 60-220
- ~ au nickel 60-185
- ~ au plomb 60-250
- ~ au silicium 60-200
- ~ au titanium 60-230
- ~ au tungstène 60-215
- ~ au vanadium 60-225
- ~ austénitique 60-420
- ~ austéno-ferritique 60-455
- ~ autotrempant 60-375
- ~ bainitique 60-435
- ~ basique 60-620
- ~ Bessemer 60-595
- ~ binaire 60-155
- ~ boursoufflé 60-585
- ~ brûlé 60-325
- ~ calmé 60-655
- ~ cementé 60-310
- ~ complexe 60-170
- ~ coque 60-755
- ~ corroyé 60-570
- ~ coulé 60-580
- ~ d'armature 60-795
- ~ damassé 60-645
- ~ de blindage 60-765
- ~ de cémentation 60-690
- ~ de construction 60-670, 60-685
- ~ de construction mécanique 60-680

acier
- ~ de construction ordinaire 60-665
- ~ de coutellerie 60-750
- ~ de décolletage 60-710
- ~ de faible trempabilité 60-380
- ~ de moulage 60-825
- ~ de nitruration 60-695
- ~ de qualité 60-140
- ~ de remplacement 60-270
- ~ de traitement 60-705
- ~ diamant 60-490
- ~ doux 60-340
- ~ du type maraging 60-460
- ~ dur 60-350
- ~ économique 60-270
- ~ écroui 60-295
- ~ effervescent 60-650
- ~ élaboré sous vide 60-640
- ~ électrique 60-625
- ~ eutectoïde 50-490
- ~ extra-doux 60-345
- ~ extra-dur 60-490
- ~ faiblement allié 60-120
- ~ ferritique 60-415
- ~ fin 60-145
- ~ fondu sous vide 60-640
- ~ forgé 60-330
- ~ fortement allié 60-130
- ~ graphitique 60-300
- ~ graphitisé 60-300
- ~ Hadfield 60-785
- ~ hautement allié 60-130
- ~ hypereutectoïde 50-500
- ~ hypoeutectoïde 50-495
- ~ inattaquable 60-535
- ~ indéformable 60-410
- ~ inoxy 60-515
- ~ inoxydable 60-515
- ~ ~ au soufre 60-525
- ~ ~ ferritique 60-520
- ~ ~ resulphuré 60-525
- ~ insensible à la surchauffe 60-470
- ~ lédéburitique 60-440
- ~ légèrement allié 60-120
- ~ magnétique 60-495
- ~ magnétiquement doux 60-500
- ~ ~ dur 60-505
- ~ mangano-silicieux 60-205
- ~ maraging 60-460
- ~ martensitique 60-430
- ~ Martin 60-610
- ~ Martin-Siemens 60-610
- ~ mi-ferritique 60-450
- ~ moulé 60-825
- ~ ~ allié 60-835
- ~ ~ non allié 60-830
- ~ mousseux 60-650
- ~ moyennement allié 60-125
- ~ nitruré 60-315
- ~ noble 60-145
- ~ non allié 60-095
- ~ non calmé 60-650
- ~ non magnétique 60-510
- ~ non trempable 60-390
- ~ non viellissable 60-405
- ~ normalisé 60-280
- ~ ordinaire 60-095, 60-135
- ~ perlitique 60-425
- ~ peu allié 60-120
- ~ peu trempant 60-380
- ~ plaqué 60-335
- ~ plombifère 60-250
- ~ poule 60-585
- ~ pour couteaux 60-750

ACIER

acier
- ~ pour dynamos 60-775
- ~ pour emboutissage profond 60-790
- ~ pour matrices 60-815
- ~ pour outils travaillant à chaud 60-810
- ~ pour outils travaillant à froid 60-805
- ~ pour ressort 60-725
- ~ pour roulements 60-715
- ~ pour roulements à billes 60-720
- ~ pour taillanderie 60-750
- ~ pour tôles de transformateur 60-770
- ~ pour transformateurs 60-770
- ~ pour travail à chaud 60-810
- ~ pour travail à froid 60-805
- ~ puddlé 60-575
- ~ quaternaire 60-165
- ~ raffiné 60-635
- ~ rapide 60-800
- ~ recuit 60-275
- ~ refondu sous vide 60-640
- ~ réfractaire 60-560
- ~ régénéré 60-280
- ~ résistant à chaud 60-565
- ~ ~ à l'usure 60-480
- ~ ~ à la corrosion 60-535
- ~ ~ à la corrosion atmosphérique 60-540
- ~ ~ au fluage 60-565
- ~ ~ au fluage aux températures élevées 60-565
- ~ ~ au revenu 60-395
- ~ ~ aux acides 60-545
- ~ ~ aux alcalis 60-550
- ~ riche en chrome 60-180
- ~ semi-austénitique 60-445
- ~ semi-calmé 60-660
- ~ semi-effervescent 60-660
- ~ semi-ferritique 60-450
- ~ soudable 60-485
- ~ sous-eutectoïde 50-495
- ~ spécial 60-115, 60-150
- ~ stabilisé 60-305
- ~ standardisé 60-265
- ~ succédané 60-270
- ~ surchauffé 60-320
- ~ susceptible de trempe 60-360
- ~ tenace à froid 60-555
- ~ ternaire 60-160
- ~ Thomas 60-600
- ~ traité 60-290
- ~ ~ sous vide 60-640
- ~ trempant 60-360
- ~ ~ à l'air 60-375
- ~ ~ à l'eau 60-365
- ~ ~ à l'huile 60-370
- ~ trempé 60-285
- ~ très trempant 60-385
- ~ vieillissable 60-400

actinium m 05-010
activation f thermique 45-275
activité f 15-280
- ~ thermodynamique 15-280

addition f principale 55-195
- ~ stabilisante 50-485

additions fpl inintentionelles 55-200
adoucissement m 70-460
affinage m 25-225
- ~ de l'alpax 25-225
- ~ du grain 40-385

affinement m du grain 40-385
affinité f 75-660
- ~ chimique 75-660
- ~ électronique 10-115

électro-affinité f 10-115
agent m antinodulisant 25-235
- ~ de cémentation 70-740

agitation f thermique 10-270
agrégat m eutectoïde 40-240
- ~ polycristallin 20-440

aiguille f de martensite 50-375
- ~ martensitique 50-375

aimantation f à saturation 75-605
- ~ de saturation 75-605

aimantation
- ~ rémanente 75-630

aladar m 65-565
alcalino-terreux mpl 55-035
alcalins mpl 55-030
alclad m 65-560
aldrey m 65-585
alitérage m 70-925
alliage m 55-160
- ~ à bas point de fusion 55-395
- ~ à base de cuivre 65-010
- ~ à base de fer 60-070
- ~ à base de nickel 65-430
- ~ à deux phases 55-335
- ~ à durcissement structural 55-385
- ~ à haute perméabilité magnétique 55-475
- ~ à haute résistance 55-420
- ~ à haute température de fusion 55-400
- ~ à laminer et à tréfiler 55-530
- ~ à monnaies 55-565
- ~ à perméabilité magnétique élevé 55-475
- ~ à pistons 55-570
- ~ à plusieurs composants 55-310
- ~ à traitement thermique 55-525
- ~ à tréfiler 55-530
- ~ à très haute résistance 55-425
- ~ à une phase 55-330
- ~ Alnico 65-530
- ~ amagnétique 55-470
- ~ anti-friction plastique 65-640
- ~ antifriction 55-555
- ~ Bahnmetall 65-685
- ~ binaire 55-295
- ~ biphasé 55-335
- ~ complexe 55-310
- ~ concentré 55-325
- ~ conducteur 55-550
- ~ cuivreux 65-010
- ~ d'addition 55-285
- ~ d'électrotypie 55-635
- ~ d'imprimerie 55-610
- ~ d'ornementation 55-605
- ~ de Babbit 65-645
- ~ de bijouterie 55-580
- ~ de cuivre 65-010
- ~ de décolletage 55-560
- ~ de deuxième fusion 55-280
- ~ de fer 60-070
- ~ de fonderie 55-535
- ~ de forge 55-530
- ~ de forge et de laminage 55-530
- ~ de Heusler 65-420
- ~ de Lipowitz 65-680
- ~ de moulage 55-535
- ~ de nickel 65-430
- ~ de première fusion 55-275
- ~ de rechargement dur 55-590
- ~ de Rose 65-675
- ~ de traitement thermique 55-525
- ~ de Wood 65-670
- ~ dentaire 55-575
- ~ dilué 55-320
- ~ dur 55-415
- ~ durci par dispersion 55-390
- ~ ~ par dispersoïdes 55-390
- ~ durcissable par précipitation 55-385
- ~ durcissant par précipitation 55-385
- ~ électrodéposé 55-290
- ~ électrolytique 55-290
- ~ eutectique 55-350
- ~ eutectoïde 55-365
- ~ extra-léger 55-410
- ~ ferromagnétique 55-455
- ~ fondu 55-150
- ~ formé par infiltration 55-670
- ~ fritté 55-655
- ~ ~ dur 55-665
- ~ fusible 55-395
- ~ hétérogène 55-340
- ~ Heusler 65-420
- ~ homogène 55-330
- ~ hypereutectique 55-360
- ~ hypereutectoïde 55-375
- ~ hypoeutectique 55-355
- ~ hypoeutectoïde 55-370

alliage
~ industriel 55-520
~ léger 55-405
~ magnétique 55-445
~ magnétiquement doux 55-460
~ ~ dur 55-465
~-mère 55-285
~ métallique 55-160
~ monétaire 55-565
~ monophasé 55-330
~ naturel 55-265
~ non-frerreux 65-005
~ ostéoplastique 55-645
~ peu dilatable 55-440
~ polyphasé 55-345
~ pour aimants permanents 55-450
~ pour contacts électriques 55-585
~ pour la coulée sous pression 55-540
~ pour le moulage sous pression 55-540
~ pour les caractères d'imprimerie 55-615
~ pseudobinaire 55-315
~ pyrophorique 55-595
~ quaternaire 55-305
~ réfractaire 55-400, 55-505
~ résistant 55-545
~ ~ à l'oxydation 55-495
~ ~ à l'usure 55-435
~ ~ à la corrosion 55-480
~ ~ au fluage 55-510
~ ~ au fluage aux températures élevées 55-510
~ ~ au ternissement 55-500
~ ~ aux acides 55-485
~ ~ aux alcalis 55-490
~ super-réfractaire 55-515
~ superplastique 55-430
~ superrésistant 55-425
~ synthétique 55-270
~ technique 55-520
~ ternaire 55-300
~ texturé 55-380
~ ultra-léger 55-410
~ Y 65-590
pré-alliage *m* 55-285
pseudo-alliage *m* 55-650
superalliage *m* 55-515
allongement *m* à la rupture 75-120
~ après rupture 75-120
~ de rupture 75-120
~ plastique de rupture 75-120
allotrope *f* 25-390
allotropie *f* 25-385
alnico *m* 65-530
aloi *m* 55-100
alpaca *m* 65-415
alpax *m* 65-565
~ raffiné 65-570
alsifer *m* 60-1165, 65-535
aludur *m* 65-580
alumag *m* 65-600
alumel *m* 65-460
aluminium *m* 05-015
~ ordinaire 65-545
~ raffiné 65-550
cupro-aluminium *m* 65-135
amalgame *m* 65-850
amas *m* de lacunes 30-125
âme *f* 70-665
amélioration *f* 70-410
américium *m* 05-020
amincissement *m* 80-355
amorçage *m* 25-215
~ de cristallisation 25-215
amorphe 20-005
analyse *f* à la microsonde 80-780
~ chimique 80-470
~ des phases par rayons X 80-585
~ dilatométrique 80-540
~ ~ absolue 80-545
~ ~ différentielle 80-550
~ fractographique 80-385

analyse
~ magnétique 80-560
~ par activation 80-765
~ par diffraction électronique 80-770
~ physico-chimique 80-490
~ radiocristallographique 80-580
~ spectrale 80-475
~ thermique 80-505
~ ~ absolue 80-510
~ ~ différentielle 80-515
~ thermomagnétique 80-565
microanalyse *f*
ancien grain *m* d'austenite 50-040
anélasticité *f* 75-050
angle *m* de désorientation 30-595
anisotropie *f* 20-445
~ magnétique 75-575
anneau *m* de Debye 80-640
anomalie *f* de structure d'acier 85-225
anticorodal *m* 65-575
antiferromagnétisme *m* 75-565
antifriction *m* 55-555
antigraphitisant *m* 50-465
antimoine *m* 05-025
~ affiné 65-665
antinodulisant *m* 25-235
antiphase *f* 40-625
aptitude *f* à former un alliage 55-175
~ à l'emboutissage profond 75-750
~ à la consolidation 75-770
~ à la surchauffe 75-760
~ à la trempe 70-020
~ au brasage 75-740
~ au grossissement du grain 75-760
arbre *m* 30-410
argent *m* 05-420
~ de monnaie 65-755
~ fin 65-760
argon *m* 05-030
arrêt *m* thermique 80-530
arsenic *m* 05-035
ascension *f* des dislocations 30-295
assemblage *m* compact 20-210
assignement *m* des indices 80-695
astate *m* 05-040
astérisme *m* 80-680
asymétrie *f* énergétique 15-235
atmosphère *f* active 70-680
~ carburante 70-695
~ controllée 70-700
~ de Cottrell 30-440
~ dislocations-impurtés 30-440
~ endothermique 70-705
~ exothermique 70-710
~ inerte 70-675
~ neutre 70-675
~ protectrice 70-690
~ réactive 70-680
atome *m* d'insertion 30-075
~ de substitution 30-070
~ déplacé 30-510
~ interstitiel 30-075
~ proche voisin 20-195
attaque *f* anodique 80-175
~ cathodique 80-180
~ chimique 80-120, 80-170
~ de joints des grains 80-145
~ de la surface des grains 80-150
~ egzagérée 80-130
~ électrolytique 80-175
~ en couleurs 80-165
~ gazeux 80-182
~ ionique 80-180
~ macrographique 80-135
~ métallographique 80-120
~ micrographique 80-140
~ par gaz ionisés 80-182

attaque
~ par oxydation 80-185
~ par pulvérisation cathodique 80-180
~ profonde 80-160
~ thermique 80-190
~ trop poussée 80-130
auréole f de ferrite 50-230
ausforming m 70-960
austempering m 70-310
austéniformage m 70-960
austénite f 50-025
~ à l'azote 70-835
~ métastable 50-265
~ primaire 50-030
~ résiduelle 50-270
~ retenue 50-270
~ transformée 50-035
nitrausténite f 70-835
austénitisation f 70-130
autodiffusion f 45-400
autoradiogramme m 80-755
autoradiographie f 80-740, 80-755
~ de contraste 80-745
~ de trace 80-750
avional m 65-610
axe m binaire 20-070
~ cristallographique 20-125
~ d'aimantation facile 75-585
~ de cristallisation primaire 25-300
~ de cristallisation secondaire 25-305
~ de macle 45-260
~ de rotation 20-065
~ de symétrie 20-065
~ de symétrie binaire 20-070
~ de symétrie quaternaire 20-080
~ de symétrie sénaire 20-085
~ de symétrie ternaire 20-075
~ privilégié d'aimantation 75-585
~ quaternaire 20-080
~ sénaire 20-085
~ ternaire 20-075
azote m 05-305

B

bain m de trempe 70-195
bainite f 50-410
~ inférieure 50-415
~ supérieure 50-420
bande f d'énergie 10-235
~ de conduction 10-260
~ de déformation 45-200
~ de ferrite libre 85-230
~ de glissement 45-190
~ de maclage 45-255
~ de macle 45-255
~ de ségrégation 85-230
~ de trempabilité 80-460
~ de valence 10-255
~ interdite 10-250
~ permise 10-240
~ pleine 10-245
~ remplie 10-245
bandes fpl de carbure 85-235
~ de Neumann 45-205
~ de Piobert 45-195
~ de Piobert-Lueders 45-195
barbe f 25-360
barreau m d'épreuve 80-435
barrière f de Lomer-Cottrell 30-450
~ de potentiel 10-110
~ énergétique 10-110
baryum m 05-045
base f d'alliage 55-185
~ du cristal 20-110
berkélium m 05-050
béryllium m 05-055

bicristal m 20-435
bilacune f 30-045
bimétal m 65-880
bismuth m 05-060
bleu m 70-390
bleuissage m 70-400
bloc m mosaïque 30-655
blocage m de dislocations 30-445
bore m 05-065
boruration f 70-900
borure m 70-905
boucle f de dislocation prismatique 30-345
~ de dissociation 30-340
~ prismatique 30-345
brasure f 65-710, 65-715
~ à l'argent 65-750
~ forte 65-720
~ tendre 65-715
braunite f 70-840
brome m 05-070
bronze m 65-095
~ à antimoine 65-165
~ à canon 65-245
~ à canons 65-245
~ à cloches 65-250
~ à faible teneur en étain 65-120
~ à graphite 65-220
~ à haute teneur en étain 65-115
~ à l'étain 65-110
~ à l'étain-plomb 65-185
~ à laminer 65-105
~ à médailles 65-260
~ à soupapes 65-275
~ à tungstène 65-170
~ antifriction 65-235
~ au béryllium 65-150
~ au cadmium 65-190
~ au chrome 65-195
~ au fer 65-175
~ au glucinium 65-150
~ au manganèse 65-155
~ au nickel 65-160
~ au plomb 65-140
~ au plomb-étain 65-185
~ au silicium 65-145
~ au vanadium 65-180
~ au zinc 65-205
~ blanc 65-230
~ conducteur 65-255
~ d'aluminium 65-135
~ d'art 65-270
~ de cloche 65-250
~ de deuxième fusion 65-285
~ de fonderie 65-100
~ de machines 65-240
~ de médailles 65-260
~ de monnaies 65-265
~ desoxydé au phosphore 65-215
~ fritté 65-225
~ malléable 65-105
~ mécanique 65-240
~ ordinaire 65-110
~ phosphoreux 65-210
~ poreux 65-225
~ sans étain 65-125
~ silicieux 65-145
~ spécial 65-130
~ téléphonique 65-255
brûlure f 85-205

C

cadmium m 05-075
cæsium m 05-080
calamine f 85-430
calcium m 05-085
silico-calcium m 60-1160
silicocalcium m 60-1160
californium m 05-090

calorisation f 70-925
canalisation f 30-500
capacité f calorifique 75-410
~ d'amortissement 75-205
~ de durcissement par trempe 70-020
caractéristique f de ductilité 75-065
~ de résistance 75-060
~ de résistance mécanique 75-060
~ mécanique 75-060
carbone m 05-095
~ combiné 50-010
~ de recuit 50-220
~ graphitique 50-005
~ libre 50-005
carbonitruration f 70-870
carbonitrure m 40-485
carburation f 70-715
~ en caisse 70-720
~ gazeuse 70-730
~ liquide 70-725
~ sélective 70-735
décarburation f superficielle 85-210
nitrocarburation f 70-875
carbure m 40-430
~ allié 40-445
~ complexe 40-475
~ de chrome 40-455
~ de fer 40-440
~ de métal 40-435
~ de niobium 40-465
~ de tungstène 40-450
~ de vanadium 40-460
~ fritté 55-665
~ métallique 40-435
~ mixte 40-475
~ simple 40-470
~ spécial 40-445
carbures mpl eutectiques 50-440
~ frittés 55-665
~ ledeburitiques 50-440
~ résiduels 40-480
cascade f de collisions 30-515
~ de déplacements 30-520
cassure f 40-670
~ à chaud 85-080
~ à froid 85-075
~ à grains fins 40-700
~ à grains gros 40-695
~ à nerfs 40-720
~ blanche 85-160
~ claire 85-155
~ conchoïdale 40-730
~ conchoïde 40-730
~ cristalline 40-685
~ de fatigue 40-745
~ ductile 40-680
~ en écailles de poisson 40-740
~ fibreuse 40-720
~ fragile 40-675
~ granulaire 40-690
~ lamellaire 40-725
~ par fatigue 40-745
~ résiliente 40-680
~ soyeuse 40-745
~ vitreuse 40-735
catalyseur m de germination 25-125
cavité f 85-100
~ d'attaque 80-205
~ octaédrique 20-330
~ tétraédrique 20-325
celtium m 05-185
cément m 70-740
~ carburant 70-740
cémentation f 70-655, 70-670
~ à la poudre 70-720
~ au carbone 70-715
~ en bain de sel 70-725
~ en caisse 70-720
~ gazeuse 70-730
~ liquide 70-725

cémentation
~ métallique 70-915
~ par l'aluminium 70-925
~ par l'azote 70-785
~ par le bore 70-900
~ par le carbone 70-715
~ par le chrome 70-920
~ par le silicium 70-910
~ par le souffre 70-895
~ par le titanium 70-945
~ par le tungstène 70-935
~ par le zinc 70-930
~ sélective 70-735
~ solide 70-720
cémentite f 50-065
~ allié 50-110
~ de séparation 50-095
~ eutectique 50-075
~ eutectoïde 50-085
~ globulaire 50-100
~ hypereutectique 50-070
~ hypereutectoïde 50-080
~ ledeburitique 50-075
~ libre 50-095
~ perlitique 50-085
~ primaire 50-070
~ proeutectique 50-070
~ proeutectoïde 50-080
~ secondaire 50-080
~ sphéroïdale 50-100
~ tertiaire 50-090
centre m de cristallisation 25-105
cérium m 05-100
ferro-cérium m 65-875
cermet m 55-675
cerrobend m 65-680
césium m 05-080
chaleur f atomique 75-370
~ critique 75-400
~ de dissolution 75-405
~ de fusion 75-380
~ de solidification 75-385
~ de sublimation 75-395
~ de transformation 75-400
~ de vaporisation 75-390
~ latente de fusion 75-380
~ ~ de vaporisation 75-390
~ moléculaire 75-375
~ spécifique 75-415
~ ~ additionnelle 25-635
~ ~ électronique 10-275
champ m coercitif 75-635
~ de contraintes d'une dislocation 30-225
changement m de phase 25-460
chauffage m 70-030
~ à cœur 70-035
~ par impulsions 70-050
prechauffage m 70-045
chlore m 05-105
chromage m thermique 70-920
chromaluminisation f 70-940
chrome m 05-110
ferro-chrome m 60-1145
chromel m 65-465
chromisation f 70-920
circuit m de Burgers 30-185
classe f cristalline 20-050
~ de symétrie 20-050
cliché m de cristal tournant 80-665
~ de Debye-Scherrer 80-635
~ de diffraction 80-610
~ de diffraction de rayons X 80-615
~ de Laue 80-620
coalescence f de la cémentite 70-515
cobalt m 05-115
coefficient m d'activité 15-285
~ d'anisotropie 75-595
~ d'autodiffusion 45-405
~ d'effet 15-300

COEFFICIENT

coefficient
- ~ d'interaction 15-295
- ~ de compressibilité 75-165
- ~ de conductibilité thermique 75-440
- ~ de contraction transversale 75-170
- ~ de diffusion 45-395
- ~ de diffusion intrinsèque 45-390
- ~ de dilatation cubique 75-460
- ~ de dilatation en volume 75-460
- ~ de dilatation linéaire 75-455
- ~ de dilatation thermique 75-450
- ~ de partage 25-260
- ~ de partage à l'équilibre 25-260
- ~ de partage réel 25-265
- ~ de Poisson 75-170
- ~ de réflexion 75-515
- ~ de striction 75-125
- ~ de température de la résistivité électrique 75-480
- ~ de transfert de carbone 70-755
- ~ élastique 75-145
- ~ linéaire de dilatation thermique 75-455

cœur m 70-665
- ~ de dislocation 30-205

cohénite f 50-115
cohérence f 30-620
cohésion f 10-310
decohésion f 40-655
collision f focalisée 30-505
collisions fpl en cascade 30-515
coloration f à chaud 80-185
- ~ thermique 80-185

combinaison f chimique 35-235
composant m 35-035
- ~ d'un alliage 55-180

composante f 35-035
composé m à fusion congruente 35-220
- ~ à fusion incongruente 35-225
- ~ à fusion non congruente 35-225
- ~ chimique 35-235
- ~ intermétallique 35-215
- ~ ~ électronique 35-280
- ~ ionique 35-230
- ~ polaire 35-230

composés mpl non stœchiométriques 35-250
- ~ stœchiométriques 35-245

composite m 55-680
composition f chimique 55-225
- ~ ~ d'un alliage 55-225
- ~ stœchiométrique 35-240

concentration f 55-230
- ~ d'équilibre 35-255
- ~ des électrons 35-285
- ~ électronique 35-285
- ~ en poids 55-235
- ~ molaire 55-240
- ~ moyenne 55-245
- ~ nominale 55-245
- ~ pondérale 55-235

condensation f de lacunes 30-115
conductibilité f électrique 75-485
- ~ thermique 75-440

supraconductibilité f 75-490
conductivité f thermique 75-440
conode f 35-695
consolidation f 45-045
- ~ sur les systèmes latents 45-180

déconsolidation f 45-535
constantan m 65-450
constante f
- ~ de diffusion 45-395
- ~ d'équilibre thermodynamique 15-335
- ~ de réseau 20-120
- ~ élastique 75-145
- ~ réticulaire 20-120

constituant m 35-035
- ~ d'un alliage 55-180
- ~ métallographique 40-010

constituant
- ~ micrographique 40-010

microconstituant m 40-010
constitution f 35-020
contraction f 75-795
contrainte f d'écoulement 45-210
- ~ critique de glissement 45-090
- ~ de cisaillement 45-085
- ~ de rupture 75-130
- ~ tangentielle 45-085

contraintes fpl de coulée 85-170
- ~ de retrait 85-175
- ~ de trempe 85-185
- ~ internes 85-020
- ~ résiduelles 85-020
- ~ thermiques 85-025

contraste m d'orientation 80-155
- ~ de phase 80-275

contrôle m gammagraphique 80-825
- ~ magnétique 80-805
- ~ magnétoscopique 80-805
- ~ non destructif 80-785
- ~ par colorants pénétrants 80-835
- ~ par ressuage 80-835
- ~ par ultra-sons 80-800
- ~ radiographique 80-705
- ~ ~ par les rayons X 80-820

coordinence f 20-185
corrodabilité f 75-675
corrosion f 85-260
- ~ à l'air ambiant 85-275
- ~ à l'eau 85-280
- ~ à l'eau de mer 85-290
- ~ aqueuse 85-280
- ~ atmosphérique 85-275
- ~ bimétallique 85-320
- ~ chimique 85-265
- ~ de contact 85-320
- ~ électrochimique 85-270
- ~ fissurante 85-300
- ~ ~ sous tension 85-385
- ~ galvanique 85-270
- ~ intercristalline 85-360
- ~ intergranulaire 85-360
- ~ interne 85-355
- ~ locale 85-340
- ~ localisée 85-340
- ~ marine 85-290
- ~ nucléaire 85-365
- ~ par contact 85-320
- ~ par fatigue 85-315
- ~ par frottement 85-325
- ~ par l'eau 85-280
- ~ par l'eau de mer 85-290
- ~ par le gaz 85-295
- ~ par piqûres 85-350
- ~ ponctuelle 85-345
- ~ radiolytique 85-365
- ~ sèche 85-295
- ~ sélective 85-305
- ~ sous contrainte 85-310
- ~ sous tension 85-310
- ~ souterraine 85-285
- ~ superficielle 85-330
- ~ uniforme 85-335

corroyage m à chaud 45-020
- ~ à froid 45-025

couche f carburée 70-780
- ~ cémentée 70-660, 70-780
- ~ de Beilby 80-090
- ~ électronique 10-155
- ~ épitaxiale 25-350
- ~ nitrurée 70-845
- ~ trempée 70-355

sous-couche f 10-160
coulabilité f 75-780
couleur f de revenu 70-385
coupe f à température constante 35-735
- ~ isotherme 35-735
- ~ métallographique 80-095
- ~ micrographique 80-095
- ~ mince 80-350

coupe
~ polie 80-095
~ pseudobinaire 35-745
~ verticale 35-740
couple m de diffusion 45-375
courbe f contrainte-déformation 80-410
~ d'écrouissage 45-070
~ d'isoactivité 15-315
~ de chauffage 80-520
~ de consolidation 45-070
~ de dilatation 80-555
~ de fatigue 80-425
~ de fluage 45-515
~ de pénétration de trempe 80-455
~ de refroidissement 80-525
~ de saturation 35-530
~ de Wöhler 80-425
~ dilatométrique 80-556
~ Jominy de pénétration de trempe 80-455
~ limite de solubilité 35-530
~ spinodale 15-190
~ tension-déformation 80-410
courbes fpl conjuguées 35-540
cran m 30-240
~ de dislocation 30-240
cri m de l'étain 45-270
criquage m 85-015
crique f 85-050
~ de trempe 85-190
microcrique f 85-045
cristal m 20-015
~ covalent 20-030
~ d'inoculation 25-130
~ dendritique 25-315
~ filiforme 25-360
~ idiomorphe 25-355
~ imparfait 30-010
~ ionique 20-025
~ maclé 45-230
~ mère 40-295
~ métallique 20-020
~ mixte 40-195
~ parfait 30-005
~ réel 30-010
~ unique 20-430
bicristal m 20-435
monocristal m 20-430
polycristal m 20-440
cristallin 20-010
cristallisation f 25-010
~ basaltique 25-065
~ équiaxe 25-075
~ mixte 25-055
~ primaire 25-050
~ secondaire 25-060
~ spontanée 25-045
électrocristallisation f 25-080
recristallisation f 45-440
transcristallisation f 25-070
cristallite f 40-360
cristallographie f 20-035
radiocristallographie f 80-580
cristaux mpl bacillaires 40-280
~ basaltiques 40-280
~ équiaxes 40-285
~ isomorphes 20-480
~ primaires 40-270
~ secondaires 40-275
croissance f dendritique 25-310
~ des cristaux 25-285
~ du grain 25-290
~ épitaxiale 25-345
~ épitaxique 25-345
crowdion m 30-470
cuivre m 05-120
~ affiné 65-020
~ ~ au feu 65-025
~ ~ thermiquement 65-025
~ arsenical 65-090
~ au cadmium 65-190
~ au chrome 65-200

cuivre
~ blister 65-015
~ brut 65-015
~ cémentaire 65-045
~ de cathode 65-030
~ de cément 65-045
~ de haute conductibilité 65-065
~ de haute conductibilité à très basse teneur en oxygène 65-070
~ électro 65-035
~ électrolytique 65-035
~ HC 65-065
~ jaune 65-290
~ noir 65-015
~ non désoxydé 65-050
~ non désoxydulé 65-050
~ OFHC 65-070
~ phosphoreux 65-055
~ pour conducteurs électriques 65-065
~ raffiné 65-020
~ rouge 65-040
~ sans oxygène 65-060
cupro-aluminium m 65-135
cupro-étain m 65-110
cupro-manganèse m 65-085
cupro-nickel m 65-400
cupronickel m 65-400, 65-410
cupro-phosphore m 65-075
cupro-plomb m 60-140
cupro-silicium m 65-080
curium m 05-125
cyanuration f 70-855
~ au bain 70-865
~ gazeuse 70-860
~ liquide 70-865
cyclage m thermique 70-520
cycle m thermique 70-970

D

damas m 60-645
decalescence f 25-490
décarburation f superficielle 85-210
décomposition f d'une dislocation 30-320
~ spinodale 25-370
déconsolidation f 45-535
décoration f 80-375
décrochement m 30-245
dédoublure f 85-405
défaut m cristallin 30-015
~ cristallographique 30-015
~ d'empilement 30-540
~ d'empilement extrinsèque 30-550
~ d'empilement intrinsèque 30-545
~ d'irradiation 30-485
~ de fonderie 85-095
~ de Frenkel 30-060
~ de surface 30-535, 85-010
~ de trempe 85-180
~ du matériau 85-005
~ du réseau cristallin 30-015
~ linéaire 30-140
~ plan 30-535
~ ponctuel 30-030
~ réticulaire 30-015
~ superficiel 85-010
défectoscopie f 80-790
défectuosité f due à l'irradiation 30-490
déformation f à chaud 45-020
~ à froid 45-025
~ élastique 45-005
~ permanente 45-010
~ plastique 45-010
dégazage m 70-550
degré m d'écrouissage 45-040
~ d'ordre 25-605
~ de corroyage 45-035

degré
~ de déformation 45-035
~ de désordre 25-610
~ de dispersion 40-335
~ de surfusion 25-250
déjettement *m* 85-200
demi-dislocation *f* 30-315
demi-métal *m* 10-370
demi-plan *m* atomique supplémentaire 30-200
~-~ supplémentaire 30-200
demi-vie *f* radioactive 75-525
démixion *f* 25-365
dendrite *f* 25-315
~ chimique 25-320
~ thermique 25-325
dénitruration *f* 70-850
densité *f* 75-345
~ d'états 10-200
~ de dislocations 30-235
~ de flux magnétique 75-610
déplacement *m* en cascade 30-520
déshydrogénation *f* 70-555
désignation *f* de l'acier 60-090
désordre *m* 25-585
désorientation *f* 30-595
déstabilisation *f* de l'austénite résiduelle 50-430
détection *f* des défauts 80-790
détermination *f* des indices 80-695
deuxième nombre *m* quantique 10-135
déviation *f* de glissement 45-145
dézincage *m* 85-380
dézingage *m* 85-380
diagramme *m*
~ anisotherme TTT 70-160
~ d'équilibre 35-480
~ d'équilibre des phases 35-480
~ d'équilibre fer-carbone 50-020
~ d'état 35-480
~ de chauffage 80-520
~ de constitution 35-480
~ de cristal tournant 80-665
~ de Debye et Scherrer 80-635
~ de diffraction 80-610
~ de diffraction à cristal tournant 80-665
~ de diffraction des électrons 80-775
~ de diffraction des rayons X 80-615
~ de diffraction électronique 80-775
~ de Laue 80-620
~ de Laüe 80-620
~ de phases 35-480
~ de poudre 80-635
~ de recristallisation 45-485
~ de refroidissement 80-525
~ de Schaeffer 60-530
~ de structure 40-265
~ de transformation en conditions isothermes 70-155
~ de transformation en conditions anisothermes 70-160
~ de transformation isoausténitique 70-155
~ de transformation isotherme 70-155
~ de transformations anisothermes 70-160
~ de transformations en refroidissement continu 70-160
~ dilatométrique 80-555
~ en retour 80-670
~ isotherme temps-température-transformation 70-155
~ ~ TTT 70-155
~ par transmission 80-675
~ ternaire d'équilibre 35-710
~ ~ prismatique 35-750
~ ~ spatial 35-750
~ tension-allongement 80-410
~ triangulaire 35-720
diamagnétique *m* 75-560

diamètre *m* atomique 10-080
~ critique 70-360
~ ~ de trempe 70-360
diffusion *f* 45-290
~ à contresens 45-340
~ accélerée 45-355
~ chimique 45-320
~ dans les tubes 45-360
~ en insertion 45-315
~ en régime stationnaire 45-345
~ en surface 45-300
~ en volume 45-295
~ intergranulaire 45-310
~ interstitielle 45-315
~ inversée 45-340
~ isotherme 45-350
~ lacunaire 45-305
~ massique 45-295
~ monophasée 45-330
~ montante 45-340
~ négative 45-340
~ par lacunes 45-305
~ polyphasée 45-325
~ stationnaire 45-345
~ thermique 45-370
~ volumique 45-295
autodiffusion *f* 45-400
électrodiffusion *f* 45-365
hétérodiffusion *f* 45-335
interdiffusion *f* 45-380
diffusivité *f* thermique 75-445
dilatométrie *f* 80-540
~ absolue 80-545
~ différentielle 80-550
dipole *m* de dislocations 30-380
direction *f* cristallographique 20-150
~ d'aimantation facile 75-585
~ de glissement 45-105
~ dense 20-220
~ la plus dense 20-220
~ privilégiée d'aimantation 75-585
disclination *f* 30-480
discordance *f* du réseau 30-615
dislocation *f* 30-145
~ ancrée 30-270
~ anguleuse 30-360
~ chargée 30-465
~ coin 30-150
~ ~ négative 30-160
~ ~ positive 30-155
~ d'arête 30-360
~ d'interface 30-580
~ de Burgers 30-165
~ de Frank 30-335
~ de joint de grains 30-580
~ de macle 30-375
~ de Shockley 30-330
~ de surstructure 30-220
~ de Taylor 30-150
~ dissociée 30-325
~ en anneau 30-340
~ en hélice 30-355
~ épinglée 30-270
~ glissile 30-265
~ hélicoïdale 30-355
~ imparfaite 30-310
~ mixte 30-180
~ parfaite 30-305
~ partielle 30-315
~ ~ de Shockley 30-330
~ pôle *f* 30-365
~ poteau *f* 30-365
~ prismatique 30-345
~ sessile 30-280
~ ~ de Frank 30-335
~ spirale 30-355
~ stair-rod 30-360
~ tringle 30-360
~ unitaire 30-215
~ vis 30-165
~ ~ à droite 30-170
~ ~ à gauche 30-175
~ ~ à pas inversé 30-175
demi-dislocation 30-315

ÉNERGIE

dislocation
superdislocation 30-220
dislocations *fpl* émissaires 30-370
~ primaires 30-250
~ secondaires 30-255
dispersité *f* 40-335
dispersoïde *m* 40-330
dissociation *f* d'une dislocation 30-320
distance *f* d'identité 20-115
~ entre lamelles 40-110
~ interatomique 20-230
~ interlamellaire 40-110
~ interparticulaire 40-425
~ interréticulaire 20-225
equidistance *f* des plans réticulaires 20-225
distortion *f* du réseau 30-025
domaine *m* 40-615
~ à deux phases 35-685
~ antiphase 40-625
~ bainitique 50-305
~ biphasé 35-685
~ critique 35-560
~ ~ de transformation 35-560
~ de fermeture 40-640
~ de transformation 35-560
~ de Weiss 40-635
~ magnétique 40-635
~ martensitique 50-300
~ monophasé 35-680
~ ordonné 40-625
~ perlitique 50-295
drasticité *f* de trempe 70-200
ductilité *f* 75-030
duralinox *m* 65-600
duralumin *m* 65-555
durcissement *m* 70-015
~ latent 45-180
~ par cémentation 70-670
~ par déformation à froid 45-050
~ par des particules 55-220
~ par dispersion 55-220
~ par dispersoïdes 55-220
~ par écrouissage 45-050
~ par explosion 45-055
~ par formation d'un alliage 55-210
~ par irradiation 30-495
~ par précipitation 70-560
~ par ségrégation 70-560
~ par solution 55-215
~ par trempe 70-175
~ par vieillissement 70-565
~ secondaire 70-405
~ structural 70-560
durée *f* d'attaque 80-125
~ d'incubation 70-150
~ de maintien 70-060
dureté *f* 75-245
~ à chaud 75-255
~ au rouge 75-260
microdureté *f* 75-250
dysprosium *m* 05-130

E

écaille *f* 85-430
échantillon *m* 80-050
~ massif 80-055
échantillonnage *m* 80-060
échauffement *m* 70-030
~ à cœur 70-035
~ par impulsions 70-050
écheveau *m* de dislocations 30-385
éclat *m* métallique 75-510
écoulement *m* plastique 45-215
~ visqueux 45-220
écrouissage *m* 45-025
~ critique 45-075

écrouissage
~ interne 45-080
effet *m* de Bauschinger 75-175
~ Hall 10-290
~ Kirkendall 45-385
~ Mössbauer 10-285
~ Snoek 30-460
~ Suzuki 30-455
einsteinium *m* 05-135
élasticité *f* 75-020
anélasticité *f* 75-050
électro-affinité *f* 10-115
électrocristallisation *f* 25-080
électrodiffusion *f* 45-365
électromigration *f* 45-365
électron *m* 65-625
~ de conduction 10-265
~ de valence 10-065
~ intérieur 10-070
~ interne 10-070
~ libre 10-055
~ lié 10-050
~ périphérique 10-065
électronégativité *f* 10-095
électrum *m* 65-810
élektron *m* 65-625
élinvar *m* 65-505
élément *m* 05-005
~ alphagène 50-455
~ antigraphitisant 50-465
~ associé 55-260
~ betagène 65-860
~ carburigène 50-435
~ d'addition 55-190
~ d'alliage 55-190
~ d'élaboration 55-260
~ de trace 55-250
~ durcissant 55-255
~ gammagène 50-450
~ graphitisant 50-460
~ majeur d'addition 55-195
~ métallique 10-360
~ nitrurigène 50-470
~ non métallique 10-365
~ résiduel 55-260
~ stabilisant 50-485
~ -trace *m* 55-250
oligo-élément *m* 55-250
éléments *mpl* accidentels 55-200
~ de transition 55-025
emboutissabilité *f* 75-750
embryon *m* 25-135
empilage *m* cristallographique 20-180
empilement *m* de dislocations 30-400
empreinte *f* 80-345
~ Baumann 80-325
~ de Baumann 80-325
énergie *f* d'activation 45-285
~ d'anisotropie 75-590
~ d'interaction 10-295
~ d'interface 30-635
~ d'ionisation 10-100
~ de défaut d'empilement 30-560
~ de faute(s) d'empilement 30-560
~ de Fermi 10-210
~ de liaison 15-090
~ de mise en ordre 25-630
~ de surface 30-640
~ disponible 15-105
~ emmagasinée 45-060
~ Fermi 10-210
~ interfaciale 30-635
~ intergranulaire 30-575
~ interne 15-095
~ ~ d'une dislocation 30-230
~ libre 15-105
~ ~ de formation 15-135
~ ~ de Gibbs 15-140
~ ~ de Helmholtz 15-105
~ ~ de mélange 15-130

ÉNERGIE

énergie libre
~ ~ d'excès 15-115
~ ~ excédentaire 15-115
~ ~ interfaciale 15-125
~ ~ partielle 15-110
~ ~ superficielle 15-120
~ liée 15-100
~ stockée 45-060
~ superficielle 30-640
enrobage m 80-065
enrobement m 80-065
ensemencement m 25-215
enthalpie f d'excès 15-145
~ de formation 15-160
~ de fusion 15-165
~ de liaison 15-170
~ de mélange 15-155
~ excédentaire 15-145
~ libre 15-140
~ partielle 15-150
entropie f configurationnelle 15-210
~ d'excès 15-220
~ de configuration 15-210
~ de formation 15-200
~ de fusion 15-225
~ de mélange 15-195
~ de vibration 15-205
~ excédentaire 15-220
~ partielle 15-215
~ vibrationnelle 15-205
épitaxie f 25-340
éponge f de fer 60-035
épreuve f Baumann 80-320
éprouvette f 80-050
équation f d'état 15-025
~ de Hildebrand 15-310
~ de Gibbs-Duhem 15-305
~ de Schrödinger 10-175
équidistance f des plans réticulaires 20-225
équilibre m chimique 15-330
~ des phases 35-015
~ hétérogène 35-015
~ instable 35-060
~ métastable 35-060
~ physico-chimique 35-015
~ stable 35-055
~ structural 40-020
~ thermique 15-325
~ thermodynamique 15-320
équivalent m de carbone 50-480
~ en carbone 50-480
erbium m 05-140
espace m k 10-205
espacement m entre particules 40-425
espaces mpl interdendritiques 40-355
essai m à blanc 70-775
~ à blanc de trempe 70-775
~ à la touche 80-465
~ à témoin de trempe 70-775
~ au nitrate mercureux 80-485
~ aux étincelles 80-440
~ Baumann 80-320
~ d'endurance 80-420
~ de choc 80-430
~ de dureté 80-415
~ de fatigue 80-420
~ de rupture 80-445
~ de traction 80-405
~ de trempabilité Jominy 80-450
~ dilatométrique 80-540
~ ~ absolu 80-545
~ ~ différentiel 80-550
~ mécanique 80-400
~ métallographique 80-035
~ non destructif 80-785
~ par le son 80-795
étain m 05-475
~ blanc 65-630
~ gris 65-635
cupro-étain m 65-110

état m d'équilibre 35-050
~ énergétique 10-120
~ métallique 10-345
~ métastable 35-070
~ quantique 10-120
~ stable 35-065
~ thermodynamique 15-015
étude f calorimétrique 80-500
~ dilatométrique différentielle 80-550
~ microfractographique 80-390
~ par diffraction électronique 80-770
~ radiographique 80-740
~ thermodynamique par calorimétrie 80-535
europium m 05-145
eutectique m 40-200
~ anormal 40-230
~ binaire 40-205
~ discontinu 40-235
~ fer-graphite 50-260
~ lamellaire 40-220
~ ledeburitique 50-255
~ phosphoreux 50-235
~ ~ ternaire 50-235
~ quaternaire 40-215
~ ternaire 40-210
pseudo-eutectique m 40-225
eutectoïde m 40-240
eutexie f 40-200
examen m au rayon X 80-570
~ fractographique 80-385
~ gammagraphique 80-825
~ macrographique 80-045
~ magnétoscopique 80-805
~ métallographique 80-035
~ micrographique 80-040
~ microscopique 80-040
~ non destructif 80-785
~ par fluorescence 80-835
~ radiocristallographique 80-580
~ radiographique 80-705, 80-740

F

face f du cristal 20-060
faciès m 20-460
~ d'un cristal 20-460
facteur m de dimension 10-395
~ de réflexion 75-515
~ de taille 10-395
~ de valence 10-405
~ dimensionnel 10-395
~ électrochimique 10-400
fatigue f 85-410
~ de contact 85-420
~ par cyclage thermique 85-425
~ sous corrosion 85-415
~ thermique 85-425
faute f d'empilement 30-540
~ d'empilement de maclage 30-555
~ d'empilement extrinsèque 30-550
~ d'empilement intrinsèque 30-545
fêlure f à chaud 85-080
~ à froid 85-075
fer m 05-220
~ affiné 60-050
~ Armco 60-010
~ au bois 60-050
~ au charbon de bois 60-050
~ au silicium 60-210
~ carbonyle 60-020
~ catalan 60-045
~ commercial 60-015
~ corroyé 60-040
~ coulé 60-055
~ électrolytique 60-025
~ en éponge 60-035
~ ex-carbonyle 60-020
~ fritté 60-065
~ météorique 60-005
~ puddlé 60-055
~ spongieux 60-035

fer
~ suédois 60-030
fermium m 05-150
ferrite f 50-045
~ au silicium 50-475
~ delta 50-050
~ libre 50-055
~ perlitique 50-060
~ proeutectoïde 50-055
silicoferrite f 50-475
ferrimagnétisme m 75-580
ferritisation f 70-645
ferro-alliages mpl 60-1135
ferro-cerium m 65-875
ferro-chrome m 60-1145
ferro-manganèse m 60-1150
ferro-silicium m 60-1140
ferromagnétique m 75-550
ferromagnétisme m 75-545
antiferromagnétisme m 75-565
ferros mpl 60-1135
ferrosilicium m 60-1140
figures fpl d'attaque 80-200
~ de pôle 80-685
finesse f du grain 40-380
fissuration f 85-015
~ par corrosion sous tension 85-385
~ saisonnière 85-385
fissure f de déformation 85-070
~ de Griffith 85-055
~ de température 85-060
~ due à la déformation 85-070
~ sous tension 85-065
microfissure f 85-045
flèche f d'emboutissage 75-755
flocons mpl 85-240
fluage m 45-495
~ accéléré 45-510
~ de transition 45-500
~ Nabarro 45-520
~ par diffusion 45-520
~ primaire 45-500
~ secondaire 45-505
~ stationnaire 45-505
~ tertiaire 45-510
~ transitoire 45-500
microfluage m 45-530
fluctuations fpl de concentration 25-100
fluidité f 75-785
fluor m 05-155
flux m 15-085
focuson m 30-505
fonction f d'état 15-035
~ d'onde 10-180
~ d'ondes 10-180
~ de Bloch 10-195
~ propre 10-190
~ thermodynamique 15-030
~ ~ d'excès 15-065
~ ~ excédentive 15-065
~ ~ intégrale 15-055
~ ~ partielle 15-060
fonte f à cœur blanc 60-1120
~ à cœur noir 60-1115
~ à graphite nodulaire 60-975
~ à graphite sphéroïdal 60-975
~ à haut carbone 60-925
~ à haute résistance 60-1065
~ à l'aluminium 60-1020
~ à matrice bainitique 60-1060
~ à matrice eutectoïde 60-1045
~ à matrice ferritique 60-1040
~ à matrice martensitique 60-1055
~ à matrice perlitique 60-1045
~ aciculaire 60-1060
~ aciérée 60-1035
~ allié 60-995

fonte
~ antifriction 60-1105
~ armée 60-1070
~ au bois 60-875
~ au charbon de bois 60-875
~ au chrome 60-1015
~ au coke 60-870
~ au manganèse 60-1030
~ au nickel 60-1025
~ au silicium 60-1010
~ austénitique 60-1050
~ bainitique 60-1060
~ Bessemer 60-845
~ blanche 60-910, 60-955
~ ~ eutectique 50-505
~ ~ hypereutectique 50-515
~ ~ hypoeutectique 50-510
~ brute 60-840
~ ~ alliée 60-935
~ crue 60-840
~ d'affinage 60-880
~ de deuxième fusion 60-945
~ de haut fourneau 60-860
~ de moulage 60-885
~ de première fusion 60-840
~ ductile 60-975
~ électrique 60-865
~ européenne 60-1120
~ faiblement alliée 60-1000
~ ferritique 60-1040
~ fine 60-1065
~ fortement alliée 60-1005
~ graphiteuse 60-925
~ grise 60-915, 60-950
~ ~ à matrice ferritique 60-1040
~ ~ à matrice perlitique 60-1045
~ ~ ferritique 60-1040
~ ~ ordinaire 60-965
~ ~ perlitique 60-1045
~ hématite 60-895
~ industrielle 60-965
~ inoculée 60-970
~ ledeburitique 50-505
~ malléable 60-1110
~ ~ à cœur blanc 60-1120
~ ~ à cœur noir 60-1115
~ ~ à matrice ferritique 60-1125
~ ~ à matrice perlitique 60-1130
~ ~ américaine 60-1115
~ ~ européenne 60-1120
~ ~ ferritique 60-1125
~ ~ perlitique 60-1130
~ martensitique 60-1055
~ Martin 60-855
~ mélangée 60-920, 60-960
~ Ni-resist 60-1090
~ nodulaire 60-975
~ non alliée 60-990
~ ordinaire 60-965
~ perlitique 60-1045
~ phosphoreuse 60-890
~ pour malléabilisation 60-905
~ réfractaire 60-1085
~ résistante 60-1065
~ ~ aux acides 60-1075
~ ~ aux alcalis 60-1080
~ semi-hématite 60-900
~ spéciale 60-995
~ spéculaire 60-930
~ sphéroïdale 60-975
~ synthétique 60-940
~ Thomas 60-850
~ trempée 60-980
~ truitée 60-920, 60-960
force f coercitive 75-635
~ de cohésion 10-315
~ de Peierls 30-300
~ de Peierls-Nabarro 30-300
~ motrice 15-070
~ thermodynamique 15-070
forêt f de dislocations 30-405
forgeabilité f 75-725
formation f d'un alliage 55-170
forme f allotropique 25-390
~ haute température 25-395

FRACTOGRAPHIE

fractographie f 80-385
microfractographie f 80-390, 80-395
fracture f intergranulaire 40-705
~ retardée 40-795
~ transcristalline 40-710
~ transgranulaire 40-710
fragilisation f 75-275
fragilité f 75-270
~ à basse température 75-295
~ à chaud 75-290
~ à froid 75-285
~ au bleu 75-300
~ au rouge 75-290
~ caustique 75-305
~ de revenu 85-220
~ due à l'hydrogène 85-390
~ Krupp 85-220
~ par l'hydrogène 85-390
~ rhéotropique 75-310
francium m 05-160
fretting m 85-325
fritté m 55-655
front m de cristallisation 25-295
~ de croissance 45-480
~ de diffusion 45-410
~ de solidification 25-295
frontière f d'antiphase 40-630
frottement m intérieur 75-210
~ interne 75-210
fugacité f 15-290
fusibilité f 75-790
fusion f 35-445
~ congruente 35-460
~ eutectique 35-470
~ hétérotherme 35-455
~ incongruente 35-465
~ isotherme 35-450
~ non congruente 35-465
~ non isotherme 35-455
~ partielle 85-215
~ ~ au recuit 85-215
~ péritectique 35-475
surfusion f 25-240

G

gadolinium m 05-165
gallium m 05-170
gammagraphie f 80-825
gauchissement m 85-200
gaz m carburant 70-745
~ d'atmosphere endothermique 70-705
~ d'atmosphere exothermique 70-710
~ d'électrons 10-045
~ électronique 10-045
~ porteur 70-685
~ support 70-685
générateur m de carbures 50-435
~ de Frank-Read 30-430
germanium m 05-175
germe m cohérent 25-155
~ critique 25-140
~ de cristallisation 25-105
~ de recristallisation 25-170
~ étranger 25-120
~ hétérogène 25-115
~ homogène 25-110
~ orienté 25-160
~ potentiel 25-135
~ propre 25-110
~ stable 25-150
~ subcritique 25-135
prégerme m 25-135
germination f 25-175
~ athermique 25-200
~ dynamique 25-205
~ hétérogène 25-185
~ homogène 25-180

germination
~ orientée 25-195
~ par frottement 25-210
~ spontanée 25-190
glissement m 45-095
~ au joint de grains 45-160
~ dévié 45-145
~ double 45-135
~ en pinceau 45-175
~ facile 45-155
~ intercristallin 45-165
~ intergranulaire 45-160
~ léger 45-155
~ multiple 45-140
~ prismatique 45-170
~ simple 45-130
globulisation f de la cémentite 70-515
glucinium m 05-055
gonflement m de la fonte 70-640
grade m de pureté 55-095
gradient m de concentration 15-080
~ de température 15-075
~ thermique 15-075
grain m 40-360
~ austénitique originel 50-040
~ d'austénite, ancien 50-040
~ fin 40-370
~, gros 40-365
~ grossier 40-365
~ maclé 45-230
sous-grain m 30-650
grandeur f d'état 15-035
~ extensive 15-040
~ intégrale 15-055
~ intensive 15-045
~ molaire 15-050
~ partielle 15-060
~ thermodynamique 15-030
~ ~ d'excès 15-065
~ ~ excédentive 15-065
~ ~ intégrale 15-055
~ ~ partielle 15-060
graphite m 50-165
~ d'écume 50-170
~ de malléabilisation 50-220
~ de recuit 50-220
~ d'écume 50-170
~ en lamelles 50-185
~ en nodules 50-210
~ en rosettes 50-200
~ étoilé 50-200
~ eutectique 50-175
~ inter-dendritique 50-205
~ lamellaire 50-185
~ nodulaire 50-210
~ primaire 50-170
~ punctiforme 50-195
~ secondaire 50-180
~ sphéroïdale 50-190
graphitisant m 50-460
antigraphitisant m 50-465
graphitisation f 50-225, 70-630
gros grain m 40-365
grosseur f du grain 80-285
grossièreté f du grain 40-375
grossissement m 80-282
~ du grain 25-290
guérison f 30-135
~ isochrone 70-495

H

hafnium m 05-185
hardénite m 50-365
hastelloy m 65-455
hélium m 05-190
hérédité f de fonte 60-985
hétérodiffusion f 45-335

hétérogénéité *f* 40-410
~ chimique 40-415
hinduminium *m* 65-605
hipernik *m* 65-475
holmium *m* 05-195
homogénéisation *f* 70-485
homogénéité *f* 40-405
inhomogénéité *f* 40-410
horizontale *f* eutectique 35-640
~ eutectoïde 35-655
~ monotectique 35-650
~ péritectique 35-645
hydrogène *m* 05-200
hydrure *m* 40-495
~ métallique 40-495
hypertrempe *f* 70-345
hystérésis *f* de la transformation 25-480
~ magnétique 75-640
~ thermique 25-480

I

imperfection *f* cristalline 30-015
~ de surface 30-535
~ linéaire 30-140
~ ponctuelle 30-030
~ réticulaire 30-015
implantation *f* d'ions 70-805
impression *f* aux sels d'argent 80-325
impureté *f* interstitielle 30-080
impuretés *fpl* 55-205
inclusions *fpl* d'oxydes 40-535
~ de laitier 40-540
~ de nitrures 40-545
~ de scories 40-540
~ de sulphures 40-530
~ endogènes 40-525
~ exogènes 40-520
~ métalliques 40-550
~ non métalliques 40-515
~ sulphurées 40-530
inconel *m* 65-445
indexation *f* 80-695
indicateur *m* radioactif 80-760
indice *m* cristallographique 20-160
~ de coordination 20-185
~ de ségrégation 40-420
~ Erichsen 75-755
indices *mpl* de Miller 20-165
~ de Miller-Bravais 20-170
indium *m* 05-205
induction *f* magnétique 75-610
~ rémanente 75-630
inhibiteur *m* de croissance du grain 40-400
inhomogénéité *f* 40-410
inoculant *m* 25-230
inoculateur *m* 25-230
inoculation *f* 25-220
inoxy *m* 60-515
inoxydabilité *f* 75-685
inoxydable *m* 60-515
insolubilité *f* 35-340
inspection *f* ultrasonique 80-800
intensité *f* d'aimantation 75-600
~ d'aimantation rémanente 75-630
~ de dislocation 30-210
interdiffusion *f* 45-380
interface *f* 30-610
~ cohérente 30-625
~ de phase 30-610
~ incohérente 30-630
interférogramme *m* 80-300

interstice *m* 20-105
interstitiel *m* 30-075
~ dissocié 30-085
intervalle *m* critique 35-560
~ de fusion 35-545
~ de solidification 35-550
~ de température de trempe 70-145
~ de transformation 35-560
invar *m* 65-500
iode *m* 05-210
ionitruration *f* 70-800
iridium *m* 05-215
iridosmium *m* 65-845
isomorphie *f* 20-475
isomorphisme *m* 20-475
isotropie *f* 20-450
quasi-isotropie *f* 20-455

J

joint *m* à grand angle 30-585
~ cohérent 30-625
~ d'angle faible 30-590
~ de flexion 30-605
~ de grain 30-565
~ de grains 30-565
~ de macle 45-265
~ de torsion 30-600
~ incoherent 30-630
~ intergranulaire 30-565

K

kanthal *m* 65-525
kovar *m* 65-520
krypton *m* 05-225

L

lacune *f* 30-035
~ de constitution 30-090
~ de Frenkel 30-065
~ de miscibilité 35-345
~ de trempe 30-100
~ en excès 30-105
~ excessive 30-105
~ octaédrique 20-330
~ réticulaire 30-035
~ tétraédrique 20-325
~ thermique 30-095
~ trempée 30-100
bilacune *f* 30-045
monolacune *f* 30-040
quadrilacune *f* 30-055
trilacune *f* 30-050
laiton *m* 65-290
~ α 65-325
~ α+β 65-330
~ β 65-335
~ à cartouches 65-380
~ à étain 65-345
~ à haute résistance 65-355
~ à l'aluminium 65-340
~ à laminer 65-295
~ amirauté 65-390
~ au manganèse 65-355
~ au nickel 65-350
~ au plomb 65-360
~ au silicium 65-365
~ biphasé 65-320
~ complexe 65-310
~ de brasure 65-740
~ de décolletage 65-370
~ de fonderie 65-300
~ de premier titre 65-325
~ de second titre 65-330
~ haute résistance 65-355
~ malléable 65-295

laiton
~ marin 65-385
~ mécanique 65-375
~ monophasé 65-315
~ ordinaire 65-305
~ pour brasage 65-740
~ pour pièces moulés 65-300
~ spécial 65-310
~ supérieur 65-320
lame ƒ mince 80-350
lamelle ƒ de cémentite 50-120
lanthane m 05-230
lautal m 65-595
lawrencium m 05-235
ledeburite ƒ 50-160
liaison ƒ chimique 10-320
~ covalente 10-330
~ d'électrovalence 10-325
~ de covalence 10-330
~ de van der Waals 10-335
~ hétéropolaire 10-325
~ homopolaire 10-330
~ ionique 10-325
~ métallique 10-340
ligne ƒ de diffraction 80-645
~ de dislocation 30-195
~ de glissement 45-185
~ de liquidus 35-515
~ de précipitation 35-530
~ de solidus 35-490
~ d'isoactivité 15-315
lignes ƒpl d'arrêts 40-750
~ d'écoulement 45-195
~ de Lueders 45-195
~ de Neumann 45-205
~ de repos 40-750
limite ƒ apparente d'élasticité 75-110
~ conventionelle d'élasticité 75-115
~ d'antiphase 40-630
~ d'écoulement 75-110
~ d'élasticité vraie 75-105
~ d'endurance 75-200
~ d'étirement 75-110
~ de fatigue 75-200
~ de fluage 75-220
~ de fluage conventionelle 75-230, 75-235
~ de fluage vraie 75-225
~ de phase 35-535
~ de proportionalité 75-100
~ de saturation 35-330
~ de solubilité 35-330
~ élastique 75-110
~ ~ conventionelle 75-115
~ interphases 30-610
liquide m de trempe 70-190
~ mère 25-095
~ résiduel 25-280
liquidus m 25-510
surface-liquidus ƒ 35-520
liséré m de cémentite 50-105
~ ferritique 50-230
lithium m 05-245
loi ƒ de Henry 15-275
~ de Raoult 15-270
~ de Vegard 35-425
~ des phases 35-010
~ des segments proportionnels 35-700
lutétium m 05-250

M

maclage m 45-245
macle ƒ 45-230
~ de croissance 45-240
~ d'écrouissage 45-235
~ de déformation 45-235
~ de recristallisation 45-240
~ de recuit 45-240
~ mécanique 45-235

macrographie ƒ 80-305
photomacrographie ƒ 80-310, 80-315
macrophotographie ƒ 80-310, 80-315
macroscopie ƒ 80-045
macrostructure ƒ 40-025
magnalium m 65-620
magnésium m 05-255
magnétoscopie ƒ 80-805
magnétostriction ƒ 75-650
maille ƒ cristalline 20-090
~ élémentaire 20-090
~ primitive 20-095
~ simple 20-095
maillechort m 65-415
~ pour brasage 65-745
maintien m 70-055
~ à température 70-055
~ en température 70-055
~ isotherme 70-055
maladie ƒ h'ydrogène 85-395
~ de Krupp 85-220
~ de l'étain 85-245
malléable m 60-1110
malléabilisation ƒ 70-625
~ par décarburation 70-635
~ par graphitisation 70-630
malleabilité ƒ 75-025
manganèse m 05-260
cupro-manganèse m 65-085
ferro-manganèse m 60-1150
silicomanganèse m 60-1155
manganin m 65-405
manganine ƒ 65-405
maraging m 70-615
marche ƒ de croissance 25-330
martempering m 70-305
martensite ƒ 50-310
~ aciculaire 50-315
~ ~ fine 50-325
~ cryptocristalline 50-330
~ cubique 50-340
~ d'écrouissage 50-360
~ de deformation 50-360
~ de revenu 50-345
~ de trempe 50-335
~ en gros grain 50-320
~ massive 50-355
~ revenue 50-345
~ secondaire 50-350
~ tétragonale 50-335
masse ƒ atomique 10-075
~ ~ relative 10-075
~ moléculaire 75-355
~ spécifique 75-345
matériau m composite 55-680
matrice ƒ 40-015
maturation ƒ 70-575, 70-585
mélange m 40-190
~ de phases 40-190
~ eutectique 40-200
~ eutectoïde 40-240
~ gazeux 35-370
~ intime 40-190
~ ~ de phases 40-190
mendélévium m 05-265
mercure m 05-270
métal m 10-355, 10-360
~ à bas point de fusion 55-060
~ à haut point de fusion 55-065
~ affiné 55-125
~ antifriction 55-555
~ Babbit 55-645
~ blanc antifriction 65-640, 65-645
~ commun 55-005
~ compact 55-155
~ de base 55-185

métal
~ de cloche 65-250
~ de Muntz 65-330
~ de zone fondue 55-140
~ déposé par évaporation 55-145
~ élaboré sous vide 55-135
~ électrolytique 55-130
~ fissile 55-080
~ fondu 55-150
~ ~ sous vide 55-135
~ fritté 55-660
~ linotype 55-625
~ monotype 55-620
~ natif 55-110
~ neuf 55-115
~ noble 55-010
~ nucléaire 55-085
~ précieux 55-010
~ pur 55-090
~ radioactif 55-075
~ réactif 55-070
~ refondu 55-120
~ réfractaire 55-065
~ Rose 65-675
~ semi-noble 55-020
~ soudable au verre 55-600
~ stéréotype 55-630
~ typographique 55-610
~ ~ pour blancs 55-640
~ ultra-pur 55-105
~ vierge 55-115
~ Wood 65-670
demi-métal m 10-370
semi-métal m 10-370
métallographie f 80-005
~ à haute température 80-015
~ électronique 80-025
~ en couleurs 80-030
~ microscopique 80-010
~ quantitative 80-020
microfractométallographie f 80-390
radiométallographie f 80-710
métalloïde m 10-365
métalloradiographie f 80-710
métallurgie f physique 10-005
~ structurale 10-005
métatectique f 40-255
métaux mpl alcalino-terreux 55-035
~ alcalins 55-030
~ de transition 55-025
~ des terres rares 55-040
~ légers 55-050
~ lourds 55-055
~ non ferreux 55-015
~ rares 55-045
méthode f à la poudre magnétique 80-810
~ à la suspension magnétique 80-815
~ de Debye et Scherrer 80-595
~ de la tangente commune 15-180
~ de Laue 80-590
~ de Laüe 80-590
~ du cristal oscillant 80-605
~ du cristal tournant 80-600
~ du diagramme en retour 80-630
~ du diagramme par transmission 80-625
~ des poudres 80-595
microanalyse f par absorption 80-735
~ par microsonde électronique 80-780
~ par sonde électronique 80-780
microconstituant m 40-010
microcrique f 85-045
microdureté f 75-250
microfissure f 85-045
microfluage m 45-530
microfractographie f 80-390, 80-395
microfractométallographie f 80-390
micrographie f 80-245
~ électronique 80-365
photomicrographie f 80-250, 80-255

microphotographie f 80-255
~ électronique 80-365
micropores mpl 85-110
microporosité f 85-115
microradiographie f 80-720, 80-725
~ par rayons X 80-730
microretassure f 85-120
microscope m métallographique 80-210
~ métallurgique 80-210
microscopie f 80-040
~ à balayage 80-360
~ à contraste de phase 80-270
~ à haute température 80-240
~ électronique 80-330
~ ~ à balayage 80-360
~ ~ à emission 80-335
~ ~ par transmission 80-340
~ ~ par transparence 80-340
~ en lumière polarisée 80-260
~ en réflexion 80-225
~ en ultraviolet 80-265
~ interférentielle 80-295
~ ionique 80-380
~ iono-ionique 80-380
~ optique 80-215
~ par ionisation 80-380
~ par réflexion 80-225
~ par transmission 80-220
~ par transparence 80-220
~ sur fond clair 80-235
~ sur fond noir 80-230
microsondage m électronique 80-780
microstructure f 40-030
migration f des joints de grain 45-490
électromigration f 45-365
milieu m cémentant 70-470
~ de trempe 70-185
~ trempant 70-185
mischmétal m 65-865
miscibilité f 35-300
~ à l'état liquide 35-305
~ à l'état solide 35-310
~ complète 35-315
~ incomplète 35-320
~ limitée 35-320
~ partielle 35-320
mise f en évidence de la microstructure 80-280
~ en ordre magnetique 25-640
mode m 20-460
modification f 25-220
modificateur m 25-230
module m d'élasticité longitudinale 75-150
~ d'élasticité transversale 75-155
~ d'Young 75-150
~ de cisaillement 75-155
~ de compression 75-160
~ de glissement 75-155
~ de rigidité 75-160
~ de scission élastique 75-155
~ de torsion 75-155
~ de volume 75-160
~ élastique 75-150
molybdène m 05-275
~ non-sag 65-870
monel m 65-435
monocristal m 20-430
monolacune f 30-040
monotectique f 40-260
montée f des dislocations 30-295
mouvement m conservateur 30-290
~ conservatif 30-290
~ non conservateur 30-285
~ non conservatif 30-285
multiplication f des dislocations 30-420

N

néodyme m 05-280
néon m 05-285
neptunium m 05-290
neutrographie f 80-830
Ni-resist m 60-1090
nichrome m 65-470
nickel m 05-295
~ ex carbonyle 65-425
~ Mond 65-425
cupro-nickel m 65-400
cupronickel m 65-400, 65-410
nicrosilal m 60-1095
nilvar m 65-500
nimonic m 65-480
niobium m 05-300
niobiure m 40-505
nital m 80-110
nitrausténite f 70-835
nitrocarburation f 70-875
nitroxydation f 70-885
nitruration f 70-785
~ en bain de sel 70-795
~ gazeuse 70-790
~ ionique 70-800
~ par bombardement ionique 70-800
~ par décharge luminescente 70-800
carbonitruration f 70-870
dénitruration f 70-850
ionitruration f 70-800
oxynitruration f 70-885
sulfonitruration f 70-890
nitrure m 70-810
~ de chrome 70-820
~ de fer 70-815
~ de titane 70-825
~ de vanadium 70-830
carbonitrure m 40-485
niveau m
~ de carbone 70-750
~ de Fermi 10-220
~ énergétique 10-150
~ d'énergie 10-150
~ Fermi 10-220
nobélium m 05-310
nœud m du réseau 20-100
~ de dislocation 30-350
nombre m atomique 75-330
~ de complexions microscopiques 15-230
~ de coordinance 20-185
~ de coordination 20-185
~ de degrés de liberté 35-030
~ de masse 75-335
~ quantique 10-125
~ ~ azimutal 10-135
~ ~ de spin 10-145
~ ~ magnétique 10-140
~ ~, premier 10-130
~ ~ principal 10-130
~ ~, quatrième 10-145
~ ~ secondaire 10-135
~ ~ total 10-130
~ ~, troisième 10-140
normalisation f 70-475
noyau m de dislocation 30-205
nuage m d'électrons 10-035
~ d'impuretés 30-440
~ de Cottrell 30-440
~ électronique 10-035
nuance f d'acier 60-085
nucléation 25-175
numéro m du grain 80-290

O

œils mpl de poisson 85-240
oligo-élément m 55-250
ombrage m 80-370
opacité f 75-505
or m 05-180
~ à titre élevé 65-795
~ affiné 65-770
~ argental 65-810
~ au titre 65-790
~ au titre légal 65-790
~ blanc 65-805
~ de coupelle 65-770
~ de joaillerie 65-780
~ de monnaie 65-775
~ dentaire 65-785
~ faux 65-815
~ fin 65-770
~ natif 65-765
~ rouge 65-800
ordination f magnétique 25-640
ordre m 25-590
~ à courte distance 25-595
~ à longue distance 25-600
~ à grande distance 25-600
~ à petite distance 25-595
désordre m 25-585
orientation f cristalline 20-415
~ cristallographique 20-415
~ d'un monocristal 80-690
~ désordonnée 20-420
~ préférentielle 20-425
~ privilégiée 20-425
désorientation f 30-595
osmium m 05-315
oxydation f à la vapeur 70-880
~ interne 70-947
nitroxydation f 70-885
oxygène m 05-320
potentiel-oxygène m 15-350
oxynitruration f 70-885

P

paire f de Frenkel 30-060
~ de semi-interstitiels 30-085
palier m thermique 80-530
palladium m 05-325
pantal m 65-615
papillons mpl 85-255
paramagnétique m 75-555
paramètre m cristallin 20-120
~ d'efficacité 15-300
~ d'ordre à courte distance 25-615
~ d'ordre à grande distance 25-620
~ d'ordre à longue distance 25-620
~ d'ordre à petite distance 25-615
~ du réseau 20-120
~ réticulaire 20-120
paroi f d'antiphase 40-630
~ de Bloch 40-645
~ de dislocations 30-395
~ de domaine 40-645
patentage m 70-415
patine f 85-375
pénétration f de trempe 70-350
période f d'identité 20-115
~ d'incubation 70-150
~ de radioactivité 75-525
~ du réseau 20-115
péritectique m 40-245
péritectoïde m 40-250
perlite f 50-125
~ coalescée 50-150
~ dégénérée 50-155

perlite
~ fine 50-135
~ globulaire 50-150
~ globularisée 50-150
~ grossière 50-140
~ lamellaire 50-130
~ ~ fine 50-135
~ ~ grossière 50-140
~ nodulaire 50-150
~ sorbitique 50-145
~ striée 50-130
perlitisation f 70-475
permalloy m 65-440
perméabilité f magnétique 75-615
~ ~ initiale 75-620
permendur m 65-490
perminvar m 65-510
peste f de l'étain 85-245
phase f 35-025
~ σ 35-275
~ blanche 85-250
~ condensée 35-145
~ d'équilibre 35-180
~ de Hume-Rothery 35-280
~ de non-équilibre 35-185
~ désordonnée 35-205
~ dispersée 40-330
~ durcissante 40-350
~ hors d'équilibre 35-185
~ intermédiaire 35-210
~ intermétallique 35-210
~ liquide restante 25-280
~ mère 35-290
~ métallique 35-150
~ métastable 35-175
~ non métallique 35-155
~ ordonnée 35-200
~ précipitée 40-305
~ proeutectique 40-340
~ proeutectoïde 40-345
~ sigma 35-275
~ stable 35-170
~ ~ à haute pression 35-165
~ ~ à haute température 35-160
~ transitoire 35-195
antiphase f 40-625
phases fpl de Laves 35-265
~ de Zintl 35-270
~ coexistantes 35-190
phénomène m activé thermiquement 45-280
~ Portevin-Le Chatelier 45-540
phonon m 10-305
phosphore m 05-330
cupro-phosphore m 65-075
phosphure m de fer 50-250
photomacrographie f 80-310, 80-315
photomicrographie f 80-250, 80-255
physique f de l'état solide 10-010
~ des métaux 10-015
~ des solides 10-010
~ du métal 10-015
pic m thermique 30-530
picral m 80-115
pièce f d'essai 80-050
plages fpl douces 85-195
plan m à assemblage compact 20-215
~ à empilement dense 20-215
~ atomique 20-175
~ cristallographique 20-155
~ d'accolement 20-470
~ de base 20-360
~ de clivage 40-665
~ de face du prisme 20-365
~ de glissement 45-100
~ de glissement dévié 45-150
~ de l'octaèdre 20-335
~ de maclage 45-250
~ de macle 45-250
~ de octaèdre 20-335

plan
~ limite 20-465
~ matrice 20-465
~ octaédral 20-335
~ octaédrique 20-335
~ prismatique 20-365
~ réticulaire 20-155
~ ~ le plus dense 20-215
plaquette f de cémentite 50-120
~ de martensite 50-370
~ martensitique 50-370
plasticité f 75-025
superplasticité f 75-055
platine m 05-335
~ iridié 65-835
~ rhodié 65-840
platinite m 65-515
platinoïdes mpl 65-820
~ légers 65-830
~ lourds 65-825
plomb m 05-240
~ affiné 65-655
~ brut 65-650
~ d'œuvre 65-650
~ dur 65-660
~ durci 65-660
cupro-plomb m 60-140
plutonium m 05-340
poids m atomique 10-075
~ moléculaire 75-355
~ spécifique 75-350
poil m 25-360
point m critique 35-555
~ ~ du supraconducteur 25-570
~ d'ancrage 30-275
~ d'attaque 80-205
~ d'ébullition 75-430
~ d'épinglage 30-275
~ d'eutexie 35-660
~ de Curie 25-560
~ de fusion 75-420
~ de Néel 25-565
~ de solidification 75-425
~ de sublimation 75-435
~ de transition 25-570, 35-555
~ de transformation 35-555
~ eutectique 35-660
~ eutectoïde 35-675
~ eutectoïdique 35-675
~ M_f 70-170
~ M_s 70-165
~ monotectique 35-670
~ péritectique 35-665
~ singulier 80-495
~ spinodal 15-185
~ triple 35-040
pointe f de déplacements 30-525
~ thermique 30-530
points mpl durs 85-145
polissage m avec attaque 80-085
~ chimique 80-080
~ électrolytique 80-075
~ mécanique 80-070
polonium m 05-345
polycristal m 20-440
polygonisation f 45-435
polymorphie f 25-385
polymorphisme m 25-385
pore m 85-100
porosité f 85-105
~ interdendritique 85-120
microporosité f 85-115
potassium m 05-350
potentiel m carbone 70-750
~ chimique 15-175
~ d'interaction 10-300
~ d'ionisation 10-105
~ électrochimique 75-715

potentiel
~-oxygène m 15-350
~ thermodynamique à pression constante 15-140
~ ~ à volume constant 15-105
pouvoir m de refroidissement 70-125
~ ralentisseur 75-535
~ refroidissant 70-125
~ refroidisseur 70-125
~ thermoélectrique 75-495
~ trempant 70-020
praséodyme m 05-355
pré-alliage m 55-285
pré-précipitation f 70-600
préchauffage m 70-045
précipitation f 40-300
~ continue 25-375
~ discontinue 25-380
pré-précipitation f 70-600
précipité m 40-305
~ cohérent 40-320
~ incohérent 40-325
~ intergranulaire 40-310
~ submicroscopique 40-315
prégerme m 25-135
prélèvement m d'échantillons 80-060
premier nombre m quantique 10-130
premier voisin m 20-195
préparation f de monocristaux 25-085
pression f de dissociation 15-345
principe m d'exclusion de Pauli 10-170
~ d'incertitude de Heisenberg 10-165
~ d'indétermination de Heisenberg 10-165
~ de la tangente commune 15-180
~ de Pauli 10-170
probabilité f thermodynamique 15-230
profondeur f d'emboutissage 75-755
~ de trempe 70-350
prométhium m 05-360
propriétés fpl chimiques 75-655
~ d'emploi 75-820
~ d'usage 75-820
~ d'utilisation 75-820
~ de fonderie 75-775
~ de résistance 75-010
~ de service 75-820
~ dépendantes de la structure 75-860
~ directement perceptibles 75-850
~ électriques 75-465
~ électrochimiques 75-705
~ extrinsèques 75-860
~ indépendantes de la structure 75-865
~ intrinsèques 75-865
~ macroscopiques 75-850
~ magnétiques 75-540
~ mécaniques 75-005
~ ~ en traction 75-015
~ métalliques 10-350
~ métallurgiques 75-855
~ nucléaires 75-520
~ optiques 75-500
~ physiques 75-315
~ technologiques 75-720
~ thermiques 75-365
~ volumétriques 75-340
protactinium m 05-365
pseudo-alliage m 55-650
pseudo-eutectique m 40-225
puits m 30-130

Q

quadrilacune f 30-055
quasi-isotropie f 20-455
quatrième nombre m quantique 10-145

R

radiocristallographie f 80-580
radiogramme m 80-715
autoradiogramme m 80-755
radiographie f 80-705, 80-715
~ par les rayons X 80-825
~ X 80-570
autoradiographie f 80-740, 80-755
métalloradiographie f 80-710
microradiographie f 80-720, 80-725
radiométallographie f 80-710
~ par rayons X 80-575
radiotraceur m 80-760
radium m 05-370
radon m 05-375
raie f de diffraction 80-645
raies fpl de surstructure 80-655
~ du réseau principal 80-650
~ principales 80-650
rapport m cristallographique 20-355
rayon m atomique 10-085
~ critique 25-145
~ ~ d'un germe 25-145
~ ionique 10-090
réactif m d'attaque 80-100
~ d'attaque en couleurs 80-105
~ métallographique 80-100
réaction f eutectoïde 35-590
~ metatectique 35-580
~ monotectique 35-575
~ monotectoïde 35-600
~ péritectique 35-570
~ peritectoïde 35-595
~ syntectique 35-585
rebondissement m 30-260
recalescence f 25-485
recristallisation f 45-440
~ de rassemblement 45-450
~ dynamique 45-465
~ primaire 45-445
~ secondaire 45-450
~ spontanée 45-455
~ tertiaire 45-460
recuit m 70-420
~ à gros grain 70-480
~ à haute température 70-480
~ à mort 70-465
~ adoucissant 70-460
~ alterné 70-520
~ au chalumeau 70-505
~ blanc 70-430
~ bleu 70-445
~ brillant 70-430
~ complet 70-465
~ d'adoucissement 70-460
~ d'affinage structural 70-475
~ de coalescence 70-510
~ de défauts 30-135
~ de détente 70-525
~ de diffusion 70-485
~ de homogénéisation 70-485
~ de malléabilisation 70-625
~ de mise en solution 70-500
~ de normalisation 70-475
~ de perlitisation 70-450
~ de recristallisation 70-490
~ de régénération 70-475
~ de relaxation 70-525
~ de sensibilisation 70-545
~ de sphéroïdisation 70-510
~ de stabilisation 70-530, 70-535
~ doux 70-460
~ en caisse 70-435
~ en pot 70-435
~ en vase close 70-435
~ graduel 70-455
~ graphitisant 70-630
~ incomplet 70-470
~ isochrone 70-495

recuit
~ isotherme 70-450
~ noir 70-440
~ oscillant 70-520
~ partiel 70-470
réfractairité f 75-690
refroidissement m 70-070
~ à l'air 70-095
~ à l'air calme 70-095
~ à l'air libre 70-095
~ à l'eau 70-100
~ à l'huile 70-105
~ au bain 70-115
~ au four 70-110
~ brusque 70-075
~ continu 70-090
~ dans le four 70-110
~ lent 70-080
~ par projection 70-085
~ rapide 70-075
surrefroidissement m 70-110
refus m 85-130
règle f d'alliage 35-700
~ d'exclusion de Pauli 10-170
~ de Hägg 35-260
~ de Hume-Rothery 35-325
~ de Konovalov 35-705
~ de Matthiessen 75-475
~ de mélange 35-700
~ de phases 35-010
~ du bras du levier 35-700
régule m 65-640
relation f d'indétermination de Heisenberg 10-165
relaxation f des contraintes 45-525
rémanence f 75-630
remise f en ordre 25-575
remonte f d'étain 85-165
remplissage m de l'espace 20-205
~ de volume 20-205
renforcement m 70-025
réplique f 80-345
repliure f 85-400
réseau m à faces centrées 20-290
~ centré 20-295
~ cristallin 20-045
~ cubique 20-300
~ ~ à corps centré 20-315
~ ~ à faces centrées 20-310
~ ~ à maille centrée 20-315
~ ~ centré 20-315
~ ~ simple 20-305
~ ~ tétraédrique 20-320
~ de Bravais 20-275
~ de carbides 50-445
~ de cémentite 50-105
~ de dislocations 30-390
~ de transition 20-405
~ de translation 20-275
~ du type diamant 20-320
~ hexagonal 20-340
~ ~ à assemblage compact 20-350
~ ~ compact 20-350
~ ~ simple 20-345
~ imparfait 30-020
~ monoclinique 20-395
~ orthorhombique 20-380
~ ~ à bases centrées 20-385
~ polaire 20-410
~ quadratique 20-370
~ réciproque 20-410
~ rhomboédrique 20-390
~ simple 20-280
~ spatial 20-045
~ tétragonal 20-370
~ triclinique 20-400
sous-réseau m 25-285
résilience f 75-190
résistance f à l'abrasion 75-265
~ à l'oxydation 75-685

resistance
~ à l'oxydation à haute température 75-690
~ à l'usure 75-265
~ à la chaleur 75-835
~ à la compression 75-075
~ à la corrosion 75-670
~ à la fatigue 75-200
~ à la flexion 75-090
~ à la torsion 75-085
~ à la traction 75-070
~ au cisaillement 75-080
~ au clivage 75-135
~ au cyclage thermique 75-830
~ au fluage 75-215, 75-220
~ au fluage aux températures élevées 75-240
~ au revenu 75-840
~ au ternissement 75-680
~ au vieillissement 75-845
~ aux acides 75-695
~ aux bases 75-700
~ aux chocs 75-185
~ aux chocs thermiques 75-825
~ aux corrosifs 75-670
~ chimique 75-665
~ durable 75-225
~ mécanique 75-040
~ ~ à chaud 75-095
~ ~ aux températures élevées 75-095
~ relative 75-140
~ thermique 75-835
superrésistance f 75-045
résistivité f 75-470
~ électrique 75-470
~ spécifique 75-470
ressuage m d'étain 85-165
restauration f 45-420
~ de l'ordre 25-575
~ dynamique 45-425
~ statique 45-430
~ thermique 45-430
retardation f élastique 75-180
retassure f 85-125
~ dispersée 85-120
microretassure f 85-120
retrait m 75-795
~ après solidification 75-810
~ avant la solidification 75-800
~ de solidification 75-805
revenu m 70-365
~ à basse température 70-370
~ à haute température 70-375
~ de détente 70-370
~ durcissant 70-590
surrevenu m 70-375, 70-595
réversion f 70-610
rhénium m 05-380
rhodium m 05-385
rouille f 85-370
rubidium m 05-390
rupture f 85-040
~ différée 40-755
~ ductile 40-680
~ fragile 40-675
~ intercristalline 40-705
~ intergranulaire 40-705
~ intragranulaire 40-710
~ par clivage 85-085
~ par decohésion 40-655
~ par fatigue 40-745
~ par glissement 40-660
~ transcristalline 40-710
ruthénium m 05-395

S

samarium m 05-400
scandium m 05-405
season-cracking m 85-385
second voisin m 20-200

SECTION

section *f* efficace d'absorption de neutrons 75-530
~ ~ de capture de neutrons 75-530
ségrégation *f* 25-400
~ aux joints de grains 25-450
~ dendritique 25-420
~ en goutelettes 25-430
~ gazeuse 25-430
~ intercristalline 25-450
~ interdendritique 25-445
~ inverse 25-440
~ majeure 25-415
~ ~ inverse 25-440
~ mineure 25-420
~ normale 25-435
~ par gravité 25-425
~ primaire 25-405
~ secondaire 25-410
sélénium *m* 05-410
semi-conducteur *m* 10-375
~-~ à impuretés 10-385
~-~ extrinsèque 10-385
~-~ intrinsèque 10-380
semi-métal *m* 10-370
sensibilisation *f* 70-545
sensibilité *f* à l'entaille 75-195
~ à l'épaisseur 75-815
~ sectionelle 75-815
série *f* continue de solutions solides 35-390
~ des forces électromotrices 75-710
~ électrochimique 75-710
sévérité *f* de trempe 70-200
shérardisation *f* 70-930
silal *m* 60-1100
microsilal *m* 60-1095
silicium *m* 05-415
cupro-silicium *m* 65-080
ferro-silicium *m* 60-1140
ferrosilicium *m* 60-1140
siliciure *m* 40-490
silico-calcium *m* 60-1160
silicocalcium *m* 60-1160
silicoferrite *f* 50-475
silicomanganèse *m* 60-1155
silicospiegel *m* 60-1155
silico-spiegel *m* 60-1155
siliciuration *f* 70-910
sillon *m* d'attaque 80-195
~ intergranulaire 80-195
similior *m* 65-395
similor *m* 65-395
site *m* du réseau 20-140
~ interstitiel 20-145
~ réticulaire 20-140
sodium *m* 05-425
solidification *f* 25-005
~ controlée 25-030
~ d'équilibre 25-015
~ de non équilibre 25-020
~ déséquilibrée 25-020
~ dirigée 25-035
~ équilibrée 25-015
~ eutectique 25-025
~ hors d'équilibre 25-020
~ orientée 25-035
~ primaire 25-050
~ secondaire 25-025
~ unidirectionnelle 25-040
solidus *m* 35-485
~ rétrograde 35-505
solubilité *f* 35-295
~ à l'état liquide 35-305
~ à l'état solide 35-310
~ continue 35-315
~ illimitée 35-315
~ mutuelle 35-300

solubilité
~ partielle 35-320
~ réciproque 35-300
~ solide 35-310
~ totale 35-315
soluté *m* 35-360
solution *f* 35-350
~ athermale 15-265
~ binaire 35-405
~ complexe 35-410
~ de continuité du métal 85-090
~ diluée 15-250
~ gazeuse 35-370
~ idéale 15-240
~ imparfaite 15-245
~ liquide 35-365
~ métallique 35-385
~ non idéale 15-245
~ réelle 15-245
~ régulière 15-255
~ semi-régulière 15-260
~ solide 35-375
~ ~ continue 35-390
~ ~ d'insertion 35-420
~ ~ de substitution 35-415
~ ~ désordonnée 35-435
~ ~ finale 35-380
~ ~ interstitielle 35-420
~ ~ ordonnée 35-430
~ ~ primaire 35-380
~ ~ saturée 35-395
~ ~ substitutionelle 35-415
~ ~ sursaturée 35-400
~ ~ terminale 35-380
solvant *m* 35-355
son *m* métallique 75-360
sondage *m* gammagraphique 80-825
~ ultra-sonore 80-800
microsondage *m* électronique 80-780
sonims *pl* 40-515
sorbite *f* 50-380
~ de revenue 50-390
~ de trempe 50-385
sorte *f* d'acier 60-080
soudabilité *f* 75-735
soudure *f* 65-710
~ à l'étain 65-725
~ au plomb 65-735
~ au zinc 65-730
~ dure 65-720
~ forte 65-720
~ tendre 65-715
soufflure *f* 85-135
soufre *m* 05-435
source *f* de dislocations 30-425
~ de Frank-Read 30-430
~ de Koehler 30-435
~ multiplicatrice de dislocation 30-425
sous-couche *f* 10-160
sous-grain *m* 30-650
sous-joint *m* 30-660
~-~ de flexion 30-605
~-~ de grains 30-660
~-~ de torsion 30-600
sous-réseau *m* 20-285
~-~ magnétique 20-485
sous-structure *f* 40-035
~-~ de solidification orientée 40-165
sous-système *m* binaire 35-715
spectrimétrie *f* X 80-700
~ à rayons X 80-700
spectrographie *f* de masse 80-480
spectrométrie *f* par fluorescence de rayons X 80-700
spéculum *m* 65-280
spek *m* 65-280
sphère *f* de Fermi 10-225

sphéroïde m 40-290
~ de graphite 50-215
sphéroïdisation f de la cémentite 70-515
sphéroïdite f 50-150
sphérolite m de graphite 50-215
sphérule f 40-290
spiegel m 60-930
silico-spiegel m 60-1155
silicospiegel m 60-1155
spiegeleisen m 60-930
spinodale f 15-190
spirale f de croissance 25-335
stabilisant m 50-485
stabilisateur m d'austénite 50-450
~ de carbure 50-465
~ de ferrite 50-455
stabilisation f 70-520, 70-535
~ de l'austénite 50-425
~ naturelle 70-540
déstabilisation f 50-430
stabilité f thermique 75-835
stéadite f 50-235
stellite f,m 65-495
striction f à la rupture 75-125
~ après rupture 75-125
~ magnétique 75-650
~ rupture 75-125
magnétostriction f 75-650
strontium m 05-430
structure f 40-005
~ à bandes 40-140
~ à domaines 40-620
~ à grains fins 40-395
~ à gros grains 40-390
~ aciculaire 40-115
~ alvéolaire 40-095
~ anomale 40-135
~ basaltique 40-125
~ biphasée 40-150
~ brute de coulée 40-050
~ cellulaire 40-095
~ colonnaire 40-125
~ cristalline 20-040
~ cristallographique 20-040
~ de bandes 40-140
~ de dislocation 30-415
~ de fonderie 40-050
~ de métal 40-005
~ de moulage 40-050
~ de Widmanstätten 40-130
~ dendritique 40-090
~ duplex 40-150
~ électronique 10-030
~ en aiguilles 40-115
~ en bandes 40-140
~ en domaines 40-620
~ eutectique 40-060
~ eutectoïde 40-075
~ fine 40-040, 40-395
~ globulisée 40-120
~ globulitique 40-120
~ granulaire 40-120
~ grossière 40-390
~ hétérogène 40-155
~ homogène 40-145
~ hypereutectique 40-065
~ hypereutectoïde 40-080
~ hypoeutectique 40-070
~ hypoeutectoïde 40-085
~ lamellaire 40-105
~ macrographique 40-025
~ magnétique 40-650
~ métallographique 40-005
~ micrographique 40-030
~ monophasée 40-145
~ mosaïque 40-170
~ polygonale 40-095
~ polyphasée 40-160
~ primaire 40-045
~ réticulaire 40-100
~ secondaire 40-055

structure
macrostructure f 40-025
microstructure f 40-030
sous-structure f 40-035
surstructure f 35-440
substance f diamagnétique 75-560
~ ferromagnétique 75-550
~ paramagnétique 75-555
substrat m 25-165
sulfinuzation f 70-890
sulfonitruration f 70-890
sulfuration f 70-895
sulfure m de fer 50-245
~ de manganèse 50-240
superalliage m 55-515
superdislocation f 30-220
superparamagnétisme m 75-570
superplasticité f 75-055
superrésistance f 75-045
supraconducteur m 10-390
supraconductibilité f 75-490
supraconduction f 75-490
surchauffe f 70-065
surface f de cassure 40-670
~ de Fermi 10-215
~ de liquidus 35-520
~ de rupture 40-670
~ de solidus 35-495
~-liquidus f 35-520
surfusion f 25-240
~ constitutionnelle 25-255
surrefroidissement m 25-245
surrevenu m 70-375, 70-595
surstructure f 35-440
survieillissement m 70-595
susceptibilité f à la corrosion 75-675
~ à la surchauffe 75-760
~ au revenu 75-765
~ de corrosion 75-675
~ magnétique 75-625
symbole m de nuance 60-090
symétrie f cristalline 20-055
syngonie f 20-235
systeme m 35-005
~ à composants multiples 35-140
~ à constituants multiples 35-140
~ à deux composants 35-120
~ à quatre composants 35-135
~ à trois composants 35-125
~ à un composant 35-115
~ binaire 35-120
~ biphasé 35-100
~ bivariant 35-085
~ cristallin 20-235
~ cristallographique 20-235
~ cubique 20-240
~ d'équilibre 35-045
~ de glissement 45-110
~ de glissement conjugué 45-125
~ de glissement primaire 45-115
~ de glissement secondaire 45-120
~ fer-carbone 50-015
~ hétérogène 35-095
~ hexagonal 20-245
~ invariant 35-075
~ métallique 55-165
~ monoclinique 20-265
~ monophasé 35-090
~ monovariant 35-080
~ orthorhombique 20-255
~ polyphasé 35-110
~ pseudobinaire 35-130
~ quadratique 20-250
~ quasibinaire 35-130
~ quaternaire 35-135
~ rhomboédrique 20-260
~ senaire 20-245
~ ternaire 35-125

systeme
- ~ tétragonal 20-250
- ~ triclinique 20-270
- ~ triphasé 35-105
- ~ unaire 35-115
- ~ unitaire 35-115
- ~ univariant 35-080

T

tache *f* de diffraction 80-660
taches *fpl* blanches 85-250
- ~ grises 85-140

taille *f* de grain 80-285
tantale *m* 05-440
tapure *f* 85-060
- ~ de trempe 85-190

taux *m* d'écrouissage 45-040
- ~ de consolidation 45-065
- ~ de corroyage 45-035
- ~ de déformation 45-035
- ~ de ségrégation 40-420
- ~ de surfusion 25-250

technétium *m* 05-445
tellure *m* 05-450
température *f* caractéristique de Debye 10-280
- ~ critique 25-570, 35-335, 35-555
- ~ d'austenitisation 70-135
- ~ d'ébullition 75-430
- ~ d'équicohésion 30-570
- ~ d'équilibre 35-690
- ~ d'eutexie 35-605
- ~ de Curie 25-560
- ~ de début de la transformation martensitique 70-165
- ~ de Debye 10-280
- ~ de décomposition 15-340
- ~ de fin de la transformation martensitique 70-170
- ~ de fusion 75-420
- ~ de liquidus 35-525
- ~ de Néel 25-565
- ~ de recristallisation 45-470
- ~ de recuit 70-425
- ~ de revenu 70-380
- ~ de solidification 75-425
- ~ de solidus 35-500
- ~ de sublimation 75-435
- ~ de transformation 35-555
- ~ de transition 25-570, 35-555, 75-280
- ~ de transition de ductilité 75-280
- ~ de transition ductile-fragile 75-280
- ~ de transition ordre-désordre 25-625
- ~ de trempe 70-140
- ~ du bleu 70-395
- ~ eutectique 35-605
- ~ eutectoïde 35-630
- ~ eutectoïdique 35-630
- ~ homologue 45-475
- ~ métatectique 35-620
- ~ monotectique 35-615
- ~ péritectique 35-610
- ~ péritectoïde 35-635
- ~ syntectique 35-625

temps *m* d'attaque 80-125
- ~ de maintien 70-060

tenacité *f* 75-035
tension *f* critique de cisaillement 45-090
- ~ ~ de glissement 45-090
- ~ d'écoulement 45-210
- ~ de cisaillement 45-085
- ~ de dissociation 15-345
- ~ de glissement 45-085
- ~ superficielle 30-645
- ~ tangentielle 45-085

tensions *fpl* de coulée 85-170
- ~ de retrait 85-175
- ~ de trempe 85-185
- ~ internes 85-020
- ~ macroscopiques 85-030
- ~ microscopiques 85-035

tenue *f* à la corrosion 75-670
- ~ au fluage 75-215

terbium *m* 05-455
terres *fpl* rares 55-040
tétragonalité *f* 20-375
texture *f* 40-555
- ~ cubique 40-605
- ~ d'écrouissage 40-560
- ~ d'étirage 40-575
- ~ de compression 40-590
- ~ de déformation 40-560
- ~ de fibre 40-565
- ~ de fonderie 40-580
- ~ de Goss 40-610
- ~ de laminage 40-570
- ~ de recristallisation 40-600
- ~ de recuit 40-600
- ~ de torsion 40-595
- ~ de traction 40-585
- ~ fibreuse 40-565

thallium *m* 05-460
théorie *f* des bandes 10-020
- ~ électronique des métaux 10-025

thermalloy *m* 65-540
thermodynamique *f* des alliages 15-010
- ~ des solides 15-005
- ~ métallurgique 15-010

thermogramme *m* 80-527
thorium *m* 05-465
thulium *m* 05-470
tirage *m* de monocristaux 25-090
titane *m* 05-480
titanisation *f* 70-945
titanure *f* 40-510
tombac *m* 65-395
traceur *m* radioactif 80-760
radiotraceur *m* 80-760
trainage *m* magnétique 75-645
traitement *m* au froid 70-340
- ~ d'adoucissement 70-460
- ~ d'affinage 25-225
- ~ d'affinage structural 70-475
- ~ d'amélioration 70-410
- ~ de cémentation 70-655
- ~ de détente 70-525
- ~ de durcissement 70-015
- ~ de graphitisation 70-630
- ~ de normalisation 70-475
- ~ de phase 25-460
- ~ de recristallisation 70-490
- ~ de relaxation 70-525
- ~ de renforcement 70-025
- ~ de stabilisation 70-535
- ~ de surface 70-010
- ~ par le froid 70-340
- ~ purement thermique 70-006
- ~ superficiel 70-010
- ~ thermique 70-005, 70-006
- ~ ~ de degazage 70-550
- ~ ~ de diffusion 70-650
- ~ ~ ordinaire 70-006
- ~ ~ prorement dit 70-006
- ~ thermochimique 70-650
- ~ thermomagnétique 70-965
- ~ thermomécanique 70-950
- ~ ~ à haute température 70-955
- ~ uniquement thermique 70-006

transcristallisation *f* 25-070
transformation *f* à chaud 45-020
- ~ à froid 45-025
- ~ à l'état solide 25-475
- ~ allotropique 25-455
- ~ athermique 25-520
- ~ austénite-martensite 50-280
- ~ avec diffusion 25-505
- ~ bainitique 50-290
- ~ congruente 25-525
- ~ continue 25-540
- ~ d'ordonnancement 25-575

transformation
~ de structure 25-500
~ discontinue 25-545
~ en masse 25-550
~ eutectique 35-565
~ eutectoïde 35-590
~ hétérogène 25-535
~ homogène 25-530
~ isotherme 25-515
~ magnétique 25-555
~ martensitique 50-280
~ ~ secondaire 50-285
~ massive 25-550
~ métatectique 35-580
~ monotectique 35-575
~ monotectoïde 35-600
~ ordre-désordre 25-580
~ par corroyage 45-015
~ péritectique 35-570
~ péritectoïde 35-595
~ perlitique 50-275
~ polymorphe 25-455
~ polymorphique 25-455
~ réversible 25-495
~ sans diffusion 25-510
~ structurale 25-500
~ syntéctique 35-585
transition f de phase 25-460
~ de phase de première espèce 25-465
~ de phase de seconde espèce 25-470
~ de première espèce 25-465
~ de seconde espèce 25-470
~ désordre-ordre 25-575
~ ordre-désordre 25-580
translation f 20-130
travail m d'extraction 10-060
~ de sortie 10-060
~ par déformation 45-015
trempabilité f 75-730
trempe f 70-075, 70-175, 70-570
~ à cœur 70-280
~ à l'air 70-215
~ à l'air soufflé 70-220
~ à la saumure 70-238
~ à l'eau 70-205
~ à l'eau salée 70-238
~ à l'huile 70-210
~ après recuit de mise en solution 70-570
~ arrêtée 70-300
~ au bain 70-245
~ au bain chaud 70-225
~ au bain de plomb 70-235
~ au chalumeau 70-290
~ au défilé 70-320
~ au défilé avec rotation 70-325
~ au plomb 70-235
~ bainitique 70-310
~ brillante 70-330
~ brutale 70-260
~ dans une solution saline 70-238
~ de l'âme 70-770
~ de lacunes 30-110
~ de proche en proche 70-320
~ de volume 70-287
~ directe 70-760
~ double 70-765
~ douce 70-265
~ du cœur 70-770
~ électrique 70-295
~ en bain de sel 70-230
~ en coquille refroidie 70-255
~ énergique 70-260
~ et revenu 70-410
~ étagée 70-305
~ ~ bainitique 70-310
~ ~ martensitique 70-305
~ incomplète 70-335
~ interrompue 70-300
~ inverse 85-150
~ isotherme 70-310
~ locale 70-275
~ localisee 70-275
~ ordinaire 70-270
~ par aspersion 70-240
~ par contact 70-250

trempe
~ par étapes 70-305
~ par immersion 70-245
~ par induction 70-295
~ partielle 70-275
~ rapide 70-260
~ secondaire 70-405
~ superficielle 70-285
~ ultra-rapide 70-085
hypertrempe f 70-345
triangulation f du système 35-725
trichite f 25-360
trilacune f 30-050
troisième nombre m quantique 10-140
tronc m de l'atome 10-040
troostite f 50-395
~ de revenu 50-405
~ de trempe 50-400
~ primaire 50-400
~ secondaire 50-405
TTM 70-950
TTMHT 70-955
tungstène m 05-485
tungstènure f 40-500

U

uranium m 05-490
usinabilité f 75-745

V

vallée f eutectique 35-730
vanadium m 05-495
variable f d'état 15-020
variance f 35-030
variété f allotropique 25-390
~ polymorphique 25-390
vecteur m d'onde 10-185
~ de base du réseau 20-135
~ de Burgers 30-190
~ de translation 20-135
védal m 65-560
veine f sombre 85-230
verrou m de Cottrell-Lomer 30-450
vieillissement m 70-575
~ accéléré 70-590
~ aprés déformation à froid 45-225
~ aprés écrouissage 45-225
~ aprés trempe 70-580
~ magnétique 70-620
~ naturel 70-585
~ poussé 70-595
survieillissement m 70-595
vitallium m 65-485
vitesse f critique de trempe 70-180
~ d'échauffement 70-040
~ de chauffage 70-040
~ de croissance 25-275
~ de croissance des germes 25-275
~ de croissance du cristal 25-275
~ de germination 25-270
~ de refroidissement 70-120
voisin m direct 20-195
premier ~ 20-195
second ~ 20-200
~ immédiat 20-195
volume m atomique 75-320
~ d'activation 30-475
~ molaire 75-325

W

whisker m 25-360

X

xénon *m* 05-500

Y

ytterbium *m* 05-505
yttrium *m* 05-510

Z

zamak *m* 65-705
zinc *m* 05-515
~ affiné par distillation 65-695
~ dur 65-700
~ rectifié 65-695
~ thermique 65-690

zircaloy *m* 65-855
zirconium *m* 05-520
zone *f* **à structure basaltique 40-180**
~ appauvrie 30-120
~ basaltique 40-180
~ centrale de cristallisation équiaxe 40-185
~ de Brillouin 10-230
~ de coordination 20-190
~ de cristallisation basaltique 40-180
~ de cristaux équiaxes 40-185
~ de diffusion 45-415
~ de Seeger 30-120
~ de trempe 40-175
~ équiaxe 40-185
~ extérieure 40-175
~ lacunaire 30-120
~ trempée 70-355
zones *fpl* **de Guinier-Preston 70-605**

SKOROWIDZ POLSKI

A

ajnsztajn *m* 05-135
aktyn *m* 05-010
aktywacja *f* cieplna 45-275
aktywność *f* ciśnieniowa 15-290
~ termodynamiczna 15-280
alclad *m* 65-560
aldrej *m* 65-585
alniko *n* 65-530
alotropia *f* 25-385
alsifer *m* 65-535
aludur *m* 65-580
alumel *m* 65-460
aluminiowanie *n* dyfuzyjne 70-925
aluminium *n*
~ hutnicze 65-545
~ pierwotne 65-545
~ rafinowane 65-550
amalgamat *m* 65-850
ameryk *m* 05-020
amorficzny 20-005
analiza *f*
~ aktywacyjna 80-765
~ chemiczna 80-470
~ cieplna 80-505
~ dylatometryczna 80-540
~ ~ bezwzględna 80-545
~ ~ prosta 80-545
~ ~ różnicowa 80-550
~ fazowa rentgenowska 80-585
~ fizykochemiczna 80-490
~ fluorescencyjna 80-700
~ ~ rentgenowska 80-700
~ kroplowa 80-465
~ magnetyczna 80-560
~ mikroskopowa 80-040
~ spektralna 80-475
~ widmowa 80-475
~ termiczna 80-505
~ ~ bezpośrednia 80-510
~ ~ różnicowa 80-515
~ ~ termodynamiczna 80-535
~ termomagnetyczna 80-565
anizotermiczny wykres *m* przemiany austenitu 70-160
anizotropia *f* 20-445
~ magnetyczna 75-575
anormalność *f* stali 85-225
antyfaza *f* 40-625
antyferromagnetyzm *m* 75-565
antygrafityzator *m* 50-465
antykorodal *m* 65-575
antymon *m* 05-025
~ rafinowany 65-665
argentan *m* 65-415
argon *m* 05-030
arsen *m* 05-035
asprężystość *f* 75-050
astat *m* 05-040
asteryzm *m* 80-680
asymetria *f* energetyczna 15-235
atmosfera *f*
~ aktywna 70-680
~ atomów obcych 30-440

atmosfera
~ Cottrella 30-440
~ egzotermiczna 70-710
~ endotermiczna 70-705
~ nawęglająca 70-695
~ obojętna 70-675
~ ochronna 70-690
~ regulowana 70-700
~ zanieczyszczeń 30-440
atom *m*
~ dyslokowany 30-510
~ międzywęzłowy 30-075
~ podstawieniowy 30-070
~ z drugiej strefy koordynacyjnej 20-200
~ z pierwszej strefy koordynacyjnej 20-195
austenit *m* 50-025
~ azotowy 70-835
~ pierwotny 50-030
~ pierwszorzędowy 50-030
~ przechłodzony 50-265
~ przemieniony 50-035
~ szczątkowy 50-270
austenityzacja *f* 70-130
austenityzowanie *n* 70-130
autoradiografia *f* 80-740
~ kontrastowa 80-745
~ śladowa 80-750
autoradiogram *m* 80-755
awional *m* 65-610
azot *m* 05-305
azotek *m* 70-810
~ chromu 70-820
~ tytanu 70-825
~ wanadu 70-830
~ żelaza 70-815
azotoaustenit *m* 70-835
azotonawęglanie *n* 70-870
azotopasywowanie *n* 70-885
azotowanie *n* 70-785
~ gazowe 70-790
~ jarzeniowe 70-800
~ jonowe 70-800
~ kąpielowe 70-795
~ z nasiarczaniem 70-890
azymutalna liczba *f* kwantowa 10-135

B

babbit *m* 65-645
babit *m* 65-645
badanie *n*
~ dylatometryczne 80-540
~ fluoroscencyjne 80-835
~ fraktograficzne 80-385
~ kalorymetryczne 80-500
~ magnetyczne 80-560
~ makroskopowe 80-045
~ metalograficzne 80-035
~ mikroskopowe 80-040
~ ~ metodą kontrastu fazowego 80-270
~ ~ w nadfiolecie 80-265
~ ~ w polu ciemnym 80-230
~ ~ w polu jasnym 80-235
~ ~ w świetle nadfioletowym 80-265
~ ~ w świetle spolaryzowanym 80-260
~ nieniszczące 80-785
~ przełomu 80-385

BADANIE

badanie
~ radiograficzne 80-705
~ rentgenograficzne 80-570
~ ultradźwiękowe 80-800
~ własności mechanicznych 80-400
~ wytrzymałościowe 80-400
bainit *m* 50-410
~ dolny 50-415
~ górny 50-420
bar *m* 05-045
bardzo drobny perlit *m* 50-400
bariera *f*
~ Cottrella-Lomera 30-450
~ energetyczna 10-110
~ potencjału 10-110
barwa *f* nalotowa 70-385
baza *f* sieci 20-110
bąbel *m* 85-130
berkel *m* 05-050
bertolidy *mpl* 35-250
beryl *m* 05-055
bezpostaciowy 20-005
biała faza *f* 85-250
białe plamy *fpl* 85-250
białe złoto *n* 65-805
biały metal *m* 65-640
bikryształ *m* 20-435
bimetal *m* 65-880
biwakans *m* 30-045
bizmut *m* 05-060
bliźniak *m* 45-230
~ deformacyjny 45-235
~ mechaniczny 45-235
~ rekrystalizacyjny 45-240
~ wyżarzania 45-240
bliźniakowanie *n* 45-245
blok *m* mozaiki 30-655
blokowanie *n* dyslokacji 30-445
błąd *m*
~ ułożenia 30-540
~ ~ bliźniaczy 30-555
~ ~ podwójny 30-550
~ ~ pojedynczy 30-545
~ ~ wewnętrzny 30-545
~ ~ zewnętrzny 30-550
bodziec *m* termodynamiczny 15-070
bor *m* 05-065
borek *m* 70-905
borowanie *n* dyfuzyjne 70-900
brak *m* rozpuszczalności 35-340
braunit *m* 70-840
brąz *m* 65-095
~ aluminiowy 65-135
~ antymonowy 65-165
~ architektoniczny 65-270
~ armatni 65-245
~ artystyczny 65-270
~ berylowy 65-150
~ bezcynowy 65-125
~ biały 65-230
~ chromowy 65-195
~ cynowo-cynkowy 65-205
~ cynowo-fosforowy 65-210
~ cynowo-ołowiowy 65-185
~ cynowy 65-110
~ dekoracyjny 65-270
~ do przeróbki plastycznej 65-105
~ dzwonowy 65- 250
~ fosforowy 65-215
~ glinowy 65-135
~ grafitowy 65-220
~ kadmowy 65-190
~ krzemowy 65-145
~ łożyskowy 65-235
~ manganowy 65-155
~ maszynowy 65-240

brąz
~ medalierski 65-260
~ monetowy 65-265
~ niklowy 65-160
~ niskocynowy 65-120
~ odlewniczy 65-100
~ ołowiowy 65-140
~ panewkowy 65-235
~ porowaty 65-225
~ przerabialny plastycznie 65-105
~ przewodowy 65-255
~ specjalny 65-130
~ spiekany 65-225
~ wolframowy 65-170
~ wtórny 65-285
~ wysokocynowy 65-115
~ zaworowy 65-275
~ zwierciadłowy 65-280
~ żelazowy 65-175
brązal *m* 65-135
brom *m* 05-070
budowa *f*
~ elektronowa 10-030
~ fazowa 35-020
~ krystalograficzna 20-040

C

cecha *f* stali 60-090
cementyt *m* 50-065
~ drugorzędowy 50-080
~ eutektoidalny 50-085
~ eutektyczny 50-075
~ kulkowy 50-100
~ pierwotny 50-070
~ pierwszorzędowy 50-070
~ przedeutektoidalny 50-080
~ stopowy 50-110
~ trzeciorzędowy 50-090
~ wolny 50-095
cer *m* 05-100
cermetal *m* 55-675
cez *m* 05-080
chlor *m* 05-105
chłodzenie *n* 70-070
~ ciągłe 70-090
~ kąpielowe 70-115
~ powolne 70-080
~ szybkie 70-075
~ ultraszybkie cieczy 70-085
~ w oleju 70-105
~ w piecu 70-110
~ w powietrzu 70-095
~ w wodzie 70-100
~ z piecem 70-110
chmura *f*
~ elektronowa 10-035
~ zanieczyszczeń 30-440
choroba *f*
~ cynowa 35-245
~ wodorowa 85-395
chrom *m* 05-110
chromansil *m* 60-260
chromel *m* 65-465
chromoaluminiowanie *n* 70-940
chromonikielina *f* 65-470
chromowanie *n* dyfuzyjne 70-920
chrzęst *m* cynowy 45-270
ciało *n*
~ rozpuszczone 35-360
~ wielokrystaliczne 20-440
ciągliwość *f* 75-030
ciecz *f*
~ chłodząca 70-190
~ hartownicza 70-190
~ macierzysta 25-095
cieniowanie *n* 80-370

cienka folia f 80-350
ciepło f
~ atomowe 75-370
~ cząsteczkowe 75-375
~ krzepnięcia 75-385
~ molowe 75-375
~ parowania utajone 75-390
~ przemiany 75-400
~ rozpuszczania 75-405
~ sublimacji 75-395
~ topnienia utajone 75-380
~ właściwe 75-415
~ ~ elektronowe 10-275
~ ~ konfiguracyjne 25-635
ciepłowytrzymałość f 75-240
ciężar m
~ atomowy 10-075
~ cząsteczkowy 75-355
~ molowy 75-355
~ właściwy 75-350
ciśnienie n dyslokacji 15-345
crowdion m 30-470
cyjanowanie n 70-855
~ gazowe 70-860
~ kąpielowe 70-865
~ niskotemperaturowe 70-875
~ wysokotemperaturowe 70-870
cykl m obróbki cieplnej 70-970
cyna f 05-475
~ biała 65-630
~ szara 65-635
cynk m 05-515
~ hutniczy 65-690
~ redestylowany 65-695
~ rektyfikowany 65-695
~ twardy 65-700
cynkowanie n dyfuzyjne 70-930
cyrkaloj m 65-855
cyrkon m 05-520
czas m
~ trawienia 80-125
~ wygrzewania 70-060
~ wytrawiania 80-125
czasowa granica f pełzania 75-235
cząstkowy współczynnik m dyfuzji 45-390
czernienie n 70-400
czterowakans m 30-055
czwarta liczba f kwantowa 10-145
czynnik m
~ elektrochemiczny 10-400
~ elektrowartościowości ujemnej 10-400
~ wartościowości względnej 10-405
~ wielkości atomu 10-395
czystość f 55-095
czysty metal m 55-090

D

daltonidy mpl 35-245
debajogram m 80-635
defekt m
~ Frenkla 30-060
~ liniowy 30-140
~ popromienny 30-485
~ powierzchniowy 30-535
~ punktowy 30-030
~ radiacyjny 30-485
~ sieci krystalicznej 30-015
~ sieciowy 30-015
~ strukturalny 30-015
~ struktury krystalicznej 30-015
defektoskopia f 80-790
~ akustyczna 80-795
~ magnetyczna 80-805
~ rentgenowska 80-820
~ ultradźwiękowa 80-800

DYSLOKACJA

dekalescencja f 25-490
dekoracja f (dyslokacji) 80-375
dekorowanie n dyslokacji 80-375
demodyfikator m 25-235
dendryt m 25-315
~ chemiczny 25-320
~ termiczny 25-325
destabilizacja f austenitu 50-430
dezorientacja f 30-595
diamagnetyk m 75-560
dipol m
~ dyslokacji 30-380
~ dyslokacyjny 30-380
diwakans m 30-045
dodatek m
~ stopowy 55-190
~ ~ główny 55-195
domena f 40-615
~ antyfazowa 40-625
~ magnetyczna 40-635
~ Weissa 40-635
~ zamykająca 40-640
domieszki fpl
~ niezamierzone 55-200
~ przypadkowe 55-200
drgania npl
~ cieplne 10-270
~ termiczne 10-270
drobnoziarnistość f 40-380
drobny perlit m 50-385
druga liczba f kwantowa 10-135
dural m 65-555
duralumin m 65-555
duraluminium n 65-555
dwuwakans m 30-045
dyfraktogram m 80-610
~ elektronowy 80-775
~ proszkowy 80-635
~ prześwietleniowy 80-675
~ rentgenowski 80-615
~ wykonany metodą promieni przechodzących 80-675
dyfuzja f 45-290
~ atomowa 45-330
~ chemiczna 45-320
~ izotermiczna 45-350
~ jednofazowa 45-330
~ kanalikowa 45-360
~ międzywęzłowa 45-315
~ objętościowa 45-295
~ po granicach ziarn 45-310
~ poprzez wakansy 45-305
~ powierzchniowa 45-300
~ przestrzenna 45-295
~ przyspieszona 45-355
~ reakcyjna 45-325
~ sieciowa 45-295
~ termiczna 45-370
~ ujemna 45-340
~ ustalona 45-345
~ wakansowa 45-305
~ własna 45-400
~ wstępująca 45-340
~ wzajemna 45-330
dylatogram m 80-555
dylatometria f 80-540
~ prosta 80-545
~ różnicowa 80-550
dysklinacja f 30-480
dyslokacja f 30-145
~ biegunowa 30-365
~ bliźniakująca 30-375
~ całkowita 30-305
~ częściowa 30-315
~ ~ Shockleya 30-330
~ częściowo zakotwiczona 30-280
~ doskonała 30-305

DYSLOKACJA

dyslokacja
~ -drzewo 30-410
~ Franka częściowo zakotwiczona 30-335
~ granic ziarn 30-580
~ helikoidalna 30-355
~ jednostkowa 30-215
~ kątowa 30-360
~ krawędziowa 30-150
~ ~ dodatnia 30-155
~ ~ nieregularna 30-345
~ ~ ujemna 30-160
~ lasu 30-410
~ mieszana 30-180
~ naładowana 30-465
~ narożnikowa 30-360
~ niedoskonała 30-310
~ o mocy jednostkowej 30-215
~ osiadła 30-280
~ poślizgowa 30-265
~ półutwierdzona 30-280
~ pryzmatyczna 30-345
~ rozciągnięta 30-325
~ rozszczepiona 30-325
~ Shockleya 30-330
~ spiralna 30-355
~ spoczynkowa 30-280
~ ślizgowa 30-265
~ śrubowa 30-165
~ ~ lewoskrętna 30-175
~ ~ prawoskrętna 30-170
~ zakotwiczona 30-270
~ złożona 30-180
dyslokacje fpl
~ emitowane 30-370
~ pierwotne 30-250
~ wtórne 30-255
dyspersja f 40-335
dyspersoid m 40-330
dysproz m 05-130
dziedziczność f żeliwa 60-985
dźwięk m metaliczny 75-360

E

efekt m kanałowy 30-500
efektywny współczynnik m rozdziału 25-265
ekstrapłaszczyzna f 30-200
elektrodyfuzja f 45-365
elektrokrystalizacja f 25-080
elektron m 65-625
~ przewodnictwa 10-265
~ rdzeniowy 10-070
~ swobodny 10-055
~ walencyjny 10-065
~ wartościowości 10-065
~ wewnętrzny 10-070
~ zewnętrzny 10-065
~ związany 10-050
elektronografia f 80-770
elektronogram m 80-775
elektronowa teoria f metali 10-025
elektrotransport m 45-365
elektroujemność f 10-095
elektrum n 65-810
elementarna komórka f sieciowa 20-090
eliminacja f defektów termiczna 30-135
elinwar m 65-505
energia f
~ aktywacji 45-285
~ anizotropii 75-590
~ błędu ułożenia 30-560
~ Fermiego 10-210
~ granic ziarn 30-575
~ granicy międzyfazowej 30-635
~ jonizacji 10-100
~ oddziaływania 10-295
~ porządkująca 25-630
~ powierzchniowa 30-640

energia
~ swobodna cząstkowa 15-110
~ ~ Gibbsa 15-140
~ ~ Helmholtza 15-105
~ ~ nadmiarowa 15-115
~ ~ powierzchni granicznej 15-125
~ ~ powierzchniowa 15-120
~ ~ resztkowa 15-115
~ ~ tworzenia roztworu 15-130
~ ~ tworzenia związku 15-135
~ wewnętrzna 15-095
~ wiązania 15-090
~ własna dyslokacji 30-230
~ zmagazynowana 45-060
~ związana 15-100
entalpia f
~ cząstkowa 15-150
~ nadmiarowa 15-145
~ resztkowa 15-145
~ swobodna 15-140
~ topnienia 15-165
~ tworzenia roztworu 15-155
~ ~ związku 15-160
~ wiązania 15-170
entropia f
~ cząstkowa 15-215
~ konfiguracyjna 15-210
~ nadmiarowa 15-220
~ pozycyjna 15-210
~ resztkowa 15-220
~ topnienia 15-225
~ tworzenia roztworu 15-195
~ ~ związku 15-200
~ wibracyjna 15-205
epitaksja f 25-340
erb m 05-140
europ m 05-145
eutektoid m 40-240
eutektyka f 40-200
~ anormalna 40-230
~ cementytowa 50-255
~ fosforowa potrójna 50-235
~ grafitowa 50-260
~ ledeburytyczna 50-255
~ nieciągła 40-235
~ płytkowa 40-220
~ poczwórna 40-215
~ podwójna 40-205
~ potrójna 40-210

F

faza f 35-025
~ σ 35-275
~ biała 85-250
~ dyspersyjna 40-330
~ elektronowa 35-280
~ macierzysta 35-290
~ metaliczna 35-150
~ metastabilna 35-175
~ międzymetaliczna 35-210
~ ~ inkongruentna 35-225
~ ~ kongruentna 35-220
~ ~ niekongruentna 35-225
~ niemetaliczna 35-155
~ nierównowagowa 35-185
~ nieuporządkowana 35-205
~ pośrednia 35-210
~ przedeutektoidalna 40-345
~ przedeutektyczna 40-340
~ przejściowa 35-195
~ rozproszona 40-330
~ równowagowa 35-180
~ sigma 35-275
~ skondensowana 35-145
~ stabilna 35-170
~ uporządkowana 35-200
~ utwardzająca 40-350
~ wydzielona 40-305
~ wysokociśnieniowa 35-165
~ wysokotemperaturowa 35-160
~ zdyspergowana 40-330

fazy *fpl*
~ Lavesa 35-265
~ termodynamicznie sprzężone 35-190
~ współistniejące 35-190
~ Zintla 35-270
ferm *m* 05-150
ferniko *n* 65-520
ferrimagnetyzm *m* 75-580
ferromagnetyk *m* 75-550
ferromagnetyzm *m* 75-545
ferryt *m* 50-045
~ eutektoidalny 50-060
~ pierwotny 50-050
~ przedeutektoidalny 50-055
~ wolny 50-055
~ wysokotemperaturowy 50-050
ferrytyczne pasmo *n* segregacyjne 85-230
ferrytyzacja *f* 70-645
figury *fpl*
~ biegunowe 80-685
~ trawienia 80-200
fizyczna granica *f* plastyczności 75-110
fizyka *f*
~ ciała stałego 10-010
~ metali 10-015
fluktuacje *fpl* stężenia 25-100
fluor *m* 05-155
fluorescencyjna spektroskopia *f* rentgenowska 80-700
fokuson *m* 30-505
folia *f* cienka 80-350
fonon *m* 10-305
fosfor *m* 05-330
fosforek *m* żelaza 50-250
fosforobrąz *m* 65-215
fotomakrografia *f* 80-310, 80-315
fotomikrografia *f* 80-250, 80-255
~ elektronowa 80-365
fraktografia *f* 80-385
frans *m* 05-160
front *m*
~ dyfuzji 45-410
~ krystalizacji 25-295
~ rozrostu 45-480
funkcja *f*
~ Blocha 10-195
~ całkowita 15-055
~ cząstkowa 15-060
~ ekstensywna 15-040
~ falowa 10-180
~ intensywna 15-045
~ molowa 15-050
~ nadmiarowa 15-065
~ resztkowa 15-065
~ stanu 15-035
~ termodynamiczna 15-030
~ ~ całkowita 15-055
~ ~ cząstkowa 15-060
~ ~ ekstensywna 15-040
~ ~ intensywna 15-045
~ ~ molowa 15-050
~ ~ nadmiarowa 15-065
~ ~ resztkowa 15-065
~ własna 10-190

G

gadolin *m* 05-165
gal *m* 05-170
gammagrafia *f* 80-825
gammaradiografia *f* 80-825
gatunek *m* stali 60-085
gaz *m*

gaz
~ elektronowy 10-045
~ nawęglający 70-745
~ nośny 70-685
generator *m* dyslokacji 30-425
generowanie *n* dyslokacji 30-420
german *m* 05-175
gęstość *f* 75-345
~ dyslokacji 30-235
~ stanów 10-200
glin *m* 05-015
glinokrzem *m* 60-1165
glinowanie *n* 70-925
głębokość *f* zahartowania 70-350
głębokotłoczność *f* 75-750
główna liczba *f* kwantowa 10-130
główna oś *f* krystalizacji 25-300
główny pierwiastek *m* stopowy 55-195
gradient *m*
~ stężenia 15-080
~ temperatury 15-075
grafit *m* 50-165
~ drugorzędowy 50-180
~ eutektyczny 50-175
~ gniazdowy 50-210
~ gwiazdkowy 50-200
~ kulkowy 50-190
~ międzydendrytyczny 50-205
~ pierwotny 50-170
~ pierwszorzędowy 50-170
~ płatkowy 50-185
~ punktowy 50-195
~ rozetkowy 50-200
~ sferoidalny 50-190
~ szumowy 50-170
~ wtórny 50-180
grafityzacja *f* 70-630, 50-225
grafityzator *m* 50-460
granica *f*
~ antyfazowa 40-630
~ bliźniaków 45-265
~ daszkowa 30-605
~ fazowa 35-535
~ koherentna 30-625
~ międzyfazowa 30-610
~ ~ koherentna 30-625
~ ~ niekoherentna 30-630
~ ~ niesprzężona 30-630
~ ~ sprzężona 30-625
~ nachylona 30-605
~ niekoherentna 30-630
~ niesprzężona 30-630
~ pełzania 75-220
~ ~ czasowa 75-235
~ plastyczności 75-110
~ ~ fizyczna 75-110
~ ~ naturalna 75-110
~ ~ umowna 75-115
~ ~ wyraźna 75-110
~ pochylenia 30-605
~ proporcjonalności 75-100
~ rozpuszczalności 35-330
~ skośna 30-605
~ skręcenia 30-600
~ skręcona 30-600
~ sprężystości 75-105
~ sprzężona 30-625
~ szerokokątowa 30-585
~ wąskokątowa 30-590
~ ziarn(a) 30-565
~ zmęczenia 75-200
graniczna temperatura *f* kruchości na zimno 75-280
gruboziarnistość *f* 40-375
grzanie *n* 70-030
~ beznalotowe 70-430
~ bezzgorzelinowe 70-435
~ ciemne 70-440

GRZANIE

grzanie
~ czyste 70-435
~ jasne 70-430
gwiazdkowość f 80-680

H

habitus m 20-460
hafn m 05-185
hantla f 30-085
hartowanie n 70-175
~ bainityczne 70-310
~ beznalotowe 70-330
~ bezpośrednie 70-760
~ czyste 70-330
~ dwukrotne 70-765
~ indukcyjne 70-295
~ izotermiczne 70-310
~ kokilowe 70-255
~ kontaktowe 70-250
~ łagodne 70-265
~ miejscowe 70-275
~ natryskiem 70-240
~ natryskowe 70-240
~ na wskroś 70-280
~ niezupełne 70-335
~ objętościowe 70-287
~ obrotowe 70-315
~ obrotowo-posuwowe 70-325
~ ostre 70-260
~ palnikowe 70-290
~ płomieniowe 70-290
~ podwójne 70-765
~ podzerowe 70-340
~ posuwowe 70-320
~ posuwowo-obrotowe 70-325
~ powierzchniowe 70-285
~ przerywane 70-300
~ przesuwowe 70-320
~ rdzenia po nawęglaniu 70-770
~ skrośne 70-280
~ stopniowe 70-305
~ w gorącej kąpieli 70-225
~ w kąpieli ołowiowej 70-235
~ w kąpieli solnej 70-230
~ w kokili chłodzonej 70-255
~ w oleju 70-210
~ w ośrodku gorącym 70-225
~ w powietrzu 70-215
~ w solance 70-238
~ w strumieniu powietrza 70-220
~ w wodzie 70-205
~ wtórne 70-405
~ z przemianą izotermiczną 70-310
~ z temperatury nawęglania 70-760
~ zanurzeniowe 70-245
~ zwykłe 70-270
hartowność f 75-730
hardenit m 50-365
hasteloj m 65-455
hel m 05-190
heterodyfuzja f 45-335
hiduminium n 65-605
hipernik m 65-475
histereza f
~ cieplna 25-480
~ magnetyczna 75-640
~ przemiany 25-480
hodowanie n monokryształów 25-085
holm m 05-195
homogenizowanie n 70-485
hydronalium n 65-600

I

igła f martenzytu 50-375
imitacja f złota 65-815
implantacja f jonów 70-805

ind m 05-205
indukcja f magnetyczna 75-610
inhibitor m rozrostu ziarna 40-400
inkludowanie n 80-065
inkonel m 65-445
intensywność f
~ chłodzenia 70-200
~ hartowania 70-200
~ namagnesowania 75-600
interferogram m 80-300
inwar m 65-500
iryd m 05-215
irydoosm m 65-845
iterb m 05-505
itr m 05-510
izoaktywa f 15-315
izomorfizm m 20-475
izotermiczny wykres m przemiany
    austenitu 70-155
izotropia f 20-450

J

jama f
~ skurczowa 85-125
~ usadowa 85-125
jamka f trawienia 80-205
jądro n dyslokacji 30-205
jednorodność f 40-405
jednostkowa komórka f sieciowa 20-090
jod m 05-210

K

kadm m 05-075
kaliforn m 05-090
kantal m 65-525
karburyzator m 70-740
kaskada f
~ przemieszczeń 30-520
~ zderzeń 30-515
katalizator m zarodkowania 25-125
kąpiel f hartownicza 70-195
kąt m
~ dezorientacji 30-595
~ wzajemnej orientacji 30-595
kierunek m
~ krystalograficzny 20-150
~ łatwego magnesowania 75-585
~ najgęściej obsadzony 20-220
~ poślizgu 45-105
kiur m 05-125
klasa f
~ krystalograficzna 20-050
~ symetrii 20-050
kobalt m 05-115
koercja f 75-635
kohenit m 50-115
koherencja f 30-620
kohezja f 10-310
kolanko n 30-245
kolejność f ułożenia płaszczyzn atomowych
    20-180
komórka f
~ prosta 20-095
~ prymitywna 20-095
~ sieciowa elementarna 20-090
~ ~ jednostkowa 20-090
kompozyt m 55-680

kondensacja f wakansów 30-115
konduktywność f 75-485
konoda f 35-695
konstantan m 65-450
kontrast m
~ fazowy 80-275
~ orientacji 80-155
kontur m Burgersa 30-185
korozja f 85-260
~ atmosferyczna 85-275
~ chemiczna 85-265
~ cierna 85-325
~ elektrochemiczna 85-270
~ gazowa 85-295
~ jądrowa 85-365
~ kontaktowa 85-320
~ lokalna 85-340
~ miejscowa 85-340
~ międzykrystaliczna 85-360
~ morska 85-290
~ naprężeniowa 85-310
~ podpowierzchniowa 85-355
~ popromienna 85-365
~ powierzchniowa 85-330
~ punktowa 85-345
~ radiacyjna 85-365
~ równomierna 85-335
~ selektywna 85-305
~ stykowa 85-320
~ szczelinowa 85-300
~ ukryta 85-355
~ wewnętrzna 85-355
~ wodna 85-280
~ wybiorcza 85-305
~ wżerowa 85-350
~ ziemna 85-285
~ zmęczeniowa 85-315
kowalność f 75-725
kowar m 65-520
kraudion m 30-470
krawędziowa pętla f dyslokacyjna 30-345
kruchość f 75-270
~ alkaliczna 75-305
~ ługowa 75-305
~ na gorąco 75-290
~ na niebiesko 75-300
~ na zimno 75-285
~ niskotemperaturowa 75-295
~ odpuszczania 85-220
~ reotropowa 75-310
~ wodorowa 85-390
krypton m 05-225
krystaliczny 20-010
krystalit m 40-360
~ podmikroskopowy 30-655
krystalizacja f 25-010
~ drugorzędowa 25-060
~ eutektyczna 25-025
~ jednokierunkowa 25-040
~ katodowa 25-080
~ kierowana 25-030
~ nierównowagowa 25-020
~ pierwotna 25-050
~ pierwszorzędowa 25-050
~ równoosiowa 25-075
~ równowagowa 25-015
~ samorzutna 25-045
~ słupkowa 25-065
~ ukierunkowana 25-035
~ wtórna 25-060
~ zorientowana 25-035
krystalografia f 20-035
kryształ m 20-015
~ bliźniaczy 45-230
~ dendrytyczny 25-315
~ doskonały 30-005
~ idiomorficzny 25-355
~ jonowy 20-025
~ macierzysty 40-295
~ metaliczny 20-020
~ nitkow(at)y 25-360

kryształ
~ pojedynczy 20-430
~ roztworu stałego 40-195
~ rzeczywisty 30-010
~ walencyjny 20-030
~ włoskowy 25-360
~ zaszczepiający 25-130
kryształy mpl
~ drugorzędowe 40-275
~ izomorficzne 20-480
~ pierwotne 40-270
~ pierwszorzędowe 40-270
~ równoosiowe 40-285
~ słupkowe 40-280
~ wtórne 40-275
krytyczna szybkość f chłodzenia 70-180
krytyczna szybkość f hartowania 70-180
krytyczne naprężenie n poślizgu 45-090
krytyczne naprężenie n styczne 45-090
krytyczny stopień m odkształcenia 45-075
krzem m 05-415
krzemek m 40-490
krzemoferryt m 50-475
krzemomangan m 60-1155
krzemowanie n dyfuzyjne 70-910
krzepnięcie n 25-005
~ eutektyczne 25-025
~ jednokierunkowe 25-040
~ kierowane 25-030
~ nierównowagowe 25-020
~ równowagowe 25-015
~ ukierunkowane 25-035
krzywa f
~ dylatometryczna 80-555
~ graniczna 35-535
~ hartowności 80-455
~ nagrzewania 80-520
~ naprężenie-odkształcenie 80-410
~ ogrzewania 80-520
~ pełzania 45-515
~ rozpuszczalności granicznej 35-530
~ spinodalna 15-190
~ stygnięcia 80-525
~ umocnienia 45-070
~ Wöhlera 80-425
~ zmęczenia 80-425
krzywe fpl termodynamicznie sprzężone 35-540
ksenon m 05-500
kujność f 75-725
kula f Fermiego 10-225
kwasoodporność f 75-695
kwazyeutektyka f 40-225
kwazyizotropia f 20-455

L

lantan m 05-230
las m dyslokacji 30-405
lauegram m 80-620
lautal m 65-595
lawina f
~ przemieszczeń 30-520
~ zderzeń 30-515
ledeburyt m 50-160
lejność f 75-780
liczba f
~ atomowa 75-330
~ koordynacyjna 20-185
~ kwantowa 10-125
~ ~ azymutalna 10-135
~ ~ czwarta 10-145
~ ~ druga 10-135
~ ~ główna 10-130
~ ~ magnetyczna 10-140
~ ~ orbitalna 10-135

LICZBA

liczba kwantowa
~ ~ pierwsza 10-130
~ ~ poboczna 10-135
~ ~ spinowa 10-145
~ ~ trzecia 10-140
~ masowa 75-335
~ mikrostanów 15-230
~ Poissona 75-170
~ porządkowa 75-330
~ stopni swobody 35-030
~ tłoczności 75-755
likwidus m 35-510
linia f
~ debajowska 80-640
~ dyfrakcyjna 80-645
~ dyslokacji 30-195
~ dyslokacyjna 30-195
~ eutektoidalna 35-655
~ eutektyczna 35-640
~ graniczna 35-535
~ likwidus(u) 35-515
~ interferencyjna 80-645
~ izoaktywności 15-315
~ monotektyczna 35-650
~ perytektyczna 35-645
~ poślizgu 45-185
~ przemiany eutektoidalnej 35-655
~ ~ eutektycznej 35-640
~ ~ monotektycznej 35-650
~ ~ perytektycznej 35-645
~ rozpuszczalności granicznej 35-530
~ solidus(u) 35-490
linie fpl
~ Lüdersa-Hartmanna 45-195
~ nadstruktury 80-655
~ Neumanna 45-205
~ płynięcia 45-195
~ podstawowe 80-650
~ spoczynkowe 40-750
lit m 05-245
lorens m 05-235
lotność f 15-290
luka f
~ czworościenna 20-325
~ międzywęzłowa 20-105
~ oktaedryczna 20-330
~ ośmiościenna 20-330
~ tetraedryczna 20-325
lut m 65-710
~ cynkowy 65-730
~ cynowy 65-725
~ łatwotopliwy 65-715
~ miękki 65-715
~ mosiężnoniklowy 65-745
~ mosiężny 65-740
~ niskotopliwy 65-715
~ ołowiany 65-735
~ srebrny 65-750
~ twardy 65-720
lutowie n 65-710
lutet m 05-250
lutowność f 75-740

Ł

łańcuch m zderzeń 30-515
ługoodporność f 75-700

M

magnal m 65-620
magnalium n 65-620
magnetostrykcja f 75-650
magnetyczna liczba f kwantowa 10-140
magnetyzacja f 75-600
~ nasycenia 75-605
magnetyzm m szczątkowy 75-630
magnez m 05-255

makrofotografia f 80-315
makrografia f 80-305
makronaprężenia npl 85-030
makrosegregacja f 25-415
makrostruktura f 40-025
makrotrawienie n 80-135
mangan m 05-260
manganin m 65-405
marka f stali 60-085
martenzyt m 50-310
~ drobnoiglasty 50-325
~ gruboiglasty 50-320
~ iglasty 50-315
~ masywny 50-355
~ odpuszczania 50-345
~ odpuszczony 50-345
~ regularny 50-340
~ skrytoiglasty 50-330
~ skrytokrystaliczny 50-330
~ tetragonalny 50-335
~ wtórny 50-350
~ zgniotowy 50-360
masa f
~ atomowa względna 10-075
~ cząsteczkowa 75-355
~ właściwa 75-345
materiał m złożony 55-680
mendelew m 05-265
metal m 10-355, 10-360
~ Auera 65-875
~ biały 65-645
~ chemicznie aktywny 55-070
~ ciekły 55-150
~ czcionkowy 55-615
~ czysty 55-090
~ elektrolityczny 55-130
~ justunkowy 55-640
~ linotypowy 55-625
~ lity 55-155
~ łatwotopliwy 55-060
~ monotypowy 55-620
~ napylany 55-145
~ nie przetapiany 55-115
~ nieszlachetny 55-005
~ niskotopliwy 55-060
~ nowy 55-115
~ oczyszczany strefowo 55-140
~ osadzany z pary 55-145
~ pierwotny 55-115
~ podstawowy 55-185
~ pospolity 55-005
~ półszlachetny 55-020
~ próżniowy 55-135
~ przetopiony 55-120
~ radioaktywny 55-075
~ rafinowany 55-125
~ reaktorowy 55-085
~ reaktywny 50-070
~ rodzimy 55-110
~ rozszczepialny 50-080
~ roztopiony 55-150
~ spiekany 55-660
~ stereotypowy 55-630
~ szlachetny 55-010
~ topiony w próżni 55-135
~ trudnotopliwy 55-065
~ ultraczysty 55-105
~ wtórny 55-120
~ wysokotopliwy 55-065
metale mpl
~ alkaliczne 55-030
~ ciężkie 50-055
~ grup przejściowych 55-025
~ lekkie 55-050
~ nieżelazne 55-015
~ przejściowe 55-025
~ rzadkie 50-045
~ ziem alkalicznych 55-035
~ ~ rzadkich 55-040
metalizowanie n dyfuzyjne 70-915
metalografia f 80-005

metalografia
~ barwna 80-030
~ elektronowa 80-025
~ ilościowa 80-020
~ mikroskopowa 80-010
~ rentgenowska 80-575
~ wysokotemperaturowa 80-015

metaloid m 10-365
metaloznawstwo n fizyczne 10-005
metatektyka f 40-255
metoda f
~ Braggów 80-600
~ Debye'a-Scherrera 80-595
~ kryształu kołysanego 80-605
~ ~ nieruchomego 80-590
~ ~ obracanego 80-600
~ ~ oscylującego 80-605
~ ~ wahliwie obracanego 80-605
~ Lauego 80-590
~ promieni przechodzących 80-625
~ ~ zwrotnych 80-630
~ proszkowa 80-595
~ mokra 80-815
~ ~ sucha 80-810
~ prześwietleniowa 80-625
~ wspólnej stycznej 15-180

miedzionikiel m 65-400
miedź f 05-120
~ anodowa 65-040
~ arsenowa 65-090
~ beztlenowa 65-060
~ o dużej przewodności elektrycznej 65-070
~ cementacyjna 65-045
~ chromowa 65-200
~ czarna 65-015
~ elektrolityczna 65-035
~ elektrotechniczna 65-065
~ fosforowa 65-075
~ hutnicza 65-025
~ kadmowa 65-190
~ katodowa 65-030
~ konwertorowa 65-015
~ krzemowa 65-080
~ manganowa 65-085
~ nie odtleniona 65-050
~ odtleniona fosforem 65-055
~ OFHC 65-070
~ przewodowa 65-065
~ rafinowana 65-020
~ ~ ogniwo 65-025
~ surowa 65-015
~ tlenowa 65-050

mieszanina f
~ eutektoidalna 40-240
~ eutektyczna 40-200
~ faz 40-190
~ gazów 35-370

międzywęźle n 20-105
miękkie plamy fpl 85-195
migracja f granic ziarna 45-490
mikroanaliza f
~ absorpcyjna 80-735
~ rentgenowska 80-780

mikrododatek m 55-250
mikrofotografia f 80-255
mikrofraktografia f 80-390, 80-395
mikrografia f 80-245
~ elektronowa 80-365

mikronaprężenia npl 85-035
mikropełzanie n 45-530
mikropęknięcie n 85-045
~ Griffitha 85-055

mikroporowatość f 85-115
mikropory mpl 85-110
mikroradiografia f 80-720
mikroradiogram m 80-725
mikrorentgenografia f 80-730
mikrosegregacja f 25-420

mikroskop m metalograficzny 80-210
mikroskopia f 80-040
~ elektronowa 80-330
~ ~ analizująca 80-360
~ ~ emisyjna 80-335
~ ~ prześwietleniowa 80-340
~ ~ rastrowa 80-360
~ ~ skaningowa 80-360
~ ~ transmisyjna 80-340
~ fazowo-kontrastowa 80-270
~ interferencyjna 80-295
~ jonowa 80-380
~ nadfioletowa 80-265
~ odbiciowa 80-225
~ optyczna 80-215
~ polaryzacyjna 80-260
~ polowo-jonowa 80-380
~ prześwietleniowa 80-220
~ refleksyjna 80-225
~ rentgenowska 80-730
~ świetlna 80-215
~ transmisyjna 80-220
~ typu odbiciowego 80-225
~ ~ prześwietleniowego 80-220
~ w nadfiolecie 80-265
~ wysokotemperaturowa 80-240

mikrostruktura f 40-030
mikrotrawienie n 80-140
mikrotwardość f 75-250
miszmetal m 65-865
mnożenie n dyslokacji 30-420
moc f dyslokacji 30-210
moduł m
~ sprężystości liniowej 75-150
~ ~ objętościowej 75-160
~ ~ podłużnej 75-150
~ ~ poprzecznej 75-155
~ ~ postaciowej 75-155
~ ~ wzdłużnej 75-150
~ ~ ściśliwości 75-160
~ Younga 75-150

modyfikacja f 25-220
modyfikator m 25-230
modyfikowanie n 25-220
~ siluminu 25-225

molibden m 05-275
~ bezwisowy 65-870

monel m 65-435
monokrystalizacja f 25-085
monokryształ m 20-430
monotektyka f 40-260
monowakans m 30-040
mosiądz m 65-290
~ α 65-325
~ alfa 65-325
~ α+β 65-330
~ aluminiowy 65-340
~ automatowy 65-370
~ β 65-335
~ cynowy 65-345
~ do przeróbki plastycznej 65-295
~ dwufazowy 65-320
~ dwuskładnikowy 65-305
~ jednofazowy 65-315
~ krzemowy 65-365
~ lutowniczy 65-740
~ łuskowy 65-380
~ manganowy 65-355
~ maszynowy 65-375
~ morski dwufazowy 65-385
~ ~ jednofazowy 65-390
~ niklowy 65-350
~ odlewniczy 65-300
~ ołowiowy 65-360
~ przerabialny plastycznie 65-295
~ specjalny 65-310
~ wanadowy 65-180
~ wieloskładnikowy 65-310
~ zwykły 65-305

motyle mpl 85-255

N

naborowywanie n 70-900
nachromowywanie n 70-920
nadplastyczność f 75-055
nadprzewodnictwo n 75-490
nadprzewodnik m 10-390
nadstop m 55-515
nadstruktura f 35-440
nadtopienie n 85-215
nadwytrzymałość f 75-045
nagrzewanie n 70-030
~ na wskroś 70-035
~ skrośne 70-035
~ tętniące 70-050
~ udarowe 70-050
nakrzemowywanie n 70-910
nalot m niebieski 70-390
namagnesowanie n 75-600
napięcie n powierzchniowe 30-645
naprężenia npl
~ cieplne 85-025
~ hartownicze 85-185
~ makroskopowe 85-030
~ mikroskopowe 85-035
~ odlewnicze 85-170
~ skurczowe 85-175
~ strefowe 85-030
~ strukturalne 85-035
~ szczątkowe 85-020
~ termiczne 85-025
~ wewnętrzne 85-020
~ własne 85-020
~ ~ drugiego rodzaju 85-035
~ ~ pierwszego rodzaju 85-030
naprężenie n
~ płynięcia plastycznego 45-210
~ poślizgu krytyczne 45-090
~ rozrywające 75-130
~ styczne 45-085
~ ~ krytyczne 45-090
~ ścinające 45-085
nasiarczanie n 70-895
nasycanie n dyfuzyjne 70-655
natężenie n koercyjne 75-635
~ namagnesowania 75-600
~ powściągające 75-635
naturalna granica f plastyczności 75-110
nawęglanie n 70-715
~ gazowe 70-730
~ kąpielowe 70-725
~ miejscowe 70-735
~ proszkowe 70-720
~ w proszkach 70-720
nawrót m 45-420, 70-610
neodym m 05-280
neon m 05-285
neptun m 05-290
neutronografia f 80-830
nichrom m 65-470
nicrosilal m 60-1095
nieciągłość f materiału 85-090
niedopasowanie f sieciowe 30-615
niejednorodność f 40-410
~ chemiczna 40-415
niemetal m 10-365
nieprzezroczystość f 75-505
niesprężystość f 75-050
nieuporządkowanie n 25-585
nikiel m 05-295
~ karbonylkowy 65-425
~ Monda 65-425
nikielina f 65-410

nimonik m 65-480
niob m 05-300
niobek m 40-505
niresist m 60-1090
nirezist m 60-1090
nital m 80-110
nitronawęglanie n 70-870
nobel m 05-310
normalizowanie n 70-475
~ rdzenia 70-770
nowe srebro n 65-415
numer m ziarna 80-290

O

objętość f
~ aktywowana 30-475
~ atomowa 75-320
~ molowa 75-325
obrabialność f 75-745
obróbka f cieplna 70-005
~ ~ zwykła 70-006
~ cieplno-chemiczna 70-650
~ ~dyfuzyjna 70-650
~ ~magnetyczna 70-965
~ ~mechaniczna 70-950
~ ~-~ niskotemperaturowa 70-960
~ ~-~ wysokotemperaturowa 70-955
~ ~-plastyczna 70-950
~ ~ niskotemperaturowa 70-960
~ ~ ~ wysokotemperaturowa 70-955
~ plastyczna 45-015
~ ~ na gorąco 45-020
~ ~ na zimno 45-025
~ podzerowa 70-340
~ powierzchniowa 70-010
~ umacniająca 70-025
obszar m
~ dwufazowy 35-685
~ jednofazowy 35-680
~ tworzenia się bainitu 50-305
~ ~ się martenzytu 50-300
~ ~ się perlitu 50-295
odazotowanie n 70-850
odbitka f
~ Baumanna 80-325
~ na siarkę 80-325
~ siarkowa 80-325
odcinek m identyczności 20-115
odcynkowanie n 85-380
odczynnik m
~ do trawienia 80-100
~ ~ ~ barwnego 80-105
~ ~ ~ kolorowego 80-105
~ metalograficzny 80-100
odgazowanie n 70-550
odkształcenie n
~ krytyczne 45-075
~ plastyczne 45-010
~ sprężyste 45-005
~ trwałe 45-010
odległość f
~ międzyatomowa 20-230
~ międzyblaszkowa 40-110
~ między cząstkami dyspersoidu 40-425
~ międzypłaszczyznowa 20-225
~ międzypłytkowa 40-110
~ średnia między cząstkami dyspersoidu 40-425
odmiana f
~ alotropowa 25-390
~ ~ wysokotemperaturowa 25-395
~ polimorficzna 25-390
odporność f
~ chemiczna 75-665
~ korozyjna 75-670

PĘKNIĘCIE

odporność
~ na alkalia 75-700
~ na działanie podwyższonej temperatury 75-835
~ na korozję 75-670
~ na matowienie korozyjne 75-680
~ na nagłe zmiany temperatury 75-825
~ na odpuszczanie 75-840
~ na pełzanie 75-215
~ na starzenie 75-845
~ na ścieranie 75-265
~ na udary cieplne 75-825
~ na uderzenia cieplne 75-825
~ na utlenianie 75-685
~ na zgorzelinowanie 75-690
~ na zmęczenie cieplne 75-830
~ na zużycie przez tarcie 75-265
~ termiczna 75-835
odprężanie n 70-525
odpuszczalność f 75-765
odpuszczanie n 70-365
~ niskie 70-370
~ wysokie 70-375
odwęglenie n
~ powierzchni 85-210
~ powierzchniowe 85-210
odwodorowywanie n 70-555
odwrotne utwardzenie n żeliwa 85-150
okres m
~ identyczności 20-115
~ inkubacji 70-150
~ inkubacyjny 70-150
~ translacji 20-115
oksydowanie n 70-400
ołów m 05-240
~ antymonowy 65-660
~ czysty 65-655
~ miękki 65-655
~ rafinowany 65-655
~ surowy 65-650
~ twardy 65-660
oporność f elektryczna właściwa 75-470
opór m właściwy 75-470
opóźnienie n
~ magnetyczne 75-645
~ sprężyste 75-180
orbitalna liczba f kwantowa 10-135
orientacja f
~ bezładna 20-420
~ chaotyczna 20-420
~ krystalograficzna 20-415
~ przypadkowa 20-420
~ uprzywilejowana 20-425
~ wyróżniona 20-425
orientowanie n monokryształu 80-690
ortęć f 65-850
osm m 05-315
osmoiryd m 65-845
osnowa f struktury 40-015
oś f
~ bliźniacza 45-260
~ główna dendrytu 25-300
~ krystalizacji główna 25-300
~ krystalograficzna 20-125
~ łatwego magnesowania 75-585
~ symetrii czterokrotna 20-080
~ ~ dwukrotna 20-070
~ ~ kryształu 20-065
~ ~ sześciokrotna 20-085
~ ~ trzykrotna 20-075
~ wtórna dendrytu 25-305
ośrodek m
~ chłodzący 70-185
~ hartowniczy 70-185
~ nawęglający 70-740
otoczka f ferrytyczna 50-230
oziębianie n 70-075

P

paczenie n się 85-200
pallad m 05-325
palladowce mpl 65-830
pantal m 65-615
paramagnetyk m 75-555
parametr m
~ oddziaływania 15-300
~ porządku bliskiego 25-615
~ ~ dalekiego 25-620
~ sieci krystalicznej 20-120
~ stanu 15-020
~ termodynamiczny 15-020
~ uporządkowania bliskiego 25-615
~ ~ dalekiego 25-620
pasemkowość f węglików 85-235
pasma n węglików 85-235
pasmo n
~ bliźniacze 45-255
~ dozwolone 10-240
~ energetyczne 10-235
~ energii wzbronionych 10-250
~ ferrytyczne 85-230
~ hartowności 80-460
~ obsadzone 10-245
~ odkształcenia 45-200
~ poślizgu 45-190
~ przewodnictwa 10-260
~ segregacyjne 85-230
~ ~ ferrytyczne 85-230
~ walencyjne 10-255
~ wzbronione 10-250
~ zapełnione 10-245
pasmowość f węglików 85-235
patentowanie n 70-415
patyna f 85-375
pełzanie n 45-495
~ dyfuzyjne 45-520
~ nieustalone 45-500
~ progresywne 45-510
~ przyspieszone 45-510
~ ustalone 45-505
permaloj m 65-440
permendur m 65-490
perminwar m 65-510
perlit m 50-125
~ bardzo cienkoblaszkowy 50-145
~ bardzo cienkopłytkowy 50-145
~ bardzo drobny 50-400
~ blaszkowy 50-130
~ drobnoblaszkowy 50-135
~ drobnopasemkowy 50-135
~ drobnopłytkowy 50-135
~ gruboblaszkowy 50-140
~ grubopasemkowy 50-140
~ grubopłytkowy 50-140
~ pasemkowy 50-130
~ płytkowy 50-130
~ sorbityczny 50-145
~ z cementytem kulkowym 50-150
~ zdegenerowany 50-155
~ ziarnisty 50-150
perlityzowanie n 70-450
perytektoid m 40-250
perytektyka f 40-245
pęcherz m gazowy 85-185
pęcznienie n żeliwa 70-640
pękanie n 85-015
~ sezonowe 85-385
pęknięcie n 85-040
~ ciągliwe 40-680
~ cieplne 85-060
~ Griffitha 85-055
~ hartownicze 85-190
~ kruche 40-675
~ łupliwe 85-085
~ na gorąco 85-080

pęknięcie
~ na zimno 85-075
~ naprężeniowe 85-065
~ opóźnione 40-755
~ plastyczne 40-680
~ poślizgowe 40-660
~ powierzchniowe 85-050
~ przy odkształcaniu 85-070
~ rozdzielcze 40-655, 85-085
~ zmęczeniowe 40-745
pętla f
~ dyslokacji 30-340
~ dyslokacyjna 30-340
~ ~ krawędziowa 30-345
pierścień m
~ debajowski 80-640
~ Debye'a 80-640
pierwiastek m
~ austenitotwórczy 50-450
~ azotkotwórczy 50-470
~ betatwórczy 65-860
~ chemiczny 05-005
~ ferrytotwórczy 50-455
~ metaliczny 10-360
~ niemetaliczny 10-365
~ półmetaliczny 10-370
~ rozdrabniający ziarno 40-400
~ rozpuszczony 35-360
~ stabilizujący 50-485
~ ~ austenit 50-450
~ ~ fazę β 65-860
~ ~ ferryt 50-455
~ stopowy 55-190
~ ~ główny 55-195
~ śladowy 55-250
~ węglikotwórczy 50-435
~ towarzyszący 55-260
~ utwardzający 55-255
pierwotny system m poślizgu 45-115
pierwsza liczba f kwantowa 10-130
pikral m 80-115
plamka f dyfrakcyjna 80-660
plamy fpl
~ białe 85-250
~ miękkie 85-195
plastyczność f 75-025
platyna f 05-335
platynit m 65-515
platynoiryd m 65-835
platynorod m 65-840
platynowce mpl 65-820
~ ciężkie 65-825
~ lekkie 65-830
pluton m 05-340
płaszczyzna f
~ atomowa 20-175
~ bazowa 20-360
~ bliźniacza 45-250
~ bliźniakowania 45-250
~ habitus 20-465
~ krystalograficzna 20-155
~ łupliwości 40-665
~ najgęściej obsadzona 20-215
~ ośmiościanu 20-335
~ podstawowa 20-360
~ pokroju 20-465
~ postaci 20-465
~ poślizgu 45-100
~ ~ poprzecznego 45-150
~ sieciowa 20-155
~ słupa 20-365
~ zrostu 20-470
płatki mpl 85-240
płynięcie n
~ lepkie 45-220
~ lepkościowe 45-220
~ plastyczne 45-215
płytka f
~ cementytu 50-120
~ martenzytu 50-370

pobieranie n próbek 80-060
poboczna liczba f kwantowa 10-135
pobranie n próbek 80-060
podatność f
~ magnetyczna 75-625
~ na korozję 75-675
podgranica f 30-660
podgrzewanie n 70-045
podłoże n 25-165
podpowłoka f elektronowa 10-160
podsieć f 20-285
~ magnetyczna 20-485
podstawa f stopu 55-185
podstruktura f 40-035
podukład m podwójny 35-715
podwarstwa f elektronowa 10-160
podziarno n 30-650
pojawienie n się kruchości 75-275
pojemność f cieplna 75-410
pokrój m kryształu 20-460
pole n naprężeń dyslokacji 30-225
polerowanie n
~ chemiczne 80-080
~ elektrolityczne 80-075
~ mechaniczne 80-070
~ z trawieniem 80-085
poligonizacja f 45-435
polikryształ m 20-440
polimorfizm m 25-385
polon m 05-345
połysk m metaliczny 75-510
por m 85-100
porowatość f 85-105
porządek m 25-590
~ bliski 25-595
~ bliskiego zasięgu 25-595
~ daleki 25-600
~ dalekiego zasięgu 25-600
postać f kryształu 20-460
poślizg m 45-095
~ łatwy 45-155
~ ołówkowy 45-175
~ po granicach ziarn 45-160
~ podwójny 45-135
~ pojedynczy 45-130
~ poprzeczny 45-145
~ prosty 45-130
~ pryzmatyczny 45-170
~ śródkrystaliczny 45-165
~ wewnątrzkrystaliczny 45-165
~ wielokrotny 45-140
~ wzdłuż granic ziarn 45-160
pot m cynowy 85-165
potas m 05-350
potencjał m
~ chemiczny 15-175
~ elektrochemiczny 75-715
~ izotermiczno-izobaryczny 15-140
~ jonizacji 10-105
~ jonizacyjny 10-105
~ oddziaływania 10-300
~ termodynamiczny pod stałym ciśnieniem 15-140
~ ~ w stałej objętości 15-105
~ tlenowy 15-350
~ węglowy 70-750
potrójna eutektyka f fosforowa 50-235
powierzchnia f
~ Fermiego 10-215
~ likwidus(u) 35-520
~ międzyfazowa 30-610
~ przełomu 40-670
~ rozdziału faz 30-610
~ solidus(u) 35-495
powinowactwo n

powinowactwo
~ chemiczne 75-660
~ elektronowe 10-115
powłoka f elektronowa 10-155
powstawanie n roztworu stałego 25-055
poziom m
~ energetyczny 10-150
~ Fermiego 10-220
pozostałość f magnetyczna 75-630
pozycja f
~ międzywęzłowa 20-145
~ węzłowa 20-140
półmetal m 10-370
półokres m rozpadu 75-525
półpłaszczyzna f nadmiarowa 30-200
półprzewodnik m 10-375
~ domieszkowy 10-385
~ niesamoistny 10-385
~ samoistny 10-380
praca f wyjścia 10-060
prawdopodobieństwo n termodynamiczne 15-230
prawo n
~ Henry'ego 15-275
~ Raoulta 15-270
prazeodym m 05-355
prężność f
~ dysocjacji 15-345
~ rozkładowa 15-345
proces m aktywowany cieplnie 45-280
promet m 05-360
promień m
~ atomowy 10-085
~ atomu 10-085
~ jonowy 10-090
~ krytyczny zarodka 25-145
protaktyn m 05-365
próba f
~ Baumanna 80-320
~ dźwiękowa 80-795
~ dźwięku 80-795
~ hartowania od czoła 80-450
~ iskrowa 80-440
~ Jominy'ego 80-450
~ metali szlachetnych 55-100
~ na siarkę 80-320
~ na zmęczenie 80-420
~ przełomu 80-445
~ rozciągania 80-405
~ rtęciowa 80-485
~ ślepego hartowania 70-775
~ twardości 80-415
~ udarności 80-430
~ zmęczeniowa 80-420
próbka f 80-050
~ masywna 80-055
~ prętowa 80-435
próg m
~ dyslokacji 30-240
~ kruchości 75-280
przechładzanie n 70-345
~ wakansów 30-110
przechłodzenie n 25-240, 25-245
~ stężeniowe 25-255
przedwydzielanie n 70-600
przegięcie n 30-245
przegrzanie n 70-065
przegrzewanie n 70-480
przegrzewność f 75-760
przekrój m
~ czynny na wychwyt neutronów 75-530
~ izotermiczny 35-735
~ pionowy 35-740
~ politermiczny 35-740
~ poziomy 35-735

przekrój
~ pseudopodwójny 35-745
~ stężeniowy 35-740
przełom m 40-670
~ aksamitny 40-715
~ biały 85-160
~ ciągliwy 40-680
~ drobnoziarnisty 40-700
~ drzewiasty 40-720
~ gruboziarnisty 40-695
~ jasny 85-155
~ jedwabisty 40-715
~ kruchy 40-675
~ krystaliczny 40-685
~ łupkowy 40-725
~ międzykrystaliczny 40-705
~ muszlowy 40-730
~ naftalinowy 40-740
~ opóźniony 40-755
~ plastyczny 40-680
~ poślizgowy 40-660
~ rozdzielczy 40-655
~ szklisty 40-735
~ śródkrystaliczny 40-710
~ warstwowy 40-725
~ włóknisty 40-720
~ ziarnisty 40-690
~ zmęczeniowy 40-745
przemiana f
~ alotropowa 25-455
~ atermiczna 25-520
~ bainityczna 50-290
~ bezdyfuzyjna 25-510
~ ciągła 25-540
~ drugiego rodzaju 25-470
~ drugiego rzędu 25-470
~ dyfuzyjna 25-505
~ eutektoidalna 35-590
~ eutektyczna 35-565
~ fazowa 25-460
~ izotermiczna 25-515
~ jednorodna 25-530
~ kongruentna 25-525
~ magnetyczna 25-555
~ martenzytyczna 50-280
~ ~ wtórna 50-285
~ masywna 25-550
~ metatektyczna 35-580
~ monotektoidalna 35-600
~ monotektyczna 35-575
~ nieciągła 25-545
~ niejednorodna 25-535
~ odwracalna 25-495
~ perlityczna 50-275
~ perytektoidalna 35-595
~ perytektyczna 35-570
~ pierwszego rodzaju 25-465
~ pierwszego rzędu 25-465
~ polimorficzna 25-455
~ porządek-nieporządek 25-580
~ strukturalna 25-500
~ syntektyczna 35-585
~ w stanie stałym 25-475
przenikalność f
~ magnetyczna 75-615
~ ~ początkowa 75-620
przepalenie n 85-205
przepływ m termodynamiczny 15-085
przeróbka f plastyczna 45-015
przerwa f
~ ciągłości materiału 85-090
~ energetyczna 10-250
przeskok m 30-265
przestarzenie n 70-595
przestrzenny wykres m równowagi układu potrójnego 35-750
przestrzeń n
~ falowa 10-205
~ k 10-205
przestrzenie fpl międzydendrytyczne 40-355
przesycanie n 70-570

przetrawienie *n* 80-130
przewężenie *n* po rozerwaniu 75-125
przewodnictwo *n* cieplne właściwe 75-440
przewodność *f*
~ cieplna 75-440
~ elektryczna właściwa 75-485
~ właściwa 75-485
przystanek *m* cieplny 80-530
~ termiczny 80-530
pseudoeutektyka *f* 40-225
pseudostop *m* 55-650
puchnięcie *n* żeliwa 70-640
punkt *m*
~ Curie 25-560
~ eutektoidalny 35-675
~ eutektyczny 35-660
~ krytyczny 35-555
~ monotektyczny 35-670
~ Néela 25-565
~ osobliwy 80-495
~ perytektyczny 35-665
~ potrójny 35-040
~ przełomowy 35-555
~ spinodalny 15-185
~ szczególny 80-495
~ zakotwiczenia 30-275

Q

quasi-eutektyka *f* 40-225
quasi-izotropia *f* 20-455

R

rad *m* 05-370
radiografia *f* 80-705
~ metali 80-710
~ promieniami gamma 80-825
radiogram *m* 80-715
radioindykator *m* 80-760
radon *m* 05-375
rdza *f* 85-370
rdzeń *m* 70-665
~ atomowy 10-040
~ dyslokacji 30-205
reakcja *f*
~ monotektoidalna 35-600
~ monotektyczna 35-575
~ perytektoidalna 35-595
~ perytektyczna 35-570
~ syntektyczna 35-585
reguła *f*
~ dźwigni 35-700
~ faz 35-010
~ Hägga 35-260
~ Hume-Rothery'ego 35-325
~ Konowałowa 35-705
~ Matthiessena 75-475
~ odcinków 35-700
~ Vegarda 35-425
~ wspólnej stycznej 15-180
rekalescencja *f* 25-485
rekrystalizacja *f* 45-440
~ dynamiczna 45-465
~ pierwotna 45-445
~ samorzutna 45-455
~ trzeciorzędowa 45-460
~ wtórna 45-450
rekrystalizowanie *n* 70-490
relaksacja *f* naprężeń 45-525
ren *m* 05-380
rentgenodefektoskopia *f* 80-820
rentgenografia *f*
~ dyfrakcyjna 80-580
~ metali 80-575

rentgenografia
~ strukturalna 80-580
rentgenogram *m*
~ dyfrakcyjny 80-615
~ Lauego 80-620
~ proszkowy 80-635
~ wykonany metodą kryształu obracanego 80-665
~ ~ ~ promieni przechodzących 80-675
~ ~ ~ ~ zwrotnych 80-670
rentgenomikroskopia *f* 80-730
rentgenowska analiza *f* fazowa 80-585
rentgenowska analiza *f* fluorescencyjna 80-700
rentgenowska *f* analiza strukturalna 80-580
replika *f* 80-345
resztka *f*
~ cieczy 25-280
~ roztworu ciekłego 25-280
rezystywność *f* 75-470.
rod *m* 05-385
rodzaj *m* stali 60-080
rowek *m* trawienny 80-195
rozciągnięcie *n* dyslokacji 30-320
rozdrobnienie *n*
~ struktury 40-385
~ ziarna 40-385
rozkład *m*
~ roztworu 25-365
~ spinodalny 25-370
rozmnażanie *n* się dyslokacji 30-420
rozpad *m*
~ roztworu 25-365
~ spinodalny 25-370
rozpuszczalnik *m* 35-355
rozpuszczalność *f* 35-295
~ ciągła 35-315
~ graniczna 35-330
~ nieograniczona 35-315
~ ograniczona 35-320
~ w stanie ciekłym 35-305
~ w stanie stałym 35-310
~ wzajemna 35-300
rozpuszczanie *n* 70-500
rozrost *m* ziarn 25-290
rozszczepienie *n* dyslokacji 30-320
roztwarzanie *n* 70-500
roztwór *m* 35-350
~ atermiczny 15-265
~ ciekły 35-365
~ doskonały 15-240
~ dwuskładnikowy 35-405
~ gazowy 35-370
~ metaliczny 35-385
~ nasycony 35-397
~ niedoskonały 15-245
~ regularny 15-255
~ rozcieńczony 15-250
~ rzeczywisty 15-245
~ semiregularny 15-260
~ stały 35-375
~ ~ ciągły 35-390
~ ~ graniczny 35-395
~ ~ międzywęzłowy 35-420
~ ~ nieograniczony 35-390
~ ~ nieuporządkowany 35-435
~ ~ pierwotny 35-380
~ ~ podstawieniowy 35-415
~ ~ podstawowy 35-380
~ ~ przesycony 35-400
~ ~ różnowęzłowy 35-415
~ ~ śródwęzłowy 35-420
~ ~ uporządkowany 35-430
~ wieloskładnikowy 35-410
rozwarstwienie *n* 85-405
równanie *n*
~ Gibbsa-Duhema 15-305

równanie
~ Hildebranda 15-310
~ Schrödingera 10-175
~ stanu 15-025

równowaga *f*
~ chemiczna 15-330
~ cieplna 15-325
~ fazowa 35-015
~ metastabilna 35-060
~ stabilna 35-055
~ strukturalna 40-020
~ termiczna 15-325
~ termodynamiczna 15-320
~ trwała 35-055

równowagowy współczynnik *m* rozdziału 25-260

równoważnik *m* węglowy 50-480

rtęć *f* 05-270

rubid *m* 05-390

ruch *m*
~ konserwatywny dyslokacji 30-290
~ niekonserwatywny dyslokacji 30-285

ruten *m* 05-395

rynna *f* eutektyczna 35-730

rzadkopłynność *f* 75-785

rzadzizna *f* 85-120

rzeczywista granica *f* pełzania 75-225

S

samar *m* 05-400

samodyfuzja *f* 45-400

schładzanie *n* 70-070

schodek *m* wzrostu 25-330

segregacja *f* 25-400
~ ciężarowa 25-425
~ dendrytyczna 25-420
~ grawitacyjna 25-425
~ kroplista 25-430
~ międzydendrytyczna 25-445
~ na granicach ziarn 25-450
~ normalna 25-435
~ odwrotna 25-440
~ pierwotna 25-405
~ prosta 25-435
~ strefowa 25-415
~ wlewka 25-415
~ wtórna 25-410

selen *m* 05-410

sezonowanie *n* 70-540

sezonowe pękanie *n* 85-385

sfera *f* Fermiego 10-225

sferoid *m* 40-290
~ grafitu 50-215

sferoidyt *m* 50-150

sferoidyzacja *f* cementytu 70-515

sferoidyzowanie *n* 70-510

sferolit *m* 40-290

siarczek *m*
~ manganu 50-240
~ żelaza(wy) 50-245

siarka *f* 05-435

siarkoazotowanie *n* 70-890

siarkowanie *n* 70-895
~ dyfuzyjne 70-895

siarkowęgloazotowanie *n* 70-892

siatka *f*
~ cementytu 50-105
~ węglików 50-445

sieć *f*
~ Bravais 20-275
~ centrowana 20-295
~ dyslokacji 30-390
~ dyslokacyjna 30-390

sieć
~ jednoskośna 20-395
~ heksagonalna 20-340
~ ~ prosta 20-345
~ ~ zwarcie wypełniona 20-350
~ ~ zwarta 20-350
~ krystalograficzna 20-045
~ odwrotna 20-410
~ płasko-centryczna 20-290
~ prosta 20-280
~ przejściowa 20-405
~ przestrzenna 20-045
~ przestrzennie centrowana 20-295
~ ~ centryczna 20-295
~ punktowa 20-275
~ regularna 20-300
~ ~ płasko-centryczna 20-310
~ ~ prosta 20-305
~ ~ przestrzennie centrowana 20-315
~ ~ ~ centryczna 20-315
~ ~ ściennie centrowana 20-310
~ ~ typu diamentu 20-320
~ romboedryczna 20-390
~ rombowa 20-380
~ ~ jednostronnie centrowana 20-385
~ ~ ~ centryczna 20-385
~ ~ o centrowanych podstawach 20-385
~ sześcienna 20-300
~ ~ centrowana 20-315
~ ~ prosta 20-305
~ ~ zwarta 20-310
~ ściennie centrowana 20-290
~ tetragonalna 20-370
~ translacyjna 20-275
~ trójskośna 20-400
~ trygonalna 20-390
~ zdefektowana 30-020

silal *m* 60-1100

silumin *n* 65-565
~ modyfikowany 65-570
~ uszlachetniony 65-570

siła *f*
~ kohezji 10-315
~ Peierlsa-Nabarro 30-300
~ spójności 10-315
~ termodynamiczna 15-070
~ termoelektryczna 75-495

skand *m* 05-405

skład *m*
~ chemiczny stopu 55-225
~ stechiometryczny 35-240

składnik *m*
~ rozpuszczony 35-360
~ stopu 55-180
~ strukturalny 40-010
~ układu 30-035

skłonność *f*
~ do gruboziarnistości 75-760
~ do korozji 75-675
~ do rozrostu ziarna 75-760

skrawalność *f* 75-745

skupienie *n* dyslokacji 30-400

skupisko *n* wakansów 30-125

skurcz *m* 75-795
~ krzepnięcia 75-805
~ przy krzepnięciu 75-805
~ w stanie ciekłym 75-800
~ w stanie stałym 75-810

solidus *m* 35-485
~ cofający się 35-505

solwus *m* 35-530

sorbit *m* 50-380
~ hartowania 50-385
~ odpuszczania 50-390

sód *m* 05-425

spawalność *f* 75-735

spektrografia *f* masowa 80-480

spektroskopia *f*
~ rentgenowska 80-700
~ ~ fluorescencyjna 80-700

SPIEK

spiek m 55-655
~ metaloceramiczny 55-675
~ metalowy 55-660
~ nasycany 55-670
~ twardy 55-665
~ węglikowy 55-665
~ żelazny 60-065
spiętrzenie n dyslokacji 30-400
spinoda f 15-190
spinowa liczba f kwantowa 10-145
spirala f wzrostu 25-335
spiż m 65-205
splot m dyslokacji 30-385
spójność f 10-310
sprężystość f 75-020
~ opóźniona 75-050
sprzężenie n 30-620
sprzężony system m poślizgu 45-125
srebro n 05-420
~ czyste 65-760
~ monetowe 65-755
~ nowe 65-415
~ rafinowane 65-760
stabilizacja f 50-425
stabilizator m 50-485
stabilizowanie n 70-530, 70-535
~ naturalne 70-540
stabilność f termiczna 75-835
stal f 60-075
~ aluminiowa 60-235
~ anormalna 60-465
~ austenityczna 60-420
~ austenityczno-ferrytyczna 60-455
~ automatowa 60-710
~ azotowana 60-315
~ bainityczna 60-435
~ bardzo miękka 60-345
~ bessemerowska 60-595
~ budowlana 60-685
~ cementowa 60-585
~ chromo-niklowa 60-190
~ chromowa 60-175
~ chromowo-niklowa 60-190
~ czteroskładnikowa 60-160
~ damasceńska 60-645
~ diamentowa 60-490
~ do azotowania 60-695
~ do budowy kotłów 60-740
~ do głębokiego ciągnienia 60-790
~ do głębokiego tłoczenia 60-790
~ do hartowania w oleju 60-370
~ ~ ~ w powietrzu 60-375
~ ~ ~ w wodzie 60-365
~ do nawęglania 60-690
~ do ulepszania cieplnego 60-705
~ drobnoziarnista 60-470
~ elektropiecowa 60-625
~ elektryczna 60-625
~ eutektoidalna 50-490
~ ferrytyczna 60-415
~ głęboko hartująca się 60-385
~ grafityzowana 60-300
~ gruboziarnista 60-475
~ Hadfielda 60-785
~ hartowana 60-285
~ hartowna 60-360
~ hartująca się 60-360
~ ~ się w powietrzu 60-375
~ ~ się w oleju 60-370
~ ~ się w wodzie 60-365
~ jakościowa 60-140
~ kobaltowa 60-240
~ konstrukcyjna 60-670
~ ~ budowlana 60-685
~ ~ maszynowa 60-680
~ konwertorowa 60-590
~ ~ świeżona tlenem 60-605
~ ~ z procesu tlenowego 60-605
~ kotłowa 60-740
~ kriotechniczna 60-655
~ krzemowa 60-200
~ ~ niskowęglowa 60-210

stal
~ kuta 60-330
~ kwasoodporna 60-545
~ kwaśna 60-615
~ ledeburytyczna 60-440
~ łożyskowa 60-715
~ ługoodporna 60-550
~ magnetyczna 60-495
~ magnetycznie miękka 60-500
~ ~ twarda 60-505
~ manganowa 60-195
~ ~ nieścieralna 60-785
~ manganowo-krzemowa 60-205
~ martenowska 60-610
~ martenzytyczna 60-430
~ ~ starzona 60-460
~ matrycowa 60-815
~ miedziowa 60-245
~ miękka 60-340
~ molibdenowa 60-220
~ mrozoodporna 60-555
~ na łożyska kulkowe 60-720
~ na łożyska toczne 60-715
~ na magnesy trwałe 60-780
~ na matryce 60-815
~ na nity 60-745
~ na noże 60-750
~ na obręcze kolejowe 60-760
~ nadeutektoidalna 50-500
~ najwyższej jakości 60-145
~ narzędziowa 60-675
~ ~ do pracy na gorąco 60-810
~ ~ do pracy na zimno 60-805
~ nawęglona 60-310
~ niehartowna 60-390
~ nie hartująca się 60-390
~ niemagnetyczna 60-510
~ nie pacząca się 60-410
~ nierdzewna 60-515
~ ~ automatowa 60-525
~ ~ ferrytyczna 60-520
~ niestopowa 60-095
~ nieuspokojona 60-650
~ niklowa 60-185
~ niskostopowa 60-120
~ niskowęglowa 60-100
~ normalizowana 60-280
~ nożowa 60-750
~ o dużej wytrzymałości 60-355
~ obręczowa 60-760
~ odporna na korozję 60-535
~ ~ na korozję atmosferyczną 60-540
~ ~ na odpuszczanie 60-395
~ ~ na przegrzanie 60-470
~ ~ na starzenie 60-405
~ ~ na ścieranie 60-480
~ ogólnego przeznaczenia 60-665
~ okrętowa 60-755
~ ołowiowa 60-250
~ oszczędnościowa 60-270
~ pancerna 60-765
~ perlityczna 60-425
~ pięcioskładnikowa 60-165
~ platerowana 60-335
~ płytko hartująca się 60-380
~ podeutektoidalna 50-495
~ półaustenityczna 60-445
~ półferrytyczna 60-450
~ półspokojona 60-660
~ prądnicowa 60-475
~ próżniowa 60-640
~ przegrzana 60-320
~ przepalona 60-325
~ przeżarzona 60-320
~ pudlarska 60-575
~ rafinowana 60-635
~ resorowa 60-725
~ samohartowna 60-375
~ skłonna do drobnoziarnistości 60-470
~ ~ do gruboziarnistości 60-475
~ spawalna 60-485
~ specjalna 60-115, 60-150
~ sprężynowa 60-725
~ srebrzanka 60-820
~ stabilizowana 60-305
~ starzejąca się 60-400
~ stopowa 60-115
~ ~ do azotowania 60-700

stal
- ~ szybkotnąca 60-800
- ~ szynowa 60-730
- ~ średniostopowa 60-125
- ~ średniowęglowa 60-105
- ~ tomasowska 60-600
- ~ transformatorowa 60-770
- ~ trójskładnikowa 60-155
- ~ twarda 60-350
- ~ tyglowa 60-630
- ~ tytanowa 60-230
- ~ ulegająca starzeniu 60-400
- ~ ulepszona cieplnie 60-290
- ~ umocniona zgniotem 60-295
- ~ uspokojona 60-655
- ~ wanadowa 60-225
- ~ węglowa 60-095
- ~ wieloskładnikowa 60-170
- ~ wolframowa 60-215
- ~ wysokiej jakości 60-145
- ~ wysokochromowa 60-180
- ~ wysokostopowa 60-130
- ~ wysokowęglowa 60-110
- ~ wyżarzona 60-275
- ~ wyższej jakości 60-140
- ~ z dodatkiem boru 60-255
- ~ z dodatkiem ołowiu 60-250
- ~ zahartowana 60-285
- ~ zasadowa 60-620
- ~ zastępcza 60-270
- ~ zaworowa 60-735
- ~ zbrojeniowa 60-795
- ~ zgrzewalna 60-485
- ~ zgrzewna 60-570
- ~ zlewna 60-580
- ~ znormalizowana 60-265
- ~ zwykła 60-095
- ~ zwykłej jakości 60-135
- ~ żaroodporna 60-560
- ~ żarowytrzymała 60-565

staliwo n 60-825
- ~ niestopowe 60-830
- ~ stopowe 60-835
- ~ węglowe 60-830

stała f
- ~ anizotropii 75-595
- ~ równowagi termodynamicznej 15-335
- ~ sieci(owa) 20-120
- ~ sprężystości 75-145

stan m
- ~ energetyczny 10-120
- ~ kwantowy 10-120
- ~ metaliczny 10-345
- ~ metastabilny 35-070
- ~ nieuporządkowania 25-585
- ~ nieuporządkowany 25-585
- ~ równowagi 35-050
- ~ stabilny 35-065
- ~ termodynamiczny 10-015
- ~ uporządkowania 25-590
- ~ uporządkowany 25-590

starzenie n 70-575
- ~ magnetyczne 70-620
- ~ martenzytu 70-615
- ~ mechaniczne 45-225
- ~ naturalne 70-585
- ~ po odkształceniu plastycznym na zimno 45-225
- ~ po przesycaniu 70-580
- ~ po zgniocie 45-225
- ~ przyspieszone 70-590
- ~ samorzutne 70-585
- ~ sztuczne 70-590

steadyt m 50-235

stellit m 65-495

stężenie n 55-230
- ~ ciężarowe 55-235
- ~ elektronowe 35-285
- ~ molowe 55-240
- ~ nominalne 55-245
- ~ równowagowe 35-255
- ~ średnie 55-245
- ~ wagowe 55-235

stop m 65-590

stop
- ~ anatomiczny 55-645
- ~ architektoniczny 55-605
- ~ Auera 65-875
- ~ automatowy 55-560
- ~ B 65-685
- ~ ciekły 55-150
- ~ czcionkowy 55-615
- ~ czteroskładnikowy 55-305
- ~ dekoracyjny 55-605
- ~ dentystyczny 55-575
- ~ do lutowania 65-710
- ~ do napawania 55-590
- ~ do przeróbki plastycznej 55-530
- ~ drukarski 55-610
- ~ ~ do galwanotypii 55-635
- ~ dwufazowy 55-335
- ~ dwuskładnikowy 55-295
- ~ dyspersyjny 55-390
- ~ elektrolityczny 55-290
- ~ eutektoidalny 55-365
- ~ eutektyczny 55-350
- ~ ferromagnetyczny 55-455
- ~ galwaniczny 55-290
- ~ galwanotypowy 55-635
- ~ Heuslera 65-420
- ~ jednofazowy 55-330
- ~ jubilerski 55-580
- ~ justunkowy 55-640
- ~ kwasoodporny 55-485
- ~ lekki 55-405
- ~ linotypowy 55-625
- ~ Lipowitza 65-680
- ~ lutowniczy 65-710
- ~ łatwotopliwy 55-395
- ~ łożyskowy 55-555
- ~ ługoodporny 55-490
- ~ magnetycznie miękki 55-460
- ~ ~ twardy 55-465
- ~ magnetyczny 55-445
- ~ metali 55-160
- ~ metalu 55-160
- ~ miedzi 65-010
- ~ Monela 65-435
- ~ monetowy 55-565
- ~ monotypowy 55-620
- ~ na magnesy trwałe 55-450
- ~ na odlewy ciśnieniowe 55-540
- ~ nadeutektoidalny 55-375
- ~ nadeutektyczny 55-360
- ~ nadplastyczny 55-430
- ~ nadwytrzymały 55-425
- ~ nasycany 55-670
- ~ naturalny 55-265
- ~ niejednofazowy 55-340
- ~ niemagnetyczny 55-470
- ~ nieżelazny 65-005
- ~ niklu 65-430
- ~ niskoprocentowy 55-320
- ~ niskotopliwy 55-395
- ~ o dużej odporności właściwej 55-545
- ~ o dużej przenikalności magnetycznej 55-475
- ~ o dużej wytrzymałości 55-420
- ~ o małej rozszerzalności cieplnej 55-440
- ~ o małym współczynniku rozszerzalności cieplnej 55-440
- ~ o podstawie miedziowej 65-010
- ~ o podstawie niklowej 65-430
- ~ o podstawie żelazowej 60-070
- ~ obrabialny cieplnie 55-525
- ~ odlewniczy 55-535
- ~ odporny na korozję 55-480
- ~ ~ na matowienie korozyjne 55-500
- ~ ~ na ścieranie 55-435
- ~ ~ na utlenianie 55-495
- ~ oporowy 55-545
- ~ panewkowy 55-555
- ~ pierwotny 55-275
- ~ piroforyczny 55-595
- ~ poczwórny 55-305
- ~ podeutektoidalny 55-370
- ~ podeutektyczny 55-355
- ~ podwójny 55-295
- ~ potrójny 55-300
- ~ przejściowy 55-285
- ~ przemysłowy 55-520
- ~ przerabialny plastycznie 55-530

stop
- ~ przewodowy 55-550
- ~ pseudopodwójny 55-315
- ~ Rose'go 65-675
- ~ steksturowany 55-380
- ~ stereotypowy 55-630
- ~ stykowy 55-585
- ~ syntetyczny 55-270
- ~ techniczny 55-520
- ~ tłokowy 55-570
- ~ trójskładnikowy 55-300
- ~ trudnotopliwy 55-400
- ~ twardy 55-415
- ~ ~ do napawania 55-590
- ~ ultralekki 55-410
- ~ umocniony dyspersyjnie 55-390
- ~ utwardzalny wydzieleniowo 55-385
- ~ wielofazowy 55-345
- ~ wieloskładnikowy 55-310
- ~ Wooda 65-670
- ~ wstępny 55-285
- ~ wtopieniowy 55-600
- ~ wtórny 55-280
- ~ wysokoprocentowy 55-325
- ~ wysokotopliwy 55-400
- ~ wysokowytrzymały 55-420
- ~ zdobniczy 55-605
- ~ zwierciadłowy 65-280
- ~ żaroodporny 55-505
- ~ żarowytrzymały 55-510
- ~ żelaza 60-070

stopień m
- ~ czystości 55-095
- ~ dyspersji 40-335
- ~ niejednorodności 40-420
- ~ nieuporządkowania 25-610
- ~ odkształcenia krytyczny 45-075
- ~ odkształcenia plastycznego 45-035
- ~ przechłodzenia 25-250
- ~ segregacji 40-420
- ~ tetragonalności 20-375
- ~ uporządkowania 25-605
- ~ wypełnienia 20-205
- ~ wzrostu 25-330
- ~ zgniotu 45-040

stopowanie n 55-170

stosunek m osi(owy) 20-355

strefa f
- ~ Brillouina 10-230
- ~ dyfuzji 45-415
- ~ dyfuzyjna 45-415
- ~ energii wzbronionych 10-250
- ~ koordynacyjna 20-190
- ~ kryształów równoosiowych 40-185
- ~ ~ słupkowych 40-180
- ~ ~ wolnych 40-185
- ~ ~ zamrożonych 40-175
- ~ rozrzedzona 30-120
- ~ zahartowania 70-355
- ~ zamrożona 40-175
- ~ zubożona 30-120

strefy fpl Guinier-Prestona 70-605

stront m 05-430

struktura f
- ~ anormalna 40-135
- ~ dendrytyczna 40-090
- ~ domenowa 40-620
- ~ drobnokrystaliczna 40-395
- ~ drobnoziarnista 40-395
- ~ dwufazowa 40-150
- ~ dyslokacyjna 30-415
- ~ dziedziczna 40-165
- ~ elektronowa 10-030
- ~ eutektoidalna 40-075
- ~ eutektyczna 40-060
- ~ globularna 40-120
- ~ globulityczna 40-120
- ~ grubokrystaliczna 40-390
- ~ gruboziarnista 40-390
- ~ iglasta 40-115
- ~ jednofazowa 40-145
- ~ komórkowa 40-095
- ~ krystal(ograf)iczna 20-040
- ~ magnetyczna 40-650

struktura
- ~ makroskopowa 40-025
- ~ metalu 40-005
- ~ mikroskopowa 40-030
- ~ mozaikowa 40-170
- ~ nadeutektoidalna 40-080
- ~ nadeutektyczna 40-065
- ~ niejednofazowa 40-155
- ~ pasemkowa 40-105
- ~ pasmowa 40-140
- ~ pierwotna 40-045
- ~ płytkowa 40-105
- ~ podeutektoidalna 40-085
- ~ podeutektyczna 40-070
- ~ podmikroskopowa 40-040
- ~ poliedryczna 40-095
- ~ siatkowa 40-100
- ~ słupkowa 40-125
- ~ w stanie lanym 40-050
- ~ wtórna 40-055
- ~ Widmannstättena 40-130
- ~ wielofazowa 40-160

studzenie n 70-080

submikrostruktura f 40-040

substancja f
- ~ rozpuszczona 35-360
- ~ zarodkotwórcza 25-125

substruktura f 40-035

superdyslokacja f 30-220

superparamagnetyzm m 75-570

surówka f 60-840
- ~ bessemerowska 60-845
- ~ biała 60-910
- ~ drzewnowęglowa 60-875
- ~ elektropiecowa 60-865
- ~ elektryczna 60-865
- ~ fosforowa 60-890
- ~ hematytowa 60-895
- ~ koksowa 60-870
- ~ martenowska 60-855
- ~ na żeliwo ciągliwe 60-905
- ~ odlewnicza 60-885
- ~ połowiczna 60-920
- ~ półhematytowa 60-900
- ~ przeróbcza 60-880
- ~ pstra 60-920
- ~ stalownicza 60-880
- ~ stopowa 60-935
- ~ syntetyczna 60-940
- ~ szara 60-915
- ~ tomasowska 60-850
- ~ wielkopiecowa 60-860
- ~ wysokowęglowa 60-925
- ~ zwierciadlista 60-930

symetria f kryształu 20-055

syngonia f 20-235

system m
- ~ poślizgu 45-110
- ~ ~ pierwotny 45-115
- ~ ~ sprzężony 45-125
- ~ ~ wtórny 45-120

szczyt m
- ~ cieplny 30-530
- ~ przemieszczeń 30-525
- ~ temperatury 30-530

szerardyzacja f 70-930

szereg m napięciowy metali 75-710

szlif m 80-095

szybkie chłodzenie n 70-075

szybkość f
- ~ chłodzenia 70-120
- ~ krystalizacji 25-275
- ~ krytyczna chłodzenia 70-180
- ~ ~ hartowania 70-180
- ~ nagrzewania 70-040
- ~ wzrostu kryształów 25-275
- ~ zarodkowania 25-270

ściana f
- ~ Blocha 40-645
- ~ dyslokacji 30-395
- ~ kryształu 20-060

ścieranie n 80-355
średnia odległość f między cząstkami dyspersoidu 40-425
średnica f
~ atomowa 10-080
~ krytyczna 70-360
środek m nawęglający 70-740
środowisko n chłodzące 70-185

T

tal m 05-460
tantal m 05-440
tarcie n wewnętrzne 75-210
technet m 05-445
tekstura f 40-555
~ ciągnienia 40-575
~ deformacji 40-560
~ deformacyjna 40-560
~ Gossa 40-610
~ odkształcenia 40-560
~ odlewu 40-580
~ pierwotna odlewu 40-580
~ regularna 40-605
~ rekrystalizacji 40-600
~ rozciągania 40-585
~ skręcania 40-595
~ sześcienna 40-605
~ ściskania 40-590
~ walcowania 40-570
~ włóknista 40-565
~ w stanie lanym 40-580
~ wyżarzania 40-600
tellur m 05-450
temperatura f
~ austenityzacji 70-135
~ austenityzowania 70-135
~ charakterystyczna Debye'a 10-280
~ Curie 25-560
~ Debye'a 10-280
~ ekwikohezyjna 30-570
~ eutektoidalna 35-630
~ eutektyczna 35-605
~ hartowania 70-140
~ homologiczna 45-475
~ końca przemiany martenzytycznej 70-170
~ kruchości na zimno graniczna 75-280
~ krytyczna 35-335, 35-555
~ krzepnięcia 75-425
~ likwidusu 35-525
~ Mr 70-170
~ Ms 70-165
~ metatektyczna 35-620
~ monotektyczna 35-615
~ Néela 25-565
~ niebieskiego nalotu 70-395
~ odpuszczania 70-380
~ perytektoidalna 35-635
~ perytektyczna 35-610
~ początku przemiany martenzytycznej 70-165
~ pojawienia się kruchości 75-280
~ ~ się nadprzewodnictwa 25-570
~ przejścia w stan kruchości 75-280
~ ~ w stan nadprzewodnictwa 25-570
~ przemiany 35-555
~ ~ magnetycznej 25-560
~ ~ martenzytycznej 70-165
~ ~ porządek-nieporządek 25-625
~ przeskoku 25-570
~ rekrystalizacji 45-470
~ rozkładu 15-340
~ równowagi 35-690
~ równowagowa 35-690
~ solidusu 35-500
~ sublimacji 75-435
~ syntektyczna 35-625
~ topnienia 75-420
~ wrzenia 75-430
~ wyżarzania 70-425
teoretyczny współczynnik m rozdziału 25-260

teoria f
~ elektronowa metali 10-025
~ pasmowa ciała stałego 10-020
terb m 05-455
termaloj m 65-540
termiczna eliminacja f defektów 30-135
termodyfuzja f 45-370
termodynamika f
~ ciała stałego 15-005
~ stopów 15-010
termogram n 80-527
tetragonalność f 20-375
tlen m 05-320
tlenoazotowanie n 70-885
tłoczność f 75-750, 75-755
tombak m 65-395
topliwość f 75-790
topnienie n 35-445
~ eutektyczne 35-470
~ heterotermiczne 35-455
~ inkongruentne 35-465
~ izotermiczne 35-450
~ kongruentne 35-460
~ nieizotermiczne 35-455
~ niekongruentne 35-465
~ perytektyczne 35-475
~ w zakresie temperatur 35-455
tor m 05-465
transkrystalizacja f 25-070
translacja f 20-130
trawienie n 80-120
~ barwne 80-165
~ chemiczne 80-170
~ elektrolityczne 80-175
~ gazowe 80-182
~ głębokie 80-160
~ granic ziarn 80-145
~ jonowe próżniowe 80-180
~ katodowe próżniowe 80-180
~ kolorowe 80-165
~ makrostruktury 80-135
~ mikrostruktury 80-140
~ pól ziarn 80-150
~ przez odpuszczanie 80-185
~ przez utlenianie 80-185
~ termiczne 80-190
trąd m cynowy 85-245
triangulacja f układu 35-725
troostyt m 50-395
~ hartowania 50-400
~ odpuszczania 50-405
trójkąt m
~ przystanków eutektycznych 40-265
~ składów 35-720
~ stężeń 35-720
trójwakans m 30-050
trzecia liczba f kwantowa 10-140
tul m 05-470
twardość f 75-245
~ na gorąco 75-255
~ w temperaturze czerwonego żaru 75-260
tworzenie n stopu 55-170
tytan m 05-480
tytanek m 40-510
tytanowanie n dyfuzyjne 70-945

U

udarność f 75-190
ujawnianie n struktury 80-280
ujawnienie n struktury 80-280
ujednorodnianie n 70-485
ujście n 30-130

UKŁAD

układ m 35-005
~ brzegowy 35-715
~ czteroskładnikowy 35-135
~ dwufazowy 35-100
~ dwuskładnikowy 35-120
~ dwuzmienny 35-085
~ heksagonalny 20-245
~ jednofazowy 35-090
~ jednoskładnikowy 35-115
~ jednoskośny 20-265
~ jednozmienny 35-080
~ krystalograficzny 20-235
~ kwazypodwójny 35-130
~ metaliczny 55-165
~ niejednofazowy 35-095
~ niezmienny 35-075
~ o dwóch stopniach swobody 35-085
~ o jednym stopniu swobody 35-080
~ poczwórny 35-135
~ podwójny 35-120
~ potrójny 35-125
~ pseudopodwójny 35-130
~ quasi-podwójny 35-130
~ regularny 20-240
~ romboedryczny 20-260
~ rombowy 20-255
~ równowagi 35-045
~ stopowy 55-165
~ sześcienny 20-240
~ tetragonalny 20-250
~ trójfazowy 35-105
~ trójskładnikowy 35-125
~ trójskośny 20-270
~ trygonalny 20-260
~ trzyskładnikowy 35-125
~ wielofazowy 35-110
~ wieloskładnikowy 35-140
~ zerozmienny 35-075
~ żelazo-węgiel 50-015

ulepszanie n cieplne 70-410
umacnianie n 70-025
~ dyspersyjne 55-220
~ przez tworzenie stopu 55-210
~ radiacyjne 30-495
~ roztworowe 55-215
~ wybuchowe 45-055
~ wydzieleniowe 70-560
~ zgniotem 45-050

umocnienie n 45-045
~ latentne 45-180

umowna granica f pełzania 75-230
~ ~ plastyczności 75-115

uplastycznianie n żeliwa 70-625
uporządkowanie n 25-575, 25-590
~ bliskie 25-595
~ bliskiego zasięgu 25-595
~ dalekie 25-600
~ dalekiego zasięgu 25-600
~ magnetyczne 25-640

uran m 05-490
uskok m dyslokacji 30-240
utajone ciepło n parowania 75-390
utajone ciepło n topnienia 75-380
utlenianie n
~ wewnętrzne 70-497
~ w parze wodnej 70-880

utwardzalność f 70-020
utwardzanie n 70-015
~ dyfuzyjne 70-670
~ dyspersyjne wydzieleniowe 70-560
~ przez odkształcanie plastyczne na zimno 45-050
~ przez starzenie 70-565
~ przez stopowanie 55-210
~ przez tworzenie stopu 55-210
~ przez zgniot 45-050
~ radiacyjne 30-495
~ roztworowe 55-215
~ wtórne 70-405
~ wybuchowe 45-055
~ wydzieleniowe 70-560
~ zgniotem 45-050

utwardzanie
~ zgniotowe 45-050

utwardzenie n żeliwa odwrotne 85-150

W

wada f
~ hartownicza 85-180
~ materiałowa 85-005
~ materiału 85-005
~ odlewnicza 85-095
~ powierzchni 85-010
~ powierzchniowa 85-010

wakans m 30-035
~ Frenkla 30-065
~ nadmiarowy 30-105
~ nierównowagowy 30-105
~ poczwórny 30-055
~ podwójny 30-045
~ pojedynczy 30-040
~ potrójny 30-050
~ przechłodzony 30-100
~ równowagowy 30-090
~ strukturalny 30-090
~ termiczny 30-095

wanad m 05-495
wapniokrzem m 60-1160
wapń m 05-085
warstewka f Beilby'ego 80-090
warstwa f
~ dyfuzyjna 70-660
~ elektronowa 10-155
~ epitaksjalna 25-350
~ naazotowana 70-845
~ nawęglona 70-780
~ zahartowana 70-355

wektor m
~ Burgersa 30-190
~ falowy 10-185
~ sieci 20-135
~ sieciowy 20-135
~ translacji 20-135

węgiel m 05-095
~ wolny 50-005
~ związany 50-010
~ żarzenia 50-220

węglik m 40-430
~ chromu 40-455
~ metalu 40-435
~ niobu 40-465
~ podwójny 40-475
~ prosty 40-470
~ specjalny 40-445
~ stopowy 40-445
~ wanadu 40-460
~ wolframu 40-450
~ złożony 40-475
~ żelaza 40-440

węgliki mpl
~ eutektyczne 50-440
~ ledeburytyczne 50-440
~ spiekane 55-665
~ szczątkowe 40-480

węglikoazotek m 40-485
węgloazotek m 40-485
węgloazotowanie n 70-855
~ gazowe 70-860
~ kąpielowe 70-865
~ niskotemperaturowe 70-875
~ wysokotemperaturowe 70-870

węgloutwardzanie n cieplne 70-670
węzeł m
~ dyslokacji 30-350
~ dyslokacyjny 30-350
~ sieci 20-100
~ sieciowy 20-100

wiązanie n
~ atomowe 10-330
~ chemiczne 10-320

WYTRZYMAŁOŚĆ

wiązanie
~ cząsteczkowe 10-335
~ elektrowalencyjne 10-325
~ heteropolarne 10-325
~ homeopolarne 10-330
~ jonowe 10-325
~ kowalencyjne 10-330
~ kowalentne 10-330
~ metaliczne 10-340
~ międzycząsteczkowe 10-335
~ van der Waalsa 10-335
wiązkość f 75-035
wielkość f
~ ekstensywna 15-040
~ intensywna 15-045
~ termodynamiczna 15-030
~ ziarna 80-285
witalium n 65-485
własności fpl
~ chemiczne 75-655
~ cieplne 75-365
~ elektrochemiczne 75-705
~ elektryczne 75-465
~ fizyczne 75-315
~ jądrowe 75-520
~ magnetyczne 75-540
~ makroskopowe 75-850
~ mechaniczne 75-005
~ ~ określane z próby rozciągania 75-015
~ metaliczne 10-350
~ metalurgiczne 75-855
~ niewrażliwe na strukturę 75-865
~ niezależne od struktury 75-865
~ objętościowe 75-340
~ odlewnicze 75-775
~ optyczne 75-500
~ technologiczne 75-720
~ użytkowe 75-820
~ wrażliwe na strukturę 75-860
~ wytrzymałościowe 75-010
~ zależne od struktury 75-860
wodorek m 40-495
wodór m 05-200
wolfram m 05-485
wolframek m 40-500
wolframowanie n dyfuzyjne 70-935
wprowadzanie n składnika stopowego 55-170
wrażliwość f
~ na działanie karbu 75-195
~ na grubość ścianki 75-815
~ na przegrzanie 75-760
wskaźnik m
~ krystalograficzny 20-160
~ plastyczności 75-065
~ wytrzymałości(owy) 75-060
wskaźniki mpl
~ Millera 20-165
~ ~-Bravais 20-170
wskaźnikowanie n 80-695
wspinanie n się dyslokacji 30-295
współczynnik m
~ aktywności 15-285
~ anizotropii 75-595
~ dyfuzji 45-395
~ ~ cząstkowy 45-390
~ odbicia 75-515
~ oddziaływania 15-295
~ Poissona 75-170
~ przejmowania węgla 70-755
~ przenoszenia węgla 70-755
~ przewodnictwa cieplnego 75-440
~ przewodzenia ciepła 75-440
~ rozdziału efektywny 25-265
~ ~ równowagowy 25-260
~ ~ teoretyczny 25-260
~ rozszerzalności cieplnej 75-450
~ ~ ~ liniowej 75-455
~ ~ ~ objętościowej 75-460
~ ~ liniowej 75-455
~ ~ objętościowej 75-460

współczynnik
~ samodyfuzji 45-405
~ sprężystości 75-145
~ ściśliwości 75-165
~ temperaturowy oporności elektrycznej 75-480
~ umocnienia 45-065
~ wypełnienia 20-205
wtórna oś f krystalizacji 25-305
wtórny system m poślizgu 45-120
wtrącenia npl
~ azotkowe 40-545
~ azotków 40-545
~ egzogeniczne 40-520
~ endogeniczne 40-525
~ metaliczne 40-550
~ niemetaliczne 40-515
~ siarczkowe 40-530
~ siarczków 40-530
~ tlenkowe 40-535
~ tlenków 40-535
~ żużlowe 40-540
wyciąganie n monokryształów 25-090
~ monokryształu 25-090
wydłużenie n
~ całkowite 75-120
~ plastyczne po rozerwaniu 75-120
~ względne 75-120
wydzielanie n 40-300
~ ciągłe 25-375
~ nieciągłe 25-380
wydzielenia npl 40-305
~ koherentne 40-320
~ na granicach ziarn 40-310
~ niekoherentne 40-325
~ podmikroskopowe 40-315
wygrzewanie n 70-055
~ izochroniczne 70-495
~ uwrażliwiające 70-545
~ wahadłowe 70-520
wykładnik m konstrukcyjny 75-140
wykres m
~ anizotermiczny przemiany austenitu 70-160
~ CTPc 70-160
~ CTPi 70-155
~ izotermiczny przemiany austenitu 70-155
~ naprężenie-odkształcenie 80-410
~ przemiany austenitu przy chłodzeniu ciągłym 70-160
~ rekrystalizacji 45-485
~ rozciągania 80-410
~ równowagi faz 35-480
~ ~ fazowej 35-480
~ ~ układu potrójnego 35-710
~ ~ ~ ~ przestrzenny 35-750
~ ~ żelazo-węgiel 50-020
~ Sauveura 40-265
~ Schaefflera 60-530
~ strukturalny 40-265
~ żelazo-węgiel 50-020
wymrażanie n 70-340
wypełnienie n zwarte 20-210
wyraźna granica f plastyczności 75-110
wysokotemperaturowa obróbka f cieplno-mechaniczna 70-955
wytrawianie n 80-120
wytrzymałość f 75-040
~ czasowa 75-235
~ długotrwała 75-235
~ dynamiczna 75-185
~ na obciążenia dynamiczne 75-185
~ na pełzanie 75-220
~ na rozciąganie 75-070
~ na skręcanie 75-085
~ na ścinanie 75-080
~ na ściskanie 75-075
~ na zginanie 75-090
~ na zmęczenie 75-200
~ na zmęczenie cieplne 75-830
~ rozdzielcza 75-135
~ trwała 75-225
~ w podwyższonej temperaturze 75-095

— 357 —

wytrzymałość
~ względna 75-140
~ zmęczeniowa 75-200
wytwarzanie n stopu 55-170
wyżarzanie n 70-420
~ beznalotowe 70-430
~ bezzgorzelinowe 70-435
~ ciemne 70-440
~ czyste 70-435
~ defektów 30-135
~ grafityzujące 70-630
~ izochroniczne 70-495
~ izotermiczne 70-450
~ jasne 70-430
~ niezupełne 70-470
~ normalizujące 70-475
~ odprężające 70-525
~ odwęglające 70-635
~ perlityzujące 70-450
~ płomieniowe 70-505
~ przeciwpłatkowe 70-555
~ przegrzewające 70-480
~ rekrystalizujące 70-490
~ rozpuszczające 70-500
~ sferoidyzujące 70-510
~ stabilizujące 70-535
~ stopniowe 70-455
~ ujednorodniające 70-485
~ uplastyczniające 70-625
~ uwrażliwiające 70-545
~ wahadłowe 70-520
~ z niebieskim nalotem 70-445
~ z przemianą izotermiczną 70-450
~ zmiękczające 70-460
~ zupełne 70-465
wzrost m
~ dendrytyczny 25-310
~ epitaksjalny 25-345
~ kruchości 75-275
~ kryształów 25-285

Z

zabielenie n 85-145
zahartowalność f 70-020
zakaz m Pauliego 10-170
zakres m
~ bainityczny 50-305
~ martenzytyczny 50-300
~ mieszanin 35-345
~ nierozpuszczalności 35-345
~ perlityczny 50-295
~ przemiany bainitycznej 50-305
~ ~ martenzytycznej 50-300
~ ~ perlitycznej 50-295
~ temperatur hartowania 70-145
~ ~ krystalizacji 35-550
~ ~ krzepnięcia 35-550
~ ~ przemiany 35-560
~ ~ topnienia 35-545
załamanie n 30-245
zanieczyszczenia npl 55-205
zanieczyszczenie n międzywęzłowe 30-080
zanik m umocnienia 45-535
zaprawa f 55-285
zaraza f cynowa 85-245
zarodek m
~ jednorodny 25-110
~ koherentny 25-155
~ krystalizacji 25-105
~ krytyczny 25-140
~ niejednorodny 25-115
~ obcy 25-120
~ podkrytyczny 25-135
~ rekrystalizacji 25-170
~ samorzutny 25-110
~ trwały 25-150
~ zorientowany 25-160
zarodkowanie n 25-175
~ atermiczne 25-200
~ dynamiczne 25-205

zarodkowanie
~ heterogeniczne 25-185
~ homogeniczne 25-180
~ jednorodne 25-180
~ niejednorodne 25-185
~ przez tarcie 25-210
~ samorzutne 25-190
~ zorientowane 25-195
zasada f
~ nieokreśloności 10-165
~ nieoznaczoności Heisenberga 10-165
~ Pauliego 10-170
~ wykluczania 10-170
zaszarzenie n 85-140
zaszczepianie n krystalizacji 25-215
zawalcowanie n 85-400
zażużlenie n 40-540
zdefektowanie n radiacyjne 30-490
zdolność f
~ chłodząca 70-125
~ chłodzenia 70-125
~ hamowania 75-535
~ spowalniania 75-535
~ stopotwórcza 55-175
~ tłumienia drgań 75-205
~ umacniania się 75-770
zdrowienie n 45-420
~ dynamiczne 45-425
~ statyczne 45-430
~ termiczne 45-430
zgład m 80-095
zgniot m 45-030
~ fazowy 45-080
~ krytyczny 45-075
~ wewnętrzny 45-080
zgorzelina f 85-430
zgrzewalność f 75-735
ziarno n
~ byłego austenitu 50-040
~ drobne 40-370
~ duże 40-365
~ grube 40-365
~ krystaliczne 40-360
zjawiska npl przedwydzieleniowe 70-600
zjawisko n
~ Bauschingera 75-175
~ Halla 10-290
~ Kirkendalla 45-385
~ Mössbauera 10-285
~ Portevina-Le Chateliera 45-540
~ Snoeka 30-460
~ Soreta 45-370
~ Suzuki 30-455
złącze n dyfuzyjne 45-375
złom m 40-670
złoto n 05-180
~ białe 65-805
~ czerwone 65-800
~ czyste 65-770
~ dentystyczne 65-785
~ dukatowe 65-795
~ jubilerskie 65-780
~ monetarne 65-775
~ monetowe 65-775
~ rafinowane 65-770
~ rodzime 65-765
~ standardowe 65-790
~ wysokokaratowe 65-795
zmęczenie n 85-410
~ cieplne 85-425
~ kontaktowe 85-420
~ korozyjne 85-415
zmiękczanie n 70-460
znacznik m promieniotwórczy 80-760
znak m stali 60-090
znal m 65-705
zniekształcenie n sieci 30-025
ząb m atomowy 10-040

ŻELIWO

związek m
~ chemiczny 35-235
~ jonowy 35-230
~ międzymetaliczny 35-215
~ ~ inkongruentny 35-225
~ ~ kongruentny 35-220
~ ~ niekongruentny 35-225
zwiększenie n wytrzymałości 70-025

Ź

źródło n
~ dyslokacji 30-425
~ Franka-Reada 30-430
~ Koehlera 30-435

Ż

żaroodporność f 75-690
żarowytrzymałość f 75-240
żelazo n 05-220
~ armco 60-010
~ dymarkowe 60-045
~ dymarskie 60-045
~ elektrolityczne 60-025
~ fryszerskie 60-050
~ gąbczaste 60-035
~ karbonylkowe 60-020
~ krzemowe 60-210
~ meteoryczne 60-005
~ meteorytowe 60-005
~ pudlarskie 60-055
~ spiekane 60-065
~ szwedzkie 60-030
~ świeżarskie 60-050
~ techniczne 60-015
~ technicznie czyste 60-015
~ zgrzewne 60-040
~ zlewne 60-060
żelazoaluminiumkrzem m 60-1165
żelazocer m 65-875
żelazochrom m 60-1145
żelazokrzem m 60-1140
~ zwierciadlisty 60-1155
żelazokrzemomangan m 60-1155

żelazomangan m 60-1150
żelazostopy mpl 60-1135
żelazowapniokrzem m 60-1160
żeliwo n 60-945
~ aluminiowe 60-1020
~ austenityczne 60-1050
~ bainityczne 60-1060
~ białe 60-955
~ ~ eutektyczne 50-505
~ ~ nadeutektyczne 50-515
~ ~ podeutektyczne 50-510
~ chromowe 60-1015
~ ciągliwe 60-1110
~ ~ amerykańskie 60-1115
~ ~ białe 60-1120
~ ~ czarne 60-1115
~ ~ europejskie 60-1120
~ ~ ferrytyczne 60-1125
~ ~ perlityczne 60-1130
~ ferrytyczne 60-1040
~ iglaste 60-1060
~ krzemowe 60-1010
~ kwasoodporne 60-1075
~ łożyskowe 60-1105
~ ługoodporne 60-1080
~ manganowe 60-1030
~ martenzytyczne 60-1055
~ modyfikowane 60-970
~ niestopowe 60-990
~ niklowe 60-1025
~ niskostopowe 60-1000
~ perlityczne 60-1045
~ połowiczne 60-960
~ przeciwcierne 60-1105
~ pstre 60-960
~ sferoidalne 60-975
~ staliste 60-1035
~ stopowe 60-995
~ szare 60-950
~ ~ ferrytyczne 60-1040
~ ~ perlityczne 60-1045
~ utwardzone 60-980
~ wysokojakościowe 60-1065
~ wysokostopowe 60-1005
~ zabielone 60-980
~ zbrojone 60-1070
~ zwykłe 60-965, 60-990
~ żaroodporne 60-1085

РУССКИЙ УКАЗАТЕЛЬ

А

абсолютный предел *m* ползучести 75-225
абсорбционный микроанализ *m* 80-735
авиональ *m* 65-610
автоионная микроскопия *f* 80-380
автоматная латунь *f* 65-370
автоматная нержавеющая сталь *f* 60-525
автоматная сталь *f* 60-710
автоматный сплав *m* 55-560
авторадиограмма *f* 80-755
авторадиография *f* 80-740
  контрастная ~ 80-745
  следовая ~ 80-750
автоэмиссионная микроскопия *f* 80-335
адмиралтейская латунь *f* 65-390
адмиралтейский металл *m* 65-390
азот *m* 05-305
азотирование *n* 70-785
  ~ в тлеющем разряде 70-800
  газовое ~ 70-790
  жидкостное ~ 70-795
  ионное ~ 70-800
азотированная сталь *f* 60-315
азотированный слой *m* 70-845
азотируемая сталь *f* 60-695
азотистый аустенит *m* 70-835
азотнауглероживание *n* 70-870
активационный объём *m* 30-475
активация *f*
  термическая ~ 45-275
активная атмосфера *f* 70-680
активность *f* 15-280
  термодинамическая ~ 15-280
актиний *m* 05-010
акустическая дефектоскопия *f* 80-795
алдрей *m* 65-585
алитирование *n* 70-925
аллотропическая модификация *f* 25-390
аллотропическая разновидность *f* 25-390
аллотропическая форма *f* 25-390
аллотропическое превращение *n* 25-455
аллотропия *f* 25-385
алмазная кубическая решётка *f* 20-320
алмазная сталь *f* 60-490
алудур *m* 65-580
алумель *m* 65-460
альдрей *m* 65-585
альклед *m* 65-560
альклэд *m* 65-560
альнико *m* 65-530
альсифер *m* 65-535
альфа+бета-латунь *f* 65-330
альфа-латунь *f* 65-325
алюдур *m* 65-580
алюмель *m* 65-460
алюминиевая бронза *f* 65-135
алюминиевая латунь *f* 65-340
алюминиевая сталь *f* 60-235

алюминиевый чугун *m* 60-1020
алюминий *m* 05-015
  ~-сырец 65-545
  катодный ~ 65-550
  первичный ~ 65-545
  рафинированный ~ 65-550
алюминирование *n*
  диффузионное ~ 70-925
амальгама *f* 65-850
американский ковкий чугун *m* 60-1115
америций *m* 05-020
амортизационная способность *f* 75-205
аморфный 20-005
анализ *m*
  дилатометрический ~ 80-540
  дифференциально-термический ~ 80-515
  дифференциальный дилатометрический ~ 80-550
  дифференциальный термический ~ 80-515
  капельный ~ 80-465
  количественный термодинамический ~ 80-535
  магнитный ~ 80-560
  металлографический ~ 80-035
  микроскопический ~ 80-040
  простой дилатометрический ~ 80-545
  простой термический ~ 80-510
  радиоактивационный ~ 80-765
  рентгеновский ~ с помощью электронного зонда 80-780
  рентгеновский спектральный ~ с помощью электронного зонда 80-780
  рентгеновский фазовый ~ 80-585
  рентгенографический фазовый ~ 80-585
  рентгеноспектральный ~ 80-700
  рентгеноструктурный ~ 80-580
  спектральный ~ 80-475
  термический ~ 80-505
  термомагнитный ~ 80-565
  физико-химический ~ 80-490
  химический ~ 80-470
  электронографический ~ 80-470
микроанализ 80-040
анатомический сплав *m* 55-645
анизотермическая диаграмма *f* превращения аустенита 70-160
анизотропия *f* 20-445
  магнитная ~ 75-575
анизотропность *f* 20-445
анодная медь *f* 65-040
аномальная структура *f* 40-135
анормальная сталь *f* 60-465
анормальность *f* стали 85-225
антикорродаль *m* 65-575
антикоррозионная устойчивость *f* 75-670
антифазная граница *f* 40-630
антифазный домен *m* 40-625
антифазовая граница *f* 40-630
антифазовый домен *m* 40-625
антиферромагнетизм *m* 75-465
антифрикционный сплав *m* 55-555
антифрикционный чугун *m* 60-1105
аргентановый припой *m* 65-745
аргон *m* 05-030
арматурная сталь *f* 60-795

армированный чугун m 60-1070
артиллерийская бронза f 65-245
астатин m 05-040
астеризм m 80-680
атермический раствор m 15-265
атермическое зарождение n 25-200
атермическое образование n 25-200
атермическое образование n зародышей 25-200
атермическое превращение n 25-520
атмосфера f
~ активная ~ 70-680
~ Коттрелла 30-440
~ примесей 30-440
~ защитная ~ 70-690
~ контролируемая ~ 70-700
~ коттрелловская ~ 30-440
~ науглероживающая ~ 70-695
~ экзотермическая ~ 70-710
~ эндотермическая ~ 70-705
атмосферная коррозия f 85-275
атмосферостойкая сталь f 60-540
атом m
~ внедрения 30-075
~ замещения 30-070
~ внедрённый ~ 30-075
~ дислоцированный ~ 30-510
~ замещённый ~ 30-070
~ междоузельный ~ 30-075
~ межузельный ~ 30-075
~ расщеплённый междоузельный ~ 30-085
~ смещённый ~ 30-510
атомная масса f 10-075
атомная плоскость f 20-175
атомная связь f 10-330
атомная теплоёмкость f 75-370
атомное число n 75-330
атомный вес m 10-075
атомный диаметр m 10-080
атомный номер m 75-330
атомный объём m 75-320
атомный радиус m 10-085
аустенизация f 70-130
аустенит m 50-025
~ азотистый ~ 70-835
~ неустойчивый ~ 50-265
~ остаточный ~ 50-270
~ первичный ~ 50-030
~ переохлаждённый ~ 50-265
~ превращённый ~ 50-035
аустенитизация f 70-130
аустенитизирование n 70-130
аустенитная сталь f 60-420
аустенитно-мартенситное превращение n 50-280
аустенитно-перлитное превращение n 50-275
аустенитно-ферритная сталь f 60-455
аустенитный чугун m 60-1050
аустенитообразующий элемент m 50-450
аусформинг m 70-960

Б

баббит m 65-640
~ высокооловянистый ~ 65-645
~ высокооловянный ~ 65-645
~ кальциевый ~ 65-685
~ оловянистый ~ 65-645
~ оловянный ~ 65-645
бабочки fpl 85-255
базис m решётки 20-110

базисная плоскость f 20-360
балл m зерна 80-290
бандажная сталь f 60-760
банметалл m 65-685
барий m 05-045
бархатистый излом m 40-715
бархатный излом m 40-715
барьер m Ломера-Коттрелла 30-450
~ потенциальный ~ 10-110
~ энергетический ~ 10-110
баумановский отпечаток m 80-325
бездиффузионное превращение n 25-510
безокислительный отжиг m 70-430
безоловянистая бронза f 65-125
безоловянная бронза f 65-125
безыгольчатый мартенсит m 50-330
бейнит m 50-410
~ верхний ~ 50-420
~ игольчатый ~ 50-415
~ нижний ~ 50-415
~ перистый ~ 50-420
бейнитирование n 70-310
бейнитная сталь f 60-435
бейнитное превращение n 50-290
белая бронза f 65-230
белое золото n 65-805
белое олово n 65-630
белосердечный ковкий чугун m 60-1120
белые пятна npl 85-250
белый антифрикционный сплав m 65-640
белый излом m 85-160
белый металл m 65-640
белый чугун m 60-910, 60-955
бериллиевая бронза f 65-150
бериллий m 05-055
беркелий m 05-050
бертоллиды mpl 35-250
бескислородная высокопроводящая медь f 65-070
бескислородная медь f 65-060
бескислородная медь f высокой проводимости 65-070
беспорядок m 25-585
беспорядочная ориентировка f 20-420
беспровесный молибден m 65-870
бессемеровская сталь f 60-595
бессемеровский чугун m 60-845
бесструктурный мартенсит m 50-330
бета-латунь f 65-335
бивакансия f 30-045
бикристалл m 20-435
биметалл m 65-880
бинарная система f 35-120
бинарный раствор m 35-405
бинарный сплав m 55-295
благородный металл m 55-010
ближайший сосед m 20-195
ближний порядок m 25-595
блок m мозаики 30-655
~ мозаичный ~ 30-655
блокирование n дислокаций 30-445
блокировка f дислокаций 30-445
блочная структура f 40-170
большеугловая граница f 30-585
бор m 05-065
борид m 70-905
борирование n 70-900

бористая сталь f 60-255
борсодержащая сталь f 60-255
браунит m 70-840
бром m 05-070
броневая сталь f 60-765
бронза f 65-095
  алюминиевая ~ 65-135
  артиллерийская ~ 65-245
  безоловянистая ~ 65-125
  безоловянная ~ 65-125
  белая ~ 65-230
  бериллиевая ~ 65-150
  ванадиевая ~ 65-180
  вольфрамовая ~ 65-170
  вторичная ~ 65-285
  высокооловянистая ~ 65-115
  высокооловянная ~ 65-115
  графитизированная ~ 65-220
  графитированная ~ 65-220
  декоративная ~ 65-270
  деформируемая ~ 65-105
  железистая ~ 65-175
  зеркальная ~ 65-280
  кадмиевая ~ 65-190
  клапанная ~ 65-275
  колокольная ~ 65-250
  кремнистая ~ 65-145
  литейная ~ 65-100
  малооловянистая ~ 65-120
  малооловянная ~ 65-120
  марганцевая ~ 65-155
  марганцовистая ~ 65-155
  машинная ~ 65-240
  медальная ~ 65-260
  металлокерамическая ~ 65-225
  монетная ~ 65-265
  низкооловянная ~ 65-120
  никелевая ~ 65-160
  оборотная ~ 65-285
  оловянистая ~ 65-110
  оловянная ~ 65-110
  оловянносвинцовистая ~ 65-185
  оловяннофосфористая ~ 65-210
  оловянноцинковая ~ 65-205
  орудийная ~ 65-245
  паспортная ~ 65-285
  подшипниковая ~ 65-235
  пористая ~ 65-225
  пушечная ~ 65-245
  свинцовая ~ 65-140
  свинцовистая ~ 65-140
  специальная ~ 65-130
  телефонная ~ 65-255
  фосфористая ~ 65-215
  хромистая ~ 65-195
  художественная ~ 65-270
бронзографит m 65-220
булавочная коррозия f 85-345
булатная сталь f 60-645
быстрое охлаждение n 70-075
быстрорежущая сталь f 60-800

В

вакансия f 30-035
~ зафиксированная закалкой 30-100
~ Френкеля 30-065
  закалённая ~ 30-100
  закалочная ~ 30-100
  избыточная ~ 30-105
  неравновесная ~ 30-105
  одиночная ~ 30-040
  структурная ~ 30-090
  тепловая ~ 30-095
  термическая ~ 30-095
бивакансия 30-045
дивакансия 30-045
моновакансия 30-040
тетравакансия 30-055
тривакансия 30-050
вакуумированная сталь f 60-640

вакуумированный металл m 55-135
валентная зона f 10-255
валентный электрон m 10-065
ванадиевая бронза f 65-180
ванадиевая латунь f 65-180
ванадиевая сталь f 60-225
ванадий m 05-495
вандерваальсов(ск)ая связь f 10-335
вариантность f 35-030
ваталлиум m 65-485
вектор m Бюргерса 30-190
~ решётки 20-135
волновой ~ 10-185
величина f зерна 80-285
  избыточная ~ 15-065
  избыточная термодинамическая ~ 15-065
  интегральная ~ 15-055
  интегральная термодинамическая ~ 15-055
  интенсивная ~ 15-045
  молярная ~ 15-050
  парциальная ~ 15-060
  парциальная термодинамическая ~ 15-060
  термодинамическая ~ 15-030
  экстенсивная ~ 15-040
веркблей m 65-650
вертикальный разрез m 35-740
верхний бейнит m 50-420
вершинная дислокация f 30-360
вес m
  атомный ~ 10-075
  удельный ~ 75-350
весовая концентрация f 55-235
весьма мягкая сталь f 60-345
вещество n
  диамагнитное ~ 75-560
  парамагнитное ~ 75-555
  растворённое ~ 35-360
  ферромагнитное ~ 75-550
взаимная диффузия f 45-380
взаимная растворимость f 35-300
взрывное упрочнение n 45-055
взятие n пробы 80-060
вибрационная энтропия f 15-205
вид m симметрии 20-050
Видманштеттова структура f 40-130
винтовая дислокация f 30-165
висмут m 05-060
включения npl нитридов 40-545
~ окислов 40-535
  металлические ~ 40-550
  неметаллические ~ 40-515
  нитридные ~ 40-545
  окисные ~ 40-535
  сернистые ~ 40-530
  сульфидные ~ 40-530
  шлаковые ~ 40-540
  экзогенные ~ 40-520
  эндогенные ~ 40-525
внедрённый атом m 30-075
внешний электрон m 10-065
внутреннее окисление n 70-945
внутреннее трение n 75-210
внутренние напряжения npl 85-020
внутренние напряжения npl второго рода 85-035
внутренние напряжения npl первого рода 85-030
внутренний наклёп m 45-080
внутренний электрон m 10-070
внутренняя газовая раковина f 85-135
внутренняя коррозия f 85-355
внутренняя энергия f 15-095

внутризёренный излом m 40-710
внутрикристаллический излом m 40-710
внутрикристаллическое скольжение n 45-165
водород m 05-200
водородная болезнь f 85-395
водородная хрупкость f 85-390
водяная закалка f 70-205
водяная коррозия f 85-280
возбудитель m зарождения 25-125
возврат m 45-420, 70-610
  динамический ~ 45-425
  термический ~ 45-425
воздушная закалка f 70-215
воздушное охлаждение n 70-095
воздушнозакаливаемая сталь f 60-375
воздушнозакаливающаяся сталь f 60-375
волновая функция f 10-180
волновой вектор m 10-185
волокнистая текстура f 40-565
волосовина f 85-045
вольфрам m 05-485
вольфрамид m 40-500
вольфрамирование n 70-935
  диффузионное ~ 70-935
вольфрамистая сталь f 60-215
вольфрамовая бронза f 65-170
вольфрамовая сталь f 60-215
воронение n 70-400
восприимчивость f к закалке 70-020
  ~ к коррозии 75-675
  ~ к упрочнению 75-770
  магнитная ~ 75-625
восходящая диффузия f 45-340
восхождение n дислокации 30-295
время n выдержки 70-060
ВТМО 70-955
вторичная бронза f 65-285
вторичная закалка f 70-405
вторичная кристаллизация f 25-060
вторичная ликвация f 25-410
вторичная рекристаллизация f 45-450
вторичная система f скольжения 45-120
вторичная структура f 40-055
вторичное мартенситное превращение n 50-285
вторичное твердение n 70-405
вторичные дислокации fpl 30-255
вторичные кристаллы mpl 40-275
вторичный графит m 50-180
вторичный мартенсит m 50-350
вторичный металл m 55-120
вторичный сплав m 55-280
вторичный цементит m 50-080
вторичный чугун m 60-945
второе квантовое число n 10-135
второй сосед m 20-200
выделение n
  непрерывное ~ 25-375
  прерывистое ~ 25-380
выделения npl 40-305
  когерентные ~ 40-320
  межзёренные ~ 40-310
  некогерентные ~ 40-325
  пограничные ~ 40-310
  ультрамикроскопические ~ 40-315
выдержка f 70-055
выносливость f 75-200

выпоты mpl 25-430
  ~ олова 85-165
выращивание n монокристаллов 25-085
вырожденный перлит m 50-155
высокий отжиг m 70-480
высокий отпуск m 70-375
высокожаропрочный сплав m 55-515
высококачественная сталь f 60-145
высококачественный чугун m 60-1065
высококоэрцитивный сплав m 55-465
высоколегированная сталь f 60-130
высоколегированный сплав m 55-325
высоколегированный чугун m 60-1005
высокооловянистая бронза f 65-115
высокооловянистый баббит m 65-645
высокооловянная бронза f 65-115
высокооловянный баббит m 65-645
высокоомный сплав m 55-545
высокоплавкий сплав m 55-400
высокопробное золото n 65-795
высокопроцентный сплав m 55-325
высокопрочная сталь f 60-355
высокопрочный сплав m 55-420
высокопрочный чугун m 60-975
высокотемпературная закалка f 70-345
высокотемпературная металлография f 80-015
высокотемпературная микроскопия f 80-240
высокотемпературная модификация f 25-395
высокотемпературная прочность f 75-095
высокотемпературная термомеханическая обработка f 70-955
высокотемпературная фаза f 35-160
высокотемпературное цианирование n 70-870
высокотемпературный отпуск m 70-375
высокотемпературный сплав m 55-510
высокотемпературный феррит m 50-050
высокоуглеродистая сталь f 60-110
высокоуглеродистый чугун m 60-925
высокохромистая сталь f 60-180
высокочастотная закалка f 70-295
высокоэлектропроводный сплав m 55-550
вытягивание n монокристаллов 25-090
вытянутая дислокация f 30-325
выявление n микроструктуры 80-280
вязкий излом m 40-680
вязкое разрушение n 40-680
вязкое течение n 45-220
вязкостное течение n 45-220
вязкость f 75-035
  ударная ~ 75-190

Г

габитус m 20-460
габитусная плоскость f 20-465
гадолиний m 05-165
газ m
  ~-носитель 70-685
  науглероживающий ~ 70-745
  цементирующий ~ 70-745
  цементующий ~ 70-745
  экзотермический ~ 70-710
  электронный ~ 10-045
  эндотермический ~ 70-705

ГРАФИТ

газ
экзогаз 70-710
эндогаз 70-705
газовая коррозия *f* 85-295
газовая ликвация *f* 25-430
газовая пустота *f* 85-135
газовая раковина *f* 85-130
газовая смесь *f* 35-370
газовая цементация *f* 70-730
газовое азотирование *n* 70-790
газовое травление *n* 80-182
газовое цианирование *n* 70-860
газообразная смесь *f* 35-370
газообразный раствор *m* 35-370
газопламенная закалка *f* 70-290
гайперник *m* 65-475
галлий *m* 05-170
гальваническая коррозия *f* 85-270
гальванотипный сплав *m* 55-635
гаммаграфия *f* 80-825
гамма-дефектоскопия *f* 80-825
гамма-лучевая дефектоскопия *f* 80-825
гантель *f* 30-085
гантельная пара *f* 30-085
гарденит *m* 50-365
гартцинк *m* 65-700
гафний *m* 05-185
гейслеров сплав *m* 65-420
гексагира *f* 20-085
гексагональная компактная решётка *f* 20-350
гексагональная плотноупакованная решётка *f* 20-350
гексагональная решётка *f* 20-340
гексагональная система *f* 20-245
гексаэдрическая решётка *f* 20-305
гелий *m* 05-190
геликоидальная дислокация *f* 30-355
гематитовый чугун *m* 60-895
генератор *m* дислокаций 30-425
генерирование *n* дислокаций 30-420
германий *m* 05-175
гетерогенная система *f* 35-095
гетерогенная структура *f* 40-155
гетерогенное зародышеобразование *n* 25-185
гетерогенное образование *n* 25-185
гетерогенное образование *n* зародышей 25-185
гетерогенное превращение *n* 25-535
гетерогенное равновесие *n* 35-015
гетерогенный зародыш *n* 25-115
гетерогенный сплав *m* 55-340
гетеродиффузия *f* 45-335
гетерополярная связь *f* 10-325
гетеротермическое плавление *n* 35-455
гидрид *m* 40-495
гидроналий *m* 65-600
гильзовая латунь *f* 65-380
гиперник *m* 65-475
гира *f* 20-065
гистерезис *m* превращения 25-480
магнитный ~ 75-640
температурный ~ 25-480
главная кристаллографическая ось *f* 25-300
главная легирующая добавка *f* 55-195

главное квантовое число *n* 10-130
главный легирующий элемент *m* 55-195
глобулярная структура *f* 40-120
глобулярный графит *m* 50-190
глобулярный перлит *m* 50-150
глубина *f* вытяжки 75-755
глубина *f* прокаливаемости 70-350
глубокое травление *n* 80-160
глубокопрокаливающаяся сталь *f* 60-385
гнездообразный графит *m* 50-210
гольмий *m* 05-195
гомеополярная связь *f* 10-330
гомогенизационный отжиг *m* 70-485
гомогенизация *f* 70-485
гомогенизирующий отжиг *m* 70-485
гомогенная область *f* 35-680
гомогенная система *f* 35-090
гомогенная структура *f* 40-145
гомогенное зародышеобразование *n* 25-180
гомогенное образование *n* зародышей 25-180
гомогенное превращение *n* 25-530
гомогенность *f* 40-405
гомогенный зародыш *m* 25-110
гомогенный сплав *m* 55-330
гомологическая температура *f* 45-475
горизонталь *f*
монотектическая ~ 35-650
перитектическая ~ 35-645
эвтектическая ~ 35-640
эвтектоидная ~ 35-655
горизонтальный разрез *m* 35-735
горячая обработка *f* давлением 45-020
горячая твёрдость *f* 75-255
горячая трещина *f* 85-080
горячеломкость *f* 75-290
градиент *m* концентрации 15-080
~ температуры 15-075
концентрационный ~ 15-080
температурный ~ 15-075
гранецентрированная решётка *f* 20-290
граница *f*
антифазная ~ 40-630
антифазовая ~ 40-630
большеугловая ~ 30-585
~ антифазных доменов 40-630
~ антифазовых доменов 40-630
~ Блоха 40-645
~ двойника 45-265
~ зерна 30-565
~ зёрен 30-565
~ кручения 30-600
~ между фазовыми областями 35-535
~ наклона 30-605
~ раздела фаз 30-610
~ растворимости 35-330
когерентная ~ 30-625
малоугловая ~ 30-590
межзёренная ~ 30-565
межфазная ~ 30-610
наклонная ~ 30-605
некогерентная ~ 30-630
фазовая ~ 35-535
субграница 30-660
субзёренная ~ 30-660
граничная кривая *f* 35-535
грань *f* кристалла 20-060
графит *m* 50-165
вторичный ~ 50-180
глобулярный ~ 50-190
гнездообразный ~ 50-210
междендритный ~ 50-205
первичный ~ 50-170

ГРАФИТ

графит
пластинчатый ~ 50-185
розеточный ~ 50-200
сфероидальный ~ 50-190
сферолитный ~ 50-190
точечный ~ 50-195
чешуйчатый ~ 50-185
шаровидный ~ 50-190
эвтектический ~ 50-175
графитизация f 50-225, 70-630
графитизированная бронза f 65-220
графитизированная сталь f 60-300
графитизирующий отжиг m 70-630
графитированная бронза f 65-220
графитистый чугун m 60-950
графитная спель f 50-170, 85-140
графитная эвтектика f 50-260
графитообразующий элемент m 50-460
грубопластинчатый перлит m 50-140
губа f 85-400
губчатое железо n 60-035

Д

давление n диссоциации 15-345
дальний порядок m 25-600
дальтониды mpl 35-245
дамасская сталь f 60-645
движение n границ (зёрен) 45-490
диффузионное ~ 30-285
консервативное ~ 30-290
неконсервативное ~ 30-285
скользящее ~ 30-290
движущая сила f 15-070
двойная закалка f 70-765
двойная ось f симметрии 20-070
двойная подсистема f 35-715
двойная система f 35-120
двойная эвтектика f 40-205
двойник m 45-230
~ деформации 45-235
~ отжига 45-240
~ рекристаллизации 45-240
~ роста 45-240
деформационный ~ 45-235
механический ~ 45-235
рекристаллизационный ~ 45-240
двойникование n 45-245
двойниковая граница f 45-265
двойниковая плоскость f 45-250
двойниковый дефект m 30-555
двойниковый дефект m упаковки 30-555
двойниковый кристалл m 45-230
двойникующая дислокация f 30-375
двойное скольжение n 45-135
двойной карбид m 40-475
двойной раствор m 35-405
двойной сплав m 55-295
двухкомпонентная система f 35-120
двухкомпонентный раствор m 35-405
двухкомпонентный сплав m 55-295
двухмерный дефект m 30-535
двухфазная латунь f 65-320
двухфазная морская латунь f 65-385
двухфазная область f 35-685
двухфазная система f 35-100
двухфазная структура f 40-150
двухфазное поле n 35-685
двухфазный сплав m 55-335
деазотирование n 70-850

дебаевская температура f 10-280
дебаевское кольцо n 80-640
дебаеграмма f 80-635
дегазирование n 70-550
дезорентация f 30-595
действительное сопротивление n разрыву 75-130
декалесценция f 25-490
декоративная бронза f 65-270
декоративный сплав m 55-605
декорирование n дислокаций 80-375
делящийся металл n 55-080
демодификатор m 25-235
дендрит m 25-315
термический ~ 25-325
химический ~ 25-320
дендритная ликвация f 25-420
дендритная структура f 40-090
дендритный кристалл m 25-315
дендритный рост m 25-310
дестабилизация f аустенита 50-430
дефект m
двойниковый ~ 30-555
двойниковый ~ упаковки 30-555
двухмерный ~ 30-535
~ кристаллической решётки 30-015
~ материала 85-005
~ решётки 30-015
~ созданный облучением 30-485
~ упаковки 30-540
~ упаковки внедрения 30-550
~ упаковки вычитания 30-545
~ Френкеля 30-060
линейный ~ 30-140
литейный ~ 85-095
нульмерный ~ 30-030
одномерный ~ 30-140
поверхностный ~ 30-535, 85-010
радиационный ~ 30-485
точечный ~ 30-030
дефектная решётка f 30-020
дефектоскопия f 80-790
акустическая ~ 80-795
гамма-~ 80-825
гамма-лучевая ~ 80-825
магнитная ~ 80-805
магнитно-порошковая ~ 80-805
нейтронная ~ 80-830
ультраакустическая ~ 80-800
ультразвуковая ~ 80-800
флуоресцентная ~ 80-835
рентгенодефектоскопия f 80-820
деформационная трещина f 85-070
деформационное разупрочнение n 45-535
деформационное старение n 45-225
деформационное упрочнение n 45-050
деформационный двойник m 45-235
деформационный мартенсит m 50-360
деформация f
критическая ~ 45-075
остаточная ~ 35-010
пластическая ~ 35-010
упругая ~ 45-005
деформируемая бронза f 65-105
деформируемая латунь f 65-295
деформируемый сплав m 55-530
деформирующее напряжение n 45-210
диаграмма f
анизотермическая ~ превращения аустенита 70-160
~ анизотермического превращения аустенита 70-160
~ железо-углерод 50-020
~ изотермического превращения аустенита 70-155
~ равновесия 35-480

диаграмма
~ равновесия фаз 35-480
~ рекристаллизация 45-485
~ состояния 35-480
~ состояния железо-углерод 50-020
~ фазового равновесия 35-480
~ фазового состояния 35-480
~ Шефлера 60-530
изотермическая ~ превращения аустенита 75-155
пространственная ~ состояния тройной системы 35-750
равновесия ~ состояния 35-480
структурная ~ 40-265
тройная ~ состояния 35-710
фазовая ~ состояния 35-480
диамагнетик m 75-560
диамагнитное вещество n 75-560
диаметр m
атомный ~ 10-080
критический ~ 70-360
дивакансия f 30-045
дивариантная система f 35-085
дигира f 20-070
дилатограмма f 80-555
дилатометрическая кривая f 80-555
дилатометрический анализ m 80-540
дилатометрическое исследование n 80-540
дилатометрия f 80-540
дифференциальная ~ 80-550
простая ~ 80-545
динамическая прочность f 75-185
динамическая рекристаллизация f 45-465
динамический возврат m 45-425
динамический отдых m 45-425
динамическое зародышеобразование n 25-205
динамическое образование n зародышей 25-205
динамная сталь f 60-775
диполь m
дислокационный ~ 30-380
дисклинация f 30-480
дислокации fpl
вторичные ~ 30-255
испущенные ~ 30-370
первичные ~ 30-250
дислокационная линия f 30-195
дислокационная петля f 30-340
дислокационная сетка f 30-390
дислокационная стенка f 30-395
дислокационная структура f 30-415
дислокационное кольцо n 30-340
дислокационное сплетение n 30-385
дислокационный диполь m 30-380
дислокационный круг m 30-340
дислокационный узел m 30-350
дислокация f 30-145
вершинная ~ 30-360
винтовая ~ 30-165
вытянутая ~ 30-325
геликоидальная ~ 30-355
двойникующая ~ 30-375
~ леса 30-410
~ Ломера-Коттрелла 30-450
~ скольжения 30-265
~ Франка 30-335
~ Шокли 30-330
единичная ~ 30-215
закреплённая ~ 30-270
заряженная ~ 30-465
зернограничная ~ 30-580
краевая ~ 30-150

дислокация
левовинтовая ~ 30-175
линейная ~ 30-150
неполная ~ 30-310
несовершенная ~ 30-310
отрицательная краевая ~ 30-160
пограничная ~ 30-580
полная ~ 30-305
положительная краевая ~ 30-155
полюсная ~ 30-365
правовинтовая ~ 30-170
призматическая ~ 30-345
растянутая ~ 30-325
расщеплённая ~ 30-325
сверхструктурная ~ 30-220
сидячая ~ 30-280
сидячая ~ Франка 30-335
скользящая ~ 30-265
смешанная ~ 30-180
составная ~ 30-180
частичная ~ 30-315
полудислокация 30-315
сверхдислокация 30-220
супердислокация 30-220
дислокационный атом m 30-510
дислокационный источник m 30-425
дислокационный лес m 30-405
диспергированная фаза f 40-330
дисперсионно-упрочнённый сплав m 55-390
дисперсионное твердение n 70-560
дисперсионное упрочнение n 70-560
дисперсионнотвердеющий сплав m 55-385
дисперсная фаза f 40-330
дисперсное упрочнение n 55-220
дисперсностное упрочнение n 70-560
дисперсность f 40-335
дисперсноупрочнённый сплав m 55-390
дисперсоид m 40-330
диспрозий m 05-130
диссоциация f дислокации 30-320
дистилляционный цинк m 65-695
дифрактограмма f 80-610
дифракционная картина f 80-610
дифракционная линия f 80-645
дифракционная рентгенограмма f 80-615
дифракционное пятно n 80-660
дифференциальная дилатометрия f 80-550
дифференциальная закалка f 70-275
дифференциально-термический анализ m 80-515
дифференциальное науглероживание n 70-735
дифференциальный дилатометрический анализ m 80-550
дифференциальный термический анализ m 80-515
диффузионная зона f 45-415
диффузионная металлизация f 70-915
диффузионная пара f 45-375
диффузионная ползучесть f 45-520
диффузионное алюминирование n 70-925
диффузионное вольфрамирование n 70-935
диффузионное движение n 30-285
диффузионное легирование n 70-655
диффузионное насыщение n 70-655
диффузионное насыщение n металлами 70-915

ДИФФУЗИОННОЕ

диффузионное превращение n 25-505
диффузионное силицирование n 70-910
диффузионное титанирование n 70-945
диффузионное хромирование n 70-920
диффузионное цинкование n 70-930
диффузионный отжиг m 70-485
диффузионный слой m 70-660
диффузионный фронт m 45-410
диффузия f 45-290
  взаимная ~ 45-380
  восходящая ~ 45-340
  ~ перемещением вакансий 45-305
  ~ по границам зёрен 45-310
  зернограничная ~ 45-310
  изотермическая ~ 45-350
  концентрационная ~ 45-320
  междуузельная ~ 45-315
  межзёренная ~ 45-310
  многофазная ~ 45-325
  неоднофазная ~ 45-325
  объёмная ~ 45-295
  однофазная ~ 45-330
  поверхностная ~ 45-300
  пограничная ~ 45-310
  реактивная ~ 45-325
  реакционная ~ 45-325
  стационарная ~ 45-345
  тепловая ~ 45-370
  термическая ~ 45-370
  трубочная ~ 45-360
  ускоренная ~ 45-355
  установившаяся ~ 45-345
  химическая ~ 45-320
  гетеродиффузия 45-335
  самодиффузия 45-400
  термодиффузия 45-370
  электродиффузия 45-365
длительная прочность f 75-235
добавка f
  главная легирующая ~ 55-195
  легирующая ~ 55-190
  микролегирующая ~ 55-250
  основная легирующая ~ 55-195
додекаэдрическая решётка f 20-310
дозародыш m 25-135
докритический зародыш m 25-135
домен m 40-615
  антифазный ~ 40-625
  антифазовый ~ 40-625
  замыкающий ~ 40-640
  магнитный ~ 40-635
доменная стенка f 40-645
доменная структура f 40-620
доменный чугун m 60-860
дополнительная удельная теплоёмкость f 25-635
доэвтектическая структура f 40-070
доэвтектический белый чугун m 50-510
доэвтектический сплав m 55-355
доэвтектоидная сталь f 50-495
доэвтектоидная структура f 40-085
доэвтектоидный сплав m 55-370
драгоценный металл m 55-010
древесно-волокнистый излом m 40-720
древесноугольный чугун m 60-875
дуктильность f 75-030
дураль f 65-555
дуралюмин m 65-555
дуралюминий m 65-555
дюраль m 65-555
дюралюмин m 65-555
дюралюминий m 65-555
дюрометрическое исследование n 80-415

Е

европейский ковкий чугун m 60-1120
европий m 05-145
единичная дислокация f 30-215
единичная ячейка f 20-090
единичное скольжение n 45-130
естественное стабилизирование n 70-540
естественное старение n 70-585

Ж

жаропрочная сталь f 60-565
жаропрочность f 75-240
жаропрочный сплав m 55-510
жаростойкая сталь f 60-560
жаростойкий сплав m 55-505
жаростойкий чугун m 60-1085
жаростойкость f 75-690
жароупорная сталь f 60-560
жароупорность f 75-690
жароупорный сплав m 55-505
жароупорный чугун m 60-1085
железная бронза f 65-175
железный сплав m 60-070
железо n 05-220
  армко ~ 60-010
  губчатое ~ 60-035
  ~ Армко 60-010
  ~ технической чистоты 60-015
  литое ~ 60-060
  карбонильное ~ 60-020
  кремнистое ~ 60-210
  кричное ~ 60-050
  металлокерамическое ~ 60-065
  метеоритное ~ 60-005
  промышленное ~ 60-015
  пудлинговое ~ 60-055
  сварочное ~ 60-040
  сернистое ~ 50-245
  сыродутное ~ 60-045
  техническое ~ 60-015
  трансформаторное ~ 60-770
  шведское ~ 60-030
  электролитическое ~ 60-025
жидкий расплав m 55-150
жидкий раствор m 35-365
жидкоплавкость f 75-785
жидкоподвижность f 75-780
жидкостная закалка f 70-245
жидкостная цементация f 70-725
жидкостное азотирование n 70-795
жидкостное цианирование n 70-865
жидкотекучесть f 75-780

З

загрязнения npl 55-205
задержанное разрушение n 40-755
закалённая вакансия f 30-100
закалённая зона f 70-355
закалённая сталь f 60-285
закалённый слой m 70-355
закаливаемость f 70-020, 75-730
закаливающаяся сталь f 60-360
закалка f 70-175, 70-570
  ~ без полиморфного превращения 70-570
  водяная ~ 70-205
  воздушная ~ 70-215
  вторичная ~ 70-405
  высокотемпературная ~ 70-345
  высокочастотная ~ 70-295

закалка
газопламенная ~ 70-290
двойная ~ 70-765
дифференциальная ~ 70-275
жидкостная ~ 70-245
~ в воде 70-205
~ в воздухе 70-215
~ в горячих средах 70-225
~ в кокилях 70-255
~ в масле 70-210
~ в охлаждаемой кокили 70-255
~ в рассоле 70-238
~ в свинцовой ванне 70-235
~ в сжатом воздухе 70-220
~ в соляной ванне 70-230
~ в соляном расплаве 70-230
~ в соляном растворе 70-238
~ в струе воздуха 70-220
~ вакансий 30-110
~ из жидкого состояния 70-085
~ методом вращения 70-315
~ методом вращения и перемещения 70-325
~ методом перемещения 70-320
~ охлаждением в ванне 70-245
~ погружением 70-245
~ расплава 70-085
~ с большой скоростью 70-260
~ с малой скоростью 70-265
~ с полиморфным превращением 70-175
~ с цементационного нагрева 70-760
~ сердцевины 70-770
~ сердцевины после цементации 70-770
~ струей воздуха 70-220
~ ТВЧ 70-295
изотермическая ~ 70-310
индукционная ~ 70-295
истинная ~ 70-570
контактная ~ 70-250
масляная ~ 70-210
местная ~ 70-275
неполная ~ 70-335
объёмная ~ 70-287
обыкновенная ~ 70-270
обычная ~ 70-270
пламенная ~ 70-290
поверхностная ~ 70-285
прерывистая ~ 70-300
пробная ~ 70-775
резкая ~ 70-260
светлая ~ 70-330
сквозная ~ 70-280
струйная ~ 70-240
струйчатая ~ 70-240
ступенчатая ~ 70-305
частичная ~ 70-335
чистая ~ 70-330
закалочная вакансия f 30-100
закалочная ванна f 70-195
закалочная жидкость f 70-190
закалочная среда f 70-185
закалочная температура f 70-140
закалочная трещина f 85-190
закалочные напряжения npl 85-185
закалочный охладитель m 70-185
закалочный порок m 85-180
закат m 85-400
закатка f 85-400
заклёпочная сталь f 60-745
закон m
~ Вегарда 35-425
~ Генри 15-275
~ Рауля 15-270
~ Рауля 15-270
закреплённая дислокация f 30-270
закрытая газовая раковина f 85-135
заливка f в оправку 80-065
замедленное разрушение n 40-755
замедляющая способность f 75-535
замещённый атом m 30-070
замыкающий домен m 40-640

запаздывающее разрушение n 40-755
запасённая энергия f 45-060
заполненная зона f 10-245
запрещённая зона f 10-250
зародыш m
гетерогенный ~ 25-115
гомогенный ~ 25-110
докритический ~ 25-135
~ кристаллизации 25-105
~ критического размера 25-140
~ рекристаллизации 25-170
когерентный ~ 25-155
критический ~ 25-140
ориентированный ~ 25-160
примесный ~ 25-120
равновесный ~ 25-140
спонтанный ~ 25-110
устойчивый ~ 25-150
дозародыш 25-135
зародышеобразование n 25-175
гетерогенное ~ 25-185
гомогенное ~ 25-180
динамическое ~ 25-205
самопроизвольное ~ 25-190
зарождение n
атермическое ~ 25-200
~ центров кристаллизации 25-175
ориентированное ~ 25-195
трибозарождение 25-210
заряженная дислокация f 30-465
затвердевание n 25-005
направленное ~ 25-035
неравновесное ~ 25-020
одноосное ~ 25-040
ориентированное ~ 25-035
управляемое ~ 25-030
эвтектическое ~ 25-025
затравка f 25-130
~ кристаллизации 25-215
затравочный кристалл m 25-130
затухающая ползучесть f 45-500
защитная атмосфера f 70-690
заэвтектическая структура f 40-065
заэвтектический белый чугун m 50-515
заэвтектический сплав m 55-360
заэвтектоидная сталь f 50-500
заэвтектоидная структура f 40-080
заэвтектоидный сплав m 55-375
зеркальная бронза f 65-280
зеркальный чугун m 60-930
зернистый излом m 40-690
зернистый перлит m 50-150
зернистый цементит m 50-100
зерно n 40-360
исходное аустенитное ~ 50-040
крупное ~ 40-365
мелкое ~ 40-370
равноосное ~ 40-285
субзерно 30-650
зернограничная дислокация f 30-580
зернограничная диффузия f 45-310
зернограничная сегрегация f 25-450
зернограничное скольжение n 45-160
золото n 05-180
белое ~ 65-805
высокопробное ~ 65-795
красное ~ 65-800
монетное ~ 65-775
одонтологическое ~ 65-785
рафинированное ~ 65-770
самородное ~ 65-765
стандартное ~ 65-790
червонное ~ 65-795
ювелирное ~ 65-780
зона f
валентная ~ 10-255

зона
 диффузионная ~ 45-415
 закалённая ~ 70-355
 запрещённая ~ 10-250
 ~ Бриллюэна 10-230
 ~ дозволенных энергий 10-240
 ~ проводимости 10-260
 ~ равноосных кристаллов 40-185
 ~ столбчатой кристаллизации 40-180
 ~ столбчатых кристаллов 40-180
 координационная ~ 20-190
 корковая ~ 40-175
 обеднённая ~ 30-120
 столбчатая ~ 40-180
 энергетическая ~ 10-235
зональная ликвация f 25-415
зонная теория f 10-020
зоны fpl Гинье-Престона 70-605
зубопротезный сплав m 55-575

И

игла f мартенсита 50-375
 мартенситная ~ 50-375
игольчатая структура f 40-115
игольчатый бейнит m 50-415
игольчатый мартенсит m 50-315
игольчатый троостит m 50-410
игрек-сплав m 65-590
идеальный кристалл m 30-005
идеальный раствор m 15-240
идиоморфный кристалл m 25-355
избирательная коррозия f 85-305
избыточная вакансия f 30-105
избыточная свободная энергия f 15-115
избыточная термодинамическая
 величина f 15-065
избыточная фаза f 40-340, 40-345
избыточная энтальпия f 15-145
избыточная энтропия f 15-220
избыточный феррит m 50-055
избыточный цементит m 50-095
излом m 30-245, 40-670, 85-040
 бархатистый ~ 40-715
 бархатный ~ 40-715
 белый ~ 85-160
 внутризёрненный ~ 40-710
 внутрикристаллический ~ 40-710
 вязкий ~ 40-680
 древесно-волокнистый ~ 40-720
 зернистый ~ 40-690
 интеркристаллический ~ 40-705
 кристаллический ~ 40-685
 крупнозернистый ~ 40-695
 крупнокристаллический ~ 40-695
 межзёренный ~ 40-705
 межкристаллитный ~ 40-705
 межкристаллический ~ 40-705
 мелкозернистый ~ 40-700
 нафталинистый ~ 40-740
 нафталиновый ~ 40-740
 раковинистый ~ 40-730
 раковинообразный ~ 40-730
 раковистый ~ 40-730
 светлый ~ 85-155
 слоистый ~ 40-725
 стекловидный ~ 40-735
 транскристаллический ~ 40-710
 тягучий ~ 40-680
 усталостный ~ 40-745
 фарфоровидный ~ 40-700, 40-715
 хрупкий ~ 40-675
 шелковистый ~ 40-715
 шиферный ~ 40-725
измельчение n зерна 40-385
измельчение n зернистости 40-385

измельчение n структуры 40-385
измельчитель m зерна 40-100
износостойкая сталь f 60-480
износостойкий сплав m 55-435
износостойкость f 75-265
износоустойчивая сталь f 60-480
износоустойчивость f 75-265
износоустойчивый сплав m 55-435
изобарно-изотермный потенциал m
 15-140
изобарный потенциал m 15-140
изоморфизм m 20-475
изоморфные кристаллы mpl 20-480
изотермическая диаграмма f
 превращения аустенита 70-155
изотермическая диффузия f 45-350
изотермическая закалка f 70-310
изотермический отжиг m 70-450
изотермический разрез m 35-735
изотермическое плавление n 35-450
изотермическое превращение n 25-515
изотопный индикатор m 80-760
изотропия f 20-450
изохорно-изотермный потенциал m
 15-105
изохрональный отжиг m 70-495
изохронный отжиг m 70-495
имитация f золота 65-815
импульсный нагрев m 70-050
инвар m 65-500
~-сталь 65-500
индексы mpl Миллера-Браве 20-170
 миллеровские ~ 20-165
индий m 05-205
индикатор m
 изотопный ~ 80-760
 радиоактивный ~ 80-760
индикационная закалка f 70-295
индицирование n 80-695
индукция f
 магнитная ~ 75-610
 остаточная ~ 75-630
 остаточная магнитная ~ 75-630
инертная атмосфера f 70-675
инконель m 65-445
инкубационный период m 70-150
инструментальная сталь f 60-675
инструментальная сталь f для горячей
 обработки 60-810
инструментальная сталь f для работы
 в холодном состоянии 60-805
инструментальная сталь f для работы
 при высоких температурах 60-810
интегральная величина f 15-055
интегральная термодинамическая
 величина f 15-055
интенсивная величина f 15-045
интенсивность f намагничения 75-600
интенсивность f охлаждения 70-200
интервал m затвердевания 35-550
 ~ кристаллизации 35-550
 ~ мартенситного превращения 50-300
 ~ плавления 35-545
 ~ температур затвердевания 35-550
 ~ температур плавления 35-545
 критический ~ 35-560
температурный ~ кристаллизации 35-550
интеркристаллитная коррозия f 85-360

КИСЛОТОУПОРНЫЙ

интеркристаллический излом *m* 40-705
интерметаллид *m* 35-215
интерметаллидная фаза *f* 35-210
интерметаллидное соединение *n* 35-215
интерметаллическая фаза *f* 35-210
интерметаллическое соединение *n* 35-215
интерференционная линия *f* 80-645
интерференционная микроскопия *f* 80-295
интерференционное пятно *n* 80-660
интерферограмма *f* 80-300
иод *m* 05-210
ионизационный потенциал *m* 10-105
ионная связь *f* 10-325
ионное азотирование *n* 70-800
ионное внедрение *n* 70-805
ионное соединение *n* 35-230
ионное травление *n* 80-180
ионный кристалл *m* 20-025
ионный радиус *m* 10-090
иридий *m* 05-215
  осмиевый ~ 65-845
  осмистый ~ 65-845
  платина-~ 65-835
  платинистый ~ 65-835
иридистый осмий *m* 65-845
искажение *n* решётки 30-025
искажённость *f* решётки 30-025
искровая проба *f* 80-440
искусственное старение *n* 70-590
испущенные дислокации *fpl* 30-370
испытание *n* без разрушения 80-785
  ~ механических свойств 80-400
  ~ на выносливость 80-420
  ~ на излом 80-445
  ~ на искру 80-440
  ~ на растяжение 80-405
  ~ на сезонное (само)растрескивание 80-485
  ~ на твёрдость 80-415
  ~ на удар 80-430
  ~ на усталость 80-420
  ~ прогностных свойств 80-400
  ~ твёрдости 80-415
  ~ ультразвуком 80-800
  калориметрическое ~ 80-500
  механическое ~ 80-400
  неразрушающее ~ 80-785
  ударное ~ 80-430
исследование *n*
  дилатометрическое ~ 80-540
  дюрометрическое ~ 80-415
  калориметрическое ~ 80-500
  макроскопическое ~ 80-045
  металлографическое ~ 80-035
  микроскопическое ~ 80-040
  радиографическое ~ 80-705
  рентгенографическое ~ 80-570
  электронно-микроскопическое ~ 80-330
истинная закалка *f* 70-570
истинный предел *m* длительной прочности 75-225
истинный предел *m* ползучести 75-225
истинный предел *m* прочности 75-130
источник *m* дислокаций 30-425
  ~ Кёлера 30-435
  ~ Франка-Рида 30-430
  дислокационный ~ 30-425
исходная фаза *f* 35-290
исходное аустенитное зерно *n* 50-040
иттербий *m* 05-505
иттрий *m* 05-510

К

кадмиевая бронза *f* 65-190
кадмий *m* 05-075
калий *m* 05-350
калифорний *m* 05-090
калоризация *f* 70-925
калориметрическое испытание *n* 80-500
калориметрическое исследование *n* 80-500
кальциевый баббит *m* 65-685
кальций *m* 05-085
канавка *f* термического травления 80-195
каналирование *n* 30-500
канталь *m* 65-525
капельная ликвация *f* 25-430
капельный анализ *m* 80-465
карандашное скольжение *n* 45-175
карбид *m* 40-430
  двойной ~ 40-475
  ~ ванадия 40-460
  ~ вольфрама 40-450
  ~ железа 40-440
  ~ металла 40-435
  ~ ниобия 40-465
  ~ хрома 40-455
  простой ~ 40-470
  сложный ~ 40-475
  специальный ~ 40-445
  спечённый ~ 55-665
карбидная сетка *f* 50-445
карбидообразующий элемент *m* 50-435
карбидостабилизирующий элемент *m* 50-465
карбиды *mpl*
  нерастворённые ~ 40-480
  остаточные ~ 40-480
  эвтектические ~ 50-440
карбонильное железо *n* 60-020
карбонильный никель *m* 65-425
карбонитрид *m* 40-485
карбюризатор *m* 70-740
касательное напряжение *n* 45-085
каскад *m* смещений 30-520
  ~ столкновений 30-515
катализатор *m* зарождения 25-125
катодная медь *f* 65-030
катодный алюминий *m* 65-550
каустическая хрупкость *f* 75-305
качественная сталь *f* 60-140
квазибинарная система *f* 35-130
квазибинарный сплав *m* 55-315
квазиизотропия *f* 20-455
квазиэвтектика *f* 40-225
квантовое состояние *n* 10-120
квантовое число *n* 10-125
кермет *m* 55-675
кипящая сталь *f* 60-650
кислая сталь *f* 60-615
кислород *m* 05-320
кислородно-конвертерная сталь *f* 60-605
кислотостойкая сталь *f* 60-545
кислотостойкий сплав *m* 55-485
кислотостойкий чугун *m* 60-1075
кислотостойкость *f* 75-695
кислотоупорная сталь *f* 60-545
кислотоупорность *f* 75-695
кислотоупорный сплав *m* 55-485

КИСЛОТОУПОРНЫЙ

кислотоупорный чугун *m* 60-1075
клапанная бронза *f* 65-275
клапанная сталь *f* 60-735
класс *m* симметрии 20-050
клубок *m* дислокаций 30-385
кобальт *m* 05-115
кобальтовая сталь *f* 60-240
ковалентная связь *f* 10-330
ковалентный кристалл *m* 20-030
кованая сталь *f* 60-330
ковар *m* 65-520
ковкий сплав *m* 55-530
ковкий чугун *m* 60-1110
ковкость *f* 75-725
когезия *f* 10-310
когенит *m* 50-115
когерентная граница *f* 30-625
когерентность *f* 30-620
когерентные выделения *npl* 40-320
когерентный зародыш *m* 25-155
коксовый чугун *m* 60-870
колебательная энтропия *f* 15-205
колебательный отжиг *m* 70-520
количественная металлография *f* 80-020
количественный термодинамический анализ *m* 80-535
колокольная бронза *f* 65-250
кольцо *n*
  дебаевское ~ 80-640
  дислокационное ~ 30-340
  ~ Дебая 80-640
комбинированный материал *m* 55-680
компактный металл *m* 55-155
композит *m* 55-680
композитный материал *m* 55-680
композиционный материал *m* 55-680
компонент *m* 35-035
  ~ сплава 55-180
  легирующий ~ 55-180
  основной ~ (сплава) 55-185
конвертерная медь *f* 65-015
конвертерная сталь *f* 60-590
конгруэнтная интерметаллическая фаза *f* 35-220
конгруэнтное интерметаллическое соединение *n* 35-220
конгруэнтное плавление *n* 35-460
конгруэнтное превращение *n* 25-525
конденсация *f* вакансий 30-115
конденсированная фаза *f* 35-145
конода *f* 35-695
консервативное движение *n* 30-290
консервативное перемещение *n* 30-290
константа *f* анизотропии 75-595
  ~ кристаллической решётки 20-120
  ~ термодинамического равновесия 15-335
  ~ упругости 75-145
  упругая ~ 75-145
константан *m* 65-450
конструкционная сталь *f* 60-670
контактная закалка *f* 70-250
контактная коррозия *f* 85-320
контактная усталость *f* 85-420
контактный сплав *m* 55-585
контраст *m*
  ориентационный ~ 80-155
  фазовый ~ 80-275
контрастная авторадиография *f* 80-745

контролируемая атмосфера *f* 70-700
контур *m* Бюргерса 30-185
конфигурационная энергия *f* 15-210
концентрационная диффузия *f* 45-320
концентрационная неоднородность *f* 40-415
концентрационное переохлаждение *n* 25-255
концентрационные флюктуации *fpl* 25-100
концентрационный градиент *m* 15-080
концентрационный треугольник *m* 35-720
концентрация *f* 55-230
  весовая ~ 55-235
  ~ по весу 55-235
  молярная ~ 55-240
  равновесная ~ 35-255
  средняя ~ 55-245
  электронная ~ 35-285
координационная зона *f* 20-190
координационная сфера *f* 20-190
координационное число *n* 20-185
корабельная сталь *f* 60-755
корковая зона *f* 40-175
корковый слой *m* 40-175
коробление *n* 85-200
корпус *m* атома 10-040
корродируемость *f* 75-675
коррозиеустойчивая сталь *f* 60-535
коррозионная стойкость *f* 75-670
коррозионная усталость *f* 85-415
коррозионная чувствительность *f* 75-675
коррозионная язва *f* 85-350
коррозионностойкая сталь *f* 60-535
коррозионностойкий сплав *m* 55-480
коррозионноустойчивая сталь *f* 60-535
коррозия *f* 85-260
  атмосферная ~ 85-275
  булавочная ~ 85-345
  внутренняя ~ 85-355
  водяная ~ 85-280
  газовая ~ 85-295
  гальваническая ~ 85-270
  избирательная ~ 85-305
  интеркристаллитная ~ 85-360
  контактная ~ 85-320
  ~ по границам зёрен 85-360
  ~ под напряжением 85-310
  ~ при трении 85-325
  локальная ~ 85-340
  межзёренная ~ 85-360
  межкристаллитная ~ 85-360
  местная ~ 85-340
  морская ~ 85-290
  питтинговая ~ 85-350
  поверхностная ~ 85-330
  подповерхностная ~ 85-355
  почвенная ~ 85-285
  равномерная ~ 85-335
  радиационная ~ 85-365
  селективная ~ 85-305
  сухая ~ 85-295
  точечная ~ 85-345
  усталостная ~ 85-315
  химическая ~ 85-265
  щелевая ~ 85-300
  электрохимическая ~ 85-270
  ядерная ~ 85-365
  язвенная ~ 85-350
котельная сталь *f* 60-740
коттрелевская атмосфера *f* 30-440
коттрелевское облако *n* 30-440
коэрцитивная сила *f* 75-635
коэффициент *m* активности 15-285
  ~ взаимодействия 15-295

КРИТИЧЕСКИЙ

коэффициент
~ диффузии 45-395
~ заполнения 20-205
~ компактности 20-205
~ линейного расширения 75-455
~ намагниченности 75-625
~ объёмного расширения 75-460
~ отражения 75-515
~ передачи 70-755
~ передачи углерода 70-755
~ Пуассона 45-170
~ распределения 25-260
~ самодиффузии 45-405
~ сжимаемости 75-165
~ теплового расширения 75-450
~ теплопроводности 75-440, 75-445
~ термического расширения 75-450
~ упаковки 20-205
~ упрочнения 45-065
парциальный ~ диффузии 45-390
равновесный ~ распределения 25-260
собственный ~ диффузии 45-390
температурный ~ удельного сопротивления 75-480
температурный ~ электрического сопротивления 75-480
температурный ~ электросопротивления 75-480
теоретический ~ распределения 25-260
термический ~ объёмного расширения 75-460
фактический ~ распределения 25-265
эффективный ~ распределения 25-265
краевая дислокация f 30-150
красная латунь f 65-395
красная медь f 65-025
красное золото n 65-800
красноломкость f 75-290
красностойкость f 75-260
красящий травитель m 80-105
краудион m 30-470
кремний m 05-415
кремнистая бронза f 65-145
кремнистая латунь f 65-365
кремнистая медь f 65-080
кремнистая сталь f 60-200
кремнистое железо n 60-210
кремнистый феррит f 50-475
кремнистый чугун m 60-1010
крепкий припой m 65-720
кривая f
граничная ~ 35-535
дилатометрическая ~ 80-555
~ Велера 80-425
~ выносливости 80-425
~ ликвидус 35-515
~ нагрева 80-520
~ нагревания 80-520
~ напряжение-деформация 80-410
~ охлаждения 80-525
~ ползучести 45-515
~ прокаливаемости 80-455
~ растворимости 35-530
~ солидус 35-490
~ упрочнения 45-070
~ усталости 80-425
спинодальная ~ 15-190
кривые fpl
сопряжённые ~ 35-540
крип m 45-495
крипоустойчивая сталь f 60-565
крипоустойчивость f 75-215
криптон m 05-225
кристалл m 20-015
двойниковый ~ 45-230
дендритный ~ 25-315
ёлочный ~ 25-315
затравочный ~ 25-130
идеальный ~ 30-005
идиоморфный ~ 25-355

кристалл
ионный ~ 20-025
ковалентный ~ 20-030
~ твёрдого раствора 40-195
материнский ~ 40-295
матричный ~ 40-295
металлический ~ 20-020
нитевидный ~ 25-360
одиночный ~ 20-430
полярный ~ 20-025
разрывной ~ 25-315
реальный ~ 30-010
совершенный ~ 30-005
бикристалл 20-435
монокристалл 20-430
поликристалл 20-440
кристаллизация f 25-010
вторичная ~ 25-060
направленная ~ 25-035
неравновесная ~ 25-020
объёмная ~ 25-075
одноосная ~ 25-040
ориентированная ~ 25-035
первичная ~ 25-050
равновесная ~ 25-015, 25-075
самопроизвольная ~ 25-045
спонтанная ~ 25-045
столбчатая ~ 25-065
управляемая ~ 25-030
эвтектическая ~ 25-025
рекристаллизация 45-440
транскристаллизация 25-070
электрокристаллизация 25-080
кристаллит m 40-360
кристаллическая ориентировка f 20-415
кристаллическая решётка f 20-045
кристаллическая система f 20-235
кристаллическая структура f 20-040
кристаллический 20-010
кристаллический излом m 40-685
кристаллическое строение n 20-040
кристаллографическая ориентированность f 20-415
кристаллографическая ориентировка f 20-415
кристаллографическая ось f 20-125
кристаллографическая плоскость f 20-155
кристаллографическая система f 20-235
кристаллографический индекс m 20-160
кристаллографическое направление n 20-150
кристаллография f 20-035
кристаллы mpl
вторичные ~ 40-275
изоморфные ~ 20-480
первичные ~ 40-270
равноосные ~ 40-285
столбчатые ~ 40-280
шестоватые ~ 40-280
критическая деформация f 45-075
критическая скорость f закалки 70-180
критическая скорость f охлаждения 70-180
критическая степень f деформации 35-075
критическая температура f 35-335, 35-555
критическая температура f хрупкости 75-280
критическая точка f 35-555
критический диаметр m 70-360
критический зародыш m 25-140
критический интервал m 35-560
критический радиус m 25-145
критический радиус m зародыша 25-145

КРИТИЧЕСКОЕ

критическое напряжение n сдвига 45-090
критическое скалывающее напряжение n 45-090
кричное железо n 60-050
крупное зерно n 40-365
крупнозернистая сталь f 60-475
крупнозернистая структура f 40-390
крупнозернистое строение n 40-390
крупнозернистость f 40-375
крупнозернистый излом m 40-695
крупноигольчатый мартенсит m 50-320
крупнокристаллическая структура f 40-390
крупнокристаллический излом m 40-695
крупнопластинчатый мартенсит m 50-320
ксенон m 05-500
кубическая гранецентрированная решётка f 20-310
кубическая объёмноцентрированная решётка f 20-315
кубическая решётка f 20-300
кубическая решётка f типа алмаз 20-320
кубическая система f 20-240
кубическая текстура f 40-605
кубическая центрированная решётка f 20-315
кубический мартенсит m 50-340
кюрий m 05-125

Л

лантан m 05-230
латентное упрочнение n 45-180
латунно-никелевый припой m 65-745
латунный припой m 65-740
латунь f 65-290
  автоматная ~ 65-370
  адмиралтейская ~ 65-390
  альфа-~ 65-325
  альфа+бета-~ 65-330
  алюминиевая ~ 65-340
  бета-~ 65-335
  ванадиевая ~ 65-180
  гильзовая ~ 65-380
  двухфазная ~ 65-320
  двухфазная морская ~ 65-385
  деформируемая ~ 65-295
  красная ~ 65-395
  кремнистая ~ 65-365
  α-~ 65-325
  α+β-~ 65-330
  β-~ 65-335
  литейная ~ 65-300
  марганцовая ~ 65-355
  марганцовистая ~ 65-355
  машинная ~ 65-375
  никелевая ~ 65-350
  однофазная ~ 65-315
  однофазная морская ~ 65-390
  оловянистая ~ 65-345
  оловянная ~ 65-345
  патронная ~ 65-380
  простая ~ 65-305
  свинцовая ~ 65-360
  свинцовистая ~ 65-360
  специальная ~ 65-310
  спецлатунь 65-310
лаутадь m 65-595
лауэграмма f 80-620
левовинтовая дислокация f 30-175
легирование n 55-170
  диффузионное ~ 70-655
  поверхностное ~ 70-655
легированная литая сталь f 60-835

легированная система f 55-165
легированная сталь f 60-115
легированный цементит m 50-110
легированный чугун m 60-935, 60-995
легируемость f 55-175
легирующая добавка f 55-190
легирующая микродобавка f 55-250
легирующий элемент m 55-190
легирующий компонент m 55-180
лёгкие металлы mpl 55-050
лёгкие платиновые металлы mpl 65-830
лёгкий сплав m 55-405
лёгкое скольжение n 45-155
легкообрабатываемая сталь f 60-710
легкоплавкий металл m 55-060
легкоплавкий припой m 65-715
легкоплавкий сплав m 55-395
ледебурит m 50-160
ледебуритная сталь f 60-440
ледебуритная эвтектика f 50-255
лес m дислокаций 30-405
дислокационный ~ 30-405
летучесть f 15-290
  термодинамическая ~ 15-290
лигатура f 55-285
ликвационная полоса f 85-230
ликвация f 25-400
  вторичная ~ 25-410
  газовая ~ 25-430
  зональная ~ 25-415
  дендритная ~ 25-420
  капельная ~ 25-430
  ~ по слитку 25-415
  ~ по удельному весу 25-425
  ~ у газового пузыря 25-430
  междендритная ~ 25-445
  нормальная ~ 25-435
  обратная ~ 25-440
  первичная ~ 25-405
  простая ~ 25-435
  прямая ~ 25-435
  макроликвация 25-415
  микроликвация 25-420
ликвидус m 35-510
линейная дислокация f 30-150
линейный дефект m 30-140
линейчатая структура f 40-165
линии fpl двойникования 45-255
  ~ Людерса-Чернова 45-195
  ~ Неймана 45-205
  основные ~ 80-650
  остановочные ~ 40-750
  сверхструктурные ~ 80-655
  сопряжённые ~ 35-540
линия f
  дислокационная ~ 30-195
  дифракционная ~ 80-645
  интерференционная ~ 80-645
  ~ дислокации 30-195
  ~ изоактивности 15-315
  ~ ликвидус 35-515
  ~ ликвидуса 35-515
  ~ монотектики 35-650
  ~ перегиба 15-190
  ~ предельной растворимости 35-530
  ~ сдвига 45-185
  ~ скольжения 45-185
  ~ солидус 35-490
  ~ солидуса 35-490
  ~ точек перегиба 15-190
  монотектическая ~ 35-650
  перитектическая ~ 35-645
  эвтектическая ~ 35-640
  эвтектоидная ~ 35-655
линотипный сплав m 55-625
литая сталь f 60-580, 60-825

литая структура f 40-050
литейная бронза f 65-100
литейная латунь f 65-300
литейная сталь f 60-825
литейные напряжения npl 85-170
литейные свойства npl 75-775
литейный дефект m 85-095
литейный порок m 85-095
литейный сплав m 55-535
литейный чугун m 60-885
литий m 05-245
литое железо n 60-060
лишняя полуплоскость f 30-200
локальная коррозия f 85-340
ломкость f
горячеломкость 75-290
красноломкость 75-290
синеломкость 75-300
теплоломкость 75-290
хладноломкость 75-285
лоуренций m 05-235
лютеций m 05-250

М

магналий m 65-620
магнетострикция f 75-650
магниевый чугун m 60-975
магний m 05-255
магнитная анизотропия f 75-575
магнитная восприимчивость f 75-625
магнитная дефектоскопия f 80-805
магнитная индукция f 75-610
магнитная область f 40-635
магнитная подрешётка f 20-485
магнитная проницаемость f 75-615
магнитная сталь f 60-495
магнитная структура f 40-650
магнитно-порошковая дефектоскопия f 80-805
магнитное квантовое число n 10-140
магнитное последействие n 75-645
магнитное превращение n 25-555
магнитное старение n 70-620
магнитное упорядочение n 25-640
магнитные свойства npl 75-540
магнитный анализ m 80-560
магнитный гистерезис m 75-640
магнитный домен m 40-635
магнитный сплав m 55-445
магнитозадерживающая сила f 75-635
магнитомягкая сталь f 60-500
магнитомягкий сплав m 55-460
магнитотвёрдая сталь f 60-505
магнитотвёрдый сплав m 55-465
макрография f 80-305
макроликвация f 25-415
макроскопические напряжения npl 85-030
макроскопические свойства npl 75-850
макроскопическое исследование n 80-045
макростроение n 40-025
макроструктура f 40-025
макрофотография f 80-310, 80-315
малолегированная сталь f 60-120
малооловянистая бронза f 65-120
малооловянная бронза f 65-120

малоуглеродистая сталь f 60-100
малоуглеродистый чугун m 60-1035
малоугловая граница f 30-590
манганин m 65-405
магранец m 05-260
сернистый ~ 50-240
марганцевая бронза f 65-155
марганцевая сталь f 60-195
марганцевый чугун m 60-1030
марганцовая латунь f 65-355
марганцовистая бронза f 65-155
марганцовистая латунь f 65-355
марганцовистая медь f 65-085
марганцовистая сталь f 60-195
марганцовистый чугун m 60-1030
марганцовокремнистая сталь f 60-205
марка f стали 60-085
марочное обозначение n 60-090
мартеновская сталь f 60-610
мартеновский чугун m 60-855
мартенсит m 50-310
безыгольчатый ~ 50-330
бесструктурный ~ 50-330
вторичный ~ 50-350
деформационный ~ 50-360
игольчатый ~ 50-315
крупноигольчатый ~ 50-320
крупнопластинчатый ~ 50-320
кубический ~ 50-340
~ деформации 50-360
~ закалки 50-335
~ наклёпа 50-360
~ отпуска 50-345
массивный ~ 50-355
мелкий ~ 50-325
мелкоигольчатый ~ 50-325
отпущенный ~ 50-345
реечный ~ 50-355
скрытокристаллический ~ 50-330
тетрагональный ~ 50-335
тонкоигольчатый ~ 50-325
мартенситная игла f 50-375
мартенситная область f 50-300
мартенситная пластинка f 50-370
мартенситная сталь f 60-430
мартенситная точка f 70-165
мартенситно-стареющая сталь f 60-460
мартенситное превращение n 50-280
мартенситный чугун m 60-1055
масляная закалка f 70-210
масс-спектрография f 80-480
масса f
атомная ~ 10-075
молекулярная ~ 75-355
удельная ~ 75-345
массивное превращение n 25-550
массивный мартенсит m 50-355
массивный образец m 80-055
массовое число n 75-335
масштабный фактор m 10-395
масштабный эффект m 10-395
материал m
комбинированный ~ 55-680
композитный ~ 55-680
композиционный ~ 55-680
металлокерамический ~ 55-675
основной ~ 25-165
материнский кристалл m 40-295
маточная фаза f 35-290
маточный расплав m 25-095
маточный раствор m 25-095
матрица f 40-015
матричный кристалл m 40-295

машинная бронза f 65-240
машинная латунь f 65-375
машиноподелочная сталь f 60-680
машиностроительная сталь f 60-680
маятниковый отжиг m 70-520
мгновенное охлаждение n 70-075
медальная бронза f 65-260
медистая сталь f 60-245
медленное охлаждение n 70-080
медноникелевый сплав m 65-400
медный сплав m 65-010
медь f 05-120
анодная ~ 65-040
бескислородная ~ 65-060
бескислородная высокопроводящая ~ 65-070
бескислородная ~ высокой проводимости 65-070
катодная ~ 65-030
конверторная ~ 65-015
красная ~ 65-025
кремнистая ~ 65-080
марганцовистая ~ 65-085
~ раскислённая фосфором 65-055
~ рафинированная в пламенной печи 65-025
~ с добавкой хрома 65-200
мышьяковистая ~ 65-090
нераскислённая ~ 65-050
рафинированная ~ 65-020
сурьмянистая ~ 65-165
фосфористая ~ 65-075
цементационная ~ 65-045
цементная ~ 65-045
чёрная ~ 65-015
черновая ~ 65-015
электролитическая ~ 65-035
электролитная ~ 65-035
электротехническая ~ 65-065
межатомное расстояние n 20-230
междендритная ликвация f 25-445
междендритная сегрегация f 25-445
междендритные полости fpl 40-355
междендритные пространства npl 40-355
междендритный графит m 50-205
междоузельный атом m 30-075
медоузлие n 20-105
междуатомное расстояние n 20-230
междуузельная диффузия f 45-315
междуузловая позиция f 20-145
междуузловое положение n 20-145
межзёренная граница f 30-565
межзёренная диффузия f 45-310
межзёренная коррозия f 85-360
межзёренное скольжение n 45-160
межзёренные выделения npl 40-310
межзёренный излом m 40-705
межкристаллитная коррозия f 85-360
межкристаллитный излом n 40-705
межкристаллический излом m 40-705
межпластиночное расстояние n 40-110
межплоскостное расстояние n 20-225
межузельный атом m 30-075
межфазная граница f 30-610
межфазная поверхность f 30-610
межфазовая энергия f 30-635
мелкий мартенсит m 50-325
мелкодисперсная фаза f 40-330
мелкое зерно n 40-370
мелкозернистая корка f 40-175
мелкозернистая сталь f 60-470
мелкозернистая структура f 40-395
мелкозернистое строение n 40-395
мелкозернистость f 40-380

мелкозернистый излом m 40-700
мелкоигольчатый мартенсит m 50-325
мелкокристаллическая структура f 40-395
мелкопластинчатый перлит m 50-135
менделевий m 05-265
местная закалка f 70-275
местная коррозия f 85-340
местная цементация f 70-735
местное науглероживание n 70-735
металл m 10-355, 10-360
адмиралтейский ~ 65-390
белый ~ 65-640
благородный ~ 55-010
вакуумированный ~ 55-135
вторичный ~ 55-120
делящийся ~ 55-080
драгоценный ~ 55-010
компактный ~ 55-155
легкоплавкий ~ 55-060
~ второго рода 10-370
~ Вуда 65-670
~ конденсированный в вакууме 55-145
~ очищенный зонной плавкой 55-140
~ Розе 65-675
~ сверхвысокой чистоты 55-105
неблагородный ~ 55-005
первичный ~ 55-115
полублагородный ~ 55-020
радиоактивный ~ 55-075
расплавленный ~ 55-150
рафинированный ~ 55-125
реакторный ~ 55-085
реакционноспособный ~ 55-070
самородный ~ 55-110
сверхчистый ~ 55-105
спечённый ~ 55-660
тугоплавкий ~ 55-065
чёрный ~ 60-070
чистый ~ 55-090
электролитический ~ 55-130
металлизация f
диффузионная ~ 70-915
термодиффузионная ~ 70-915
металлическая связь f 10-340
металлическая система f 55-165
металлическая фаза f 35-150
металлические включения npl 40-550
металлические свойства npl 10-350
металлический блеск m 75-510
металлический звук m 75-360
металлический кристалл m 20-020
металлический раствор m 35-385
металлический сплав m 55-160
металлический элемент m 10-360
металлическое состояние n 10-345
металловедение n 10-005
физическое ~ 10-005
металлографический анализ m 80-035
металлографический микроскоп m 80-210
металлографическое исследование n 80-035
металлография f 80-005
высокотемпературная ~ 80-015
количественная ~ 80-020
рентгеновская ~ 80-575
цветная ~ 80-030
электронномикроскопическая ~ 80-025
металлоид m 10-365
металлокерамика f 55-675
металлокерамическая бронза f 65-225
металлокерамический материал m 55-675
металлокерамический сплав m 55-655
металлокерамическое железо n 60-065

МОЛЕКУЛЯРНЫЙ

металломикроскоп *m* 80-210
металломикроскопия *f* 80-010
металлорадиография *f* 80-710
металлофизика *f* 10-015
металлургические свойства *npl* 75-855
металлы *mpl*
  лёгкие ~ 55-050
  лёгкие платиновые ~ 65-830
  ~ переходных групп 55-025
  ~ платиновой группы 65-820
  переходные ~ 55-025
  платиновые ~ 65-820
  рассеянные ~ 55-045
  редкие ~ 55-045
  редкоземельные ~ 55-040
  тяжёлые ~ 55-055
  тяжёлые платиновые ~ 65-825
  цветные ~ 55-015
  щёлочноземельные ~ 55-035
  щелочные ~ 55-030
метастабильная фаза *f* 35-175
метастабильное равновесие *n* 35-060
метастабильное состояние *n* 35-070
метатектика *f* 40-255
метатектическая реакция *f* 35-580
метатектическая температура *f* 35-620
метатектическое превращение *n* 35-580
метеоритное железо *n* 60-005
метод *m* вращающегося кристалла 80-600
  ~ Дебая 80-595
  ~ качающегося кристалла 80-605
  ~ колебательного кристалла 80-605
  ~ Лауэ 80-590
  ~ магнитных порошков 80-810
  ~ магнитных суспензий 80-815
  ~ на просвет 80-625
  ~ неподвижного кристалла 80-590
  ~ обратной съёмки 80-630
  ~ просвечивания 80-625
  порошковый ~ 80-595
механическая полировка *f* 80-070
механическая прочность *f* 75-040
механическая смесь *f* 40-190
механическая смесь *f* фаз 40-190
механические свойства *npl* 75-005, 75-015
механические свойства *npl*
  определённые испытанием на растяжение 70-015
механический двойник *m* 45-235
механическое испытание *n* 80-400
механическое старение *n* 45-225
миграция *f* границ (зёрен) 45-490
  термомиграция 45-370
  электромиграция 45-365
микроанализ *m* 80-040
микрография *f* 80-245
микродобавка *f* 55-250
  легирующая ~ 55-250
микролегирующая добавка *f* 55-250
микролегирующий элемент *m* 55-250
микроликвация *f* 25-420
микронапряжения *npl* 85-035
микроползучесть *f* 45-530
микропористость *f* 85-115
микропоры *fpl* 85-110
микропустоты *fpl* 85-110
микрорадиограмма *f* 80-725
микрорадиография *f* 80-720
микрорентгенография *f* 80-730
микроскоп *m*
  металлографический ~ 80-210
  металломикроскоп 80-210
микроскопический анализ *m* 80-040

микроскопическое исследование *n* 80-040
микроскопия *f*
  автоионная ~ 80-380
  автоэмиссионная ~ 80-335
  высокотемпературная ~ 80-240
  интерференциальная ~ 80-295
  оптическая ~ 80-215
  отражательная ~ 80-225
  поляризационная ~ 80-260
  просвечивающая ~ 80-220
  просвечивающая электронная ~ 80-340
  растровая электронная ~ 80-360
  рентгеновская ~ 80-730
  светлопольная ~ 80-235
  световая ~ 80-215
  сканирующая ~ 80-360
  сканирующая электронная ~ 80-360
  темнопольная ~ 80-230
  температурная ~ 80-240
  трансмиссионная ~ 80-220
  ультрафиолетовая ~ 80-265
  фазоконтрастная ~ 80-270
  электронная ~ 80-330
  эмиссионная электронная ~ 80-335
  металломикроскопия 85-010
микросоставляющая *f* 40-010
микростроение *n* 40-030
микроструктура *f* 40-030
микроструктурное равновесие *n* 40-020
микротвёрдость *f* 75-250
микротрещина *f* 85-045
микрофотография *f* 80-250, 80-255
  электронная ~ 80-365
микрофрактограмма *f* 80-395
микрофрактография *f* 80-390
микрошлиф *m* 80-095
микроэлемент *m* 55-250
миллеровские индексы *mpl* 20-165
мишметалл *m* 65-865
многокомпонентная система *f* 35-140
многокомпонентная сталь *f* 60-170
многокомпонентный раствор *m* 35-410
многокомпонентный сплав *m* 55-310
многократное скольжение *n* 45-140
многокристаллическое тело *n* 20-440
многофазная диффузия *f* 45-325
многофазная система *f* 35-110
многофазная структура *f* 40-160
многофазный сплав *m* 55-345
множественное скольжение *n* 45-140
модификатор *m* 25-230
модификация *f*
  аллотропическая ~ 25-390
  высокотемпературная ~ 25-395
  полиморфная ~ 25-390
модифицирование *n* 45-220, 25-225
модифицирование *n* силумина 25-225
модифицированный силумин *m* 65-570
модифицированный чугун *m* 60-970
модуль *m* объёмного сжатия 75-160
  ~ всестороннего сжатия 75-160
  ~ объёмной упругости 75-160
  ~ поперечной упругости 75-155
  ~ продольной упругости 75-150
  ~ сдвига 75-155
  ~ сжимаемости 75-160
  ~ Юнга 75-150
мозаичная область *f* 30-655
мозаичная структура *f* 40-170
мозаичный блок *m* 30-655
молекулярная масса *f* 75-355
молекулярная связь *f* 10-335
молекулярная теплоёмкость *f* 75-375
молекулярный вес *m* 75-355

молибден *m* 05-275
беспровесный ~ 65-870
непровисающий ~ 65-870
молибденовая сталь *f* 60-220
мольный объём *m* 75-325
молярная величина *f* 15-050
молярная концентрация *f* 55-240
мондникель *m* 65-425
монель *m* 65-435
~-металл 65-435
монетная бронза *f* 65-265
монетное золото *n* 65-775
монетное серебро *n* 65-755
монетный сплав *m* 55-565
моновакансия *f* 30-040
моновариантная система *f* 35-080
моноклинная решётка *f* 20-395
моноклинная система *f* 20-265
монокристалл *m* 20-430
монотектика *f* 40-260
монотектическая горизонталь *f* 35-650
монотектическая линия *f* 35-650
монотектическая реакция *f* 35-575
монотектическая температура *f* 35-615
монотектическая точка *f* 35-670
монотектическое превращение *n* 35-575
монотектоидная реакция *f* 35-600
монотектоидное превращение *n* 35-600
монотипный сплав *m* 55-620
морская коррозия *f* 85-290
мощность *f* дислокации 30-210
мышьяк *m* 05-035
мышьяковистая медь *f* 65-090
мягкая сталь *f* 60-340
мягкие пятна *npl* 85-195
мягкий припой *m* 65-715
мягкий свинец *m* 65-655

Н

нагартованная сталь *f* 60-295
нагартовка *f* 45-030
нагрев *m* 70-030
импульсный ~ 70-050
предварительный ~ 70-045
сквозной ~ 70-035
нагревание *n* 70-030
наиболее плотноупакованная плоскость *f* 20-215
наклёп *m* 45-030
внутренний ~ 45-080
фазовый ~ 45-080
наклонная граница *f* 30-605
накопленная энергия *f* 45-060
намагниченность *f* 75-600
~ насыщения 75-605
остаточная ~ 75-630
наплавляемый твёрдый сплав *m* 55-590
направление *n*
кристаллографическое ~ 20-150
~ лёгкого намагничивания 75-585
~ легчайшего намагничивания 75-585
~ плотнейшей упаковки 20-220
~ скольжения 45-105
направленная кристаллизация *f* 25-035
направленное затвердевание *n* 25-035
напряжение *n*
деформирующее ~ 45-210
касательное ~ 45-085
критическое ~ сдвига 45-090

напряжение
критическое скалывающее ~ 45-090
~ сдвига 45-085
~ течения 45-210
сдвиговое ~ 45-085
скалывающее ~ 45-085
разрушающее ~ 75-130
сдвиговое ~ 45-085
напряжения *npl*
внутренние ~ 85-020
внутренние ~ второго рода 85-035
внутренние ~ первого рода 85-030
закалочные ~ 85-185
литейные ~ 85-170
макроскопические ~ 85-030
~ второго рода 85-035
~ первого рода 85-030
~ I рода 85-030
объёмные ~ 85-030
остаточные ~ 85-020
собственные ~ 85-020
структурные ~ 85-035
температурные ~ 85-025
тепловые ~ 85-025
термические ~ 85-025
усадочные ~ 85-175
эпитаксиальные ~ 25-345
микронапряжения 85-035
нарушение *n* сплошности материала 85-090
наследственно крупнозернистая сталь *f* 60-475
наследственно мелкозернистая сталь *f* 60-470
наследственность *f* чугуна 60-985
насыщение *n*
диффузионное ~ 70-655
диффузионное ~ металлами 70-915
насыщенный твёрдый раствор *m* 35-395
натрий *m* 05-425
науглероженная сталь *f* 60-310
науглероживание *n* 70-715
дифференциальное ~ 70-735
местное ~ 70-735
~ в порошке 70-720
~ газами 70-730
~ порошком 70-720
науглероживающая атмосфера *f* 70-695
науглероживающий газ *m* 70-745
нафталинистый излом *m* 40-740
нафталиновый излом *m* 40-740
начальная магнитная проницаемость *f* 75-620
начальная проницаемость *f* 75-620
неблагородный металл *m* 55-005
неглубоко прокаливающаяся сталь *f* 60-380
негомогенная система *f* 35-095
негомогенность *f* 40-410
негомогенный сплав *m* 55-340
недеформирующаяся сталь *f* 60-410
незакаливающаяся сталь *f* 60-390
неидеальный раствор *m* 15-245
нейзильбер *m* 65-415
нейтронная дефектоскопия *f* 80-830
нейтронография *f* 80-830
некогерентная граница *f* 30-630
некогерентные выделения *npl* 40-325
неконгруэнтная интерметаллическая фаза *f* 35-225
неконгруэнтное интерметаллическое соединение *n* 35-225
неконгруэнтное плавление *n* 35-465

неконсервативное движение n 30-285
неконсервативное перемещение n 30-285
некорродирующий сплав m 55-480
нелегированная сталь f 60-095
нелегированный чугун m 60-990
немагнитная сталь f 60-510
немагнитный сплав m 55-470
неметалл m 10-365
неметаллическая фаза f 35-155
неметаллические включения npl 40-515
неметаллический элемент m 10-365
неограниченная растворимость f 35-315
неограниченный твёрдый раствор m 35-390
неодим m 05-280
неоднородная система f 35-095
неоднородность f 40-410
  концентрационная ~ 40-415
  химическая ~ 40-415
неоднофазная диффузия f 45-325
неон m 05-285
неотожжённый ковкий чугун m 60-905
неповреждающий контроль m 80-785
неполная дислокация f 30-310
неполная закалка f 70-335
неполный отжиг m 70-470
непрерывное выделение n 25-375
непрерывное охлаждение n 70-090
непрерывное превращение n 25-540
непровисающий молибден m 65-870
непрозрачность f 75-505
нептуний m 05-290
неравновесная вакансия f 30-105
неравновесная кристаллизация f 25-020
неравновесная фаза f 35-185
неравновесное затвердевание n 25-020
неразрушающее испытание n 80-785
нераскислённая медь f 65-050
нерастворённые карбиды mpl 40-480
нерастворимость f 35-340
нерафинированный цинк m 65-690
нержавеющая сталь f 60-515
несвязанный углерод m 50-005
несвязанный электрон m 10-055
несмешиваемость f 35-340
несовершенная дислокация f 30-310
несовершенный кристалл m 30-010
несогласованность f решётки 30-615
несоответствие n решётки 30-615
несплошная эвтектика f 40-235
нестареющая сталь f 60-405
нестационарная ползучесть f 45-500
неупорядоченная фаза f 35-205
неупорядоченное состояние n 25-585
неупорядоченный твёрдый раствор m 35-435
неупрочняющее скольжение n 45-155
неупругость f 75-050
неустановившаяся ползучесть f 45-500
неустойчивая фаза f 35-175
неустойчивое равновесие n 35-060
неустойчивый аустенит m 50-265
нижний бейнит m 50-415
низкий отжиг m 70-525
низкий отпуск m 70-370
низколегированная сталь f 60-120

низколегированный сплав m 55-320
низколегированный чугун m 60-1000
низкооловянная бронза f 65-120
низкопроцентный сплав m 55-320
низкотемпературная обработка f 70-340
низкотемпературная термомеханическая обработка f 70-960
низкотемпературная хрупкость f 75-295
низкотемпературное цианирование n 70-875
низкотемпературный отпуск m 70-370
низкоуглеродистая сталь f 60-100
никелевая бронза f 65-160
никелевая латунь f 65-350
никелевая сталь f 60-185
никелевый сплав m 65-430
никелевый чугун m 60-1025
никелин m 65-410
никель m 05-295
  карбонильный ~ 65-425
никросилал m 60-1095
нимоник m 65-480
ниобид m 40-505
ниобий m 05-300
нирезист m 60-1090
нитал m 80-110
нитевидный кристалл m 25-360
нитраллой m 60-700
нитрид m 70-810
  ~ ванадия 70-830
  ~ железа 70-815
  ~ титана 70-825
  ~ хрома 70-820
нитридные включения npl 40-545
нитридообразователь m 50-470
нитридообразующий элемент m 50-470
нитрирование n 70-785
нитрированный слой m 70-845
нитрооксидирование n 70-885
нитроцементация f 70-870
нихром m 65-470
нобелий m 05-310
новое серебро n 65-415
ножевая сталь f 60-750
номер m
  атомный ~ 75-330
  ~ зерна 80-290
нонвариантная система f 35-075
нормализационный отжиг m 70-475
нормализация f 70-475
нормализованная сталь f 60-280
нормальная ликвация f 25-435
нормальная сегрегация f 25-435
НТМО 70-960
нульмерный дефект m 30-030

О

обеднённая зона f 30-120
обезводороживание n 70-555
обезуглероживание n
  ~ поверхности 85-210
  поверхностное ~ 85-210
обезуглероживающий отжиг m 70-635
обесцинкование n 85-380
облако n
  коттрелловское ~ 30-440
  ~ Коттрелла 30-440

облако
~ примесей 30-440
~ примесных атомов 30-440
область *f*
гомогенная ~ 35-680
двухфазная ~ 35-685
магнитная ~ 40-635
мартенситная ~ 50-300
мозаичная ~ 30-655
~ бейнитного превращения 50-305
~ гомогенности 35-680
~ закалки 70-145
~ несмешиваемости 35-345
~ плавления 35-545
~ превращения 35-560
однофазная ~ 35-680
перлитная ~ 50-295
промежуточная ~ 50-305
облик *m* кристалла 20-460
оболочка *f*
основная ~ 10-155
электронная ~ 10-155
подоболочка 10-160
оборотная бронза *f* 65-285
обрабатываемость *f* резанием 75-745
обработка *f*
высокотемпературная термомеханическая ~ 70-955
горячая ~ давлением 45-020
низкотемпературная ~ 70-340
низкотемпературная термомеханическая ~ 70-960
~ водяным паром 70-880
~ давлением 45-015
~ на аустенит 70-130
~ на твёрдый раствор 70-500
~ холодом 70-340
одинарная ~ 70-415
поверхностная ~ 70-010
собственно термическая ~ 70-006
термическая ~ 70-005
термомагнитная ~ 70-965
термомеханическая ~ 70-950
термохимическая ~ 70-650
упрочняющая ~ 70-025
химико-термическая ~ 70-650
холодная ~ давлением 45-025
термообработка 70-005
образец *m* 80-050, 80-435
массивный ~ 80-055
прутковый ~ 80-435
образование *n*
атермическое ~ 25-200
атермическое ~ зародышей 25-200
гетерогенное ~ 25-185
гетерогенное ~ зародышей 25-185
гомогенное ~ зародышей 25-180
динамическое ~ зародышей 25-205
~ зародышей 25-175
~ твёрдого раствора 25-055
самопроизвольное ~ зародышей 25-190
зародышеобразование 25-175
обратимое превращение *n* 25-495
обратная ликвация *f* 25-440
обратная рентгенограмма *f* 80-670
обратная решётка *f* 20-410
обратный отбел *m* чугуна 85-150
обыкновенная закалка *f* 70-270
обыкновенный чугун *m* 60-965, 60-990
обычная закалка *f* 70-270
обычный чугун *m* 60-965
объём *m*
активационный ~ 30-475
атомный ~ 75-320
мольный ~ 75-325
объёмная диффузия *f* 45-295
объёмная закалка *f* 70-280
объёмная кристаллизация *f* 25-075
объёмное заполнение *n* 20-205

объёмноцентрированная решётка *f* 20-295
объёмные напряжения *npl* 85-030
объёмные свойства *npl* 75-340
объёмный фактор *m* 10-395
ограниченная растворимость *f* 35-320
одинарная обработка *f* 70-415
одинарное скольжение *n* 45-130
одиночная вакансия *f* 30-040
одиночное скольжение *n* 45-130
одиночный кристалл *m* 20-430
одновариантная система *f* 35-080
однокомпонентная система *f* 35-115
однократное скольжение *n* 45-130
одномерный дефект *m* 30-140
одноосная кристаллизация *f* 25-040
одноосное затвердевание *n* 25-040
однородность *f* 40-405
односторонне-гранецентрированная ромбическая решётка *f* 20-385
однофазная диффузия *f* 45-330
однофазная латунь *f* 65-315
однофазная морская латунь *f* 65-390
однофазная область *f* 35-680
однофазная система *f* 35-090
однофазная структура *f* 40-145
однофазное поле *n* 35-680
однофазный сплав *m* 55-330
одонтологический сплав *m* 55-575
одонтологическое золото *n* 65-785
окалина *f* 85-430
окалиностойкая сталь *f* 60-560
окалиностойкий сплав *m* 55-505
окалиностойкость *f* 75-690
окисление *n* внутреннее 70-947
окислительный потенциал *m* 15-350
окисные включения *npl* 40-535
оксидное травление *n* 80-185
октаэдрическая пора *f* 20-330
октаэдрическая пустота *f* 20-330
октаэдрическая решётка *f* 20-315
олово *n* 05-475
белое ~ 65-630
серое ~ 65-635
оловянистая бронза *f* 65-110
оловянистая латунь *f* 65-343
оловянистый баббит *m* 65-645
оловянная бронза *f* 65-110
оловянная латунь *f* 65-345
оловянная чума *f* 85-245
оловянносвинцовистая бронза *f* 65-185
оловяннофосфористая бронза *f* 65-210
оловянноцинковая бронза *f* 65-205
оловянный баббит *m* 65-645
оловянный припой *m* 65-725
оловянный треск *m* 45-270
оловянный хруст *m* 45-270
оплавление *n* 85-215
определение *n* прокаливаемости торцовым методом 80-450
оптическая микроскопия *f* 80-215
оптические свойства *npl* 75-500
орбитальное квантовое число *n* 10-135
ориентационный контраст *m* 80-155
ориентирование *n* монокристалла 80-690
ориентированная кристаллизация *f* 25-035

ориентированное зарождение n 25-195
ориентированное затвердевание n 25-035
ориентированный зародыш m 25-160
ориентировка f
  беспорядочная ~ 20-420
  кристаллическая ~ 20-415
  кристаллографическая ~ 20-415
  преимущественная ~ 20-425
(орто)ромбическая решётка f 20-380
орудийная бронза f 65-245
осевая единица f 20-115
осмиевый иридий m 65-845
осмий m 05-315
  иридистый ~ 65-845
осмирид m 65-845
осмистый иридий m 65-845
основа f сплава 55-185
~ структуры 40-115
основная легирующая добавка f 55-195
основная оболочка f 10-155
основная сталь f 60-620
основной компонент m (сплава) 55-185
основной легирующий элемент m 55-195
основной материал m 25-165
основные линии fpl 80-650
особая точка f 80-495
особо мягкая сталь f 60-345
остановочные линии fpl 40-750
остаточная деформация f 35-010
остаточная индукция f 75-630
остаточная магнитная индукция f 75-630
остаточная намагниченность f 75-630
остаточное относительное сужение n 75-125
остаточное относительное удлинение n 75-120
остаточные карбиды mpl 40-480
остаточные напряжения npl 85-020
остаточный аустенит m 50-270
остаточный расплав m 25-280
остов m атома 10-040
остуживание n 70-080
ось f
  главная кристаллографическая ~ 25-300
  двойная ~ симметрии 20-070
  кристаллографическая ~ 20-125
  ~ второго порядка дендрита 25-305
  ~ двойникования 45-260
  ~ лёгкого намагничивания 75-585
  ~ легчайшего намагничивания 75-585
  ~ первого порядка дендрита 25-300
  ~ симметрии второго порядка 20-070
  ~ симметрии кристалла 20-065
  ~ симметрии третьего порядка 20-075
  ~ симметрии четвёртого порядка 20-080
  ~ симметрии шестого порядка 20-085
  четверная ~ симметрии 20-080
  шестерная ~ симметрии 20-085
отбел m 85-145
  обратный ~ чугуна 85-150
отбелённый чугун m 60-980
отбор m проб 80-060
отдых m 45-420
  динамический ~ 45-425
  статический ~ 45-425
отжиг m 70-420
  безокислительный ~ 70-430
  высокий ~ 70-480
  гомогенизационный ~ 70-485
  гомогенизирующий ~ 70-485
  графитизирующий ~ 70-630
  диффузионный ~ 70-485
  изотермический ~ 70-450

отжиг
  изохрональный ~ 70-495
  изохронный ~ 70-495
  колебательный ~ 70-520
  маятниковый ~ 70-520
  неполный ~ 70-470
  низкий ~ 70-525
  нормализационный ~ 70-475
  обезуглероживающий ~ 70-635
  ~ в магнитном поле 70-965
  ~ в ящиках 70-435
  ~ дефектов 30-135
  ~ для снятия напряжений 70-525
  ~ на крупное зерно 70-480
  перлитизирующий ~ 70-450
  пламенный ~ 70-505
  полный ~ 70-465
  рекристаллизационный ~ 70-490
  светлый ~ 70-430
  синий ~ 70-445
  смягчающий ~ 70-460
  стабилизирующий ~ 70-535
  ступенчатый ~ 70-455
  сфероидизирующий ~ 70-510
  циклический ~ 70-520
  чёрный ~ 70-440
  чистый ~ 70-435
  ясный ~ 70-430
относительная прочность f 75-140
относительное сужение n 75-125
относительное удлинение n 75-120
отношение n осей 20-355
отожжённая сталь f 60-275
отпечаток m
  баумановский ~ 80-325
  ~ Баумана 80-325
отпуск m 70-365
  высокий ~ 70-375
  высокотемпературный ~ 70-375
  низкий ~ 70-370
  низкотемпературный ~ 70-370
отпускная хрупкость f 85-220
отпущенный мартенсит m 50-345
отражательная микроскопия f 80-225
отрицательная краевая дислокация f 30-160
отрыв m 40-655
оттенение n 80-370
охлаждающая способность f 70-125
охлаждающая среда f 70-185
охлаждение n 70-070
  быстрое ~ 70-075
  воздушное ~ 70-095
  мгновенное ~ 70-075
  медленное ~ 70-080
  непрерывное ~ 70-090
  ~ в ванне 70-115
  ~ в воде 70-100
  ~ в масле 70-105
  ~ в печи 70-110
  ~ в спокойном воздухе 70-095
  ~ на воздухе 70-095
  ~ с печью 70-110
  резкое ~ 70-075
охрупчивание n 75-275

П

палладий m 05-325
пантал m 65-615
пара f
  гантельная ~ 30-085
  диффузионная ~ 45-375
  ~ Френкеля 30-060
парамагнетик m 75-555
парамагнитное вещество n 75-555
параметр m ближнего порядка 25-615
~ взаимодействия 15-300

параметр
~ дальнего порядка 25-620
~ решётки 20-120
~ состояния 15-020
термодинамический ~ 15-020
термодинамический ~ состояния 15-020
парциальная величина f 15-060
парциальная свободная энергия f 15-110
парциальная термодинамическая величина f 15-060
парциальная энтальпия f 15-150
парциальная энтропия f 15-215
парциальный коэффициент m диффузии 45-390
паспортная бронза f 65-285
патентирование n 70-415
патина f 85-375
патронная латунь f 65-380
паяемость f 75-740
первичная кристаллизация f 25-050
первичная ликвация f 25-405
первичная ползучесть f 45-500
первичная рекристаллизация f 45-445
первичная система f скольжения 45-115
первичная структура f 40-045
первичные дислокации fpl 30-250
первичные кристаллы mpl 40-270
первичный алюминий m 65-545
первичный аустенит m 50-030
первичный графит m 50-170
первичный металл m 55-115
первичный сплав m 55-275
первичный твёрдый раствор m 35-380
первичный феррит m 50-050
первичный цементит m 50-070
первичный чугун m 60-840
первое квантовое число n 10-130
первый ближайший сосед m 20-195
перегиб m 30-245
перегрев m 70-065
перегреваемость f 75-760
перегретая сталь f 60-320
пережжённая сталь f 60-325
пережог m 85-205
перелёт m 30-260
переменная f состояния 15-020
перемещение n
консервативное ~ 30-290
неконсервативное ~ 30-285
перенос m
термоперенос 45-370
электроперенос 45-365
переохлаждение n 25-240, 25-245
концентрационное ~ 25-255
переохлаждённый аустенит m 50-265
переползание n дислокаций 30-295
перестарение n 70-595
пересыщенный твёрдый раствор m 35-400
перетравление n 80-130
переход m второго рода 25-470
~ первого рода 25-465
полиморфный ~ 25-455
фазовый ~ 25-460
фазовый ~ второго рода 25-470
фазовый ~ первого рода 25-465
переходная ползучесть f 45-500
переходная решётка f 20-405
переходная фаза f 35-195

переходные металлы mpl 55-025
период m
инкубационный ~ 70-150
~ инертности 70-150
~ полураспада 75-525
~ решётки 20-115
~ трансляции 20-115
подготовительный ~ 70-150
перистый бейнит m 50-420
перитектика f 40-245
перитектическая горизонталь f 35-645
перитектическая линия f 35-645
перитектическая реакция f 35-570
перитектическая температура f 35-610
перитектическая точка f 35-665
перитектическое плавление n 35-475
перитектическое превращение n 35-570
перитектоид m 40-250
перитектоидная реакция f 35-595
перитектоидная температура f 35-635
перитектоидное превращение n 35-595
перлит m 50-125
вырожденный ~ 50-155
глобулярный ~ 50-150
грубопластинчатый ~ 50-140
зернистый ~ 50-150
мелкопластинчатый ~ 50-135
пластинчатый ~ 50-130
сорбитовый ~ 50-145
сорбитообразный ~ 50-145
точечный ~ 50-150
перлитизирующий отжиг m 70-450
перлитная область f 50-295
перлитная сталь f 60-425
перлитно-графитный чугун m 60-1045
перлитное превращение n 50-275
перлитный ковкий чугун m 60-1130
перлитный чугун m 60-1045
пермаллой m 65-440
пермендур m 65-490
перминвар m 65-510
петля f
дислокационная ~ 30-340
~ дислокации 30-340
призматическая ~ 30-345
призматическая ~ дислокаций 30-345
пик m смещений 30-525
температурный ~ 30-530
тепловой ~ 30-530
термический ~ 30-530
пикрал m 80-115
пирофорный сплав m 55-595
питтинг m 80-350
питтинговая коррозия f 85-350
плавкость f 75-790
плавление n 35-445
гетеротермическое ~ 35-455
изотермическое ~ 35-450
конгруэнтное ~ 35-460
неконгруэнтное ~ 35-465
перитектическое ~ 35-475
~ в интервале температур 35-455
эвтектическое ~ 35-470
плакированная сталь f 60-335
пламенная закалка f 70-290
пламенный отжиг m 70-505
пластинка f
мартенситная ~ 50-370
~ мартенсита 50-370
~ цементита 50-120
пластинчатая структура f 40-105
пластинчатая эвтектика f 40-220
пластинчатый графит m 50-185

пластинчатый перлит *m* 50-130
пластическая деформация *f* 45-010
пластическое течение *n* 45-215
пластическое упрочнение *n* 45-050
пластичность *f* 70-025
сверхпластичность 75-055
платина *f* 05-335
~-иридий 65-835
платинистый иридий *m* 65-835
платинит *m* 65-515
платиновые металлы *mpl* 65-820
платинородий *m* 65-840
плёнка *f*
 тонкая ~ 80-350
 эпитаксиальная ~ 25-350
плоскость *f*
 атомная ~ 20-175
 базисная ~ 20-360
 габитусная ~ 20-465
 двойниковая ~ 45-250
 кристаллографическая ~ 20-155
 наиболее плотноупакованная ~ 20-215
 ~ базиса 20-360
 ~ габитуса 20-465
 ~ грани призм 20-365
 ~ двойникования 45-250
 ~ октаэдра 20-335
 ~ поперечного скольжения 45-150
 ~ раскола 40-665
 ~ решётки 20-155
 ~ скалывания 40-665
 ~ скола 40-665
 ~ скольжения 45-100
 ~ спайности 40-665
 ~ срастания 20-470
 ~ сроста 20-470
 призматическая ~ 20-365
плотная упаковка *f* 20-210
плотность *f* 75-345
 ~ дислокации 30-235
 ~ магнитного потока 75-610
 ~ состояний 10-200
 ~ упаковки 20-205
плутоний *m* 05-340
побочное квантовое число *n* 10-135
поверхностная диффузия *f* 45-300
поверхностная закалка *f* 70-285
поверхностная коррозия *f* 85-330
поверхностная обработка *f* 70-010
поверхностная трещина *f* 85-050
поверхностная энергия *f* 30-640
поверхностное легирование *n* 70-655
поверхностное натяжение *n* 30-645
поверхностное обезуглероживание *n* 85-210
поверхностный дефект *m* 30-535, 85-010
поверхностный порок *m* 85-010
поверхность *f*
 межфазная ~ 30-610
 ~ излома 40-670
 ~ ликвидус 35-520
 ~ ликвидуса 35-520
 ~ раздела 30-610
 ~ раздела фаз 30-610
 ~ солидус 35-495
 ~ солидуса 35-495
 ~ Ферми 10-215
повышение *n* хрупкости 75-275
поглотитель *m* 30-130
пограничная дислокация *f* 30-580
пограничная диффузия *f* 45-310
пограничное скольжение *n* 45-160
пограничные выделения *npl* 40-310
подготовительный период *m* 70-150

подгруппа *f* 10-160
 электронная ~ 10-160
подложка *f* 25-165
подоболочка *f* 10-160
 электронная ~ 10-160
подогрев *m* 70-045
подогревание *n* 70-045
подповерхностная коррозия *f* 85-355
подрешётка *f* 20-285
 магнитная ~ 20-485
подуровень *f* 10-160
 электронный ~ 10-160
подшипниковая бронза *f* 65-235
подшипниковая сталь *f* 60-715
подшипниковый сплав *m* 55-555
позиция *f*
 межузловая ~ 20-145
 узловая ~ 20-140
показатель *m* пластичности 75-065
 ~ прочности 75-060
поле *n*
 двухфазное ~ 35-685
 однофазное ~ 35-680
 ~ напряжений дислокации 30-225
ползучепрочность *f* 75-220
ползучесть *f* 45-495
 диффузионная ~ 45-520
 затухающая ~ 45-160
 нестационарная ~ 45-500
 неустановившаяся ~ 45-500
 первичная ~ 45-500
 переходная ~ 45-500
 прогрессирующая ~ 45-510
 стационарная ~ 45-505
 ускоренная ~ 45-510
 установившаяся ~ 45-505
 микроползучесть 45-530
полигонизация *f* 45-435
поликристалл *m* 20-440
поликристаллическое тело *n* 20-440
полиморфизм *m* 25-385
полиморфная модификация *f* 25-390
полиморфное превращение *n* 25-455
полиморфный переход *m* 25-455
полировка *f*
 механическая ~ 80-070
 ~ травлением 80-085
 химическая ~ 80-080
 электролитическая ~ 80-075
 электрополировка 80-075
политермический разрез *m* 35-740
полиэдрическая структура *f* 40-095
полная дислокация *f* 30-305
полная растворимость *f* 35-315
полное относительное удлинение *n* 75-120
полный отжиг *m* 70-465
половинчатый чугун *m* 60-920, 60-960
положение *n*
 междуузловое ~ 20-145
 узловое ~ 20-140
положительная краевая дислокация *f* 30-155
полоса *f*
 ликвационная ~ 85-230
 ~ двойникования 45-255
 ~ деформации 45-200
 ~ прокаливаемости 80-460
 ~ скольжения 45-190
 энергетическая ~ 10-235
полосчатая структура *f* 40-140
полуаустенитная сталь *f* 60-445
полублагородный металл *m* 55-020

ПОЛУГЕМАТИТОВЫЙ

полугематитовый чугун *m* 60-900
полудислокация *f* 30-315
полуметалл *m* 10-370
полупроводник *m* 10-375
  примесный ~ 10-385
  собственный ~ 10-380
полуспокойная сталь *f* 60-660
полуферритная сталь *f* 60-450
полюсная дислокация *f* 30-365
полюсные фигуры *fpl* 80-685
поляризационная микроскопия *f* 80-260
полярная связь *f* 10-325
полярное соединение *n* 35-230
полярный кристалл *m* 20-025
поперечное скольжение *n* 45-145
пора *f* 85-100
  октаэдрическая ~ 20-330
  тетраэдрическая ~ 20-325
пористая бронза *f* 65-225
пористость *f* 85-105
  усадочная ~ 85-120
  микропористость 85-115
порог *m* 30-240
  ~ рекристаллизации 45-470
порок *m*
  закалочный ~ 85-180
  литейный ~ 85-095
  поверхностный ~ 85-015
  ~ материала 85-005
порошковая рентгенограмма *f* 80-635
порошковый метод *m* 80-595
порошковый сплав *m* 55-655
порошкограмма *f* 80-635
поршневой сплав *m* 55-570
порядок *m* 25-590
  ближний ~ 25-595
  дальний ~ 25-600
  ~ упаковки 20-180
последействие *n*
  магнитное ~ 75-645
  упругое ~ 75-180
последовательность *f* упаковки 20-180
постоянная *f* решётки 20-120
  ~ упругости 75-145
  упругая ~ 75-145
потенциал *m*
  изобарно-изотермный ~ 15-140
  изобарный ~ 15-140
  изохорно-изотермный ~ 15-105
  ионизационный ~ 10-105
  окислительный ~ 15-350
  ~ взаимодействия 10-300
  ~ ионизации 10-105
  углеродный ~ 70-750
  химический ~ 15-175
  электрохимический ~ 75-715
потенциальный барьер *m* 10-110
поток *m* 15-085
почвенная коррозия *f* 85-285
появление *n* хрупкости 75-275
правило *n* Коновалова 35-705
  ~ Матисена 75-475
  ~ Маттиссена 75-475
  ~ общей касательной 15-180
  ~ отрезков 35-700
  ~ рычага 35-700
  ~ фаз 35-010
  ~ Хегга 35-260
  ~ Юм-Розери 35-325
правовинтовая дислокация *f* 30-170
празеодим *m* 05-355
превращение *n*
  аллотропическое ~ 25-455
  атермическое ~ 25-520

превращение
  аустенитно-мартенситное ~ 50-280
  аустенитно-перлитное ~ 50-275
  бездиффузионное ~ 25-510
  бейнитное ~ 50-290
  вторичное мартенситное ~ 50-285
  гетерогенное ~ 25-535
  гомогенное ~ 25-530
  диффузионное ~ 25-505
  изотермическое ~ 25-515
  конгруэнтное ~ 25-525
  магнитное ~ 25-555
  мартенситное ~ 50-280
  массивное ~ 25-550
  метатектическое ~ 35-580
  монотектическое ~ 35-575
  монотектоидное ~ 35-600
  непрерывное ~ 25-540
  обратимое ~ 25-495
  перитектическое ~ 35-570
  перитектоидное ~ 35-595
  перлитное ~ 50-275
  полиморфное ~ 25-455
  ~ в твёрдом состоянии 25-475
  ~ первого рода 25-465
  ~ порядок-беспорядок 25-580
  прерывистое ~ 25-545
  промежуточное ~ 50-290
  синтетическое ~ 35-585
  структурное ~ 25-500
  фазовое ~ 25-460
  фазовое ~ второго рода 25-470
  фазовое ~ первого рода 25-465
  эвтектическое ~ 35-565
  эвтектоидное ~ 35-590
превращённый аустенит *m* 50-035
предварительный нагрев *m* 70-045
предвыделение *n* 70-600
предел *m*
  абсолютный ~ ползучести 75-225
  истинный ~ длительной прочности 75-225
  истинный ~ ползучести 75-225
  истинный ~ прочности 75-130
  ~ выносливости 75-200
  ~ ограниченной длительной прочности 75-230
  ~ пластичности 75 110
  ~ ползучести 75-220
  ~ пропорциональности 75-100
  ~ прочности при растяжении 75-070
  ~ растворимости 35-330
  ~ текучести 75-110
  ~ упругости 75-105
  ~ усталости 75-200
  условный ~ длительной прочности 75-230
  условный ~ ползучести 75-230
  условный ~ текучести 75-115
  физический ~ текучести 75-110
предельная растворимость *f* 35-330
предельный чугун *m* 60-880
преимущественная ориентировка *f* 20-425
прерывистая закалка *f* 70-300
прерывистое выделение *n* 25-380
прерывистое превращение *n* 25-545
прививка *f* кристаллизации 25-215
придавание *n* твёрдости 70-015
призматическая дислокация *f* 30-345
призматическая петля *f* 30-345
призматическая петля *f* дислокаций 30-345
призматическая плоскость *f* 20-365
призматическое скольжение *n* 45-170
примеси *fpl* 55-205
  случайные ~ 55-200
примесный зародыш *m* 25-120
примесный полупроводник *m* 10-385
примесь *f* внедрения 30-080
примитивная элементарная ячейка *f* 20-095

примитивная ячейка f 20-095
принцип m исключения Паули 10-170
~ несовместимости 10-170
~ Паули 10-170
припой m 65-710
аргентановый ~ 65-745
крепкий ~ 65-720
латунно-никелевый ~ 65-745
латунный ~ 65-740
легкоплавкий ~ 65-715
мягкий ~ 65-715
оловянный ~ 65-725
свинцовый ~ 65-735
серебряный ~ 65-750
твёрдый ~ 65-720
тугоплавкий ~ 65-720
цинковый ~ 65-730
природный сплав m 55-265
проба f 55-100
искровая ~ 80-440
~ благородного металла 55-100
~ на звучность 80-795
~ на излом 80-445
~ на серу 80-320
пробельный сплав m 55-640
пробная закалка f 70-775
проводниковый сплав m 55-550
проводность f
удельная ~ 75-485
удельная электрическая ~ 75-485
прогрев m 70-035
прогрессирующая ползучесть f 45-510
продолжительность f травления 80-125
прозвучивание n 80-800
прокаливаемость f 75-730
промежуточная область f 50-305
промежуточная фаза f 35-210
промежуточное превращение n 50-290
промежуточный сплав m 55-285
прометий m 05-360
промышленное железо n 60-015
промышленный сплав m 55-520
проницаемость f
магнитная ~ 75-615
начальная магнитная ~ 75-620
начальная ~ 75-620
пропитанный сплав m 55-670
пропитываемый сплав m 55-670
просвечивающая микроскопия f 80-220
просвечивающая электронная микроскопия f 80-340
простая гексагональная решётка f 20-345
простая дилатометрия f 80-545
простая кубическая решётка f 20-305
простая латунь f 65-305
простая ликвация f 25-435
простая решётка f 20-280
простая сталь f 60-095
простое скольжение n 45-130
простой дилатометрический анализ m 80-545
простой карбид m 40-470
простой термический анализ m 80-510
пространственная диаграмма f состояния тройной системы 35-750
пространственная решётка f 20-045
пространство n волновых векторов 10-205
протактиний m 05-365
процесс m выделения 40-300
термически активированный ~ 45-280
термически активируемый ~ 45-280

процесс
термоактивируемый ~ 45-280
прочностные свойства npl 75-010
прочность f 75-040
высокотемпературная ~ 75-095
динамическая ~ 75-185
длительная ~ 75-235
механическая ~ 75-040
относительная ~ 75-140
~ на изгиб 75-090
~ на кручение 75-085
~ на растяжение 75-070
~ на сжатие 75-075
~ на скручивание 75-085
~ на срез 75-080
~ при повышенных температурах 75-095
~ при растяжении 70-070
сверхвысокая ~ 75-045
термоциклическая ~ 75-830
ударная ~ 75-185
удельная ~ 75-140
усталостная ~ 75-200
циклическая ~ 75-200
жаропрочность 75-240
ползучепрочность 75-220
сверхпрочность 75-045
теплопрочность 75-095
проэвтектическая фаза f 40-340
проэвтектоидная фаза f 40-345
проэвтектоидный феррит m 50-055
проэвтектоидный цементит m 50-080
пружинная сталь f 60-725
прутковый образец m 80-435
прямая ликвация f 25-435
псевдобинарная система f 35-130
псевдобинарный разрез m 35-745
псевдобинарный сплав m 55-315
псевдосплав m 55-650
пудлинговая сталь f 60-575
пудлинговое железо n 60-055
пустота f 85-100
газовая ~ 85-135
октаэдрическая ~ 20-330
тетраэдрическая ~ 20-325
путь m
свободный ~ 40-425
средний свободный ~ между частицами 40-425
пушечная бронза f 65-245
пятна npl
белые ~ 85-250
мягкие ~ 85-195
пятно n
дифракционное ~ 80-660
интерференционное ~ 80-660
~ травления 80-205

Р

работа f выхода 10-060
рабочий свинец m 65-650
равновесие n
гетерогенное ~ 35-015
метастабильное ~ 35-060
микроструктурное ~ 40-020
неустойчивое ~ 35-060
стабильное ~ 35-055
тепловое ~ 15-325
термическое ~ 15-325
термодинамическое ~ 15-320
устойчивое ~ 35-055
фазовое ~ 30-015
химическое ~ 15-330
равновесная диаграмма f состояния 35-480
равновесная концентрация f 35-255
равновесная коррозия f 85-335

равновесная кристаллизация f 25-015
равновесная система f 35-045
равновесная температура f 35-690
равновесная фаза f 35-180
равновесное состояние n 35-050
равновесные зёрна npl 40-285
равновесные кристаллы mpl 40-285
равновесный зародыш m 25-140
равновесный коэффициент m распределения 25-260
равноосная кристаллизация f 25-075
радиационная дефектность f 30-490
радиационная коррозия f 85-365
радиационное повреждение n 30-490
радиационное упрочнение n 30-495
радиационный дефект m 30-485
радий m 05-370
радиоавтограф m 80-755
радиоавтография f 80-740
радиоактивационный анализ m 80-765
радиоактивный индикатор m 80-760
радиоактивный металл m 55-075
радиограмма f 80-715
  микрорадиограмма 80-725
радиографическое исследование n 80-705
радиография f 80-705
  ~ металлов 80-710
  авторадиография 80-740
  металлорадиография 80-710
  микрорадиография 80-720
радиус m
  атомный ~ 10-085
  ионный ~ 10-090
  критический ~ 25-145
  критический ~ зародыша 25-145
радон m 05-375
разбавленная эвтектика f 40-230
разбавленный раствор m 15-250
разбавленный сплав m 55-320
разбивка f концентрационного треугольника 35-725
разгаростойкость f 75-830
размер m зерна 80-285
размерный фактор m 10-395
размерный эффект m 10-395
размножение n дислокаций 30-420
разориентировка f 30-595
разрез m
  вертикальный ~ 35-740
  горизонтальный ~ 35-735
  изотермический ~ 35-735
  политермический ~ 35-740
  псевдобинарный ~ 35-745
разрушающее напряжение n 75-130
разрушение n 85-040
  вязкое ~ 40-680
  задержанное ~ 40-755
  замедленное ~ 40-755
  запаздывающее ~ 40-755
  усталостное ~ 40-745
  хрупкое ~ 40-675
разрыв m 85-040
  раскалывающий ~ 85-085
  сдвигающий ~ 40-660
разрывной кристалл m 25-315
разупорядочение n 25-580
раковина f
  внутренняя газовая ~ 85-135
  газовая ~ 85-130
  закрытая газовая ~ 85-135
  усадочная ~ 85-125

раковинистый излом m 40-730
раковинообразный излом m 40-730
раковистый излом m 40-730
раскалывающий разрыв m 85-085
распад m раствора 25-365
  ~ твёрдого раствора 25-365
  спинодальный ~ 25-370
  эвтектический ~ 35-565
  эвтектоидный ~ 35-590
расплав m 55-150
  жидкий ~ 55-150
  маточный ~ 25-095
  остаточный ~ 25-280
расплавленный металл m 55-150
рассеянные металлы mpl 55-045
расслоение n 85-405
  ~ жидкого раствора 25-365
  ~ раствора 25-365
расслой m 85-405
расстояние n
  межатомное ~ 20-230
  междуатомное ~ 20-230
  межпластиночное ~ 40-110
  межплоскостное ~ 20-225
раствор m 35-350
  атермический ~ 15-265
  бинарный ~ 35-405
  газообразный ~ 35-370
  двойной ~ 35-405
  двухкомпонентный ~ 35-405
  жидкий ~ 35-365
  идеальный ~ 15-240
  маточный ~ 25-095
  металлический ~ 35-385
  многокомпонентный ~ 35-410
  насыщенный твёрдый ~ 35-395
  неидеальный ~ 15-245
  неограниченный твёрдый ~ 35-390
  неупорядоченный твёрдый ~ 35-435
  первичный твёрдый ~ 35-380
  пересыщенный твёрдый ~ 35-400
  разбавленный ~ 15-250
  реальный ~ 15-245
  регулярный ~ 15-255
  семирегулярный ~ 15-260
  совершенный ~ 15-240
  спиртовой ~ азотной кислоты 80-110
  спиртовой ~ пикриновой кислоты 80-115
  твёрдый ~ 35-375
  твёрдый ~ внедрения 35-420
  твёрдый ~ замещения 35-415
  упорядоченный твёрдый ~ 35-430
растворённое вещество n 35-360
растворённый элемент m 35-360
растворимость f 35-295
  взаимная ~ 35-300
  неограниченная ~ 35-315
  ограниченная ~ 35-320
  полная ~ 35-315
  предельная ~ 35-330
  ~ в жидком состоянии 35-305
  ~ в твёрдом состоянии 35-310
  частичная ~ 35-320
растворитель m 35-355
растрескивание n 85-015
  сезонное ~ 85-385
растровая электронная микроскопия f 80-360
растянутая дислокация f 30-325
расщепление n дислокации 30-320
расщеплённая дислокация f 30-325
расщеплённый междоузельный атом m 30-085
рафинированная медь f 65-020
рафинированная сталь f 60-635
рафинированная сурьма f 65-665
рафинированное золото n 65-770

рафинированное серебро n 65-760
рафинированный алюминий m 65-550
рафинированный металл m 55-125
рафинированный свинец m 65-655
реактив m для травления 80-100
травящий ~ 80-100
реактивная диффузия f 45-325
реакторный металл m 55-085
реакционная диффузия f 45-325
реакционноспособный металл m 55-070
реакция f
  метатектическая ~ 35-580
  монотектическая ~ 35-575
  монотектоидная ~ 35-600
  перитектическая ~ 35-570
  перитектоидная ~ 35-595
  синтетическая ~ 35-585
  эвтектическая ~ 35-565
реальный кристалл m 30-010
реальный раствор m 15-245
регулярный раствор m 15-255
редистиллированный цинк m 65-695
редкие металлы mpl 55-045
редкоземельные металлы mpl 55-040
реечный мартенсит m 50-355
режим m термической обработки 70-970
~ термообработки 70-970
резкая закалка f 70-260
резкое охлаждение n 70-075
резкость f закалки 70-200
рекалесценция f 25-485
рекристаллизационный двойник m 45-240
рекристаллизационный отжиг m 70-490
рекристаллизация f 45-440
  вторичная ~ 45-450
  динамическая ~ 45-465
  первичная ~ 45-445
  ~ обработки 45-445
  самопроизвольная ~ 45-455
  собирательная ~ 45-450
  третичная ~ 45-460
релаксация f напряжений 45-525
рельсовая сталь f 60-730
реманенц m 75-630
рений m 05-380
рентгеноанализ m 80-570
рентгеновская металлография f 80-575
рентгеновская микроскопия f 80-730
рентгеновская спектроскопия f 80-700
рентгеновская флуоресцентная спектроскопия f 80-700
рентгеновский анализ m с помощью электронного зонда 80-780
рентгеновский спектральный анализ m с помощью электронного зонда 80-780
рентгеновский фазовый анализ m 80-585
рентгенограмма f
  дифракционная ~ 80-615
  обратная ~ 80-670
  порошковая ~ 80-635
  ~ вращения 80-665
  ~ Дебая 80-635
  ~ снятая на просвет 80-675
рентгенографирование n 80-570
рентгенографический фазовый анализ m 80-585
рентгенографическое исследование n 80-570
рентгенография f металлов 80-575
  структурная ~ 80-580

рентгенография
  микрорентгенография 80-730
рентгенодефектоскопия f 80-820
рентгеноспектральный анализ m 80-700
рентгеноструктурный анализ m 80-580
реостатный сплав m 55-545
реотропная хрупкость f 75-310
реплика f 80-345
рессорная сталь f 60-725
ретроградный солидус m 35-505
рефлексограмма f 80-610
решётка f
  алмазная кубическая ~ 20-320
  гексагональная компактная ~ 20-350
  гексагональная плотноупакованная ~ 20-350
  гексагональная ~ 20-340
  гексаэдрическая ~ 20-305
  гранецентрированная ~ 20-290
  додекаэдрическая ~ 20-310
  кристаллическая ~ 20-045
  кубическая гранецентрированная ~ 20-310
  кубическая объёмноцентрированная ~ 20-315
  кубическая ~ 20-300
  кубическая ~ типа алмаза 20-320
  кубическая центрированная ~ 20-315
  моноклинная ~ 20-395
  обратная ~ 20-410
  объёмноцентрированная ~ 20-295
  односторонне-гранецентрированная ромбическая ~ 20-385
  октаэдрическая ~ 20-315
  орторомбическая ~ 20-380
  переходная ~ 20-405
  простая гексагональная ~ 20-345
  простая кубическая ~ 20-305
  простая ~ 20-280
  пространственная ~ 20-045
  ~ Браве 20-275
  ромбическая базоцентрированная ~ 20-385
  ромбическая ~ 20-380
  ромбоэдрическая ~ 20-390
  тетрагональная ~ 20-370
  трансляционная ~ 20-270
  триклинная ~ 20-440
  центрогранная кубическая ~ 20-310
  центрогранная ~ 20-290
  подрешётка 20-285
  сверхрешётка 35-440
ржавчина f 85-370
родий m 05-385
розеточный графит m 50-200
ромбическая базоцентрированная решёка f 20-385
ромбическая решётка f 20-380
ромбическая система f 20-255
ромбоэдрическая решётка f 20-390
ромбоэдрическая система f 20-260
рост m
  дендритный ~ 25-310
  ~ зерна 25-290
  ~ кристаллов 25-285
  ~ чугуна 70-640
  эпитаксиальный ~ 25-345
ртуть f 05-270
рубидий m 05-390
рутений m 05-395
ряд m напряжений 75-710
рядовая сталь f 60-135

С

самарий m 05-400
самодиффузия f 45-400
самозакаливающаяся сталь f 60-375
самопроизвольная кристаллизация f 25-045

САМОПРОИЗВОЛЬНАЯ

самопроизвольная рекристаллизация f 45-455
самопроизвольное зародышеобразование n 25-190
самопроизвольное образование n зародышей 25-190
самораcтрескивание n сезонное 85-385
самородное золото n 65-765
самородный металл m 55-110
свариваемая сталь f 60-485
свариваемость f 75-735
сваривающая сталь f 60-485
сварочная сталь f 60-570
сварочное железо n 60-040
сверхвысокая прочность f 75-045
сверхвысокопрочный сплав m 55-425
сверхдислокация f 30-220
сверхлёгкий сплав m 55-410
сверхпластичность f 75-055
сверхпластичный сплав m 55-430
сверхпроводимость f 75-490
сверхпроводник m 10-390
сверхпроводность f 75-490
сверхпрочность f 75-045
сверхпрочный сплав m 55-425
сверхрешётка f 35-440
сверхструктура f 35-440
сверхструктурная дислокация f 70-220
сверхструктурные линии fpl 80-655
сверхтвёрдый сплав m 55-415
сверхчистый металл m 55-105
светлая закалка f 70-330
светловина f 85-230
светлопольная микроскопия f 80-235
светлосердечный ковкий чугун m 60-1120
светлый излом m 85-155
светлый отжиг m 70-430
световая микроскопия f 80-215
свинец m 05-240
  мягкий ~ 65-655
  рабочий ~ 65-650
  рафинированный ~ 65-655
  сурьмянистый ~ 65-660
  твёрдый ~ 65-660
  черновой ~ 65-650
свинцовая бронза f 65-140
свинцовая латунь f 65-360
свинцовистая бронза f 65-140
свинцовистая латунь f 65-360
свинцовистая сталь f 60-250
свинцовокальциевонатриевый сплав m 65-685
свинцовый припой m 65-735
свободная поверхностная энергия f 15-120
свободная энергия f Гельмгольца 15-105
свободная энергия f Гиббса 15-140
свободная энергия f образования 15-135
свободная энергия f поверхности раздела 15-125
свободная энергия f смешения 15-130
свободная энтальпия f 15-140
свободный путь m 40-425
свободный углерод m 50-005
свободный феррит m 50-055
свободный электрон m 10-055
свойства npl
  литейные ~ 75-775

свойства
  магнитные ~ 75-540
  макроскопические ~ 75-850
  металлические ~ 10-350
  металлургические ~ 75-855
  механические ~ 75-005, 75-015
  механические ~ определённые испытанием на растяжение 75-015
  объёмные ~ 75-340
  оптические ~ 75-500
  прочностные ~ 75-010
  служебные ~ 75-820
  структурно нечувствительные ~ 75-865
  структурночувствительные ~ 76-860
  тепловые ~ 75-365
  термические ~ 75-365
  технологические ~ 75-720
  физические ~ 75-315
  химические ~ 75-655
  эксплуатационные ~ 75-820
  электрические ~ 75-465
  электрофизические ~ 75-465
  электрохимические ~ 75-705
  ядерные ~ 75-520
связанная энергия f 15-100
связанный углерод m 50-010
связанный электрон m 10-050
связь f
  атомная ~ 10-330
  вандерваальсов(ск)ая ~ 10-335
  гетерополярная ~ 10-325
  гомеополярная ~ 10-330
  ионная ~ 10-325
  ковалентная ~ 10-330
  металлическая ~ 10-340
  молекулярная ~ 10-335
  полярная ~ 10-325
  ~ Ван-дер-Ваальса 10-335
  электровалентная ~ 10-325
сдвиг m 45-095
  единичный ~ 45-130
сдвигающий разрыв m 40-660
сдвиговое напряжение n 45-085
сегрегаты mpl 40-305
сегрегация f 25-400
  зернограничная ~ 25-450
  междендритная ~ 25-445
  нормальная ~ 25-435
  ~ на границах зёрен 25-450
  ~ по границам зёрен 25-450
сезонная хрупкость f 85-385
сезонное растрескивание n 85-385
сезонное самораcтрескивание n 85-385
селективная коррозия f 85-305
селен m 05-410
семирегулярный раствор m 15-260
сенсибилизация f 70-545
сера f 05-435
сердцевина f 70-665
серебристая сталь f 60-820
серебро n 05-420
  монетное ~ 65-755
  новое ~ 65-415
  рафинированное ~ 65-760
серебрянка f 60-820
серебряный припой m 65-750
сернистое железо n 50-245
сернистые включения npl 40-530
сернистый марганец m 50-240
серое олово n 65-635
сероуглеродоазотирование n 70-892
серый чугун m 60-915, 60-950
сетка f
  дислокационная ~ 30-390
  карбидная ~ 50-445
  ~ дислокаций 30-390
  ~ карбидов 50-445

сетка
цементитная ~ 50-105
сетчатая структура f 40-100
сечение n поглощения нейтронов 75-530
эффективное ~ захвата нейтронов 75-530
сигма-фаза f 35-275
сидячая дислокация f 30-280
сидячая дислокация f Франка 30-335
сила f
движущая ~ 15-070
коэрцитивная ~ 75-635
магнитозадерживающая ~ 75-535
~ Пайерльса 30-300
~ Пайерльса-Набарро 30-300
~ связи 10-315
~ сцепления 10-315
термоэлектродвижущая ~ 75-495
силал m 60-1100
силикоалюминий m 60-1165
силикокальций m 60-1160
силикомарганец m 60-1155
силикошпигель m 60-1155
силицид m 40-490
силицирование n
диффузионное ~ 70-910
силумин m 65-565
модифицированный ~ 65-570
симметрия f кристалла 20-055
сингония f 20-235
сингулярная точка f 80-495
синеломкость f 75-300
синение n 70-400
синий налёт m 70-390
синий отжиг m 70-445
синий цвет m побежалости 70-390
синтетическая реакция f 35-585
синтетическая температура f 35-625
синтетическое превращение n 35-585
синтетический сплав m 55-270
синтетический чугун m 60-940
система f 35-005
бинарная ~ 35-120
вторичная ~ скольжения 45-120
гексагональная ~ 20-245
гетерогенная ~ 35-095
гомогенная ~ 35-090
двойная ~ 35-120
двухкомпонентная ~ 35-120
двухфазная ~ 35-100
дивариантная ~ 35-085
квазибинарная ~ 35-130
кристаллическая ~ 20-235
кристаллографическая ~ 20-235
кубическая ~ 20-240
легированная ~ 55-165
металлическая ~ 55-165
многокомпонентная ~ 35-140
многофазная ~ 35-120
моновариантная ~ 35-080
моноклинная ~ 20-265
негомогенная ~ 35-095
неоднородная ~ 35-095
нонвариантная ~ 35-075
одновариантная ~ 35-080
однокомпонентная ~ 35-115
однофазная ~ 35-090
первичная ~ скольжения 45-115
псевдобинарная ~ 35-130
равновесная ~ 35-045
ромбическая ~ 20-255
ромбоэдрическая ~ 20-260
~ железо-углерод 50-015
~ скольжения 45-110
сопряжённая ~ скольжения 45-125
тетрагональная ~ 20-250
трёхкомпонентная ~ 35-125
трёхфазная ~ 35-105

система
тригональная ~ 20-260
триклинная ~ 20-270
тройная ~ 35-125
унарная ~ 35-115
четверная ~ 35-135
четырёхкомпонентная ~ 35-135
скалывающее напряжение n 45-085
скандий m 05-405
сканирующая микроскопия f 80-360
сканирующая электронная
  микроскопия f 80-360
сквозная закалка f 70-280
сквозной нагрев m 70-035
склонность f к перегреву 75-760
скол m 85-085
скольжение n 45-095
внутрикристаллическое ~ 45-165
двойное ~ 45-135
единичное ~ 45-130
зернограничное ~ 45-160
карандашное ~ 45-175
лёгкое ~ 45-155
межзёренное ~ 45-160
многократное ~ 45-140
множественное ~ 45-140
неупрочняющее ~ 45-155
одинарное ~ 45-130
одиночное ~ 45-130
однократное ~ 45-130
пограничное ~ 45-160
поперечное ~ 45-145
призматическое ~ 45-170
простое ~ 45-130
~ по границам 45-160
~ по границам зёрен 45-160
сложное ~ 45-140
скользящая дислокация f 30-265
скользящее движение n 30-290
скопление n вакансий 30-125
скопление n дислокаций 30-400
скорость f
критическая ~ закалки 70-180
критическая ~ охлаждения 70-180
~ зародышеобразования 25-270
~ зарождения центров кристаллизации 25-270
~ кристаллизации 25-275
~ нагрева 70-040
~ нагревания 70-040
~ охлаждения 70-120
~ роста кристаллов 25-275
скрытая теплота f испарения 75-390
скрытая теплота f парообразования 75-390
скрытая теплота f плавления 75-380
скрытое упрочнение n 45-180
скрытокристаллический мартенсит m 50-330
слаболегированная сталь f 60-120
следовая авторадиография f 80-750
слепок m 80-345
словолитный сплав m 55-615
сложное скольжение n 45-140
сложный карбид m 40-475
слоистый излом m 40-725
слой m
азотированный ~ 70-845
диффузионный ~ 70-660
закалённый ~ 70-355
корковый ~ 40-175
нитрированный ~ 70-845
~ Бейльби 80-090
~ Бильби 80-090
цементированный ~ 70-780
цементованный ~ 70-780
эпитаксиальный ~ 25-350

служебные свойства *npl* 75-820
случайные примеси *fpl* 55-200
смесь *f*
  газовая ~ 35-370
  газообразная ~ 35-370
  механическая ~ 40-190
  механическая ~ фаз 40-190
  эвтектическая ~ 40-200
  эвтектоидная ~ 40-240
смешанная дислокация *f* 30-180
смещённый атом *m* 30-510
смягчающий отжиг *m* 70-460
собирательная рекристаллизация *f* 45-450
собственная функция *f* 10-190
собственная энергия *f* дислокации 30-230
собственно термическая обработка *f* 70-006
собственные напряжения *npl* 85-020
собственный коэффициент *m* диффузии 45-390
собственный полупроводник *m* 10-380
совершенный кристалл *m* 30-005
совершенный раствор *m* 15-240
соединение *n*
  интерметаллидное ~ 35-215
  интерметаллическое ~ 35-215
  ионное ~ 35-230
  конгруэнтное интерметаллическое ~ 35-220
  неконгруэнтное интерметаллическое ~ 35-225
  полярное ~ 35-230
  химическое ~ 35-235
  электронное ~ 35-280
солидус *m* 35-485
  ретроградный ~ 35-505
соответственная температура *f* 45-475
соотношение *n* неопределённости 10-165
соотношение *n* неопределённости Гейзенберга 10-165
сопротивление *n*
  действительное ~ разрыву 75-130
  ~ изгибу 75-090
  ~ износу 75-265
  ~ истиранию 75-265
  ~ коррозии 75-670
  ~ кручению 75-085
  ~ окислению 75-685
  ~ отрыву 75-135
  ~ ползучести 75-215
  ~ растяжению 75-070
  ~ сдвигу 75-080
  ~ сжатию 75-075
  ~ срезу 75-080
  ~ удару 75-185
  ~ усталости 75-200
  удельное ~ 75-470
  удельное электрическое ~ 75-470
сопряжённая система *f* скольжения 45-125
сопряжённые кривые *fpl* 35-540
сопряжённые линии *fpl* 35-540
сопутствующий элемент *m* 55-260
сорбит *m* 50-380
  ~закалки 50-385
  ~ отпуска 50-390
сорбитизация *f* 70-415
сорбитовый перлит *m* 50-145
сорбитообразный перлит *m* 50-145
сорт *m* стали 60-080
сосед *m*
  ближайший ~ 20-195
  второй ~ 20-200
  первый ближайший ~ 20-195
состав *m*
  стехиометрический ~ 35-240

состав
  химический ~ 55-225
  химический ~ сплава 55-225
составляющая *f* сплава 55-180
  структурная ~ 40-010
  микросоставляющая 40-010
составная дислокация *f* 30-180
состояние *n*
  квантовое ~ 10-120
  металлическое ~ 10-345
  метастабильное ~ 35-070
  неупорядоченное ~ 25-585
  равновесное ~ 35-050
  ~ равновесия 35-050
  стабильное ~ 35-065
  термодинамическое ~ 15-015
  упорядоченное ~ 25-590
  устойчивое ~ 35-065
  энергетическое ~ 10-120
сосуществующие фазы *fpl* 35-190
спеканный сплав *m* 55-655
спектральный анализ *m* 80-475
спектрография *f*
  масс-~ 80-480
спектроскопия *f*
  рентгеновская ~ 80-700
  рентгеновская флуоресцентная ~ 80-700
специальная латунь *f* 65-310
специальная сталь *f* 60-115, 60-150
специальная фаза *f* 65-130
специальный карбид *m* 40-445
спецлатунь *f* 65-310
спечённый карбид *m* 55-665
спечённый металл *m* 55-660
спечённый сплав *m* 55-655
спечённый твёрдый сплав *m* 55-665
спиновое квантовое число *n* 10-145
спинод *m* 15-185
спиноддаль *f* 15-190
спинодальная кривая *f* 15-190
спинодальная точка *f* 15-185
спинодальный распад *m* 25-370
спиральная ступенька *f* роста 25-335
спиртовой раствор *m* азотной кислоты 80-110
спиртовой раствор *m* пикриновой кислоты 80-115
сплав *m* 55-160
  автоматный ~ 55-560
  анатомический ~ 55-645
  антифрикционный ~ 55-555
  белый антифрикционный ~ 65-640
  бинарный ~ 55-295
  вторичный ~ 55-280
  высокожаропрочный ~ 55-515
  высококоэрцитивный ~ 55-465
  высоколегированный ~ 55-325
  высокоомный ~ 55-545
  высокоплавкий ~ 55-400
  высокопроцентный ~ 55-325
  высокопрочный ~ 55-420
  высокотемпературный ~ 55-510
  высокоэлектропроводный ~ 55-550
  гальванотипный ~ 55-635
  гейслеров ~ 65-420
  гетерогенный ~ 55-340
  гомогенный ~ 55-330
  двойной ~ 55-295
  двухкомпонентный ~ 55-295
  двухфазный ~ 55-335
  декоративный ~ 55-605
  деформируемый ~ 55-530
  дисперсионно-упрочнённый ~ 55-390
  дисперсионнотвердеющий ~ 55-385
  дисперсноупрочнённый ~ 55-390
  доэвтектический ~ 55-355
  доэвтектоидный ~ 55-370

СТАЛЬ

сплав
  жаропрочный ~ 55-510
  жаростойкий ~ 55-505
  жароупорный ~ 55-505
  железный ~ 60-070
  заэвтектический ~ 55-360
  заэвтектоидный ~ 55-375
  зубопротезный ~ 55-575
  игрек-~ 65-590
  износостойкий ~ 55-435
  износоустойчивый ~ 55-435
  квазибинарный ~ 55-315
  кислотостойкий ~ 55-485
  кислотоупорный ~ 55-485
  ковкий ~ 55-530
  контактный ~ 55-585
  коррозионностойкий ~ 55-480
  лёгкий ~ 55-405
  легкоплавкий ~ 55-395
  линотипный ~ 55-630
  литейный ~ 55-535
  магнитный ~ 55-445
  магнитомягкий ~ 55-460
  магнитотвёрдый ~ 55-465
  медноникелевый ~ 65-400
  медный ~ 65-010
  металлический ~ 55-160
  металлокерамический ~ 55-655
  многокомпонентный ~ 55-310
  многофазный ~ 55-345
  монетный ~ 55-565
  монотипный ~ 55-620
  наплавляемый твёрдый ~ 55-590
  негомогенный ~ 55-340
  некорродирующий ~ 55-480
  немагнитный ~ 55-470
  низколегированный ~ 55-320
  низкопроцентный ~ 55-320
  никелевый ~ 65-430
  однофазный ~ 55-330
  одонтологический ~ 55-575
  окалиностойкий ~ 55-505
  первичный ~ 55-275
  пирофорный ~ 55-595
  подшипниковый ~ 55-555
  порошковый ~ 55-655
  поршневой ~ 55-570
  природный ~ 55-265
  пробельный ~ 55-640
  проводниковый ~ 55-550
  промежуточный ~ 55-285
  промышленный ~ 55-520
  пропитанный ~ 55-670
  пропитываемый ~ 55-670
  псевдобинарный ~ 55-315
  разбавленный ~ 55-320
  реостатный ~ 55-545
  сверхвысокопрочный ~ 55-425
  сверхлёгкий ~ 55-410
  сверхпластичный ~ 55-430
  сверхпрочный ~ 55-425
  сверхтвёрдый ~ 55-415
  свинцовокальциевонатриевый ~ 65-685
  синтетический ~ 55-270
  словолитный ~ 55-615
  спеканный ~ 55-655
  спечёный ~ 55-655
  спечённый твёрдый ~ 55-665
  ~ Вуда 65-670
  ~ Гейслера 65-420
  ~ для зубных протезов 55-575
  ~ для контактов 55-585
  ~ для литья под давлением 55-540
  ~ для наварки 55-590
  ~ для наплавки 55-590
  ~ для постоянных магнитов 55-450
  ~ для спаев со стеклом 55-600
  ~ для спайки со стеклом 55-600
  ~ игрек 65-590
  ~ карбидов металлов 55-665
  ~ на медной основе 65-010
  ~ на никелевой основе 65-430
  ~ первой плавки 55-275
  ~ Розе 65-675
  ~ с высокой магнитной проницаемостью 55-475
  ~ с высокой проницаемостью 55-475

сплав
  ~ с малым коэффициентом термического расширения 55-440
  ~ с низким тепловым расширением 55-440
  ~ стойкий к окислению 55-495
  ~ стойкий к тускнению 55-500
  стереотипный ~ 55-630
  твёрдый ~ 55-415
  текстурованный ~ 55-380
  термически обрабатываемый ~ 55-525
  термообрабатываемый ~ 55-525
  термостойкий ~ 55-510
  технический ~ 55-520
  типографский ~ 55-610
  трёхкомпонентный ~ 55-300
  тройной ~ 55-300
  тугоплавкий ~ 55-400
  уай-~ 65-590
  ультралёгкий ~ 55-410
  ферромагнитный ~ 55-455
  цветной ~ 65-005
  цинкоалюминиевый ~ 65-705
  чёрный ~ 60-070
  четверной ~ 55-305
  шрифтовой ~ 55-615
  щелочестойкий ~ 55-490
  щелочеупорный ~ 55-490
  эвтектический ~ 55-350
  эвтектоидный ~ 55-365
  электролитический ~ 55-290
  электроосаждённый ~ 55-290
  электротипный ~ 55-635
  электроэрозионностойкий ~ 55-585
  ювелирный ~ 55-580
  псевдосплав 55-650
  суперсплав 55-515
сплавление n 55-170
сплавляемость f 35-305
сплетение n дислокаций 30-385
спокойная сталь f 60-655
спонтанная кристаллизация f 25-045
спонтанный зародыш m 25-110
способность f
  амортизационная ~ 75-205
  замедляющая ~ 75-535
  охлаждающая ~ 70-125
  ~ к глубокой вытяжке 75-750
  ~ к отпуску 75-765
  ~ к упрочнению 75-770
  ~ сплавляться 55-175
среда f
  закалочная ~ 70-185
  охлаждающая ~ 70-185
среднелегированная сталь f 60-125
среднеуглеродистая сталь f 60-105
средний свободный путь m между частицами 40-425
средняя концентрация f 55-245
срез m 40-660
сродство n 75-660
  химическое ~ 75-660
  электронное ~ 10-115
  электросродство 10-115
стабилизатор m
  β-~ 65-860
стабилизация f 70-530
  ~ аустенита 50-425
стабилизирование n 70-530, 70-535
  естественное ~ 70-540
стабилизированная сталь f 60-305
стабилизирующая присадка f 50-485
стабилизирующий отжиг m 70-535
стабильная фаза f 35-170
стабильное равновесие n 35-055
стабильное состояние n 35-065
сталистый чугун m 60-1035
сталь f 60-075
  автоматная нержавеющая ~ 60-525

СТАЛЬ

сталь
автоматная ~ 60-710
азотированная ~ 60-315
азотируемая ~ 60-695
алмазная ~ 60-490
алюминиевая ~ 60-235
анормальная ~ 60-465
арматурная ~ 60-795
атмосферостойкая ~ 60-540
аустенитная ~ 60-420
аустенито-ферритная ~ 60-455
бандажная ~ 60-760
бейнитная ~ 60-435
бессемеровская ~ 60-595
борсодержащая ~ 60-255
броневая ~ 60-765
булатная ~ 60-645
быстрорежущая ~ 60-800
вакуумированная ~ 60-640
ванадиевая ~ 60-225
весьма мягкая ~ 60-345
воздушнозакаливаемая ~ 60-375
воздушнозакаливающаяся ~ 60-375
вольфрамистая ~ 60-215
вольфрамовая ~ 60-215
высококачественная ~ 60-145
высоколегированная ~ 60-130
высокопрочная ~ 60-355
высокоуглеродистая ~ 60-110
высокохромистая ~ 60-180
глубокопрокаливающаяся ~ 60-385
графитизированная ~ 60-300
дамасская ~ 60-645
динамная ~ 60-475
доэвтектоидная ~ 50-495
жаропрочная ~ 60-565
жаростойкая ~ 60-560
жароупорная ~ 60-560
закалённая ~ 60-285
закаливающаяся ~ 60-360
заклёпочная ~ 60-745
заэвтектоидная ~ 50-500
износостойкая ~ 60-480
износоустойчивая ~ 60-480
инвар-~ 65-500
инструментальная ~ 60-675
инструментальная ~ для горячей обработки 60-810
инструментальная ~ для работы в холодном состоянии 60-805
инструментальная ~ для работы при высоких температурах 60-810
качественная ~ 60-140
кипящая ~ 60-650
кислая ~ 60-615
кислородно-конвертерная ~ 60-605
кислотостойкая ~ 60-545
кислотоупорная ~ 60-545
клапанная ~ 60-735
кобальтовая ~ 60-240
кованая ~ 60-330
конвертерная ~ 60-590
конструкционная ~ 60-670
корабельная ~ 60-755
коррозиеустойчивая ~ 60-535
коррозионностойкая ~ 60-535
коррозионноустойчивая ~ 60-535
котельная ~ 60-740
кремнистая ~ 60-200
крипоустойчивая ~ 60-565
крупнозернистая ~ 60-475
ледебуритная ~ 60-440
легированная литая ~ 60-835
легированная ~ 60-115
легкообрабатываемая ~ 60-710
литая ~ 60-580, 60-825
литейная ~ 60-825
магнитная ~ 60-495
магнитомягкая ~ 60-500
магнитотвёрдая ~ 60-505
малолегированная ~ 60-120
малоуглеродистая ~ 60-100
марганцевая ~ 60-195
марганцовистая ~ 60-195
марганцовокремнистая ~ 60-205
мартеновская ~ 60-610
мартенситная ~ 60-430

сталь
мартенситно-стареющая ~ 60-460
машиноподелочная ~ 60-680
машиностроительная ~ 60-680
медистая ~ 60-245
мелкозернистая ~ 60-470
многокомпонентная ~ 60-170
молибденовая ~ 60-220
мягкая ~ 60-340
нагартованная ~ 60-295
наследственно крупнозернистая ~ 60-475
наследственно мелкозернистая ~ 60-470
науглероженная ~ 60-310
неглубоко прокаливающаяся ~ 60-380
недеформирующаяся ~ 60-410
незакаливающаяся ~ 60-390
нелегированная ~ 60-095
немагнитная ~ 60-510
нержавеющая ~ 60-515
нестареющая ~ 60-405
никелевая ~ 60-185
низколегированная ~ 60-120
низкоуглеродистая ~ 60-100
ножевая ~ 60-750
нормализованная ~ 60-280
окалиностойкая ~ 60-560
основная ~ 60-620
особо мягкая ~ 60-345
отожжённая ~ 60-275
перегретая ~ 60-320
пережжённая ~ 60-325
перлитная ~ 60-425
плакированная ~ 60-335
подшипниковая ~ 60-715
полуаустенитная ~ 60-445
полуспокойная ~ 60-660
полуферритная ~ 60-450
простая ~ 60-095
пружинная ~ 60-725
пудлинговая ~ 60-575
рафинированная ~ 60-635
рельсовая ~ 60-730
рессорная ~ 60-725
рядовая ~ 60-135
самозакаливающаяся ~ 60-375
свариваемая ~ 60-485
сваривающаяся ~ 60-485
сварочная ~ 60-570
свинцовистая ~ 60-250
серебристая ~ 60-820
слаболегированная ~ 60-120
специальная ~ 60-115, 60-150
спокойная ~ 60-655
среднелегированная ~ 60-125
среднеуглеродистая ~ 60-105
стабилизированная ~ 60-305
~ высокого качества 60-145
~ Гадфильда 60-785
~ глубокой штамповки 60-790
~ для глубокой вытяжки 60-790
~ для горячего деформирования и прессформ 60-810
~ для постоянных магнитов 60-780
~ для работы при высоких температурах 60-810
~ для цементации 60-690
~ закаливаемая в воде 60-365
~ закаливаемая в масле 60-370
~ закаливающаяся в воде 60-365
~ закаливающаяся в масле 60-370
~-заменитель 60-270
~ легированная одним компонентом 60-155
~ небольшой прокаливаемости 60-380
~-нитраллой 60-700
~ общего назначения 60-665
~ обыкновенного качества 60-135
~ обычного качества 60-135
~ повышенного качества 60-140
~ с глубокой прокаливаемостью 60-385
~ с неглубокой прокаливаемостью 60-380
~ торгового качества 60-135
~ устойчивая против отпуска 60-395
стандартная ~ 60-265
стареющая ~ 60-400
строительная ~ 60-585
судовая ~ 60-755
судостроительная ~ 60-755

сталь
 твёрдая ~ 60-350
 теплопрочная ~ 60-565
 теплостойкая ~ 60-395
 термическая улучшенная ~ 60-290
 тигельная ~ 60-630
 титанистая ~ 60-230
 титановая ~ 60-230
 томасовская ~ 60-600
 томлёная ~ 60-585
 трансформаторная ~ 60-770
 тройная ~ 60-160
 углеродистая литая ~ 60-830
 углеродистая ~ 60-095
 улучшаемая ~ 60-705
 успокоенная ~ 60-655
 ферритная нержавеющая ~ 60-520
 ферритная ~ 60-415
 хладостойкая ~ 60-555
 хладоустойчивая ~ 60-555
 хромистая ~ 60-175
 хромовая ~ 60-175
 хромоникелевая ~ 60-190
 цементированная ~ 60-310
 цементируемая ~ 60-690
 цементная ~ 60-585
 цементуемая ~ 60-690
 четверная ~ 60-165
 шарикоподшипниковая ~ 60-720
 штамповая ~ 60-815
 щелочестойкая ~ 60-550
 щелочеупорная ~ 60-550
 эвтектоидная ~ 50-490
 электрическая ~ 60-625
 электросталь 60-625

стандартная сталь f 60-265

стандартное золото n 65-790

старение n 70-575
 деформационное ~ 45-225
 естественное ~ 70-585
 искусственное ~ 70-590
 магнитное ~ 70-620
 механическое ~ 45-225
 ~ мартенсита 70-615
 ~ после закалки 70-580
 ~ при нагреве 70-590
 ускоренное ~ 70-590
 перестарение 70-595

стареющая сталь f 60-400

статический отдых m 45-430

стационарная диффузия f 45-345

стационарная ползучесть f 45-505

стедит m 50-235

стекловидный излом m 40-735

стеллит m 65-495

стенка f
 дислокационная ~ 30-395
 доменная ~ 40-645
 ~ Блоха 40-645
 ~ дислокации 30-395
 ~ домена 40-645

степень f беспорядка 25-610
 ~ вариантности 35-030
 ~ дендритной ликвации 40-420
 ~ деформации 45-035
 ~ дисперсности 40-335
 ~ компактности 20-205
 ~ наклёпа 45-040
 ~ неоднородности 40-420
 ~ переохлаждения 25-250
 ~ пластической деформации 45-035
 ~ порядка 25-605
 ~ тетрагональности 20-375
 ~ упорядочения 25-605
 ~ упорядоченности 25-605
 ~ чистоты 55-095

стереотипный сплав m 55-630

стехиометрический состав m 35-240

стойкость f
 коррозионная ~ 75-670
 ~ к тепловым ударам 75-825

стойкость
 ~ к термической усталости 75-830
 ~ к тускнению 75-680
 термическая ~ 75-835
 химическая ~ 75-665
 жаростойкость 75-690
 износостойкость 75-265
 кислотостойкость 75-695
 красностойкость 75-260
 окалиностойкость 75-690
 теплостойкость 75-835
 термостойкость 75-835
 щелочестойкость 75-695

сток m 30-130

столбчатая зона f 40-180

столбчатая кристаллизация f 25-065

столбчатая структура f 50-125

столбчатые кристаллы mpl 40-280

строение n
 кристаллическое ~ 20-040
 крупнозернистое ~ 40-390
 мелкозернистое ~ 40-395
 фазовое ~ 35-020
 электронное ~ 10-030
 макростроение 40-025
 микростроение 40-030

строительная сталь f 60-685

стронций m 05-430

строчечная структура f 40-140

строчечность f карбидов 85-235

струйная закалка f 70-240

струйчатая закалка f 70-240

структура f 40-005
 аномальная ~ 40-135
 блочная ~ 40-170
 видманштеттова ~ 40-130
 вторичная ~ 40-055
 гетерогенная ~ 40-155
 глобулярная ~ 40-120
 гомогенная ~ 40-145
 двухфазная ~ 40-150
 дендритная ~ 40-090
 дислокационная ~ 30-415
 доменная ~ 40-620
 доэвтектическая ~ 40-070
 доэвтектоидная ~ 40-085
 заэвтектическая ~ 40-065
 заэвтектоидная ~ 40-080
 игольчатая ~ 40-115
 кристаллическая ~ 20-040
 крупнозернистая ~ 40-390
 крупнокристаллическая ~ 40-390
 линейчатая ~ 40-165
 литая ~ 40-050
 магнитная ~ 40-650
 мелкозернистая ~ 40-395
 мелкокристаллическая ~ 40-395
 многофазная ~ 40-160
 мозаичная ~ 40-170
 однофазная ~ 40-145
 первичная ~ 40-045
 пластинчатая ~ 40-105
 полиэдрическая ~ 40-095
 полосчатая ~ 40-140, 40-165
 сетчатая ~ 40-100
 столбчатая ~ 40-125
 строчечная ~ 40-140
 ~ металла 40-005
 тонкая ~ 40-040
 эвтектическая ~ 40-060
 эвтектоидная ~ 40-075
 электронная ~ 10-030
 ячеистая ~ 40-095
 макроструктура 40-025
 микроструктура 40-030
 сверхструктура 35-440
 субструктура 40-035

структурная вакансия f 30-090

структурная диаграмма f 40-265

структурная рентгенография f 80-580

структурная составляющая f 40-010

СТРУКТУРНО

структурно нечувствительные свойства *npl* 75-865
структурно свободный цементит *m* 50-095
структурное превращение *n* 25-500
структурночувствительные свойства *npl* 75-860
структурные напряжения *npl* 85-035
ступенчатая закалка *f* 70-305
ступенчатый отжиг *m* 70-455
ступенька *f* 30-240
спиральная ~ роста 25-335
~ роста 25-330
субграница *f* 30-660
субзёренная граница *f* 30-660
субзерно *n* 30-650
субструктура *f* 40-035
судовая сталь *f* 60-755
судостроительная сталь *f* 60-755
сужение *n* относительное 75-125
сульфид *m* железа 50-245
~ марганца 50-240
сульфидирование *n* 70-895
сульфидные включения *npl* 40-530
сульфоазотирование *n* 70-890
сульфоцианирование *n* 70-892
сульфуризация *f* 70-895
супердислокация *f* 30-220
суперпарамагнетизм *m* 75-570
суперсплав *m* 55-515
сурьма *f* 05-025
рафинированная ~ 65-665
сурьмянистая медь *f* 65-165
сурьмянистый свинец *m* 65-660
сухая коррозия *f* 85-295
сфера *f*
координационная ~ 20-190
~ Ферми 10-225
сфероидальный графит *m* 50-190
сфероидальный цементит *m* 50-100
сфероидизация *f* 70-510
~ цементита 70-515
сфероидизирующий отжиг *m* 70-510
сфероидизованный чугун *m* 60-975
сферолит *m* 40-290
~ графита 50-215
сферолитный графит *m* 50-190
сходственная температура *f* 45-475
съёмка *f* на просвет 80-675
сыродутное железо *n* 60-045

Т

таллий *m* 05-460
тантал *m* 05-440
твёрдая сталь *f* 60-350
твёрдая цементация *f* 70-720
твердение *n*
вторичное ~ 70-405
дисперсионное ~ 70-560
твёрдорастворное упрочнение *n* 55-215
твёрдость *f* 75-245
горячая ~ 75-255
~ при красном калении 75-260
микротвёрдость 75-250
твёрдый припой *m* 65-720
твёрдый раствор *m* 35-375
твёрдый раствор *m* внедрения 35-420
твёрдый раствор *m* замещения 35-415

твёрдый свинец *m* 65-660
твёрдый сплав *m* 55-415
текстура *f* 40-555
волокнистая ~ 40-565
кубическая ~ 40-605
~ волочения 40-575
~ вытяжки 40-585
~ Госса 40-610
~ деформации 40-560
~ кристаллизации 40-580
~ кручения 40-595
~ металла отливки 40-580
~ отжига 40-600
~ отливки 40-580
~ прокатки 40-570
~ растяжения 40-585
~ рекристаллизации 40-600
~ сжатия 40-590
текстурованный сплав *m* 55-380
телефонная бронза *f* 65-255
теллур *m* 05-450
тело *n*
многокристаллическое ~ 20-440
поликристаллическое ~ 20-440
темнопольная микроскопия *f* 80-230
температура *f*
гомологическая ~ 45-475
дебаевская ~ 10-280
закалочная ~ 70-140
критическая ~ 35-335, 35-555
критическая ~ хрупкости 75-280
метатектическая ~ 35-620
монотектическая ~ 35-615
перитектическая ~ 35-610
перитектоидная ~ 35-635
равновесная ~ 35-690
синтетическая ~ 35-625
соответственная ~ 45-475
сходственная ~ 45-475
~ аустенитизации 70-135
~ вязко-хрупкого перехода 75-280
~ Дебая 10-280
~ закалки 70-140
~ затвердевания 75-425
~ кипения 75-430
~ конца мартенситного превращения 70-170
~ Кюри 25-560
~ ликвидуса 35-525
~ магнитного превращения 25-560
~ начала затвердевания 35-525
~ начала мартенситного превращения 70-165
~ начала плавления 35-500
~ начала рекристаллизации 45-470
~ Нееля 25-565
~ Нэля 25-565
~ отжига 70-425
~ отпуска 70-380
~ охрупчивания 75-280
~ перехода 75-280
~ перехода в сверхпроводящее состояние 25-570
~ перехода в хрупкое состояние 75-280
~ плавления 75-420
~ превращения 35-555
~ равновесия 35-690
~ равносвязи 30-570
~ разложения 15-340
~ разупорядочения 25-625
~ рекристаллизации 45-470
~ синего цвета побежалости 70-395
~ сублимации 75-435
характеристическая ~ 10-280
характеристическая ~ Дебая 10-280
эвтектическая ~ 35-605
эвтектоидная ~ 35-630
эквикогезивная ~ 30-570
температурная микроскопия *f* 80-240
температурные напряжения *npl* 85-025
температурный гистерезис *m* 25-480
температурный градиент *m* 15-175
температурный интервал *m* кристаллизации 35-550

температурный коэффициент *m* удельного сопротивления 75-480
температурный коэффициент *m* электрического сопротивления 75-480
температурный коэффициент *m* электросопротивления 75-480
температурный пик *m* 30-530
температуропроводность *f* 75-445
теоретический коэффициент *m* распределения 25-260
теория *f*
зонная ~ 10-020
электронная ~ металлов 10-025
тепловая вакансия *f* 30-095
тепловая диффузия *f* 45-370
тепловая трещина *f* 85-060
тепловая хрупкость *f* 75-290
тепловое окрашивание *n* 80-185
тепловое равновесие *n* 15-325
тепловое травление *n* 80-190
тепловое травление *n* на воздухе 80-185
тепловой пик *m* 30-530
тепловые колебания *npl* 10-270
тепловые напряжения *npl* 85-025
тепловые свойства *npl* 75-365
теплоёмкость *f* 75-410
атомная ~ 75-370
дополнительная удельная ~ 25-635
молекулярная ~ 75-375
удельная ~ 75-415
электронная ~ 10-275
теполомкость *f* 75-290
теплопроводность *f* 75-440
теплопрочная сталь *f* 60-565
теплопрочность *f* 75-240
теплостойкая сталь *f* 60-395
теплостойкость *f* 75-835
теплота *f*
скрытая ~ испарения 75-390
скрытая ~ парообразования 75-390
скрытая ~ плавления 75-380
~ затвердевания 75-385
~ испарения 75-390
~ парообразования 75-390
~ плавления 75-380
~ превращения 75-400
~ растворения 75-405
~ сублимации 75-395
удельная ~ 75-415
теплоустойчивость *f* 75-835
тербий *m* 05-455
термаллой *m* 65-540
термическая активация *f* 45-275
термическая вакансия *f* 30-095
термическая диффузия *f* 45-370
термическая обработка *f* 70-005
термическая остановка *f* 80-530
термическая площадка *f* 80-530
термическая стойкость *f* 75-835
термическая трещина *f* 85-060
термическая усталость *f* 85-425
термическая устойчивость *f* 75-835
термически активированный процесс *m* 45-280
термически активируемый процесс *m* 45-280
термически обрабатываемый сплав *m* 55-525
термически улучшенная сталь *f* 60-290
термические напряжения *npl* 85-025

термические свойства *npl* 75-365
термический анализ *m* 80-505
термический возврат *m* 45-425
термический дендрит *m* 25-325
термический коэффициент *m* объёмного расширения 75-460
термический пик *m* 30-530
термическое возбуждение *n* 45-275
термическое равновесие *n* 15-325
термическое травление *n* 80-190
термическое улучшение *n* 70-410
термоактивируемый процесс *m* 45-280
термодинамика *f* сплавов 15-010
~ твёрдого тела 15-005
термодинамическая активность *f* 15-280
термодинамическая величина *f* 15-030
термодинамическая вероятность *f* 15-230
термодинамическая летучесть *f* 15-290
термодинамическая переменная *f* 15-020
термодинамическая функция *f* 15-030
термодинамический параметр *m* 15-020
термодинамический параметр *m* состояния 15-020
термодинамическое равновесие *n* 15-320
термодинамическое состояние *n* 15-015
термодиффузионная металлизация *f* 70-915
термодиффузия *f* 45-370
термограмма *f* 80-527
термомагнитная обработка *f* 70-965
термомагнитный анализ *m* 80-565
термомеханическая обработка *f* 70-950
термомиграция *f* 45-370
термообрабатываемый сплав *m* 55-525
термообработка *f* 70-005
термоперенос *m* 45-370
термостойкий сплав *m* 55-510
термостойкость *f* 75-835
термохимическая обработка *f* 70-650
термоциклирование *n* 70-520
термоциклическая прочность *f* 75-830
термоэлектродвижущая сила *f* 75-495
тетравакансия *f* 30-055
тетрагира *f* 20-080
тетрагональная решётка *f* 20-370
тетрагональная система *f* 20-250
тетрагональность *f* 20-375
тетрагональный мартенсит *m* 50-335
тетраэдрическая пора *f* 20-325
тетраэдрическая пустота *f* 20-325
технеций *m* 05-445
технические свойства *npl* 75-720
технический сплав *m* 55-520
техническое железо *n* 60-015
течение *n*
вязкое ~ 45-220
вязкостное ~ 45-220
пластическое ~ 45-215
тигельная сталь *f* 60-630
типографский сплав *m* 55-610
титан *m* 05-480
титанид *m* 40-510
титанирование *n* 70-945
диффузионное ~ 70-945
титанистая сталь *f* 60-230
титановая сталь *f* 60-230
ТМО 70-965

томасовская сталь *f* 60-600
томасовский чугун *m* 60-850
томлёная сталь *f* 60-585
томление *n* 70-625
~ чугуна 70-625
томпак *m* 65-395
тонкая плёнка *f* 80-350
тонкая структура *f* 40-040
тонкая фольга *f* 80-350
тонкоигольчатый мартенсит *m* 50-325
торий *m* 05-465
точечная коррозия *f* 85-345
точечный графит *m* 50-195
точечный дефект *m* 30-030
точечный перлит *m* 50-150
точка *f*
   критическая ~ 35-555
   мартенситная ~ 70-165
   монотектическая ~ 35-670
   особая ~ 80-495
   перитектическая ~ 35-665
   сингулярная ~ 80-495
   спинодальная ~ 15-185
   ~ закрепления 30-275
   ~ затвердевания 75-425
   ~ кипения 75-430
   ~ Кюри 25-560
   ~ M_K 70-170
   ~ M_H 70-165
   ~ Нееля 26-565
   ~ перегиба 15-185
   ~ плавления 75-420
   тройная ~ 35-040
   эвтектическая ~ 35-660
   эвтектоидная ~ 35-675
травитель *m* 80-100
   красящий ~ 80-105
травление *n* 80-120
   газовое ~ 80-182
   глубокое ~ 80-160
   ионное ~ 80-180
   оксидное ~ 80-185
   тепловое ~ 80-190
   тепловое ~ на воздухе 80-185
   термическое ~ 80-190
   ~ для макроскопического исследования 80-135
   ~ для микроскопического исследования 80-140
   ~ зёрен 80-150
   ~ ионной бомбардировкой 80-180
   ~ окислением на воздухе 80-185
   ~ по границам зёрен 80-145
   ~ цветами побежалости 80-185
   химическое ~ 80-170
   цветное ~ 80-165
   электролитическое ~ 80-175
   перетравление 80-130
   электротравление 80-175
травящий реактив *m* 80-100
транскристаллизация *f* 25-070
транскристаллический излом *m* 40-710
трансляционная решётка *f* 20-275
трансляция *f* 20-130
трансмиссионная микроскопия *f* 80-220
трансформаторная сталь *f* 60-770
трансформаторное железо *n* 60-770
третичная рекристаллизация *f* 45-460
третичный цементит *m* 50-090
третье квантовое число *n* 10-140
треугольник *m*
   концентрационный ~ 35-720
   ~ Гиббса-Розебоома 35-720
   ~ Гаманна 40-265
трёхкомпонентная система *f* 35-125
трёхкомпонентный сплав *m* 55-300
трёхфазная система *f* 35-105

трещина *f*
   горячая ~ 85-080
   деформационная ~ 85-070
   закалочная ~ 85-190
   поверхностная ~ 85-050
   тепловая ~ 85-060
   термическая ~ 85-060
   ~ вследствие внутренних напряжений 85-065
   ~ Гриффиса 85-055
   ~ Гриффитса 85-055
   холодная ~ 85-075
   микротрещина 85-045
трещинообразование *n* 85-015
трибозарождение *n* 25-210
тривакансия *f* 30-050
тригира *f* 20-075
тригональная система *f* 20-260
триклинная решётка *f* 20-400
триклинная система *f* 20-270
тройная диаграмма *f* состояния 35-710
тройная ось *f* симметрии 20-075
тройная система *f* 35-125
тройная сталь *f* 60-160
тройная точка *f* 35-040
тройная эвтектика *f* 40-210
тройной сплав *m* 55-300
троостит *m* 50-395
   игольчатый ~ 50-410
   ~ закалки 50-400
   ~ отпуска 50-405
трубочная диффузия *f* 45-360
тугоплавкий металл *m* 55-065
тугоплавкий припой *m* 65-720
тугоплавкий сплав *m* 55-400
тулий *m* 05-470
тягучесть *f* 75-030
тягучий излом *m* 40-680
тяжёлые металлы *mpl* 55-055
тяжёлые платиновые металлы *mpl* 65-825

У

уай-сплав *m* 65-590
увеличение *n* 80-282
углерод *m* 05-095
   несвязанный ~ 50-005
   свободный ~ 50-005
   связанный ~ 50-010
   ~ отжига 50-220
углеродистая литая сталь *f* 60-830
углеродистая сталь *f* 60-095
углеродный потенциал *m* 70-750
углеродный эквивалент *m* 50-480
углероазотирование *n* 70-875
угол *m* разориентировки 30-595
ударная вязкость *f* 75-190
ударная прочность *f* 75-185
ударное испытание *f* 80-430
ударное упрочнение *n* 45-055
удельная масса *f* 75-345
удельная проводность *f* 75-485
удельная прочность *f* 75-140
удельная теплоёмкость *f* 75-415
удельная теплота *f* 75-415
удельная электрическая проводность *f* 75-485
удельная электропроводность *f* 75-485
удельное сопротивление *n* 75-470

ФАЗА

удельное электрическое сопротивление *n* 75-470
удельное электросопротивление *n* 75-470
удельный вес *m* 75-350
удлинение *n*
  остаточное относительное ~ 75-120
  относительное ~ 75-120
  полное относительное ~ 75-120
узел *m* дислокации 30-350
  ~ кристаллической решётки 20-100
  ~ решётки 20-100
  дислокационный ~ 30-350
узловая позиция *f* 20-140
узловое положение *n* 20-140
укрупнение *n* зерна 25-290
улучшаемая сталь *f* 60-705
ультраакустическая дефектоскопия *f* 80-800
ультразвуковая дефектоскопия *f* 80-800
ультралёгкий сплав *m* 55-410
ультрамикроскопические выделения *npl* 40-315
ультрафиолетовая микроскопия *f* 80-265
унарная система *f* 35-115
упорядочение *n* 25-575
  магнитное ~ 25-640
упорядоченная фаза *f* 35-200
упорядоченное состояние *n* 25-590
упорядоченный твёрдый раствор *m* 35-430
управляемая кристаллизация *f* 25-030
управляемое затвердевание *n* 25-030
упрочнение *n* 45-045, 70-015, 70-025
  взрывное ~ 45-055
  деформационное ~ 45-050
  дисперсионное ~ 70-560
  дисперсное ~ 55-220
  дисперсностное ~ 70-560
  латентное ~ 45-180
  пластическое ~ 45-050
  радиационное ~ 30-495
  скрытое ~ 45-180
  твёрдорастворное ~ 55-215
  ударное ~ 45-055
  ~ взрывом 45-055
  ~ дисперсными частицами 55-220
  ~ легированием 55-210
  ~ наклёпом 45-050
  ~ облучением 30-495
  ~ при старении 70-565
  ~ растворением 55-215
  фазовое ~ 45-080
упрочнитель *m* 40-350, 55-255
упрочняемость *f* 75-770
упрочняющая обработка *f* 70-025
упрочняющая фаза *f* 40-350
упрочняющий элемент *m* 55-255
упругая деформация *f* 45-005
упругая константа *f* 75-145
упругая постоянная *f* 75-145
упругое последействие *n* 75-180
упругость *f* 75-020
  ~ диссоциации 15-345
уравнение *n* Гиббса-Дюгема 15-305
  ~ Гильдебранда 15-310
  ~ состояния 15-025
  ~ Шредингера 10-175
уран *m* 05-490
уровень *m* Ферми 10-220
  энергетический ~ 10-150
ус *m* 25-360
усадка *f* 75-795
  ~ в жидком состоянии 75-800

усадка
  ~ в твёрдом состоянии 75-810
  ~ до затвердевания 75-800
  ~ при затвердевании 75-805
усадочная пористость *f* 85-120
усадочная раковина *f* 85-125
усадочная рыхлота *f* 85-120
усадочные напряжения *npl* 85-175
усик *m* 25-360
ускоренная диффузия *f* 45-355
ускоренная ползучесть *f* 45-510
ускоренное старение *n* 70-590
условный предел *m* длительной прочности 75-230
условный предел *m* ползучести 75-230
условный предел *m* текучести 70-115
успокоенная сталь *f* 60-655
усталостная коррозия *f* 85-315
усталостная прочность *f* 75-200
усталостное разрушение *n* 40-745
усталостный излом *m* 40-445
усталость *f* 85-410
  контактная ~ 85-420
  коррозионная ~ 85-415
  термическая ~ 85-425
установившаяся диффузия *f* 45-345
установившаяся ползучесть *f* 45-505
устойчивая фаза *f* 35-170
устойчивое равновесие *n* 35-055
устойчивое состояние *n* 35-065
устойчивость *f*
  антикоррозионная ~ 75-670
  термическая ~ 75-835
  ~ против отпуска 75-840
  ~ против старения 75-845
  химическая ~ 75-665
  износоустойчивость 75-265
  криоустойчивость 75-215
  теплоустойчивость 75-835
устойчивый зародыш *m* 25-150
утонение *n* 80-355

Ф

фаза *f* 35-025
  высокотемпературная ~ 35-160
  диспергированная ~ 40-330
  дисперсная ~ 40-330
  избыточная ~ 40-340, 40-345
  интерметаллидная ~ 35-210
  интерметаллическая ~ 35-210
  исходная ~ 35-290
  конгруэнтная интерметаллическая ~ 35-220
  конденсированная ~ 35-145
  маточная ~ 35-290
  мелкодисперсная ~ 40-330
  металлическая ~ 35-150
  метастабильная ~ 35-175
  неконгруэнтная интерметаллическая ~ 35-225
  неметаллическая ~ 35-155
  неравновесная ~ 35-185
  неупорядоченная ~ 35-205
  неустойчивая ~ 35-175
  переходная ~ 35-195
  промежуточная ~ 35-210
  проэвтектическая ~ 40-340
  проэвтектоидная ~ 40-345
  равновесная ~ 35-180
  сигма-~ 35-275
  стабильная ~ 35-170
  упорядоченная ~ 35-200
  упрочняющая ~ 40-350
  устойчивая ~ 35-170
  ~ устойчивая при высоком давлении 35-165
  ~ Юм-Розери 35-280
  электронная ~ 35-280

ФАЗОВАЯ

фазовая граница *f* 35-535
фазовая диаграмма *f* состояния 35-480
фазовое превращение *n* 25-460
фазовое превращение *n* второго рода 25-470
фазовое превращение *n* первого рода 25-465
фазовое равновесие *n* 30-015
фазовое строение *n* 35-020
фазовое упрочнение *n* 45-080
фазовый контраст *m* 80-275
фазовый наклёп *m* 45-080
фазовый переход *m* 25-460
фазовый переход *m* второго рода 25-470
фазовый переход *m* первого рода 25-465
фазоконтрастная микроскопия *f* 80-270
фазы *fpl*
сосуществующие ~ 35-190
~ Лавеса 35-265
~ Цинтля 35-270
фактический коэффициент *m* распределения 25-265
фактор *m*
масштабный ~ 10-395
объёмный ~ 10-395
размерный ~ 10-395
~ анизотропии 75-595
электрохимический ~ 10-400
фарфоровидный излом *m* 40-715
фермий *m* 05-150
ферянико *m* 65-520
ферримагнетизм *m* 75-580
феррит *m* 50-045
высокотемпературный ~ 50-050
избыточный ~ 50-055
кремнистый ~ 50-475
первичный ~ 50-050
проэвтектоидный ~ 50-055
свободный ~ 50-055
~ перлита 50-060
эвтектоидный ~ 50-060
ферритизация *f* 70-645
ферритная нержавеющая сталь *f* 60-520
ферритная сталь *f* 60-415
ферритно-графитный чугун *m* 60-1040
ферритно-графитовый чугун *m* 60-1040
ферритный венец *m* 50-230
ферритный ковкий чугун *m* 60-1125
ферритный чугун *m* 60-1040
ферритообразующий элемент *m* 50-455
ферромагнетизм *m* 75-545
ферромагнетик *m* 75-550
ферромагнитное вещество *n* 75-550
ферромагнитность *f* 75-545
ферромагнитный сплав *m* 55-455
ферромарганец *m* 60-1150
ферросилиций *m* 60-1140
ферросплавы *mpl* 60-1135
феррохром *m* 60-1145
феррроцерий *m* 65-875
фигуры *fpl*
полосные ~ 80-685
~ травления 80-200
физика *f* металлов 10-015
~ твёрдого тела 10-010
металлофизика 10-015
физико-химический анализ *m* 80-490
физические свойства *npl* 75-315
физический предел *m* текучести 75-110
физическое металловедение *n* 10-005

флокены *mpl* 85-240
флуоресцентная дефектоскопия *f* 80-835
флюктуации *fpl*
концентрационные ~ 25-100
~ концентрации 25-100
~ состава 25-100
фокусирующее столкновение *n* 30-505
фокусон *m* 30-505
фонон *m* 10-305
форма *f* кристалла 20-460
фосфид *m* железа 50-250
фосфор *m* 05-330
фосфористая бронза *f* 65-215
фосфористая медь *f* 65-075
фосфористая эвтектика *f* 50-235
фосфористый чугун *m* 60-890
фотография *f*
макрофотография 80-310, 80-315
микрофотография 80-250, 80-255
фрактография *f* 80-385
франций *m* 05-160
фронт *m*
диффузионный ~ 45-410
~ диффузии 45-410
~ кристаллизации 25-295
~ роста 45-480
фтор *m* 05-155
фугитивность *f* 15-290
функция *f*
волновая ~ 10-180
собственная ~ 10-190
термодинамическая ~ 15-030
~ Блоха 10-195
~ состояния 15-035

X

характеристическая температура *f* 10-280
характеристическая температура *f* Дебая 10-280
хастеллой *m* 65-455
хидуминий *m* 65-605
химико-термическая обработка *f* 70-650
химическая диффузия *f* 45-320
химическая коррозия *f* 85-265
химическая неоднородность *f* 40-415
химическая полировка *f* 80-080
химическая связь *f* 10-320
химическая стойкость *f* 75-665
химическая устойчивость *f* 75-665
химические свойства *npl* 75-655
химический анализ *m* 80-470
химический дендрит *m* 25-320
химический потенциал *m* 15-175
химический состав *m* 55-225
химический состав *m* сплава 55-225
химический элемент *m* 05-005
химическое равновесие *n* 15-330
химическое соединение *n* 35-235
химическое сродство *n* 75-660
химическое травление *n* 80-170
хладноломкость *f* 75-285
хладостойкая сталь *f* 60-555
хладоустойчивая сталь *f* 60-555
хлор *m* 05-105
холодная обработка *f* давлением 45-025
холодная трещина *f* 85-075
хром *m* 05-110
~-никель 65-470

ЧУГУН

хромансиль m 60-260
хромель m 65-465
хромистая бронза f 65-195
хромистая сталь f 60-175
хромистый чугун m 60-1015
хромоалитирование n 70-940
хромоалюминирование n 70-940
хромовая сталь f 60-175
хромоникелевая сталь f 60-190
хрупкий излом m 40-675
хрупкое разрушение n 40-675
хрупкость f 75-270
  водородная ~ 85-390
  каустическая ~ 75-305
  низкотемпературная ~ 75-295
  отпускная ~ 85-220
  реотропная ~ 75-310
  сезонная ~ 85-385
  тепловая ~ 75-290
  ~ отпуска 85-220
  щелочная ~ 75-305
художественная бронза f 65-270

Ц

цвет m
  синий ~ побежалости 70-390
  ~ побежалости 70-385
цветная металлография f 80-030
цветное травление n 80-165
цветной сплав m 65-005
цветные металлы mpl 55-015
цезий m 05-080
цементационная медь f 65-045
цементация f 70-670, 70-715
  газовая ~ 70-730
  жидкостная ~ 70-725
  местная ~ 70-735
  твёрдая ~ 70-720
цементированная сталь f 60-310
цементированный слой m 70-780
цементируемая сталь f 60-690
цементирующий газ m 70-745
цементит m 50-065
  вторичный ~ 50-080
  зернистый ~ 50-100
  избыточный ~ 50-095
  легированный ~ 50-110
  первичный ~ 50-070
  проэвтектоидный ~ 50-080
  структурно-свободный ~ 50-095
  сфероидальный ~ 50-100
  третичный ~ 50-090
  ~ перлита 50-085
  шаровидный ~ 50-100
  эвтектический ~ 50-075
цементитная пластинка f 50-120
цементитная сетка f 50-105
цементитная эвтектика f 50-255
цементная медь f 65-045
цементная сталь f 60-585
цементованный слой m 70-780
цементуемая сталь f 60-690
цементующий газ m 70-745
центр m кристаллизации 25-105
~ рекристаллизации 25-170
центрогранная кубическая решётка f 20-310
центрогранная решётка f 20-290
церий m 05-100
церробенд m 65-680

цианирование n 70-855
  высокотемпературное ~ 70-870
  газовое ~ 70-860
  жидкостное ~ 70-865
  низкотемпературное ~ 70-875
циклическая прочность f 75-200
циклический отжиг m 70-520
цинк m 05-515
  дистилляционный ~ 65-695
  нерафинированный ~ 65-690
  редистиллированный ~ 65-695
  черновой ~ 65-690
цинкоалюминиевый сплав m 65-705
цинковый припой m 65-730
циркаллой m 65-855
цирконий m 05-520

Ч

частичная дислокация f 30-315
частичная закалка f 70-335
частичная растворимость f 35-320
червонное золото n 65-795
чёрная медь f 65-015
черновая медь f 65-015
черновой свинец m 65-650
черновой цинк m 65-690
черносердечный ковкий чугун m 60-1115
чёрный металл m 60-070
чёрный отжиг m 70-440
чёрный сплав m 60-070
четверная ось f симметрии 20-080
четверная система f 35-135
четверная сталь f 60-165
четверная эвтектика f 40-215
четверной сплав m 55-305
четвёртое квантовое число n 10-145
четырёхкомпонентная система f 35-135
чешуйчатый графит m 50-185
число n
  атомное ~ 75-330
  второе квантовое ~ 10-135
  главное квантовое ~ 10-130
  квантовое ~ 10-125
  координационное ~ 20-185
  магнитное квантовое ~ 10-140
  массовое ~ 75-335
  орбитальное квантовое ~ 10-135
  первое квантовое ~ 10-130
  побочное квантовое ~ 10-135
  спиновое квантовое ~ 10-145
  третье квантовое ~ 10-140
  четвёртое квантовое ~ 10-145
  ~ степеней свободы 35-030
чистая закалка f 70-330
чистый металл m 55-090
чистый отжиг m 70-435
чувствительность f
  коррозионная ~ 75-675
  ~ к надрезам 75-195
  ~ к перегреву 75-760
  ~ к толщине стенки 75-815
чугаль m 60-1020
чугун m
  алюминиевый ~ 60-1020
  американский ковкий ~ 60-1115
  антифрикционный ~ 60-1105
  армированный ~ 60-1070
  аустенитный ~ 60-1050
  белосердечный ковкий ~ 60-1120
  белый ~ 60-910, 60-955
  бессемеровский ~ 60-845
  вторичный ~ 60-945
  высококачественный ~ 60-1065
  высоколегированный ~ 60-1005

чугун
 высокопрочный ~ 60-975
 высокоуглеродистый ~ 60-925
 гематитовый ~ 60-895
 графитический ~ 60-950
 доменный ~ 60-860
 доэвтектический белый ~ 50-510
 древесноугольный ~ 60-875
 европейский ковкий ~ 60-1120
 жаростойкий ~ 60-1085
 жароупорный ~ 60-1085
 заэвтектический белый ~ 50-515
 зеркальный ~ 60-930
 кислотостойкий ~ 60-1075
 кислотоупорный ~ 60-1075
 ковкий ~ 60-1110
 коксовый ~ 60-870
 кремнистый ~ 60-1010
 легированный ~ 60-935, 60-995
 литейный ~ 60-885
 магниевый ~ 60-975
 малоуглеродистый ~ 60-1035
 марганцевый ~ 60-1030
 марганцовистый ~ 60-1030
 мартеновский ~ 60-855
 мартенситный ~ 60-1055
 модифицированный ~ 60-970
 нелегированный ~ 60-990
 неотожжённый ковкий ~ 60-905
 низколегированный ~ 60-1000
 никелевый ~ 60-1025
 обыкновенный ~ 60-965, 60-990
 обычный ~ 60-965
 отбелённый ~ 60-980
 первичный ~ 60-840
 перлитно-графитный ~ 60-1045
 перлитный ковкий ~ 60-1130
 перлитный ~ 60-1045
 половинчатый ~ 60-920, 60-960
 полугематитовый ~ 60-900
 предельный ~ 60-880
 светлосердечный ковкий ~ 60-1120
 серый ~ 60-915, 60-950
 синтетический ~ 60-940
 сталистый ~ 60-1035
 сфероидизированный ~ 60-975
 томасовский ~ 60-850
 ферритно-графитный ~ 60-1040
 ферритно-графитовый ~ 60-1040
 ферритный ковкий ~ 60-1125
 ферритный ~ 60-1040
 фосфористый ~ 60-890
 хромистый ~ 60-1015
 черносердечный ковкий ~ 60-1115
 ~ с игольчатой структурой 60-1060
 ~ с шаровидным графитом 60-975
 щелочестойкий ~ 60-1080
 щелочеупорный ~ 60-1080
 эвтектический белый ~ 50-505
 электрочугун 60-865

Ш

шарикоподшипниковая сталь f 60-720
шаровидный графит m 50-190
шаровидный цементит m 50-100
шведское железо n 60-030
шелковистый излом m 40-715
шерардизация f 70-930
шестерная ось f симметрии 20-085
шестоватые кристаллы mpl 40-280
шиферный излом m 40-725
шлаковины fpl 40-540
шлаковые включения npl 40-540
шлиф m 80-095
 микрошлиф 80-095
шрифтовой сплав m 55-615
штамповая сталь f 60-815
штампуемость f 75-750

Щ

щелевая коррозия f 85-300
щелочестойкая сталь f 60-550
щелочестойкий сплав m 55-490
щелочестойкий чугун m 60-1080
щелочестойкость f 75-700
щелочеупорная сталь f 60-550
щелочеупорность f 75-700
щелочеупорный сплав m 55-490
щелочеупорный чугун m 60-1080
щелочная хрупкость f 75-305
щёлочноземельные металлы mpl 55-035
щелочные металлы mpl 55-030

Э

эвтектика f 40-200
 графитная ~ 50-260
 двойная ~ 40-205
 ледебуритная ~ 50-255
 несплошная ~ 40-235
 пластинчатая ~ 40-220
 разбавленная ~ 40-230
 тройная ~ 40-210
 фосфористая ~ 50-235
 цементитная ~ 50-255
 четверная ~ 40-215
 квазиэвтектика 40-225
эвтектическая горизонталь f 35-640
эвтектическая кристаллизация f 25-025
эвтектическая линия f 35-640
эвтектическая реакция f 35-565
эвтектическая смесь f 40-200
эвтектическая структура f 40-060
эвтектическая температура f 35-605
эвтектическая точка f 35-660
эвтектические карбиды mpl 50-440
эвтектический белый чугун m 50-505
эвтектический графит m 50-175
эвтектический распад m 35-565
эвтектический сплав m 55-350
эвтектический цементит m 50-075
эвтектическое затвердевание n 25-025
эвтектическое плавление n 35-470
эвтектическое понижение n 35-730
эвтектическое превращение n 35-565
эвтектоид m 40-240
эвтектоидная горизонталь f 35-655
эвтектоидная линия f 35-655
эвтектоидная смесь f 40-240
эвтектоидная сталь f 50-490
эвтектоидная структура f 40-075
эвтектоидная температура f 35-630
эвтектоидная точка f 35-675
эвтектоидное превращение n 35-590
эвтектоидный распад m 35-590
эвтектоидный сплав m 55-365
эвтектоидный феррит m 50-060
эйнштейний m 05-135
эквикогезивная температура f 30-570
экзогаз m 70-710
экзогенные включения npl 40-520
экзотермическая атмосфера f 70-710
экзотермический газ m 70-710
эксплуатационные свойства npl 75-820
экстенсивная величина f 15-040
экстраплоскость f 30-200
электрическая сталь f 60-625

электрические свойства *npl* 75-465
электровалентная связь *f* 10-325
электродиффузия *f* 45-365
электрокристаллизация *f* 25-080
электролитическая медь *f* 65-035
электролитическая полировка *f* 80-075
электролитический металл *m* 55-130
электролитический сплав *m* 55-290
электролитическое железо *n* 60-025
электролитическое травление *n* 80-175
электролитная медь *f* 65-035
электромиграция *f* 45-365
электрон *m* 65-625
  валентный ~ 10-065
  внешний ~ 10-065
  внутренний ~ 10-070
  несвязанный ~ 10-055
  свободный ~ 10-055
  связанный ~ 10-050
  ~ проводимости 10-265
электронная концентрация *f* 35-285
электронная микроскопия *f* 80-330
электронная микрофотография *f* 80-365
электронная оболочка *f* 10-155
электронная подгруппа *f* 10-160
электронная подоболочка *f* 10-160
электронная структура *f* 10-030
электронная теория *f* металлов 10-025
электронная теплоёмкость *f* 10-275
электронная фаза *f* 35-280
электронно-микроскопическое исследование *n* 80-330
электронное соединение *n* 35-280
электронное средство *n* 10-115
электронное строение *n* 10-030
электронномикроскопическая металлография *f* 80-025
электронномикроскопический снимок *m* 80-365
электронный газ *m* 10-045
электронный подуровень *m* 10-160
электронограмма *f* 80-775
электронографирование *n* 80-770
электронографический анализ *m* 80-770
электронография *f* 80-770
электроосаждённый сплав *m* 55-290
электроотрицательность *f* 10-095
электроперенос *m* 45-365
электрополировка *f* 80-075
электропроводность *f*
  удельная ~ 75-485
электросродство *n* 10-115
электросталь *f* 60-625
электротехническая медь *f* 65-065
электротипный сплав *m* 55-635
электротравление *n* 80-175
электрофизические свойства *npl* 75-465
электрохимическая коррозия *f* 85-270
электрохимические свойства *npl* 75-705
электрохимический потенциал *m* 75-715
электрохимический фактор *m* 10-400
электрочугун *m* 60-865
электроэрозионностойкий сплав *m* 55-585
электрум *m* 65-810
элемент *m* 05-005
  аустенитообразующий ~ 50-450
  главный легирующий ~ 55-195

элемент
  графитообразующий ~ 50-460
  карбидообразующий ~ 50-435
  карбидостабилизирующий ~ 50-465
  легирующий ~ 55-190
  металлический ~ 10-360
  микролегирующий ~ 55-250
  неметаллический ~ 10-365
  нитридообразующий ~ 5Р-470
  основной легирующий ~ 55-195
  растворённый ~ 35-360
  сопутствующий ~ 55-260
  упрочняющий ~ 55-255
  ферритообразующий ~ 50-455
  химический ~ 05-005
элементарная кристаллическая ячейка *f* 20-090
элементарная ячейка *f* 20-090
элинвар *m* 65-505
эмбрион *m* 25-135
эмиссионная электронная микроскопия *f* 80-335
эндогаз *m* 70-705
эндогенные включения *npl* 40-525
эндотермическая атмосфера *f* 70-705
эндотермический газ *m* 70-705
энергетическая асимметрия *f* 15-235
энергетическая зона *f* 10-235
энергетическая полоса *f* 10-235
энергетическая щель *f* 10-250
энергетический барьер *m* 10-110
энергетический стимул *m* 15-070
энергетический уровень *m* 10-150
энергетическое состояние *n* 10-120
энергия *f*
  внутренняя ~ 15-095
  запасённая ~ 45-060
  избыточная свободная ~ 15-115
  межфазовая ~ 30-635
  накопленная ~ 45-060
  парциальная свободная ~ 15-110
  поверхностная ~ 30-640
  свободная поверхностная ~ 15-120
  свободная ~ Гельмгольца 15-105
  свободная ~ Гиббса 15-140
  свободная ~ образования 15-135
  свободная ~ поверхности раздела 15-125
  свободная ~ смешения 15-130
  связанная ~ 15-100
  собственная ~ дислокации 30-230
  ~ активации 45-285
  ~ анизотропии 75-590
  ~ взаимодействия 10-295
  ~ границ зёрен 30-575
  ~ дефекта упаковки 30-560
  ~ дефектов упаковки 30-560
  ~ ионизации 10-100
  ~ межфазовой границы 30-635
  ~ связи 15-090
  ~ упорядочения 25-630
  ~ Ферми 10-210
энтальпия *f*
  избыточная ~ 15-145
  парциальная ~ 15-150
  свободная ~ 15-140
  ~ образования 15-160
  ~ плавления 15-165
  ~ связи 15-170
  ~ смешения 15-155
энтропия *f*
  вибрационная ~ 15-205
  избыточная ~ 15-220
  колебательная ~ 15-205
  конфигурационная ~ 15-210
  парциальная ~ 15-215
  ~ образования 15-200
  ~ плавления 15-225
  ~ смешения 15-195
эпиграмма *f* 80-670

эпитаксиальная плёнка f 25-350
эпитаксиальное наращивание n 25-345
эпитаксиальный рост m 25-345
эпитаксиальный слой m 25-350
эпитаксия f 25-340
эрбий m 05-140
эффект m
  масштабный ~ 10-395
  размерный ~ 10-395
  ~ Баушингера 75-175
  ~ каналирования 30-500
  ~ Киркендалла 45-385
  ~ Киркендолла 45-385
  ~ Мёссбауэра 10-285
  ~ относительной валентности 10-405
  ~ Протевена-Ле Шателье 45-540
  ~ Снука 30-460
  ~ Соре 45-370
  ~ Сузуки 30-455
  ~ Холла 10-290
  ~ электроотрицательной валентности 10-400
эффективное сечение n захвата нейтронов 75-530
эффективный коэффициент m распределения 25-265

Ю

ювелирное золото n 65-780
ювелирный сплав m 55-580

Я

явление n предвыделения 70-600
ядерная коррозия f 85-365
ядерные свойства npl 75-520
ядро n дислокации 30-205
язвенная коррозия f 85-350
ямка f травления 80-205
ясный отжиг m 70-430
ячеистая структура f 40-095
ячейка f
  единичная ~ 20-090
  примитивная элементарная ~ 20-095
  примитивная ~ 20-095
  элементарная кристаллическая ~ 20-090
  элементарная ~ 20-090